Principles of Engineering

PROJECT LEAD THE WAY

PLTW

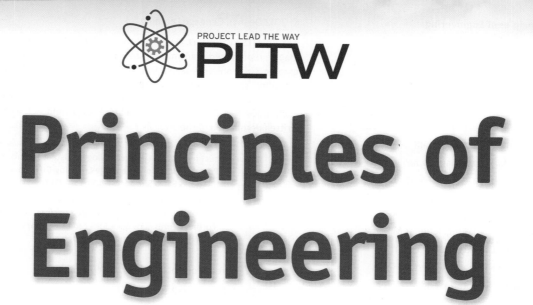

Principles of Engineering

Brett A. Handley
Wheatland-Chili Central School District

David M. Marshall
West Irondequoit Central School District

Craig Coon
Brockport Central High School

DELMAR
CENGAGE Learning™

Australia • Brazil • Japan • Korea • Mexico • Singapore • Spain • United Kingdom • United States

Principles of Engineering
Brett A. Handley, David M. Marshall,
Craig Coon

Vice President, Career, Education, and
    Training Editorial: Dave Garza

Director of Learning Solutions: Sandy Clark

Senior Acquisitions Editor: James DeVoe

Managing Editor: Larry Main

Product Manager: Mary Clyne

Development: iD8 Publishing Services

Editorial Assistant: Cris Savino

Vice President, Career, Education, and
    Training Marketing: Jennifer McAvey

Marketing Director: Deborah Yarnell

Marketing Manager: Katherine Hall

Production Director: Wendy Troeger

Production Manager: Mark Bernard

Content Project Manager: David Plagenza

Art Director: Bethany Casey

For product information and technology assistance, contact us at
**Professional Group Cengage Learning Customer
& Sales Support, 1-800-354-9706**

For permission to use material from this text or product,
submit all requests online at **cengage.com/permissions.**
Further permissions questions can be e-mailed to
permissionrequest@cengage.com.

Library of Congress Control Number: 2010932532

ISBN-13: 978-1-435-42836-2

ISBN-10: 1-435-42836-6

Delmar
5 Maxwell Drive
Clifton Park, NY 12065-2919
USA

Cengage Learning is a leading provider of customized learning solutions with office locations around the globe, including Singapore, the United Kingdom, Australia, Mexico, Brazil and Japan. Locate your local office at: **international. cengage.com/region**

Cengage Learning products are represented in Canada by Nelson Education, Ltd.

For your lifelong learning solutions, visit **delmar.cengage.com**

Visit our corporate website at **cengage.com.**

Printed in the United States of America
1 2 3 4 5 6 7 15 14 13 12 11

BRIEF CONTENTS

TABLE OF CONTENTS

Engineering is by no means a new profession. It has been practiced since the dawn of human civilization. The methods that engineers employ continue to change with the evolution and integration of more powerful and robust technologies. However, the basic *principles* of engineering (i.e., concepts, scientific laws, and mathematical formulas) remain the same.

Principles of Engineering is a new textbook for teachers who want to help their high school students understand what engineering is at a level that is appropriate to their knowledge of mathematics and science. It is also designed to inspire students to explore career pathways in engineering and technology. By presenting the principles and concepts that engineers and design professionals use to shape our modern, human-designed world, *Principles of Engineering* will help students develop the technological literacy and problem-solving skills that are invaluable in first-year postsecondary engineering and engineering-technology programs.

 ## PRINCIPLES OF ENGINEERING AND PROJECT LEAD THE WAY, INC.

This text resulted from a partnership forged between Delmar Cengage Learning and Project Lead The Way, Inc. in February 2006. As a not-for-profit foundation that develops curriculum for engineering, Project Lead The Way, Inc. provides students with the rigorous, relevant, reality-based knowledge they need to pursue engineering or engineering technology programs in college.

The Project Lead The Way© curriculum developers strive to make mathematics and science relevant for students by building hands-on, real-world projects in each course. To support Project Lead The Way's© curriculum goals and to support all teachers who want to develop project- and problem-based programs in engineering and engineering technology, Delmar Cengage Learning is developing a complete series of texts to complement all of Project Lead the Way'© nine courses:

Gateway to Technology
Introduction to Engineering Design
Principles of Engineering
Digital Electronics
Aerospace Engineering
Biotechnical Engineering
Civil Engineering and Architecture
Computer Integrated Manufacturing
Engineering Design and Development

To learn more about Project Lead The Way's© ongoing initiatives in middle school and high school, please visit www.pltw.org.

HOW THIS TEXT WAS DEVELOPED

The development of Delmar's Project Lead the Way© series began with a focus group that brought together experienced teachers and curriculum developers from a broad range of engineering disciplines. Two important themes emerged from that discussion: (1) Teachers need a single resource that fits the way they teach engineering today, and (2) teachers want an engaging, interactive resource to support project- and problem-based learning.

For years, teachers have struggled to fit conventional textbooks to STEM-based curricula for engineering. Most high school teachers have collected stacks of thick textbooks in their search for the right mix of materials to teach engineering concepts in a high school setting. *Principles of Engineering* finally answers that need with an interactive text organized around the principles and applications of engineering. For the first time, teachers will be able to choose a single text that addresses the challenges and individuality of project- and problem-based learning while presenting sound coverage of the essential concepts and techniques used in the engineering design process.

This book is unique in that it was written by a team of authors with experience teaching engineering concepts to high school students. The authors have taken great care to create a principles text that high school students can understand without sacrificing rigor. Every chapter was thoroughly reviewed by an engineering professor to ensure accurate and valid engineering content. *Principles of Engineering* supports project- and problem-based learning through the follwing.

▶ Creating an unconventional, show-don't-tell pedagogy that is driven by *concepts*, not traditional textbook content. Concepts are mapped at the beginning of each chapter, and clearly identified as students navigate the chapter.

▶ Reinforcing major concepts with Applications, Projects, and Problems based on real-world examples.

▶ Providing a text rich in features designed to bring architecture to life in the real world. Case studies, Career Profiles, Boxed Articles highlighting human achievements, and resources for extended learning will show students how engineers and architects develop career pathways, work through failures, and innovate to continuously improve the success and quality of their projects.

▶ Reinforcing the text's interactivity with an exciting design that invites students to participate in a journey through the history and current practice of civil engineering and architecture.

Organization

The topics in this text are presented in an order that builds in rigor. The topics have been carefully chosen and developed to support and complement Project Lead the Way's© curriculum for *Principles of Engineering*.

Features

Teachers want an interactive text that keeps students interested in the story behind the engineered products and solutions that shape our world. This text delivers that story with plentiful boxed articles and illustrations of current technology. Here are some examples of how this text is designed to keep students engaged in a journey through the design process.

▶ **Case Studies** allows students to explore the process design teams use to create new technologies.

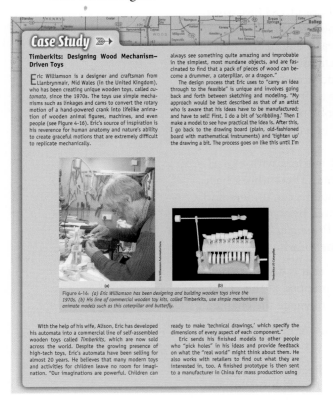

Case Study ▶▶→

Timberkits: Designing Wood Mechanism–Driven Toys

Eric Williamson is a designer and craftsman from Llanbrynmair, Mid Wales (in the United Kingdom), who has been creating unique wooden toys, called *automata*, since the 1970s. The toys use simple mechanisms such as linkages and cams to convert the rotary motion of a hand-powered crank into lifelike animation of wooden animal figures, machines, and even people (see Figure 4-16). Eric's source of inspiration is his reverence for human anatomy and nature's ability to create graceful motions that are extremely difficult to replicate mechanically.

always see something quite amazing and improbable in the simplest, most mundane objects, and are fascinated to find that a pack of pieces of wood can become a drummer, a caterpillar, or a dragon."

The design process that Eric uses to "carry an idea through to the feasible" is unique and involves going back and forth between sketching and modeling. "My approach would be best described as that of an artist who is aware that his ideas have to be manufactured; and have to sell! First, I do a bit of 'scribbling.' Then I make a model to see how practical the idea is. After this, I go back to the drawing board (plain, old-fashioned board with mathematical instruments) and 'tighten up' the drawing a bit. The process goes on like this until I'm

Figure 4-16: (a) Eric Williamson has been designing and building wooden toys since the 1970s. (b) His line of commercial wooden toy kits, called Timberkits, use simple mechanisms to animate models such as this caterpillar and butterfly.

With the help of his wife, Alison, Eric has developed his automata into a commercial line of self-assembled wooden toys called *Timberkits*, which are now sold across the world. Despite the growing presence of high-tech toys, Eric's automata have been selling for almost 20 years. He believes that many modern toys and activities for children leave no room for imagination. "Our imaginations are powerful. Children can

ready to make 'technical drawings,' which specify the dimensions of every aspect of each component."

Eric sends his finished models to other people who "pick holes" in his ideas and provide feedback on what the "real world" might think about them. He also works with retailers to find out what they are interested in, too. A finished prototype is then sent to a manufacturer in China for mass production using

Case Study ▶▶→

(continued)

traditional and computer-aided manufacturing methods. Step-by-step assembly instructions are provided with each kit, which are accompanied by hand-sketched isometric and oblique pictorial drawings. One example of Eric's Timberkit models is the T-REX (see Figure 4-17).

crank into a swinging motion that sends the left leg forward and backward. The right arm and bottom jaw are part of a four-bar linkage system that controls the opening and closing of the mouth of the T-REX. As the right arm rotates upward, the jaw appears to close.

Eric and his wife hope to inspire children, both young and old, to make their own automata using

Figure 4-17: The parts of the T-REX Timberkit are animated through a series of interconnected linkage mechanisms and cams.

The T-REX assembly is driven by a hand-powered crank mechanism that is attached to a plate cam (see Figure 4-18). The plate cam is in turn connected to the toy's left leg via a linkage rod and pin joint. The rod changes the continuous circular motion of the

whatever resources are available. "We want people to make their own models, and we hope ours will be the starting point for that process." To learn more about Timberkits, visit www.timberkits.com.

Figure 4-18: (a) A modified version of the slider and crank is used to animate the toy's left leg. (b) The motion of T-REX's jaw is controlled by a four-bar linkage system.

▶ **Boxed Articles** highlight fun facts and points of interest on the road to new and better products.

Point of Interest
A Bridge to the Future

Not all of Leonardo's inventions were practical, and many were before their time. For example, in 1502, Leonardo produced a drawing for a single-span, 720-foot bridge crossing the Bosphorus River. The project was part of a civil engineering contract with Sultan Bajazet II of Constantinople, in what is now Turkey. The bridge was never built—Leonardo's ideas for its construction were about 300 years ahead of generally accepted engineering practices. Nearly 500 years later, Leonardo's simple but powerful drawing drew the attention of Norwegian artist Vebjørn Sand. Sand was inspired by the drawing to create the Leonardo Project to bring Leonardo's vision to life. Five hundred years after Leonardo created these drawings, the Leonardo Bridge was constructed and opened for pedestrians and bicyclists on October 31, 2001, in Ås, Norway. Sand is currently considering several sites in the United States for the next Leonardo bridge project. You can read more about the Leonardo Project at www.vebjorn-sand.com and view a slide show of its construction at www.leonardobridgeproject.org.

Source: www.vebjorn-sand.com. Accessed 1/26/2010.

► **Your Turn** activities reinforce text concepts with skill-building activities.

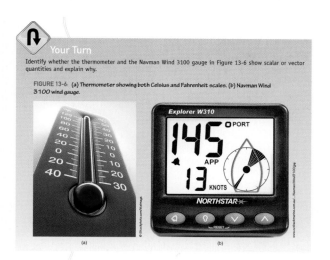

Your Turn

Identify whether the thermometer and the Navman Wind 3100 gauge in Figure 13-6 show scalar or vector quantities and explain why.

FIGURE 13-6: (a) Thermometer showing both Celsius and Fahrenheit scales. (b) Navman Wind 3100 wind gauge.

(a)   (b)

► **Off-Road Exploration** provides links to extended learning with resources for additional reading and research.

OFF-ROAD EXPLORATION

To get a better idea of how the design process is used in real-world challenges, watch the films *The Launch: A Product Is Born* and *ABC News Nightline Report: The Deep Dive* (July 13, 1999).

► **Key Terms** are defined throughout the text to help students develop a reliable lexicon for the study of engineering.

Linkage:

an assembly of rigid mechanical components within a mechanism that are linked together for the purpose of transmitting force and controlling motion.

► **STEM** connections show examples of how science and math principles are used to solve problems in engineering and technology.

There are fixed amounts of each stable chemical element on Earth. These elements are cycled through different mediums that act as reservoirs. For example, the carbon dioxide in air is absorbed by the leaves of a pineapple fruit in Hawaii, which uses the carbon atoms to form its fruit pulp. When you consume the pineapple, your body converts the carbon in the pineapple into organic polymer molecules such as new skin or hair. This cycling of matter between the various reservoirs is referred to as the *geochemical cycle*.

▶ **Career Profiles** provide role models and inspiration for students to explore career pathways in engineering.

▶ **Design Briefs** and **Engineers' Notebooks** are included where relevant to model effective communication techniques.

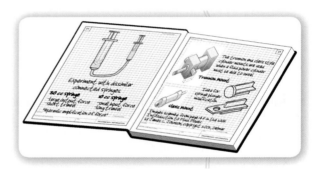

Design Brief

Client:	Genesee Valley Power Authority
Designer:	Kristin Haas
Problem Statement:	Dangerous conditions exist during the manipulation of low-level nuclear waste, wherein workers cannot risk direct contact with the objects that they must handle.
Design Statement:	Design a one-of-a-kind hydraulic-powered robot arm that will move one ping-pong ball-sized object from a given location in a work environment and drop it down a 1 ½" I.D. tube. The ping-pong ball represents the sensitive material. The work environment represents a space in which dangerous conditions exist that prevent human beings from making direct contact with the material.
Constraints:	1. 3-week deadline
	2. $100 budget
	3. Maximum weight =+/N10 lb
	4. No human contact with anything inside the work environment during the operation of the robot arm
	5. Fully assembled, the solution must fit within an 11" × 9" × 17" space (interior of a standard photocopier paper box)
	6. The solution must interface with two ¼-20 threaded inserts on the base of the test environment that are spaced 3 inches apart

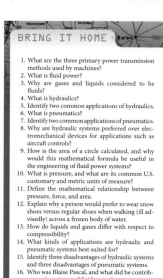

BRING IT HOME

1. What are the three primary power transmission methods used by machines?
2. What is fluid power?
3. Why are gases and liquids considered to be fluids?
4. What is hydraulics?
5. Identify two common applications of hydraulics.
6. What is pneumatics?
7. Identify two common applications of pneumatics.
8. Why are hydraulic systems preferred over electromechanical devices for applications such as aircraft controls?
9. How is the area of a circle calculated, and why would this mathematical formula be useful in the engineering of fluid power systems?
10. What is pressure, and what are its common U.S. customary and metric units of measure?
11. Define the mathematical relationship between pressure, force, and area.
12. Explain why a person would prefer to wear snow shoes versus regular shoes when walking (ill advisedly) across a frozen body of water.
13. How do liquids and gases differ with respect to compressibility?
14. What kinds of applications are hydraulic and pneumatic systems best suited for?
15. Identify three disadvantages of hydraulic systems and three disadvantages of pneumatic systems.
16. Who was Blaise Pascal, and what did he contribute to the science of fluid power?
17. What was the first hydraulic-powered machine to utilize Pascal's law, and who was responsible for its invention?
18. What are standards, and why do engineers develop them?
19. What are schematic symbols, and why are they used to communicate the design of fluid power systems?
20. What common components are found within both hydraulic and pneumatic systems?
21. What is the difference between a working line and a pilot line?
22. How would an engineer tell the difference between connected and nonconnected fluid power lines where the lines cross on a schematic diagram?
23. What is the purpose of a filter in a fluid power circuit?
24. What are the two most common types of linear actuators in fluid power systems, and how do they differ from each other?
25. What is the difference between a shutoff valve and a check valve, and what are these valves used for?
26. Under what conditions might an engineer use a shuttle valve in a fluid power circuit?
27. What is the purpose of a flow control valve?
28. Sketch the schematic symbols for a filter, single-acting cylinder, double-acting cylinder, shutoff valve, check valve, shuttle valve, and flow control valve.
29. What is the purpose of a reservoir in a hydraulic circuit?
30. What kind of mechanical device is used to generate pressure in a hydraulic system?
31. What is a prime mover, and is it needed in a fluid power system?
32. What is the purpose of a pressure relief valve in a hydraulic circuit?
33. What is the function of a directional control valve (DCV), and how are the ports on a hydraulic DCV identified?
34. Sketch the schematic symbols for a hydraulic reservoir, fixed-displacement pump, pressure relief valve, and a four-way tandem center condition DCV.
35. Why is ambient atmosphere put through a conditioning process before it is used in a pneumatic system?
36. What kind of mechanical device is used to generate the air pressure in a pneumatic system?
37. What is the function of a pneumatic pressure regulator?
38. Why is a lubricator used in a pneumatic system?
39. Where is compressed air stored before it is put to use in a pneumatic system?
40. Why might an engineer substitute a simple schematic symbol for a compressed air source in place of the schematic symbols used to identify all the conditioning devices that air must travel through before it is used in a pneumatic system?
41. Sketch the schematic symbols for an air compressor, an air filter outfitted with a manual drain, a pressure regulator, a flow meter, a lubricator, an air receiver tank, and a generic air supply.

▶ **Bring it Home:** Questions or activities are provided at the end of each chapter that are designed to reinforce and tax a student's understanding of the information that they have learned in the chapter.

▶ **Extra Mile**: *An Engineering Design Challenge* or *major exploration* at the end of each chapter provides extended learning opportunities for students who want an additional challenge.

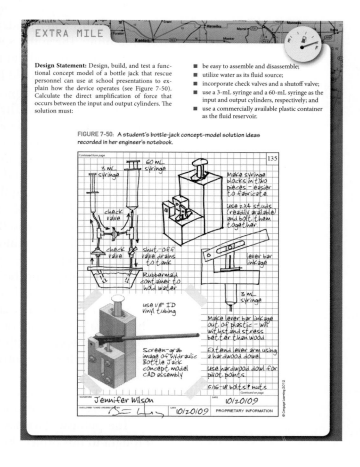

EXTRA MILE

Design Statement: Design, build, and test a functional concept model of a bottle jack that rescue personnel can use at school presentations to explain how the device operates (see Figure 7-50). Calculate the direct amplification of force that occurs between the input and output cylinders. The solution must:

■ be easy to assemble and disassemble;
■ utilize water as its fluid source;
■ incorporate check valves and a shutoff valve;
■ use a 3-mL syringe and a 60-mL syringe as the input and output cylinders, respectively; and
■ use a commercially available plastic container as the fluid reservoir.

FIGURE 7-50: A student's bottle-jack concept-model solution ideas recorded in her engineer's notebook.

Supplements

A complete supplements package accompanies this text to help instructors implement 21st-century strategies for teaching engineering design:

▶ A **Student Workbook** reinforces text concepts with practice exercises and hands-on activities.

▶ An **Instructor's E-Resource** includes solutions to text and workbook problems, instructional outlines and helpful teaching hints, a STEM mapping guide, PowerPoint presentations, and computerized testing options.

HOW *PRINCIPLES OF ENGINEERING* SUPPORTS STEM EDUCATION

Mathematics and science are the languages we use to communicate ideas about engineering and technology. It would be difficult to find even a single paragraph in this text that does not discuss science, technology, engineering, or mathematics. The authors have taken the extra step of showing the links that bind Mathematics and science to engineering and technology. The STEM icon shown here highlights passages throughout this text that explain how engineers and other technical professionals use Mathematics and science principles to support successful designs. In addition, the Instructor's E-Resource contains a STEM mapping guide to this career cluster.

ACKNOWLEDGMENTS

A text like *Principles of Engineering* could not be produced without the patient support of family and friends and the valuable contributions of a dedicated educational community. The authors wish to acknowledge several individuals for their support and patience throughout this process.

Brett Handley would like to recognize his parents, wife, and daughter for their patience, understanding, and unconditional support during the long road to this first edition textbook. Thanks are also given to his students, colleagues, and administrators who were never at a loss for encouraging words. Brett's efforts on this book are dedicated in memory of his grandmother, Laura Cornwall, who passed away shortly after the completion of the first draft.

Dave Marshall would like to thank his wife Melanie for her help and his two boys for understanding that sometimes the book came first. Special thanks to his students and his school's Tech Team for inspiring him with new ideas. Dave would like to acknowledge and thank Project Lead The Way® for the opportunity to teach, write curriculum, and interact with peers, all of which has enabled him to become a better teacher.

The prototype for the Project Lead The Way© Principles of Engineering course was developed in the late 1980s by our friend and retired colleague, Paul Kane, from Shenendehowa High School in Clifton Park, New York. Thanks for "leading the way," Paul! Had it not been for you, this book might never have been developed.

Many people contributed to the textbook beyond the named authors. We are very grateful for the contributions made by Mary Pat Shaffer, Allyson Powell, Dawn Jacobson, Kent Williams, and Lindsay Schmonsees. A special thank you is extended to our Project Manager, Mary Clyne, for her professionalism, patience, persistence, guidance, and understanding.

The publisher wishes to acknowledge the invaluable wisdom and experience brought to this project by our focus group and review panel.

Focus Group

Connie Bertucci, Victor High School, Victor, NY
Omar Garcia, Kearny High School, San Diego, CA
Brett Handley, Wheatland-Chili Middle/High School, Scottsville, NY
Donna Matteson, State University of New York at Oswego
Curt Reichwein, North Penn High School, Lansdale, PA
George Reluzco, Mohonasen High School, Rotterdam, NY
Mark Schroll, Program Coordinator, The Kern Family Foundation
Lynne Williams, Coronado High School, Colorado Springs, CO

Review Panel

Andy Brendel, Triad High School, Troy, IL
Kyle Buck, East Lake High School, Tarpon Springs, FL
Nick Havlik, Brookfield East High School, Brookfield, WI
Jason Huber, South Milwaukee High School, South Milwaukee, WI
David Lynch, Memorial High School, Madison, WI
Michael Martin, Martin Luther King High School, Riverside, CA

Tim Newton, Union-Endicott High School, Endicott, NY

Rick W. Orr, Weber State University, Ogden, UT

Dara M. Randerson, Oswego East High School, Oswego, IL

Daniel Spak, Firestone High School, Akron, OH

George Streit, Edgar High School, Edgar, WI

Peter Tucker, Triad High School, Troy, IL

The authors wish to recognize the extraordinary efforts of two reviewers:

Pamela Lottero Perdue, Towson University, Towson, Maryland

Barry Witte, South Colonie Central School District, Colonie, New York

Pam Lottero Perdue brought her perspective as an engineer, university professor, former POE Master Teacher, and teacher of young children to her thorough review of every chapter in this text. Barry Witte brought many years of experience in curriculum development and classroom instruction, along with a keen eye for detail to his review. In their generosity with time and talent, these two reviewers have demonstrated their commitment to quality engineering education. We are truly grateful for their contributions.

Technical Edit

The authors and the publisher extend special thanks to our patient and thorough technical editor: Professor Aly El-Iraki, Marine Engineering and Naval Architecture Department, Faculty of Engineering, Alexandria University, Egypt.

The publisher also thanks our special consultant for this series Aaron Clark, North Carolina State University, Raleigh, NC.

The publisher also especially thanks Project Lead The Way's© Director of Engineering Curriculum, Sam Cox, and Associate Director of Engineering Curriculum, Wes Terrell, for reviewing chapters at the manuscript stage.

ABOUT THE AUTHORS

Brett Handley is a Technology Education teacher at Wheatland-Chili Middle School/High School, a Project Lead The Way (PLTW) school. Mr. Handley currently teaches PLTW's Gateway to Technology program for middle school, and PLTW's high school courses in IED, CIM, and POE. He has served as a PLTW Master Teacher for the IED and POE courses and is a former PLTW Associate Director of Curriculum. Mr. Handley holds a BS in K–12 Technology Education and an MS in Professional Studies with a concentration in Engineering Education.

David Marshall is a Technology Education teacher at West Irondequoit School District, a Project Lead The Way School. Mr. Marshall currently teaches Principles of Engineering, Digital Electronics, Design and Drawing for Production, and Gateway to Technology. He helped to develop the most recent revision of the POE curriculum. Mr. Marshall has served as a POE Master Teacher for 15 years and continues to instruct from across the United States each summer. He holds a BS in K–12 Technology Education and an MS degree in Technology Education.

Craig Coon teaches Technology Education at the Brockport Central School District, a Project Lead The Way School. Mr. Coon has served as a POE Master Teacher for 10 years and has spent his summers instructing and collaborating with educators from across the country. He helped to develop the POE curriculum that is now being taught nationwide. Mr. Coon currently teaches Principles of Engineering and Engineering Design and Development. He holds a BS degree in K–12 Technology Education and an MS in Integrating Technology in the Classroom.

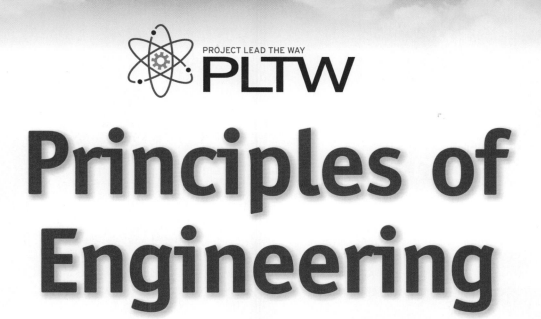

Principles of Engineering

CHAPTER 1
Overview and History of Engineering

Before You Begin

Think about these questions as you study the concepts in this chapter.

1. When did engineering begin?

2. What were some of the first engineering designs?

3. Who were the important pioneers in the field of engineering, and what did they contribute?

4. What are the first steps to becoming an engineer today?

5. How have advances in technology spurred growth in engineering fields?

Are you a problem solver? Are you curious about how things work? Do you enjoy looking for ways to make things better and more efficient? If you answered "yes" to any of these questions, maybe you should consider a career in the field of engineering.

An engineer applies knowledge of science and math to solve problems. Many people confuse the work of engineers with the work of scientists. As a student of engineering, you should learn how to define these separate disciplines. Scientists ask and answer questions, discovering new knowledge about how the world works. Engineers use the knowledge that scientists discover to solve problems and create technologies that satisfy human wants and needs.

People have been studying how the world works and using their powerful minds to improve the quality of life since prehistoric times. Prehistoric men and women used their ingenuity to survive hunger, climate conditions, and the occasional attack from enemies. In order to hunt, farm, fish, and fight, these people needed to think like engineers do today. In fact, many simple tools developed by prehistoric humans, such as the wedge, lever, and wheel, are still used in engineering today (Figure 1-1).

The pace of engineering developments quickened around 3000 B.C.E. The Egyptians built canals to control flooding and provide irrigation to feed their growing population. The Chinese invented paper, enabling architects to sketch masterpieces of the Middle Ages and allowing world explorers to develop navigational maps. Modern civilization—its sophistication, its diversity, and its vast reach—was built on many small improvements achieved throughout human history. In the course of about 15,000 years, humans have advanced from living in caves to inhabiting an international space station (Figure 1-2).

This chapter will take a brief look at the long history of engineering and show you some of the key moments in its astonishing evolution. The great scientist Isaac Newton once wrote to a friend, "If I have seen further

© iStockphoto.com /.rest.

Figure 1-1: *Archeologists can only guess at the tools and technology that prehistoric engineers used to construct Stonehenge in England 5,000 years ago. The builders probably used some of the same simple tools that we use in engineering today such as wheels, wedges, and levers.*

Courtesy of NASA.

Figure 1-2: *Mission specialist James Newman waves at a camera during an extravehicular activity (EVA) performed at the International Space Station.*

Engineer:

a person trained and skilled in the design and development of technological solutions to human problems.

than others, it is by standing upon the shoulders of giants." This chapter will introduce you to some of those giants—people who enjoyed solving problems, were curious about how things worked, and wanted to make them work more efficiently. As you read this chapter, you will see how their achievements still affect the quality of your daily life.

Engineering touches every aspect of our daily living. It helps us communicate, work, travel, and stay healthy. It is such a large discipline that it is broken into a few major categories: civil, electrical, chemical, and mechanical engineering. These general disciplines are further divided into specialized areas such as biomechanical, aerospace, and computer engineering. In this chapter, we will explore many of the career opportunities available to today's engineering students.

THE MESOPOTAMIANS

Thousands of years ago, the land between the Tigris and Euphrates Rivers was called Mesopotamia. Today, this land is within the country of Iraq (Figure 1-3). The invention of the wheel is said to have taken place in this area, along with the wheeled cart. In the southern part of Mesopotamia at the beginning of recorded history, Sumerians built canals, walls, and temples. These were the first known engineering accomplishments.

Astronomy was also explored in Mesopotamia. The oldest astrolabe, dated to around 2,000 B.C.E, was found there. An astrolabe (see Figure 1-4) is an instrument used to measure the altitude of the sun or stars. The Mesopotamian astrolabe was created by engraving three concentric circles divided into 12 sections

Point of Interest
The Code of Hammurabi

The Code of Hammurabi, inscribed in cuneiform script, is still studied in law schools. Modern scholars have translated portions of the code as follows:

229. If a builder build a house for someone and does not construct it properly, and the house which he built fall in and kill its owner, then that builder shall be put to death.
230. If it kill the son of the owner, the son of that builder shall be put to death.
231. If it kill a slave of the owner, then he shall pay slave for slave to the owner of the house.

232. If it ruins goods, he shall make compensation for all that has been ruined, and inasmuch as he did not construct properly this house which he built and it fell, he shall re-erect the house from his own means.
233. If a builder builds a house for someone, even though he has not yet completed it, if then the walls seem toppling, the builder must make the walls solid from his own means.

Adapted from translation by L. W. King. Source: http://avalon.law.yale.edu/ancient/hamframe.asp, © 2008 Lillian Goldman Law Library, 127 Wall Street, New Haven, CT 06511; accessed 3/17/2010.

on clay tablets. Astrolabe technology was used to solve astronomical or navigational problems and wasn't replaced by a more precise instrument until the Middle Ages.

The best-known rulers in Mesopotamia at this time were the Assyrians and the Babylonians. Hammurabi was one of Babylonia's great kings. He created what some say was the first building code, providing a system for administering penalties for poor construction. The "Code of Hammurabi" established a set of building ethics, reinforced the importance of quality, and specified consequences for breaking the code.

FIGURE 1-3a and b: The Mesopotamians lived and worked in present-day Iraq, occupying the land between the Tigris and Euphrates Rivers. They achieved some of the earliest known engineering accomplishments.

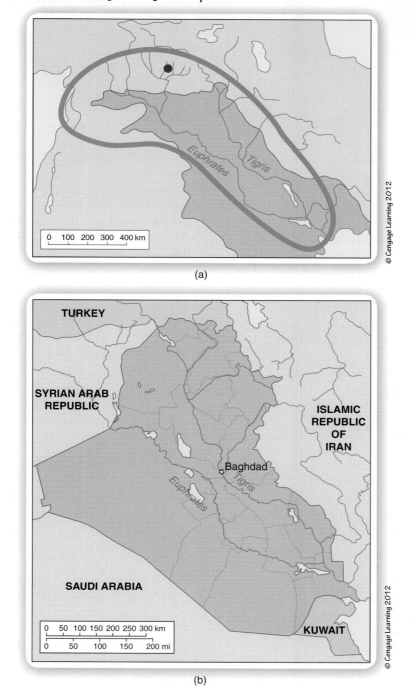

(a)

(b)

FIGURE 1-4: Mesopotamians used clay tablets to make astronomical measurements; this ancient instrument is known as an *astrolabe*. Although different materials were used in later centuries, this instrument wasn't replaced by a newer technology until the Middle Ages.

Horizon

The people of Mesopotamia developed ways to control floods and irrigate fields. Evidence of these irrigation systems is still present today. The remains of the Nahrwan Canal, a canal system that paralleled the Tigris, is one example. The canal was 400 feet wide and extended about 200 miles.

Around 700 B.C.E., the Assyrians completed the first public water supply. They constructed a freshwater canal system that extended from the mountains of Tas to the existing Khosr River. The river then carried the water to Ninevah, a total of 45 miles. In Jerwan, an aqueduct was built from cut stone, designed to carry the freshwater over an existing stream. The cut stone aqueduct was 863 feet long, 68 feet wide, and 28 feet at its highest point. The channel through which the water flowed was 50 feet wide and 5 feet deep. The channel was lined with concrete and is the first known use of concrete for construction.

THE EGYPTIANS

The Egyptians used the natural wealth of the Nile River to support their complex technological society. Each year, the Nile rose in the spring and flooded the valley. When the flood waters receded during the dry season, the river left behind a new layer of silt. The silt was filled with rich nutrients that made the Nile River valley an especially fertile place for growing food. By developing a controlled irrigation system, the Egyptians were able to use the rich soil in the valley to grow food during the dry season.

The Egyptians' ability to adapt to the Nile's flooding cycle was a key to their success. With an ample food supply, Egyptians could focus on social and cultural development. We can still see evidence of their complex culture in their construction projects.

The Egyptians used many different building styles during their 2,000-year history, but they are best known for the pyramids. Three famous examples are still standing at Giza (Figure 1-5). The largest one is known as the Great Pyramid

Irrigation:
the application of water to crop-producing land through artificial means.

FIGURE 1-5: The Greeks considered the Great Pyramid at Giza to be one of the seven wonders of the ancient world. It is the only one of those structures that remains intact.

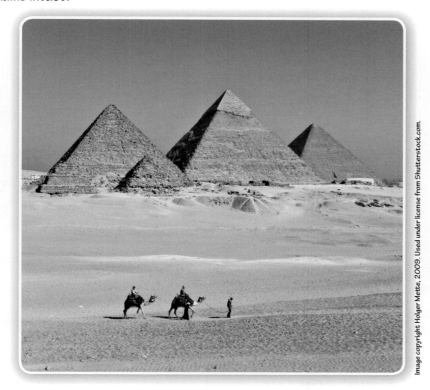

Image copyright Holger Mette, 2009. Used under license from Shutterstock.com.

or the Pyramid of Cheops. It stands 481 feet high and covers 13 acres at its base. It is amazing to consider that these architectural wonders were built without power tools.

In ancient Egypt, the king chose a trusted individual to serve as "chief of works." An expert in general construction, the chief advised the king on plans for irrigation, flood control, and surveying. In today's world, an analogy would be the appointment of the secretary of the U.S. Department of Transportation by the president of the United States. This trusted individual leads the organization whose mission is to "[s]erve the United States by ensuring a fast, safe, efficient, accessible and convenient transportation system that meets our vital national interests and enhances the quality of life of the American people, today and into the future." In the United States, the appointment is approved by Congress.

THE GREEKS

The Greek landscape forced its inhabitants to live near the sea. By focusing their efforts on harbor construction and ship building, the Greeks became the navigational leaders of their time. They were the first to construct a lighthouse, built at the port of Alexandria in what is now Egypt around 300 B.C.E. (Figure 1-6). The Pharos lighthouse stood 370 feet tall and was named one of the seven wonders of the ancient world.

During the rule of Pericles, Athens' leaders sought to make their city the most beautiful metropolis. Pericles rounded up the best artists, architects, and builders to create well-known shrines, statues, and structures. They used simple machines, timber frames, and hand-powered lifts (similar to those used today) to construct the columns, beams, and pillars that we still identify with classical Greek architecture.

FIGURE 1-6: The Greeks built the Pharos lighthouse at the port of Alexandria about 2,300 years ago and called it one of the seven wonders of the ancient world.

© Cengage Learning 2012

FIGURE 1-7: Roman engineers focused on building functional structures such as aqueducts, which carried water to municipalities miles away from the source. The Segovia aqueduct in what is now Spain carried water from the Frío River 10 miles away.

Image copyright Jose Ignacio Soto, 2010. Used under license from Shutterstock.com.

THE ROMANS

The Romans had realistic engineering goals. They kept things simple in design and form but massive in scale. They were less concerned with artistic appeal than with the function of the structures they built; see Figure 1-7.

The civil engineers of ancient Rome focused on public works projects built by slaves. The Romans are known for their bath houses, arenas, roadways, temples, and public forums.

Some of the greatest Roman engineering feats are the new construction methods and techniques they invented. They developed cement and numerous construction machines such as the treadmill hoist, pile driver, and the wooden bucket wheel (Figure 1-8).

THE MIDDLE AGES

Historians describe the time after the fall of the Roman Empire as the Middle Ages. This was a slow time for engineering developments. Even so, we do attribute some advances in structural design to medieval inventors. They also developed energy-efficient machines and power-saving devices.

Many useful inventions that are still a part of our daily life came from China during the Middle Ages. These inventions include gun powder, papermaking, iron casting, and textiles.

ENGINEERING PIONEERS

Many advances were made in science and technology during the 15th, 16th, and 17th centuries that held tremendous promise for the industrial development of the world. In the next section, we will introduce a few of this era's important pioneers

© Cengage Learning 2012

in science and technology and explain how their achievements remain relevant to the study and practice of engineering today.

Technology That Continues to Improve

During the late Middle Ages, Johann Gutenberg (1398–1468) invented the movable type mold (Figure 1-9). Gutenberg is credited with the first printed book in A.D. 1450.

Gutenberg's printing press was the spark that ignited the flame of widespread communication. His invention came at a time when societies felt a growing need for a convenient way to share information. People still have this need today, and the printing industry remains a large business even as methods for sharing information have adapted to a digital environment.

Various forms of the printing press already existed at the time of Gutenberg's invention. As early as the 11th century, the Chinese had developed a set of movable letters from a baked mixture of clay and glue. The letter pieces were stuck onto a plate where impressions could be taken by pressing paper onto the stationary letters. Because the letters were glued to the plate, the plate could be heated and the letters removed and resituated. The process was not used in any form of mass production, but it did form a basis for the future invention of Gutenberg's printing press. In Europe, would-be printers had attempted to create presses that could produce cheap playing

FIGURE 1-9: Johann Gutenberg's printing press allowed societies to share a growing body of information about science and engineering.

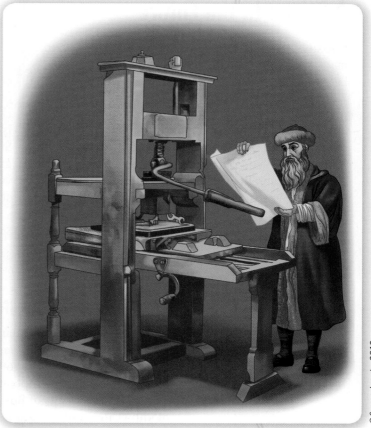

© Cengage Learning 2012

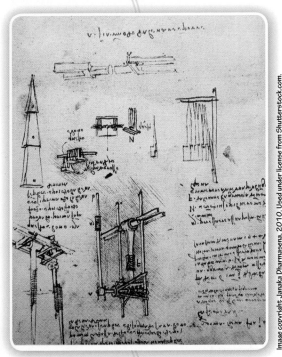

cards and frivolous items, but Gutenberg saw the practical need for a press that could provide common people with reading material. Having the ability to print and store information in written form allowed for the fast spread of knowledge. By 1500, books on chemistry, mining, metallurgy, and many other science and engineering topics were in print.

Today, there are many printing methods available to address commercial and consumer needs. In your own home, you probably have an ink-jet printer that sprays tiny drops of ink onto paper. Or maybe you have a laser printer, which uses static electricity to force toner to stick to the paper until it is fused by heated rollers. A variety of methods exist today to print newspapers, packaging, and even t-shirts.

Technology Before Its Time

Leonardo da Vinci (1452–1519), the Renaissance painter of the *Mona Lisa*, used his artistic talent to observe and illustrate how machines work. With the knowledge he gained, Leonardo was able to make mechanical improvements and invent new machines. That is exactly what engineers do today. Leonardo's inventions included musical instruments, hydraulic pumps, reversible crank mechanisms, finned mortar shells, and a steam cannon; see Figure 1-10.

Point of Interest
A Bridge to the Future

Not all of Leonardo's inventions were practical, and many were before their time. For example, in 1502, Leonardo produced a drawing for a single-span, 720-foot bridge crossing the Bosphorus River. The project was part of a civil engineering contract with Sultan Bajazet II of Constantinople, in what is now Turkey. The bridge was never built—Leonardo's ideas for its construction were about 300 years ahead of generally accepted engineering practices. Nearly 500 years later, Leonardo's simple but powerful drawing drew the attention of Norwegian artist Vebjørn Sand. Sand was inspired by the drawing to create the Leonardo Project to bring Leonardo's vision to life. Five hundred years after Leonardo created these drawings, the Leonardo Bridge was constructed and opened for pedestrians and bicyclists on October 31, 2001, in Ås, Norway. Sand is currently considering several sites in the United States for the next Leonardo bridge project. You can read more about the Leonardo Project at www.vebjorn-sand.com and view a slide show of its construction at www.leonardobridgeproject.org.

Source: www.vebjorn-sand.com. Accessed 1/26/2010.

Connections Between Science and Engineering

Many scientific laws are named for the notable scientists who discovered them. Boyle's law, for example, was named after Robert Boyle, who lived in Ireland and England from 1627 until 1691. Boyle's law states that if the volume of a gas is decreased at constant temperature, the pressure increases proportionally. This law, along with the results of Boyle's other gas experiments, formed an important body of knowledge underlying many pneumatic inventions.

Robert Hooke (1635–1703) worked for Robert Boyle. Among Hooke's many engineering achievements, he created the air pump on which Boyle's experiments could be conducted. Hooke and Boyle's working relationship provides an excellent example of how science and engineering minds work together. Much of Boyle's work on gases might have been inspired by, if not strongly based on, work carried out by Hooke's work with clocks, springs, and gases.

Hooke published an important book titled *Micrographia*, which became the foundation for the science of **microscopy**. Hooke's use of the microscope to investigate a previously invisible world led to more fascinating discoveries and new inventions. For example, Hooke observed through the microscope that hairs from the beard of a goat would bend when dry and straighten when wet. This understanding led to his invention of the **hygrometer**, a device we still use today to measure water vapor content in the air (Figure 1-11).

Microscopy:
the use of a microscope for investigation.

Hygrometer:
an instrument used to measure the water vapor content in the air.

FIGURE 1-11: Hygrometers are used to measure humidity and help control indoor environmental air quality. These instruments still rely on principles Hooke discovered more than 300 years ago while studying goat hairs under a microscope.

© Andreas Rehl/IStockphoto.com.

Connections Between Math and Engineering

Math is an indispensable tool in the engineer's toolbox of skills. Engineers and scientists rely on calculus to solve complex problems. Although many mathematicians all over the world contributed to the development of calculus, Gottfried Wilhelm Leibniz (1646–1716) and Isaac Newton (1642–1727) are both credited its invention (Figure 1-12).

There are two branches of calculus. *Differential calculus* determines the slope or steepness of a complicated curve. For example, we can calculate the slope of a mountain that might become a new ski area or determine the speed of a new roller coaster using differential calculus.

FIGURE 1-12: Sir Isaac Newton is one of the most influential mathematicians and scientists in history. His invention of calculus enabled engineers to solve problems that regular mathematics could not address.

© iStockphoto.com/GeorgiosArt.

Integral calculus helps us find the area or the volume of a complex figure. For example, we can use integral calculus to determine the amount of water needed to fill an unusually shaped pool.

THE AGE OF TRANSPORTATION

Since the invention of the wheeled cart, engineers have been continuously improving how people travel. In the 1760s, James Watt's working model of a steam engine led indirectly to the age of transportation.

Steam Engines as Power for Transportation

Collaborating with Matthew Boulton, a well-known manufacturer, Watt produced hundreds of steam engines. These early machines had little to do with transportation. The Boulton and Watt engines were used in England during the 1800s to pump water out of mines and to drive machinery that helped power textile mills and iron works.

Experimentation led to advances in steam engine technology. These advances made steam engines useful for transportation in applications such as steam boats and locomotives. The first commercially run steam boat, *The Clermont*, was created by Robert Fulton in 1807 and ran between New York City and Albany on the Hudson River. Fulton's first trip took 32 hours.

In 1823, Englishman George Stephenson demonstrated the feasibility of a steam-powered railroad transportation system. His invention sparked the growth of railway systems everywhere. By the end of the Civil War, the United States had 35,000 miles of railway; that increased to more than 190,000 miles by 1900.

Canals as Conduits for Transportation

Other forms of transportation systems were also growing quickly. Extensive canal systems were being constructed in England from 1780 through 1900 and in the United States during the first half of the 19th century (see Figure 1-13).

FIGURE 1-13: Important U.S. Canals, 1800–1850
Several important canals improved transportation in the United States in the first half of the 19th century.

Waterway	Constructed	Connects
Erie Canal	1817–1825	New York state cities between Buffalo and Albany
Ohio and Erie Canal	1828–1836	Cleveland to Portsmouth on the Ohio River
Chesapeake and Ohio Canal	1828–1850	Washington, D.C., to Cumberland, Maryland

Although the commercial use of these canals declined in the 20th century, we still rely on the Panama Canal to facilitate international trade. The Panama Canal was one of the most important engineering projects of the 20th century. Finished in 1914, the canal is 50 miles long, 110 feet wide, and 70 feet deep. It reduced the distance ships had to travel by about 5,000 miles when sailing from the East Coast of the United States to the West Coast (Figure 1-14).

The Need for Better Roads for Transportation

The evolution of the automobile in the early 1900s transformed our society. The number of autos on the road exploded in 1904 when Henry Ford implemented his vision of the assembly line to mass produce high-quality, affordable vehicles (Figure 1-15). By the end of the 20th century, 90 percent of U.S. homes had automobiles.

FIGURE 1-14: The Panama canal was a 50-mile project that could shorten a ship's journey from the U.S. East Coast to West Coast by 5,000 miles.

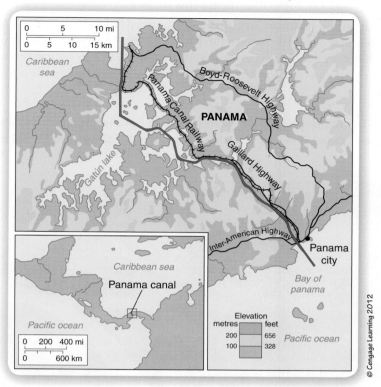

FIGURE 1-15: Henry Ford's innovations to assembly-line technology allowed his company to produce a high-quality, affordable product. The Model T remains a symbol of mass production.

As cars became more affordable and people wanted to travel farther, it was necessary to build better roads. The era of road building was first influenced by John MacAdam (Scotland, 1756–1836), who devised a method for compacting broken and crushed stone. MacAdam used his methods of construction to pave 180 miles of turnpike. MacAdam's Scottish compatriot, Thomas Telford, used large, flat stones set on edge and wedged them together to form a solid, flat surface. Telford's innovation created smoother, more durable roadways.

The Federal Aid Highway Act of 1956 authorized federal funding to build more than 45,500 miles of paved roads in the United States. The total cost of the resulting interstate highway system was estimated in 1991 to be $128.9 billion (Figure 1-16). Engineers still play an important role in maintaining existing roads and building new ones.

FIGURE 1-16: Mass-produced automobiles had a transforming effect on U.S. society, leading indirectly to the development of the interstate highway system.

Taking to the Sky

At the beginning of the 20th century, many inventors were focused on a single goal: heavier-than-air flight. Brothers Orville and Wilbur Wright first achieved success in 1903 with a 12-second, 120-foot flight (Figure 1-17). In 1905, Charles and Gabriel Voisin started the first aircraft company in France, building custom planes. The first American aircraft company was formed in 1911 by Glenn Curtiss in Hammondsport, New York. Curtiss switched from manufacturing motorcycles to manufacturing airplanes and is the inventor of the seaplane. Today, air transportation makes long-distance travel practical and affordable.

Moving Underground

Engineers first envisioned the Channel Tunnel, or "Chunnel," in the early 1800s. Several attempts to construct an underwater passage from England to France failed before the Chunnel opened in 1994. Construction started in 1988 with 11 boring machines tunneling their way toward each other from France to Britain and from Britain to France. Think of the mathematical calculations of speed and distance that had to be so accurate that the two opposite tunnels could join at the exact same point successfully. Today, high-speed trains travel this modern marvel of engineering.

Courtesy of NASA.

THE AGE OF ELECTRICITY

The fundamental nature of electricity was first studied by physicists in the early 1800s. Countless scientists have worked to develop **electricity** as a source of power. Over the years, engineers have harnessed electrical energy to solve many problems.

Can you imagine what your life would be like without batteries? Just consider how many batteries are at work in your home right now, supplying power to watches, cell phones, remote controls, and other portable electronics (Figure 1-18). When you use battery-powered devices, you are using electricity in the form of direct current (DC) (Figures 1-19 a and b). In direct current, electrons move in the same direction.

Alessandro Volta invented the first electric battery in 1827. To honor his contribution to science and technology, Volta's name was chosen to describe the electromotive force that moves electric current: the volt.

In alternating current (AC), electrons move forward and backward at extremely high speeds. AC is the form of electricity that power supply companies deliver to U.S. homes. Household appliances such as refrigerators, clothes dryers, and televisions plug directly into wall sockets and use alternating current (Figures 1-19c and d).

Electricity:

the transfer of energy through the flow of electrons along a conductor.

FIGURE 1-18: What would your life be like without batteries?

Image copyright monkeybusiness, 2010. Used under license from Shutterstock.com.

FIGURE 1-19: Products that rely on batteries for power use direct current to drive internal motors. An (a) electric toothbrush and (b) car starter both use DC. A household appliance that plugs in such as (c) a hair dryer or (d) a refrigerator uses alternating current to drive motors.

(a)

(b)

(c)

(d)

Nikola Tesla played a major role in deciding how we use AC today. Tesla's inventions influenced how electricity is generated, transmitted, and converted to mechanical power today. He invented the AC induction motor and designed the Niagara Falls Power Station, which was completed in 1895 and lit up Broadway in New York City about 400 miles away.

We all know Thomas Edison as the inventor of the incandescent bulb (Figure 1-20), but he also holds an astounding 1,093 U.S. patents. Edison developed many of the principles of mass production and is credited with creating the first industrial research laboratory. Electric power generation and the distribution of electricity to factories and homes were first commercially implemented in New York City thanks to Edison's work.

FIGURE 1-20: Thomas Edison's 1,093 patents cover a wide range of familiar and unfamiliar endeavors including electric lighting systems, the phonograph, movies, batteries, automobiles, cement, rubber, the telegraph, the telephone, and a method for preserving fruit.

Courtesy of Library of Congress/Brady-Handy Photograph Collection.

(a)

© iStockphoto.com/Felix Möckel.

(b)

Inventors and innovators in the 20th century took full advantage of the power of electricity to create numerous applications that would vastly improve the quality of life. In fact, we use so much electricity that we are now creating technologies to help us use energy more efficiently. ENERGY STAR is a joint program of the U.S. Environmental Protection Agency and the U.S. Department of Energy. The ENERGY STAR mission is to help consumers save money and to protect the environment by using energy-efficient products and practices. This program encourages manufacturers to design energy-efficient household appliances and home-building materials such as windows and lighting. ENERGY STAR challenges engineers to design new products that use less energy without sacrificing comfort or quality.

Your Turn

When the Burj Khalifa opened in downtown Dubai, United Arab Emirates, in January 2010, it took the title of world's tallest building. The Burj stands 2,625 feet tall and holds more than 160 stories (Figure 1-21). Investigate the world's tallest building on the Internet and identify some of the challenges and solutions the numerous engineers needed to solve during this 6-year project.

FIGURE 1-21: The Burj Khalifa is the tallest building in the world, a towering 2,625 feet.

Engineers have improved the quality of life by designing systems for supplying clean water, developing genetically modified crops, and creating better medicines and medical procedures. By improving the human condition, engineers play a vital role in our society. Let's take a closer look at some important engineering disciplines and at the kind of education and training that students need to become engineers.

ENGINEERING SOCIETIES

Engineering first achieved formal recognition as a profession in the late 19th century. John Smeaton was the first person to call himself a civil engineer. Hoping to attract others with similar interests, Smeaton created an engineering society in 1771. He realized that such an organization would help validate the profession. In 1881, the Institution of Civil Engineers was formed, and Thomas Telford, the great road builder, was elected as its first president. Soon after that, the Institute of Mechanical Engineers was formed with George Stephenson, the man associated with the introduction of steam-powered rail, as its first president.

By 1908, civil, mechanical, electrical, chemical, mining, and metallurgical engineering all had societies in the United States (Figure 1-22).

FIGURE 1-22: Engineering Societies in the United States

1852	American Society of Civil Engineers
1871	American Institute of Mining, Metallurgical, and Petroleum Engineers
1880	American Society of Mechanical Engineers
1884	Institute of Electrical and Electronics Engineers
1908	American Institute of Chemical Engineers

PREPARING FOR THE FIELD OF ENGINEERING

When should you start thinking seriously about your education and how it can shape your future? Now! High school is a great place to prepare for a future in engineering. Many schools offer engineering programs or STEM—science, technology, engineering, and mathematics—classes. STEM classes often provide collaboration between these discplines that reinforces the connections between them (Figure 1-23). Your guidance department can help you identify the courses that will best prepare you for the field of engineering.

To seek employment as an engineer, you will need to earn a bachelor of science (BS) degree from a college or university. This degree might take four or five years to complete, depending on where you choose to study. Spend some time researching which college is right for you. Plan a visit to the school to ask professors questions. More important, ask the students some questions. Ask them why they are pursuing a degree in engineering and what made them choose the school they are attending.

Engineering programs are accredited through the Accreditation Board of Engineering & Technology (ABET), an organization that helps standardize requirements among colleges and universities

After receiving your degree, you will most likely do one of two things. You might enter the workforce within your area of specialization, or you might pursue a secondary degree in your area of specialization. In many states, a test is required for you to become a licensed engineer. Some colleges offer a co-op or work-study program. These programs give students valuable real-world experience that can be helpful in landing that first job. Many times, a student co-op leads to a full-time job.

FIGURE 1-23: STEM classes help students see how their studies of math and science can support the projects they do in engineering and technology.

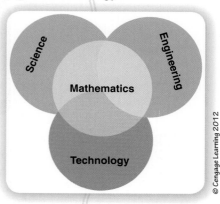

© Cengage Learning 2012

WHAT DO ENGINEERS DO?

As you study engineering principles and their applications in this textbook, you will discover that engineers have many possible career paths that lead to a wide variety of engineering disciplines. The major engineering categories are civil, electrical, chemical, and mechanical. These disciplines are so broad that some engineers pursue a specialty within the category; others blend two categories together. For example, aerospace engineering is considered a subcategory of mechanical engineering, whereas an industrial engineer blends electrical and mechanical disciplines together. Let's take a look at what some engineers do.

Aerospace Engineer

Aerospace engineering is concerned with engineering applications in the area of aeronautics (the science of air flight) and astronautics (the science of space flight). Aerospace engineering deals with flight of every kind: balloon flight, sailplanes, propeller- and jet-powered aircraft, missiles, rockets, and satellites, as well as advanced interplanetary concepts such as ion-propulsion rockets and solar-wind vehicles (Figure 1-24). Aerospace engineering will be responsible for making future interplanetary travel possible. The challenge is to produce vehicles that can traverse the long distances of space in ever-shorter periods of time. New propulsion systems will require that engineers venture into areas never before imagined.

Agricultural Engineering

At the turn of the 20th century, a large majority of the working population was engaged in **agriculture**. As farming efficiency increased and people left farming

Agriculture:
the activity of producing crops or raising animals.

FIGURE 1-24: **This aerospace engineer is inspecting the blades of an industrial wind tunnel.**

for jobs in industry, fewer farmers were needed. Contemporary farmers are able to feed about 10 times as many families as their ancestors did 100 years ago. Today, less than 5 percent of the U.S. population works on farms. How are fewer farmers able to feed a larger population? The application of new technologies and engineering to agricultural practices has vastly increased farmers' efficiency and productivity. Agricultural engineers design machinery to plant, fertilize, and harvest food crops. They design structures for crop storage and develop methods to conserve soil and water. They even produce new seeds and plants that are genetically engineered to resist pests (Figure 1-25).

FIGURE 1-25: **Some agricultural engineers use genetic engineering to improve crop resiliency.**

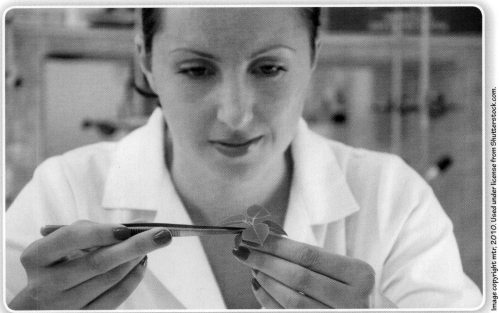

FIGURE 1-26: *Farmers today rely heavily on chemical engineering.*

Chemical Engineering

Chemistry has been studied for centuries, but chemical engineering began its rise to prominence shortly after World War I. Chemical engineering uses the science of chemistry to change the composition or properties of materials and make them more useful for industrial processes. Chemical engineers are involved in manufacturing polymers, drugs, paints, pesticides, cosmetics, and food (Figure 1-26). They work in oil refining, paper products and textiles manufacturing, and the extraction of metals from ores. Many specialty areas within chemical engineering are growing rapidly.

Civil Engineering

Civil engineers provide communities with structures such as buildings, bridges, dams, and roads (Figure 1-27). Civil engineers are concerned with planning, designing, and supervising the construction of structures. Consider the challenges that civil engineers have as they design and build structures that can withstand earthquakes and extreme weather.

FIGURE 1-27: *The Oresund Bridge between Denmark and Sweden is the world's longest single bridge carrying both road and railway traffic. Civil engineers worked on this project from 1991 to 1999.*

Today, civil engineering focuses on structures and structural systems such as bridges, skyscrapers, tunnels, canals, rapid-transit systems, highway systems, recreational facilities, industrial plants, dams, nuclear power plants, railroad lines, harbors, and off-shore oil and gas facilities. When civil engineers improve the movement of people on roadways, they are also protecting society and the environment. Reducing traffic can reduce toxic emissions from individual vehicles, and improvements in highway design can minimize the number of accidents. The civil engineer is always at the center of discussions when municipalities are planning new buildings and transportation systems.

Computer Hardware Engineering

Large, medium, and small organizations alike all depend on computers to manage inventories, create invoices, record sales, and communicate with customers. The amount of data storage possible makes the computer indispensable. Computers of the past were the size of a room; today's computers have been downsized to the size of your fist with computing speeds that were unthinkable just a year ago. The world has become almost completely computer centered, and the computer's role should only increase, thanks to engineers who focus their efforts on the research, design, and testing of computer hardware components (Figure 1-28).

Electrical Engineering

Electrical engineers solve problems using electrical principles. These principles are relatively new to us—Georg Ohm and Michael Faraday were only beginning to build our understanding of this science in the 19th century. Our first electrical engineers included inventors such Thomas Edison, who supplied DC to New York City residents in 1882, and Guglielmo Marconi, who made the first wireless radio transmission in 1896. Imagine how many alterations there have been to take us from the first electronic television using cathode ray tubes in 1931 to today's liquid crystal display (LCD) and plasma flat-screen TVs (Figure 1-29a and b). Every household gadget has been transformed by electrical engineers.

Industrial Engineering

Industrial engineering is a growing branch of the engineering family. Industrial engineers study engineering principles and techniques that can help make manufacturing processes more efficient and more profitable. These innovators are skilled at finding ways to do more with less. Their efforts can help improve customer service and product quality, make workplaces safer, and make the work easier. Industrial engineers must design, install, and improve systems that integrate people, materials, and equipment to efficiently produce goods (Figure 1-30). They must coordinate their understanding of the physical and social sciences with the activities of workers to design areas in which workers will produce the best results.

FIGURE 1-29: From the (a) cathode ray tubes of the 1950s to (b) today's flat screens, television technology has enjoyed a dramatic evolution.

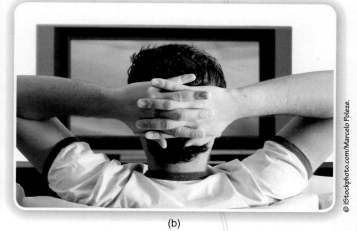

(a)

(b)

FIGURE 1-30: Industrial designers consider the efficiency of manufacturing process and the safety of industry workers.

Mechanical Engineering

Automobiles, engines, heating and air-conditioning systems, gas and steam turbines, air and space vehicles, trains, ships, servomechanisms, transmission mechanisms, radiators, **mechatronics**, and pumps are a few of the systems and devices that require knowledge of mechanical engineering. Mechanical engineering deals with power, its generation, and its application. Power affects the rate of change or "motion" of something. This can be a change in temperature or a change in motion because of outside stimulus.

Mechanical engineers use their knowledge of how things move to accomplish a variety of tasks. The breadth of study required to become a mechanical engineer allows students to diversify into many other engineering areas. The major specialty areas of mechanical engineering are applied mechanics, control, design, engines and power plants, energy, fluids, lubrication, materials, pressure vessels and piping, transportation and aerospace, and heating, ventilation, and air conditioning (HVAC). Are you interested in continuing to shape the world to become a better place? If you are up to a challenge and your curiosity can be harnessed to solve problems, consider a career in engineering.

Mechatronics:

the study of the combination of computer engineering, electronic engineering, and mechanical engineering.

SUMMARY

- Our world has been shaped by people from ancient civilizations who thought and behaved like engineers.

- Modern civilization was built on many small improvements throughout human history to satisfy human wants and needs.

- Mesopotamians invented the wheel, established a building code, and developed ways to control floods and irrigate fields.

- The Nile River's cyclical flooding and the use of irrigation allowed the Egyptians to develop highly productive agricultural systems. An abundant food supply gave the ancient Egyptians freedom to develop a complex technological culture and build great structures such as the pyramids.

- The coastal-dwelling Greeks became navigational leaders and built the first lighthouse.

- The Greeks valued aesthetics in building projects. They used simple machines, timber frames, and hand-powered lifts to build classical works of architecture.

- The Romans valued function over form in building projects. They were the first to use cement to construct public works projects.

- Medieval inventors advanced structural design and created energy-efficient machines and power-saving devices.

- Gunpowder and paper were introduced by the Chinese during the Middle Ages.

- Many important pioneers in science contributed to our current knowledge about engineering. The work of these pioneers often shows the connections between engineers, scientists, and mathematicians.

- Over the course of 200 years, engineers have improved transportation by harnessing the power of steam to move boats and trains and by mastering heavier-than-air flight. They have decreased travel time by building canals, better roads, and tunnels.

- Alessandro Volta invented the first electric battery, and his name was chosen to describe the electromotive force that moves electric current: the volt.

- Battery-powered devices use electricity in the form of direct current (DC), whereas household appliances plug directly into wall sockets and use alternating current (AC).

- Engineering societies provide an opportunity for people with like interests and skills to share ideas.

- Preparing for the field of engineering starts in high school by taking courses that introduce problem solving using science, math, and engineering concepts.

- The field of engineering is divided into the broad categories of civil, electrical, chemical, and mechanical engineering. Engineers often pursue specialties within these broad categories.

1. Describe the difference between a scientist and an engineer.
2. What type of standards did the Code of Hammurabi establish, and how are they used today?
3. What important agricultural invention did the Mesopotamians and Egyptians develop?
4. How did the landscape in Greece affect the ancient Greeks' development of technology?
5. Describe two aspects of ancient Roman engineering that distinguish it from ancient Greek engineering.
6. What civilization invented gunpowder, and when?
7. What invention allowed scholars to spread their knowledge more quickly during the 15th, 16th, and 17th centuries?
8. What were the first uses of the steam engine?
9. How did the steam engine adapt to use in transportation systems?
10. What major construction projects were completed during the 19th and 20th centuries that led to better and faster travel?
11. Name some important figures in the history of flight.
12. What did Alessandro Volta invent to provide direct current to today's portable electronic devices?
13. What did Nikola Tesla do that influenced the way alternating current is used today?
14. Who was first person to call himself a civil engineer? How did he help bring formal recognition to the profession?
15. When should you start researching engineering as a potential career path?
16. Name the four broad branches of engineering.
17. What specific engineering discipline interests you the most? Why?

EXTRA MILE

Identify an engineer in your community and conduct an informational interview. Your objective is to:

- Discover why he or she wanted to be an engineer and when.
- Summarize the person's education.
- Determine who the engineer works for and what his or her responsibilities include.
- Define the skills needed for their job.
- Inquire what advice the engineer would give to an aspiring engineer.

ENGINEERING TIME LINE

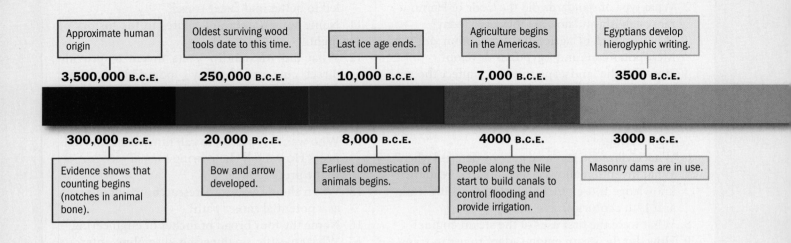

Approximate human origin	Oldest surviving wood tools date to this time.	Last ice age ends.	Agriculture begins in the Americas.	Egyptians develop hieroglyphic writing.
3,500,000 B.C.E.	**250,000 B.C.E.**	**10,000 B.C.E.**	**7,000 B.C.E.**	**3500 B.C.E.**

300,000 B.C.E.	**20,000 B.C.E.**	**8,000 B.C.E.**	**4000 B.C.E.**	**3000 B.C.E.**
Evidence shows that counting begins (notches in animal bone).	Bow and arrow developed.	Earliest domestication of animals begins.	People along the Nile start to build canals to control flooding and provide irrigation.	Masonry dams are in use.

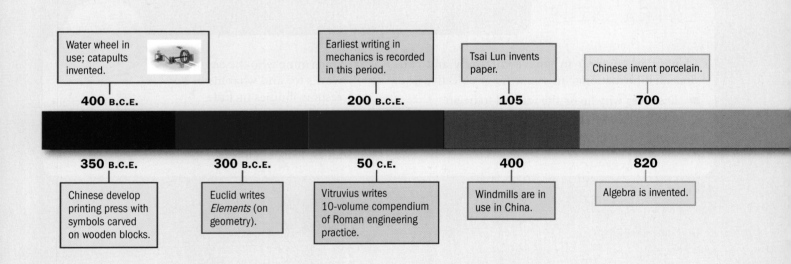

Water wheel in use; catapults invented.		Earliest writing in mechanics is recorded in this period.	Tsai Lun invents paper.	Chinese invent porcelain.
400 B.C.E.		**200 B.C.E.**	**105**	**700**

350 B.C.E.	**300 B.C.E.**	**50 C.E.**	**400**	**820**
Chinese develop printing press with symbols carved on wooden blocks.	Euclid writes *Elements* (on geometry).	Vitruvius writes 10-volume compendium of Roman engineering practice.	Windmills are in use in China.	Algebra is invented.

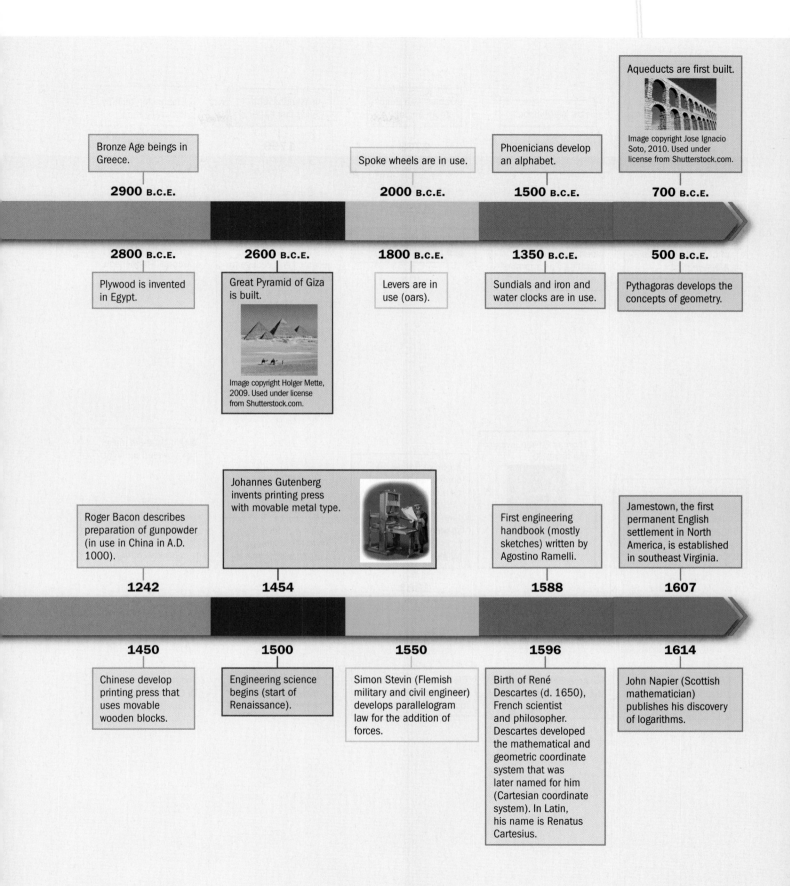

Bronze Age beings in Greece.

2900 B.C.E.

Spoke wheels are in use.

2000 B.C.E.

Phoenicians develop an alphabet.

1500 B.C.E.

Aqueducts are first built.

Image copyright Jose Ignacio Soto, 2010. Used under license from Shutterstock.com.

700 B.C.E.

2800 B.C.E.

Plywood is invented in Egypt.

2600 B.C.E.

Great Pyramid of Giza is built.

Image copyright Holger Mette, 2009. Used under license from Shutterstock.com.

1800 B.C.E.

Levers are in use (oars).

1350 B.C.E.

Sundials and iron and water clocks are in use.

500 B.C.E.

Pythagoras develops the concepts of geometry.

Johannes Gutenberg invents printing press with movable metal type.

Roger Bacon describes preparation of gunpowder (in use in China in A.D. 1000).

1242

1454

First engineering handbook (mostly sketches) written by Agostino Ramelli.

Jamestown, the first permanent English settlement in North America, is established in southeast Virginia.

1588

1607

1450

Chinese develop printing press that uses movable wooden blocks.

1500

Engineering science begins (start of Renaissance).

1550

Simon Stevin (Flemish military and civil engineer) develops parallelogram law for the addition of forces.

1596

Birth of René Descartes (d. 1650), French scientist and philosopher. Descartes developed the mathematical and geometric coordinate system that was later named for him (Cartesian coordinate system). In Latin, his name is Renatus Cartesius.

1614

John Napier (Scottish mathematician) publishes his discovery of logarithms.

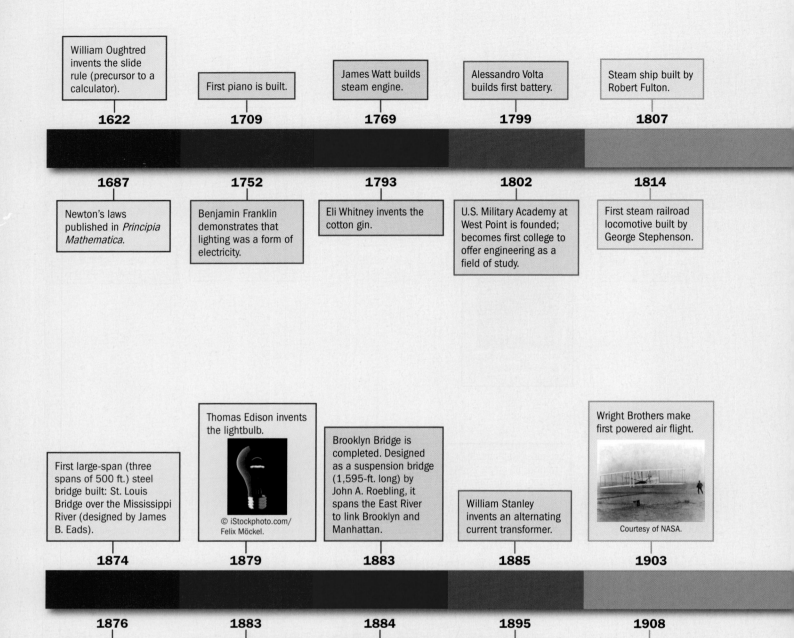

William Oughtred invents the slide rule (precursor to a calculator).

1622

First piano is built.

1709

James Watt builds steam engine.

1769

Alessandro Volta builds first battery.

1799

Steam ship built by Robert Fulton.

1807

1687

Newton's laws published in *Principia Mathematica*.

1752

Benjamin Franklin demonstrates that lighting was a form of electricity.

1793

Eli Whitney invents the cotton gin.

1802

U.S. Military Academy at West Point is founded; becomes first college to offer engineering as a field of study.

1814

First steam railroad locomotive built by George Stephenson.

First large-span (three spans of 500 ft.) steel bridge built: St. Louis Bridge over the Mississippi River (designed by James B. Eads).

Thomas Edison invents the lightbulb.

© iStockphoto.com/ Felix Möckel.

Brooklyn Bridge is completed. Designed as a suspension bridge (1,595-ft. long) by John A. Roebling, it spans the East River to link Brooklyn and Manhattan.

William Stanley invents an alternating current transformer.

Wright Brothers make first powered air flight.

Courtesy of NASA.

1874

1879

1883

1885

1903

1876

Alexander Graham Bell invents the telephone.

1883

The American Society of Mechanical Engineers is founded.

1884

The American Institute of Electrical Engineers is founded.

1895

Wilhelm Roentgen discovers X-rays.

1908

The American Institute of Chemical Engineers is founded.

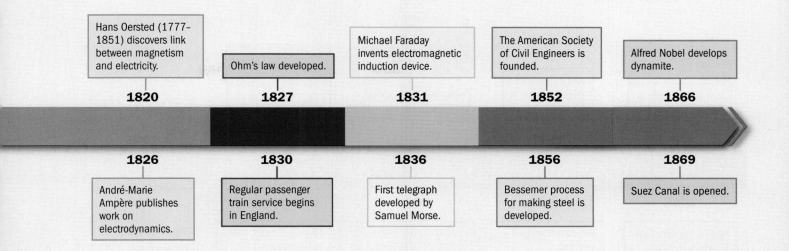

1820 — Hans Oersted (1777–1851) discovers link between magnetism and electricity.

1827 — Ohm's law developed.

1831 — Michael Faraday invents electromagnetic induction device.

1852 — The American Society of Civil Engineers is founded.

1866 — Alfred Nobel develops dynamite.

1826 — André-Marie Ampère publishes work on electrodynamics.

1830 — Regular passenger train service begins in England.

1836 — First telegraph developed by Samuel Morse.

1856 — Bessemer process for making steel is developed.

1869 — Suez Canal is opened.

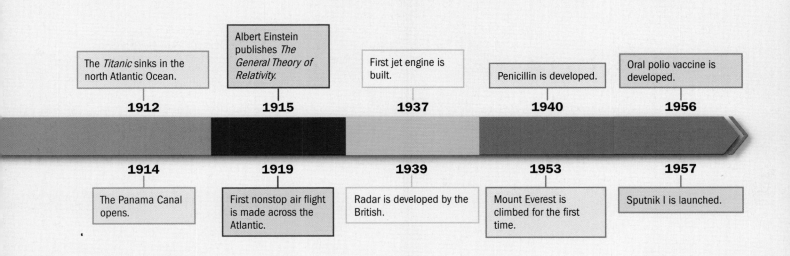

1912 — The *Titanic* sinks in the north Atlantic Ocean.

1915 — Albert Einstein publishes *The General Theory of Relativity.*

1937 — First jet engine is built.

1940 — Penicillin is developed.

1956 — Oral polio vaccine is developed.

1914 — The Panama Canal opens.

1919 — First nonstop air flight is made across the Atlantic.

1939 — Radar is developed by the British.

1953 — Mount Everest is climbed for the first time.

1957 — Sputnik I is launched.

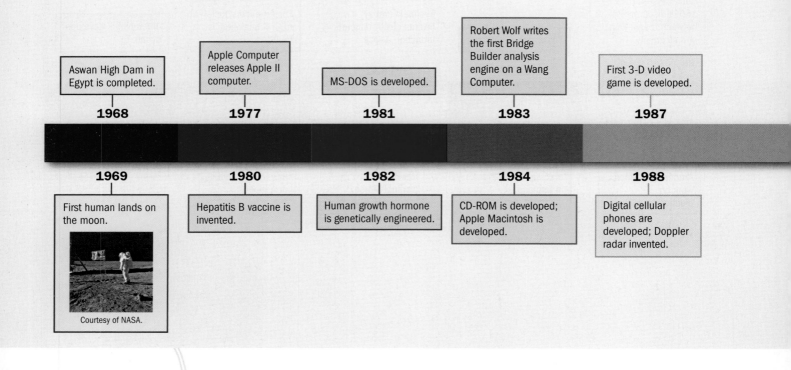

1968 — Aswan High Dam in Egypt is completed.

1977 — Apple Computer releases Apple II computer.

1981 — MS-DOS is developed.

1983 — Robert Wolf writes the first Bridge Builder analysis engine on a Wang Computer.

1987 — First 3-D video game is developed.

1969 — First human lands on the moon.

Courtesy of NASA.

1980 — Hepatitis B vaccine is invented.

1982 — Human growth hormone is genetically engineered.

1984 — CD-ROM is developed; Apple Macintosh is developed.

1988 — Digital cellular phones are developed; Doppler radar invented.

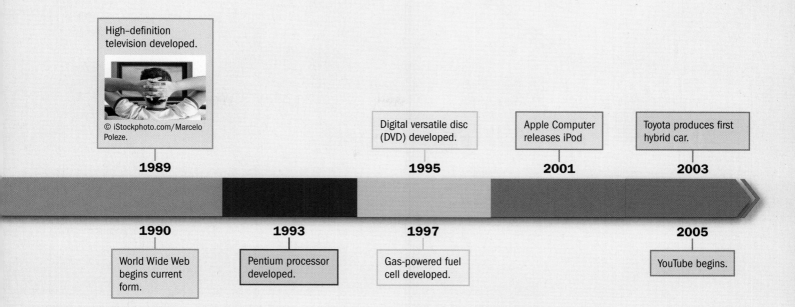

High–definition television developed.

© iStockphoto.com/Marcelo Poleze.

1989

Digital versatile disc (DVD) developed.

1995

Apple Computer releases iPod

2001

Toyota produces first hybrid car.

2003

1990

World Wide Web begins current form.

1993

Pentium processor developed.

1997

Gas-powered fuel cell developed.

2005

YouTube begins.

CHAPTER 2
Design Tools

Menu | START LOCATION | DISTANCE | END LOCATION

Before You Begin
Think about these questions as you study the concepts in this chapter:

 1 How does a design process guide the engineer from a loosely understood problem to a highly refined solution?

 2 What brainstorming techniques do engineers use to generate ideas during a design process?

 3 What kinds of sketches do engineers use to record their initial ideas?

 4 What kinds of computer-based modeling methods do engineers use to develop their solutions?

 5 How and why do engineers use analytical tools?

 6 How do engineers communicate and present their ideas?

Engineers understand that science, technology, and mathematics are not only interrelated bodies of knowledge but also the most fundamental tools of engineering itself. Indeed, the chapters of this textbook are extensions and explorations of this general idea. Engineers also employ other tools and skills that are common across all engineering disciplines. These include:

▷ design processes,

▷ brainstorming,

▷ sketching,

▷ modeling,

▷ analysis tools, and

▷ presentation and communication tools.

Unlike the typical formula-based problems that are often encountered in mathematics and science classes (and which have definite, correct answers), a design problem will have more than one solution. The process of developing a solution to a design problem is not nearly as cut and dry as solving a formula-based problem. However, there are methods that engineers and other technical professionals employ to guide them through the task. These methods are known as *design processes*.

Engineers are able to accomplish more when they operate as a part of a well-focused team because no one individual holds the answers to all the questions that arise during a design project (see Figure 2-1). Teams of engineers use various brainstorming methods throughout a design process to generate questions and stimulate ideas. A team that is capable of generating large numbers of ideas has a greater chance of developing a workable solution.

Figure 2-1: *Teams use a process called brainstorming to generate questions and stimulate ideas.*

Figure 2-2: *Creative ideas occur throughout a design project and must be recorded before they are lost. Designers often find themselves sketching out their ideas using whatever writing materials are available at the time.*

© Cengage Learning 2012

The importance of technical sketching to the process of design cannot be overemphasized. Quickly capturing a preliminary idea through various technical sketching methods is far more efficient than trying to record an idea through the use of a computer. Creative ideas often occur at random times throughout a design project and in places that do not allow for access to the necessary computer tools. It is, therefore, unlikely that an idea will last long enough to be inputted into a computer if it is not first sketched in sufficient detail before being lost to the depths of memory (see Figure 2-2).

Virtual tools, such as computer-aided design (CAD) programs, are used to take an idea to the next level of detail once it has been recorded and sufficiently worked out on paper. Finite element analysis (FEA) methods are also used on CAD models to observe the effects that simulated forces will have on a design. CAD models can be made into tangible mock-ups through the use of several different rapid prototyping methods. They may also be imported into computer-aided manufacturing (CAM) programs for the purpose of generating machine-tool control programs.

Once a solution has been modeled, it is put through a series of tests that generate design and performance data. Data from both virtual and actual tests is compiled, summarized, and communicated in the form of graphs and charts. This information is then used to drive the improvement or redesign of a solution.

An engineer will be called on to update the client or a manager on her progress at various stages of a design project. Sometimes this requires the development of a digital presentation, such as a PowerPoint slide show. If the presentation is not designed properly, the medium of the digital presentation will overshadow the message that is supposed to be communicated. This can be avoided by employing simple and consistent layouts and backgrounds, large fonts and graphics, and subtle transitions between images, text, and slides. Effective communication is also dependent on preparation, such as rehearsing and checking the equipment to make sure everything is in working order.

DESIGN PROCESS

Of all the different tools that engineers use across the spectrum of engineering disciplines, the design process is arguably the most universal. Engineers and other design professionals use various design processes to help focus and guide their efforts through the complex and often monumental task of moving from a loosely understood problem to a highly refined solution. Though there is no engineering standard, a **design process** can be thought of as a systematic problem-solving method for generating and developing ideas into solutions.

Many design processes are depicted as a linear sequence of steps. This can be misleading because problem-solving endeavors are not always carried out in a linear fashion. Even the most talented designers will concede that the nature of the process involves taking a few steps backward once they realize they have taken an idea too far in

Design process:

a systematic problem-solving method for generating and developing ideas into solutions.

a wrong direction. Recognizing this, they find it easier to develop an initial understanding of the process by breaking it down into a series of steps. This chapter will focus on a 12-step design process (see Figure 2-3).

STEP 1 ▶ *Define the Problem*

A successful design engineer has the ability to analyze a less-than-ideal situation, figure out the root of the problem, and clearly define that problem in a concise manner. This may seem simple, but it is a skill that can take years to develop. Many times a client will approach an engineer with a problem that is loosely defined. Here, too, the abilities previously mentioned are of great value because the problem may be presented in a way that requires refinement through careful and thoughtful listening, questioning, and interpretation.

At this point in the design process, an engineer will begin to formulate the details of a design brief. Simply defined, a **design brief** is a concise information tool that summarizes the most important information about a design project. It serves as an agreement between the engineer and the client, and it is used as a standard for assessing a solution's validity at any point in the development process. Figure 2-4 shows an example of a design brief. Bear in mind that all of the parts of a design brief may not be identified at this step in the design process.

The information contained in a design brief will identify:

▶ the person or entity who serves as the client;

▶ the individual(s) or group who is to be held accountable for the design service;

▶ the client's problem, need, or want;

▶ the challenge that will result in an engineering solution, and the degree to which that solution must be realized; and

▶ any limitations within which the engineer must operate throughout the design process.

The **problem statement** is written to clearly and concisely identify a client's or target consumer's problem, need, or want. It is important to understand the difference between a *problem* and a *challenge*. A problem does not require a person to act. It merely identifies a situation that is in need of remedy. If your boss or teacher tells you to design something, he or she is presenting you with a challenge, not a problem.

The challenge is summarized within the **design statement**, which follows the problem statement in a design brief. A design statement challenges the engineer to take action to solve the problem. It describes what a design solution should do without describing how to solve the problem. The design statement also identifies the degree to which the solution must be realized. Is the final solution expected to be a one-of-a-kind design or will it be mass produced? Simply put, the design statement tells you what you are supposed to do; the problem statement helps you understand *why* you are doing it.

It is also important to understand that there are no industrial standards associated with the format of a design brief, just as there are no industrial standards associated with the steps of a design process. Both are organizational tools that help the design engineer achieve a desired result.

FIGURE 2-3: This 12-step design process serves as a problem-solving approach and guide for anyone who is given the task of generating ideas and then developing them into a working solution.

From Rogers, Wright, and Yates, Gateway to Engineering, © 2010 Delmar Learning, a part of Cengage Learning, Inc.

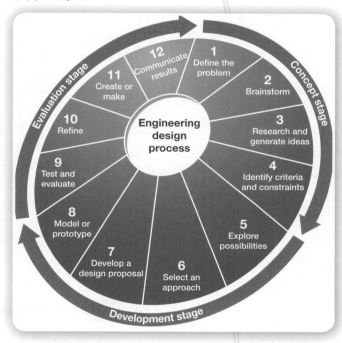

Design brief:

a concise information tool that summarizes the most important information about a design project, serves as an agreement between the engineer and the client, and is used as a standard for assessing a solution's validity at any point in the development process.

FIGURE 2-4: Typical design brief.

Design Brief

Client:	Genesee Valley Power Authority
Designer:	Kristin Haas
Problem Statement:	Dangerous conditions exist during the manipulation of low-level nuclear waste, wherein workers cannot risk direct contact with the objects that they must handle.
Design Statement:	Design a one-of-a-kind hydraulic-powered robot arm that will move one ping-pong ball-sized object from a given location in a work environment and drop it down a 1 ½" I.D. tube. The ping-pong ball represents the sensitive material. The work environment represents a space in which dangerous conditions exist that prevent human beings from making direct contact with the material.
Constraints:	1. 3-week deadline
	2. $100 budget
	3. Maximum weight =+/N10 lb
	4. No human contact with anything inside the work environment during the operation of the robot arm
	5. Fully assembled, the solution must fit within an 11" × 9" × 17" space (interior of a standard photocopier paper box)
	6. The solution must interface with two ¼-20 threaded inserts on the base of the test environment that are spaced 3 inches apart

STEP 2 ▶ *Brainstorm*

Once the problem and challenge have been defined, a design team will conduct a meeting of minds to openly discuss the problem, identify existing knowledge that team members have that relates to the problem, generate questions that need to be answered, and define areas of focus that will be assigned to various members of the team for research. This step in the design process is called *brainstorming*.

Brainstorming:
any technique that is used by a design team to generate ideas.

Brainstorming can include any technique that a design team uses to generate ideas. Because of the nature of creativity, brainstorming is often repeated during other steps in the design process. Design teams use different types of brainstorming techniques. Some are purely oral. Others mostly involve writing down ideas in the form of two- or three-word concepts. Still other forms of brainstorming involve sketching out massive numbers of ideas, sharing, and explaining them on the spur of the moment. It is important to note that ideas are not evaluated or judged at this point. Brainstorming methods will be explored in greater detail later in this chapter.

STEP 3 ▶ *Research and Generate Ideas*

Research:
the systematic study of materials and sources in order to establish facts and reach new conclusions.

Contrary to some engineering stereotypes, engineers do not know everything. However, some of the qualities of a successful engineer are an insatiable passion for knowledge and an understanding of the critical role that *research* plays in the development of creative solutions. Research is the systematic study of materials and sources in order to establish facts and reach new conclusions. At some point in your education, you have probably performed research as part of a writing assignment for a school report.

Some psychologists believe that creative ideas result from the mind's ability to recognize patterns and establish connections between seemingly random pieces of information that are stored in the brain. If this theory is true, then teams of people who learn as much as they can about the topics associated with a particular problem will have a better chance of generating creative ideas that lead to a workable solution.

Once the team has clearly defined the challenge, generated questions, and established topics that need to be explored, it is the responsibility of the team members to fan out and search for the answers. Research often involves gathering and sifting through preexisting information. This is known as *secondary research*. You perform secondary research when you do a keyword search on a website such as Google. Books, films, patents, professional journals, periodicals, and encyclopedias are other common sources of secondary research. Engineers also perform another type of research called *primary research*. Primary research involves generating original information through interviews, surveys, experiments, and observations. Though it takes more effort and is usually more time consuming, primary research will often provide the most useful information for developing creative ideas.

At this stage of the design process, it is critical to keep a detailed record of the information that you learn by using an engineer's notebook. An **engineer's notebook** is a type of journal that serves as both an archival record of engineering research and a repository for solution ideas that are generated throughout the design process (see Figure 2-5). The engineer's notebook is a place for recording important book titles, useful website URL addresses, and the names and contact information of people who have expert knowledge about topics that relate to the problem. Keeping an engineer's notebook handy at all times is the best way to ensure that there is a place to document ideas as they occur during research efforts.

Like brainstorming, research is something that an engineer will perform multiple times throughout a design process. Realizing this, design teams must establish tight time deadlines for initial research or this step in the design process will drag on at the expense of the entire project. Once the research deadline occurs, the members of the design team will meet again to share the jewels of information that they have learned. Again, having an engineer's notebook handy not only allows designers to recall important information accurately to their team members but also gives them a place to jot down sketches and ideas that occur as teammates share their knowledge.

> **Engineer's notebook:**
>
> a type of journal that serves as an archival record of engineering research and as a repository for solution ideas that are generated throughout a design process.

FIGURE 2-5: An engineer's notebook is a type of journal that serves as a repository for information and ideas throughout a design project. The contents of an engineer's notebook often include written and sketched ideas, summaries of work sessions, printouts of images, references to important resources, and write-ups of tests and experiment results.

From Johnson, Introduction to Fluid Power, © 2002 Delmar Learning, a part of Cengage Learning, Inc.

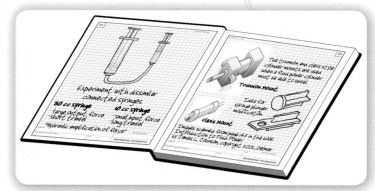

| STEP 4 | *Identify Criteria and Constraints* |

Limitations are often spelled out at the beginning of design projects that are undertaken in school for educational purposes. In the real world of open-ended problems, however, this is not often the case. Having researched the problem and developed a clear understanding of the challenge, the team and the client will come together once again to identify and agree on limits within which an acceptable solution must be developed. These limits will then be added to the design brief as a list of criteria and constraints.

A design project will drag on indefinitely if limits are not imposed because there is always room for improvements. **Criteria** are specific standards against which a design will be judged acceptable or unacceptable. For example, a solution may be deemed unacceptable if it cannot support a specific load or fit within a specific size container. **Constraints** are general limits that are imposed on a design project such as project deadlines, budgets, materials, or manufacturing processes. It is important that the criteria and constraints be identified in list form because new limitations may be imposed or realized as the design process unfolds.

STEP 5 *Explore Possibilities*

Exploring possibilities involves returning to the initial solution ideas and sketches that were recorded in the engineer's notebook during step 2 (brainstorming) and step 3 (research and generating ideas) and building from them. A different type of brainstorming is used during this stage of the process, wherein each team member sketches "big picture" ideas and shares them with the group. This technique allows the group to play off each person's thoughts. Members of the group will listen to the ideas expressed and expand on them, combine them, and sometimes twist them in completely unexpected directions. A visual record of the team's ideas will take shape as the sketches are plastered across the walls of the meeting space. The goal is to generate a critical mass of sketches from which the group will be able to identify a handful of the most promising ideas for more detailed exploration.

Team members are then given the task of identifying two or three ideas that they believe have the greatest potential, recognizing the limitations that were set forth in the previous step. This task can be accomplished by giving each team member two or three small Post-it notes to place on their selections. The sketches that get the most votes are then assigned to various team members so that the ideas can be worked out in greater detail. These team members will generate more sketches (see Figure 2-6) and might even develop CAD models if the designers are proficient and have enough time to do so. The team, and sometimes the client, then meets to conduct a formal **critique** of the detailed ideas. This meeting provides an

FIGURE 2-6: Detailed sketches of the selected ideas are made for presentation and discussion during a formal critique.

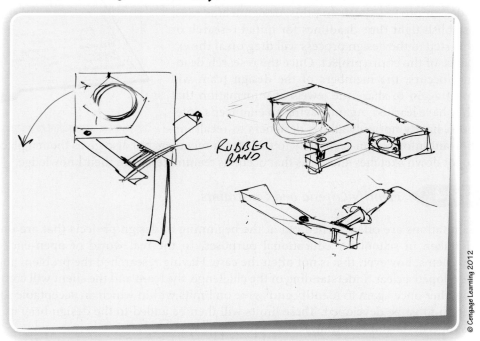

RUBBER BAND

opportunity for each team member to voice his or her opinion and share feedback in the form of constructive criticism.

STEP 6 ▶ *Select an Approach*

Realizing that there is more than one solution to a design problem, how does one decide which solution is the best? The answer is often based on trade-offs that the client or designer is willing to make between a solution's functional, structural, aesthetic, economic, and environmental qualities.

Sometimes we develop a fondness or attachment to our ideas that serves to block our ability to make objective and logical decisions. This is known as being "married to an idea," and it can ultimately lead to the demise of a design project. Two ways to prevent this from happening include (1) giving the client the responsibility for selecting one of the ideas and (2) using a tool called a *decision matrix* to help the team come to consensus on one idea. A **decision matrix** is a chart that designers use to quantify their opinions of two or more design ideas by assessing each idea according to a series of important considerations.

To create a decision matrix, the team will meet again to conduct a brainstorming session. The purpose of this session is to identify a list of considerations against which the solution ideas will be assessed. Size, function, simplicity, manufacturability, budget, safety, and development time are major considerations in most design challenges. In some cases, it is appropriate to evaluate whether or not an idea meets a specific criterion. When all of the necessary considerations have been agreed to, they are added to the decision matrix across the top row. Each solution idea is identified in the left column of the decision matrix (see Figure 2-7).

> **Decision matrix:**
>
> a chart used by designers to quantify their opinions of two or more design ideas by assessing each idea according to a series of important considerations.

FIGURE 2-7: Each member of a design team will use a decision matrix to rank all of the preliminary ideas according to important design and manufacturing considerations. The totals from each team member's decision matrix are then averaged to determine which idea has the best chance of being developed successfully.

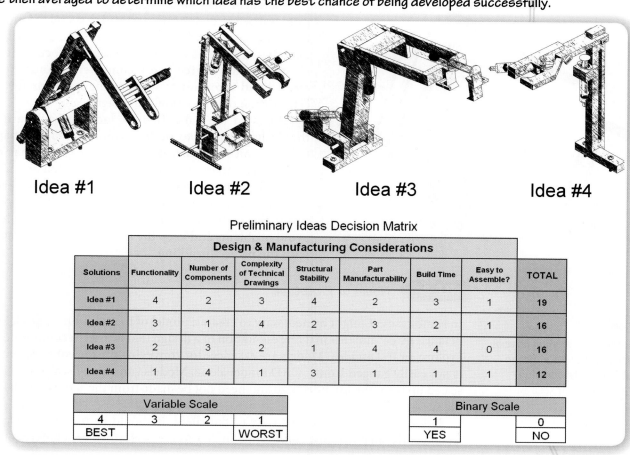

Idea #1   Idea #2   Idea #3   Idea #4

Preliminary Ideas Decision Matrix

Solutions	Functionality	Number of Components	Complexity of Technical Drawings	Structural Stability	Part Manufacturability	Build Time	Easy to Assemble?	TOTAL
Idea #1	4	2	3	4	2	3	1	19
Idea #2	3	1	4	2	3	2	1	16
Idea #3	2	3	2	1	4	4	0	16
Idea #4	1	4	1	3	1	1	1	12

Design & Manufacturing Considerations

Variable Scale			
4	3	2	1
BEST			WORST

Binary Scale	
1	0
YES	NO

Two number scales may be used, one variable and one binary. The variable scale can include any range of values that is more than two. The example decision matrix shown in Figure 2-7 uses a four-value variable scale. In this case, a score of 4 is awarded to the idea that does the best job of meeting an important consideration, whereas a score of 1 is given to the idea that lacks the most in that category. A binary scale is used in place of a variable scale when the ideas are assessed according to a specific criterion or a question. Therefore, 0 equates to "No" and 1 is synonymous with "Yes." Some considerations may have greater bearing on the final decision. For example, the team may decide that structural stability is twice as important as manufacturability. In such cases, a multiplication factor may be used.

Each member of the design team will fill out the decision matrix to determine a total point score for each preliminary idea. The totals from each team member's decision matrix are then averaged to determine which idea received the highest score according to the group. The hope is that the idea that garnishes the most points will have the greatest potential for development into a successful working solution. This kind of technique helps to ensure that important decisions are not made as a result of individual pride or emotional attachment to any one idea.

STEP 7 ▶ Develop a Design Proposal

Having selected an idea to work out to completion, the design team must now develop the idea into a solution. This requires the creation of documentation in the form a design proposal. A design proposal usually consists of CAD part models and assemblies (see Figure 2-8), assembly drawings, and dimensioned multiview drawings of each manufactured component. If a solution involves electronics or fluid power components, then one or more schematic diagrams will also be needed.

The client may be brought in to review the design proposal before giving the go-ahead to spend the money that is needed to build and test a prototype. In most situations, the goal of a design proposal is to document the solution to the degree that a competent manufacturer could build a prototype without the need for further clarification on the part of the design team.

This step may also involve looping backward in the design process to perform more research related to mechanical, electrical, or material issues that are not clearly understood. Major decisions that are made from this point forward have the potential to generate more problems down the road. Thinking strategically about the path that each decision will take will help avoid problems and keep the design solution on target with the project deadline.

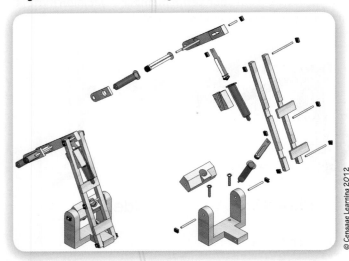

FIGURE 2-8: **The assembly and exploded assembly CAD models show how the design's components fit together to form a working solution.**

© Cengage Learning 2012

STEP 8 ▶ Model or Prototype

Model:

a detailed three-dimensional representation of a design that is used to communicate, explore, or test an idea.

In some cases, it might be necessary to create a model of the solution. A **model** is a detailed three-dimensional representation of a design that is used to communicate, explore, or test an idea. Models are excellent for conveying the idea of scale, which is not easily gauged from a CAD model alone. Models can also be scaled down or scaled up from the intended size and need not be made from the same materials as the final product.

Before a solution is turned over to the client, it should be tested under controlled and actual working conditions. Such tests are performed on a different type of model called a *prototype*. A **prototype** is a one-of-a-kind working model of a solution that is developed for testing purposes. The prototype is manufactured according to the information contained in the design proposal.

Prototype:

a one-of-a-kind working model of a solution that is developed for testing purposes.

STEP 9 ▶ *Test and Evaluate*

Once a prototype of the solution has been created, the testing phase of the design process is ready to begin. This often involves the client working with the engineering design team to determine what tests will be used to measure the quality of the design solution. Testing may include applying forces to structural components or an entire assembly to ensure that the design can withstand the loads to which it will be subjected. If a test produces data, then that data must be recorded, organized, and analyzed. A computer **spreadsheet** program such as Microsoft Excel is often used for this purpose. This step in the process also includes tests that measure a solution's functionality. If all goes according to plan, then the solution will satisfy the design statement and constraints as identified in the design brief.

STEP 10 ▶ *Refine*

Any issues or shortcomings that are identified through testing and analysis are discussed among the design team. At this point, the team may need to backtrack to an earlier step in the design process in order to move forward with the development of the solution.

All ideas for changes to the design should be sketched out in an engineer's notebook and accompanied by explanations that identify why the changes are necessary. Once the ideas have been worked out to an appropriate degree of detail on paper, the designer will work out the refinements in greater detail by modifying and updating the existing CAD models and technical drawings.

STEP 11 ▶ *Create the Final Solution*

When all of the necessary modifications have been identified, implemented, tested, and evaluated, the final solution that will be delivered to the client is now ready for fabrication. As mentioned earlier, the design statement contained in the design brief will identify whether the final solution is to be a custom (one-of-a-kind) design, a limited-production item, or a mass-produced product.

Designs that involve mass production often require additional efforts from other professionals such as manufacturing engineers, systems engineers, and tool and die makers. Their work may involve the design of special processes, machines, or tooling that speeds production and minimizes dimensional variation between like parts.

STEP 12 ▶ *Communicate the Results*

The results of a design process are often communicated to a manager or client in the form of a technical report, digital presentation (see Figure 2-9), and demonstration of the actual working solution. This step in the process may also involve the creation of instruction manuals, graphic advertisements, and patent documents.

Brainstorming Ideas

We noted earlier that brainstorming is a technique that design teams use to generate ideas during a design process. Brainstorming combines individual brain power with human interaction to create the best possible solution to a design problem. It provides a great opportunity for team building and is a vital step in any design process. The word *brainstorming* was first coined in the 1920s by an advertising executive named Alex Osborn. Osborn wrote many books on creative thinking

and operated on the philosophy that, in order for ideas to form, one must have the opportunity to think.

Brainstorming allows team members to think creatively without judgments or constraints. Good brainstorming practices create a relaxed environment that encourages everyone to participate. Ideas do not have to make sense, and participants need not be overly concerned with the issues of practicality or cost. Participants should neither feel embarrassed to share their ideas nor pressured to conform to the expectations of others. Brainstorming gives team members the freedom to voice any solution that comes to mind. Once a team begins brainstorming, the initial ideas will spark new ideas, creating a chain reaction.

Group versus Individual Brainstorming

Brainstorming can be done in groups or as individuals. Both approaches have advantages. Sometimes individual and group brainstorming sessions are used in combination to maximize the benefits of each method. Individual brainstorming can yield better ideas, because 100% of the participant's attention is focused on generating ideas without the distraction of listening to others (see Figure 2-10).

Group brainstorming, on the other hand, takes advantage of the wider experience that the diverse members of a group provide. Listening to ideas from a different perspective can jump-start an individual's ideas or help him or her break through a mental block. Group brainstorming is a great way to demonstrate that everyone has something to offer, and it helps everyone feel involved in defining the end solution.

Guidelines for a Traditional Group Brainstorming Session

Ideally, a traditional group brainstorming session involves five to seven individuals who have diverse backgrounds. The group is seated around a large round or square table so that participants can

FIGURE 2-10: Individual brainstorming can yield better ideas because 100% of the participant's attention is focused on generating ideas without the distraction of listening to others.

see each other. It is important that the environment be comfortable and free from distractions. Everyone has a generous stack of paper on which to jot notes and make sketches as they wait their turn to speak. As the ideas will need to be voted on and narrowed down, it is common practice to limit each piece of paper to one idea. Sketching with a dark marker makes it easier for other participants to see the idea as it is being held up and explained.

Group efforts often require the guidance of a facilitator. This person may participate in the brainstorming session but is also responsible for directing the session, keeping the group focused, and encouraging the quieter members to participate. At the onset of the session, the group facilitator should remind the participants of the well-established **guidelines** that are designed to help the group work cohesively in order to achieve the best possible results from a brainstorming session. These guidelines are:

▶ Stay focused on the task at hand.

▶ Do not judge or criticize ideas but feel free to ask questions and provide positive comments.

▶ Don't limit yourself to only rational ideas.

▶ Build off the ideas of the other participants.

▶ Only one person should present their idea at a time.

Once the group facilitator has clearly identified the problem along with any important criteria that the solution must adhere to, the group should start sketching its ideas as they come. When someone has an idea, he or she should hold up a sketch for everyone to see, explain it, and affix it to a large wall that serves as a collection point for the ideas. After the first idea is presented, the other participants should present their ideas as the opportunity arises. Following the above listed guidelines, being polite and waiting your turn to talk will help maintain a comfortable atmosphere that encourages creative thought. It is important to understand that brainstorming is supposed to be a fun and engaging process (see Figure 2-11). If team members aren't having fun, then the results of the process will suffer. The

FIGURE 2-11: Brainstorming should be a fun and engaging process.

group facilitator will monitor the group's pace and set a time limit for the remainder of the session when she recognizes that group is losing steam.

Other Methods of Brainstorming

One way to overcome the initial inertia of a brainstorming session is to begin by asking questions. Participants can form questions beginning with "Can you . . . ," "What if you . . . ," and "Are there ways to . . . " to generate ideas. Other methods of brainstorming have been developed to stimulate the mind and trigger ideas. Two such techniques include the stepladder method and the reverse-brainstorming process.

STEPLADDER METHOD The stepladder method enables involvement from everyone in a group. It is especially useful for groups that have a mix of both introverts (quiet people) and extroverts (outgoing people). This method guarantees that everyone's ideas are heard.

Step 1. Clearly define the problem and criteria.

Step 2. Allow group members time to create ideas individually.

Step 3. Two group members come together to share their ideas and brainstorm new ideas that develop as a result of the sharing process.

Step 4. A third member is added who presents his or her ideas to the other two members and then listens to the ideas from the original two members. After all of the ideas have been shared, time is allowed to brainstorm new ideas.

Step 5. Repeat step 4 by adding one new group members at a time until all members have joined and shared their ideas.

REVERSE BRAINSTORMING The reverse-brainstorming process indirectly generates solutions to a problem. Whereas a traditional brainstorming process is designed to generate ideas that solve a problem, the reverse-brainstorming process focuses on the cause of the problem. Once potential causes are identified, solutions for each of the causes are then discussed. Reverse brainstorming is a good method to use when a problem statement is very broad. The following is an example of the reverse-brainstorming process:

Step 1. Clearly define the problem. Example: Students are dissatisfied with the school cafeteria.

Step 2. Reverse the problem by stating it in the form of a question. Example: What could cause the students to be dissatisfied with the school cafeteria?

Step 3. Generate ideas to answer the question. Examples:

> ▶ Food is cold.
>
> ▶ Noise level is too high.
>
> ▶ Lines are long.
>
> ▶ Favorite items run out early.
>
> ▶ Menu options are limited.
>
> ▶ Cafeteria is not clean.

Step 4. Generate ideas to alleviate the causes of the problem. Examples:

> ▶ Install warming trays or heat lamps.
>
> ▶ Play music over the loudspeaker.
>
> ▶ Use swipe cards to speed up the checkout process.

▶ Obtain feedback through surveys to determine students' favorite foods.

▶ Increase the number and accessibility of garbage receptacles.

Wrapping Up a Brainstorming Session

Once the brainstorming session is finished and all of the team's ideas are on display, the next step involves figuring out which idea to go with. The most promising solutions are rarely obvious immediately after a brainstorming session. The following steps are designed to help a team determine which idea should be picked for development into a workable solution.

Step 1. Start by eliminating duplicate ideas.

Step 2. Give everyone time to study and understand all of the ideas.

Step 3. Ask each member to identify his or her three favorite ideas by placing a small Post-it note or sticker on the sketch.

Step 4. Tally the votes to determine the top choice.

SKETCHING

Brainstorming activities often involve sketching ideas on paper so that they may be shared and explained to other members of a team. A sketch is a rough, handmade drawing that represents the main features of an object or scene. The act of sketching is a skill that serves to bridge the artist and the engineer. Being able to sketch well is critical because of the fleeting nature of ideas. To the engineer, sketching is a form of visual communication that is used to quickly record and convey ideas before they are lost. Once recorded, a sketch may evolve into an elaborate computer-aided design model or a detailed technical drawing. This text will briefly explore two general categories of sketches: pictorials and multiviews.

Sketch:
a rough, handmade drawing that represents the main features of an object or scene.

Pictorial Sketch

A pictorial is a type of drawing that gives the illusion of three dimensions by showing an object's height, width, and depth in a single view. Pictorial sketches show three mutually perpendicular faces on a prismatic object and are often drawn in isometric (see Figure 2-12), oblique, or perspective format. The object

FIGURE 2-12: This pictorial sketch of a robot arm gripper plate idea gives the illusion of a three-dimensional object on a two-dimensional surface.

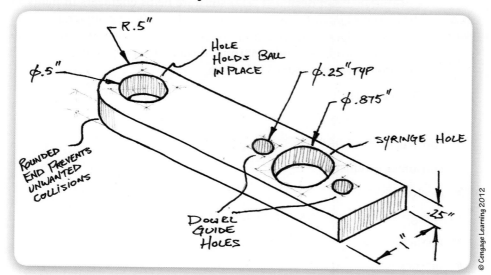

© Cengage Learning 2012

FIGURE 2-13: This two-view multiview sketch shows the front and right side views of the robot arm gripper plate. Each view shows what the object looks like with a 90° rotation of its adjacent view. This type of sketch employs the use of different line conventions such as object lines, hidden lines, and center lines to help the viewer image what the object looks like in her mind.

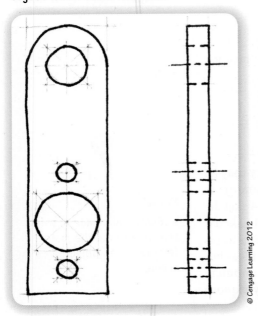

© Cengage Learning 2012

that is portrayed in a pictorial sketch is usually easier to understand because we perceive the world around us in three dimensions. Adding notes to the sketch makes it easier for others to understand aspects of the idea that may not be obvious by the drawing alone. Placing dimensions on the sketch helps the viewer establish a sense of scale. Becoming a good pictorial sketcher takes a lot of practice, but a good sketch can communicate more information in less time, which is critical during a brainstorming session.

Multiview Sketch

Some ideas are too complicated to be sketched in pictorial format. In such cases, a multiview sketch may be the easiest option. A multiview is a type of drawing that portrays an object as a series of two or more two-dimensional views arranged in a specific pattern (see Figure 2-13). Dimensions are usually added to a multiview drawing to communicate geometric information about an object's overall height, width, and depth, as well as information about the sizes and locations of features such as holes. The most common views given in a multiview sketch are the top, front, and side views. The top view communicates an object's width and depth. The front view shows the object's width and height. A side view shows an object's height and depth. Width describes an object's side-to-side dimension. Height is the measure of an object from top to bottom. An object's depth is measured from front to back.

MODELING IDEAS

After an idea has been sketched, designers usually create a detailed drawing. Not long ago, design teams created meticulously detailed drawings by hand using little more than pencil and paper. Their final drawings were then copied as blueprints using a blueprint machine, resulting in a finished drawing called a *blueprint*. Computers have changed the way these final drawings are created.

OFF-ROAD EXPLORATION

To learn more about the differences between isometric, oblique, and perspective pictorials along with the standards that are associated with multiview drawings, explore *Visualization, Modeling, and Graphics for Engineering Design* by Dennis Lieu and Sheryl Sorby, Cengage, Delmar Learning, 2009.

Advantages and Applications of Computer-Aided Design

Computer-aided design (CAD) software helps engineers in many industries design and manufacture products, including skyscrapers, suspension bridges, tunnels, automobiles, boats, aircraft, spacecraft, trains, consumer products, and electronic devices. CAD systems are used in just about every industry, including:

▶ aerospace

▶ architecture

▶ construction

▶ consumer products

▶ electronics

▶ energy

▶ medical

▶ plastics

► packaging

► transportation

The use of CAD software has revolutionized the process of designing products. The driving force behind the development of tools and technologies such as CAD has always been competition. To win business, companies must produce superior designs more efficiently and at a lower cost than their competitors. The user-friendly interface of most CAD software enables the user to draw complex shapes easily and in a fraction of the time it would take to draw it by hand (see Figure 2-14). Changes can be made quickly to CAD models, which will in turn update their respective technical drawings. Also, digital CAD files can be instantly e-mailed to other engineers and manufacturers around the world.

Another advantage of CAD systems is their ability to store models of off-the-shelf parts that are frequently used in engineering designs. Libraries of regularly used parts can be downloaded and shared with other designers. Assembly models are generated by combining specialized components with existing component models and positioning them as required (see Figure 2-15).

CAD models resemble the final product to a degree of geometric precision that is beyond that of a physical prototype. Realistic images and animations of the virtual models may be used by marketing personnel to produce promotional materials such as video advertisements and sales brochures. They may be used in conjunction with market research to determine whether the product is even worth producing. Also, CAD model images are used in the development of instructional use manuals.

FIGURE 2-14: (a) CAD programs give the engineer the ability to develop a solution in detail within a virtual environment. (b) Dimensioned technical drawings are then created from the CAD models, which communicate all of the geometric information that is necessary to fabricate the object.

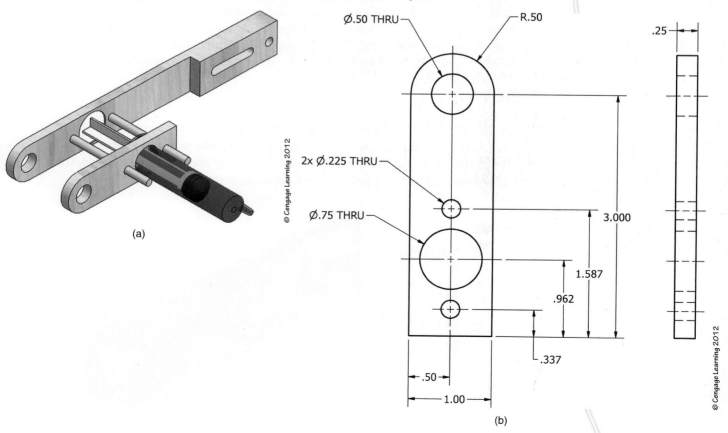

FINITE ELEMENT ANALYSIS Mathematical information that used to take engineers hours to calculate by hand is now available at the click of a button. Given a 3-D CAD model's dimensions in combination with an assigned material, a modern CAD system will calculate the object's surface area, volume, and weight as well as identify its center of gravity and other physical characteristics. Many 3-D CAD programs now have the ability to perform structural analysis on a component or assembly through a process called **finite element analysis (FEA)**. Simply speaking, FEA is a computer-based design analysis tool that allows the user to apply virtual forces and pressures to a 3-D CAD model in order to determine deflections, stress concentrations, and other effects. Using FEA, the engineer need not wait for the development and testing of a prototype to determine if the geometry is in need of further development (see Figure 2-16). This type of virtual testing ultimately saves time and money.

FIGURE 2-16: FEA programs give the engineer the ability to simulate the effects that forces will have on an object such as deflection and stress concentrations. The red areas show the highest levels of stress concentration, thus identifying where the part may first experience structural failure.

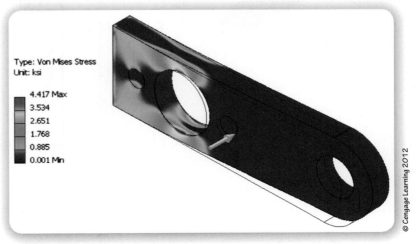

Type: Von Mises Stress
Unit: ksi

4.417 Max
3.534
2.651
1.768
0.885
0.001 Min

© Cengage Learning 2012

COMPUTER-AIDED MANUFACTURING CAD systems are used in conjunction with computer-aided manufacturing (CAM) software to create machine tool control programs that take the form of G&M code. G&M code programs are used to autonomously operate computer numerical control (CNC) machines such as lathes, mills, water jets, laser cutters, and welding equipment (see Figure 2-17). A typical design begins by producing a virtual model of the part using a CAD system. The virtual model is then imported into the CAM system. In the case of milling machines and lathes, the CAM software performs tool path analysis for both

FIGURE 2-17: (a) A CAD model is imported into a CAM program for the purpose of tool-path analysis and G&M code creation. (b) Once created, the code is then downloaded to a CNC machine where the actual part is created.

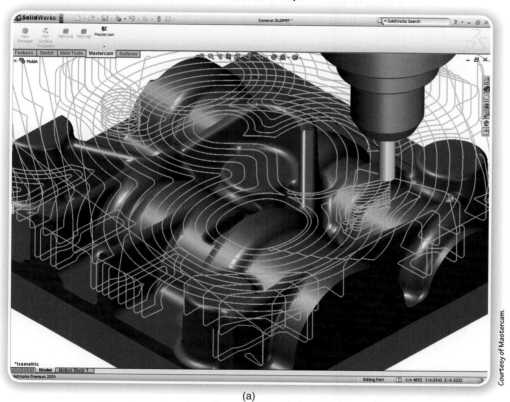

(a)

(b)

rough and finished cut operations. The sequence of machining operations is then converted to G&M code and sent to the appropriate machine(s) for manufacturing the part.

RAPID PROTOTYPING Creating a physical model of a design solution through traditional manufacturing methods or CAD, CAM, or CNC systems still requires the support of skilled craftspeople, which slows down the product-development process. Starting in the late 1980s, a new type of technology called *rapid prototyping* was developed, which has given the engineer the ability to convert CAD models directly into physical models without the skills, tools, and assistance of the traditional craftsperson. Rapid prototyping (RP) is a collection of CAD data–driven physical model construction technologies that use additive manufacturing processes. All RP processes begin with a 3-D CAD model, which is sliced into a series of ultrathin layers. The geometric information for each layer is then sent to a machine where the object is built one layer at a time.

RP technologies offer the following advantages over conventional manufacturing methods that are usually based on subtractive processes:

▶ RP technologies are able to produce complex models more quickly and easily than traditional manufacturing methods.

▶ RP models can be made from a wide selection of polymers, ceramics, and metals.

▶ Some RP technologies offer cost-effective alternatives for low-volume production of actual finished products.

Designers have several different types of RP methods to choose from. The four most common methods include:

▶ stereolithography,

▶ fused deposition modeling,

▶ selective laser sintering, and

▶ powder binder printing.

Stereolithography Stereolithography (SLA) is the oldest of the RP technologies. The SLA process builds a solid plastic model by using an ultraviolet laser to harden a photosensitive liquid polymer one layer at a time. An SLA machine consists of a build platform that is suspended in a large vat of a light-sensitive liquid polymer. Before the first layer is built, a leveling blade called the *sweeper* deposits an even layer of the liquid polymer over the build platform. An ultraviolet laser beam is sent through lenses that focus the beam onto a scanning mirror. The scanning mirror precisely steers the laser beam to trace individual layers onto the build surface, which hardens the photosensitive liquid polymer. After the first layer is complete, an elevator lowers the build platform two to three thousandths of an inch (0.002–0.003 in) and the process starts over again. This cycle continues until the entire part is created (see Figure 2-18). The hardened part is removed from the vat, drained, and then baked in an ultraviolet oven to cure the material. Finished SLA polymer components can be made to especially high dimensional accuracy and detail.

Fused Deposition Modeling The SLA process was soon followed by another RP technology called fused deposition modeling (FDM). The FDM process involves extruding a plastic or wax material through a nozzle that traces the cross-sectional geometry of a part layer by layer (Figure 2-19). Most FDM machines supply build

FIGURE 2-18: (a) The SLA process builds a solid plastic model by using an ultraviolet laser to harden a photosensitive liquid polymer one layer at a time. (b) The model of the automobile tire rim was made using the SLA process.

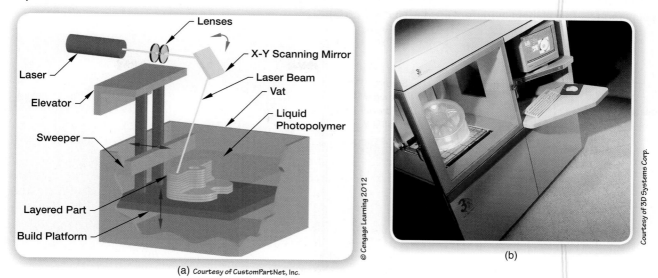

(a) Courtesy of CustomPartNet, Inc.
© Cengage Learning 2012
(b)
Courtesy of 3D Systems Corp.

FIGURE 2-19: (a) The FDM process involves extruding a plastic or wax material through a nozzle that traces the cross-sectional geometry of a part layer by layer. (b) Some FDM build materials display flexural properties.

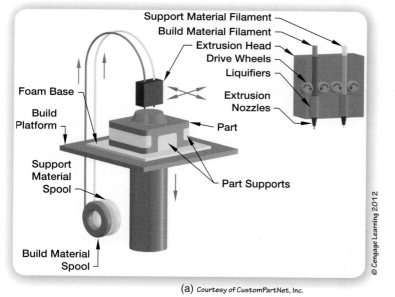

(a) Courtesy of CustomPartNet, Inc.
© Cengage Learning 2012
(b)
Courtesy of Stratasys.

and support materials through large spools of thin filament that are contained in removable cartridges. The filaments are extruded through a nozzle that contains heaters to raise the temperature of the material to just their melting points. The material hardens almost immediately after flowing from the nozzle and bonds to the layer below. After each layer is finished, the platform lowers, and the extrusion nozzle deposits another layer. FDM parts can be made from a variety of plastics and specialized waxes.

Selective Laser Sintering The selective laser sintering (SLS) process was invented at the University of Texas in Austin by Carl Deckard and his colleagues in 1987. The SLS process is closely similar to stereolithography. SLS uses a moving laser beam to trace the shape of each layer onto a powdered polymer material (see Figure 2-20).

FIGURE 2-20: (a) The SLS process uses a moving laser beam to trace the shape of each layer onto a powdered polymer or metal material, which sinters the powder together to form the part. (b) SLS models are structurally stronger than SLA models, and in some cases the process may be used to create finished parts rather than display models (air plenum manufactured with PA 2210).

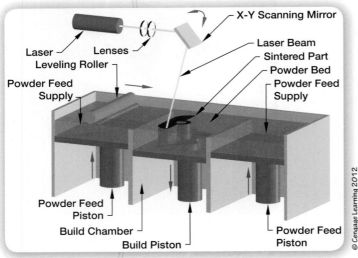

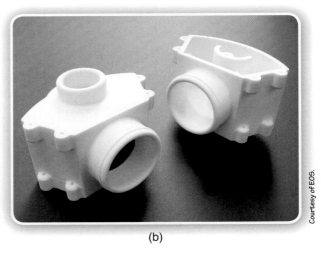

(a) Courtesy of CustomPartNet, Inc.

© Cengage Learning 2012

Courtesy of EOS.

(b)

In a process called *sintering*, the energy from the laser beam melts the powdered material and fuses it together. Before the next layer is sintered, a new layer of powder is deposited on top of each solidified layer. A similar process called direct metal laser sintering (DMLS) uses powdered metal as the build material.

Powder-Binder Printing Like the SLS process, another RP technology called *powder-binder printing* builds models using a powdered medium. In this case, the powder is made from either starch or plaster. Unlike the SLS process, powder-binder printers use a liquid adhesive instead of a laser to fuse the powder together. An ink-jet print head, similar to those found inside ink-jet printers, deposits the liquid adhesive to targeted regions of the powder bed to form one thin cross-sectional layer of the part (see Figure 2-21). After each layer is built, the build

FIGURE 2-21: The powder-binder printing process forms a part by spraying liquid adhesive to targeted regions of a powder bed (a). It is the only process that can generate a multi-colored model at one time (b).

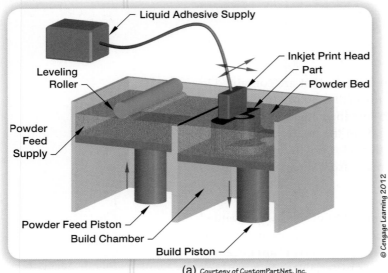

(a) Courtesy of CustomPartNet, Inc.

© Cengage Learning 2012

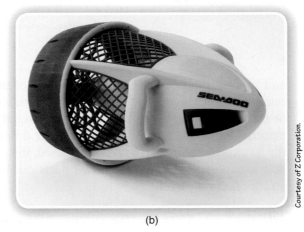

Courtesy of Z Corporation.

(b)

platform is lowered and a fresh layer of powder is added and leveled. After the part is completed, the loose powder surrounding the part is brushed or blown away. Parts made using the powder-binder printing process are often coated with a sealant to improve their strength and surface finish.

ANALYSIS PROCEDURE

By definition, engineering is the application of mathematics and science. As such, engineers find themselves solving scientific and mathematical problems that are not that different in nature from the problems that you have probably encountered in your science and mathematics classes. When solving such problems, engineers utilize a systematic analysis procedure that can be broken down into a series of steps:

Step 1. State the problem: A well-written problem statement should clearly, concisely, and logically explain the problem to be solved. All information needed to solve the problem as well as the desired outcome should be included. What is known and what is unknown should be clearly stated. A well-written problem statement is an important step to achieving a valid solution.

Step 2. Create a diagram: The saying "a picture is worth a thousand words" is certainly true when it comes to solving engineering problems. A diagram can be a sketch, drawing, or schematic of the item or system being analyzed. The diagram should include all given information from the problem statement.

Step 3. Make assumptions: Most engineering problems require the use of assumptions to develop a solution. Assumptions allow engineers to simplify a complex problem by eliminating potential variables. The experienced engineer knows what assumptions are safe to make versus those that may cause a disaster.

Step 4. Choose equations: All physical systems can be described using mathematical relationships. For example, Ohm's law describes the relationship between voltage, current, and resistance in an electrical circuit. It is important to choose the proper equation for the system being analyzed.

Step 5. Calculations: Substitute all given quantities into the equation to find the solution. Use a calculator or a computer to perform mathematical operations. Be sure to label the solution with proper units and the correct number of significant figures.

Step 6. Check solution: Ask yourself, "Does this answer make sense?" Does your answer align with the desired outcome of the problem statement? Double-check each step of your computations. The more engineering problems you solve, the better your ability to spot errors will become.

Step 7. Discuss results: After your solution has been thoroughly checked, discuss your answer with others to explain what the result of the analysis means.

PRESENTATION TOOLS

The final step in a design process involves communicating the results. A team's confidence in its solution is not always enough to warrant the go-ahead for full-scale implementation. It is often the case that a solution must be presented to an organization's decision makers in order to receive approval for the

© iStockphoto.com/Baris Simsek.

necessary financial backing (see Figure 2-22). Therefore, success depends not only on the quality of the solution but also on the effectiveness of the team's presentation.

Communicating an idea successfully depends on using the right tools and being prepared. Most designers today use digital formats to make important presentations. The following section will identify guidelines for creating effective digital presentations and address important tips for situations that involve public speaking.

Digital Presentations

The most popular software program used today to create and deliver digital presentations is Microsoft® PowerPoint®. A good PowerPoint slide show should be eye-catching, well-organized, and informative. Although PowerPoint is a powerful tool, it is important to remember that the most dynamic element in a presentation is the presenter. As such, the message is only as good as the messenger who delivers it.

A lot more time is needed to prepare for a presentation than to deliver it. Between research, generating graphics, writing, revision, and rehearsal, the development of a digital presentation can average an hour or more per slide. Taking this into consideration as part of the overall project plan will increase the likelihood of creating a presentation that achieves the team's underlying goal of moving a design solution into full development.

KNOW YOUR AUDIENCE A presentation's effectiveness will depend on how well the presenters understand their audience. Audiences differ in their knowledge base, which should influence the depth to which technical concepts and details are addressed. A good presentation should focus only on what the audience needs to know. Details can be addressed during a question-and-answer session at the end of the presentation.

CHARACTERISTICS OF EFFECTIVE DIGITAL PRESENTATIONS A good presentation should use as few slides as possible to convey the message. It should begin with a title slide, which is followed by an outline of the presentation's objectives. The end of the presentation should include a summary slide, followed by a slide that solicits questions and comments from the audience. It is also good practice to give the audience any necessary contact information so that communication can take place after the presentation is over.

How the information is presented is just as important as the information itself. The following guidelines should be considered when laying out the composition of each slide:

▶ Leave a border of space around the information contained on a slide, much like a frame around a picture.

▶ Keep the information on the slides to a minimum by using bulleted statements rather than complete sentences (except for direct quotes).

▶ Use a consistent and simple background, and make sure that it contrasts appropriately with the color of the text.

▶ Use graphics to help explain technical concepts and ideas.

▶ Use font sizes that are large enough to be read from across the room; use even larger font sizes for headings.

▶ Use sounds, transitions, and animation effects consistently and sparingly.

▶ Leave some white space in the overall composition so that the viewer is not overwhelmed by too much text and too many graphics.

Even after you've used your software's spell-checking tools, it is critical that you proofread your presentation carefully. Minor errors in spelling or grammar are easily noticed by the audience.

PRACTICE The more you practice what you want to say, the more confident you will be when you are saying it. Being confident is important when you want people to listen to and agree with an idea. Practicing out loud several times can free you from reading from notes, which results in a more natural presentation. Although you need to rehearse many times, you don't need to memorize everything you want to say. Your words can be different each time you rehearse. Focus on the *message* you want to convey and ensure that the message is the same every time you rehearse. To avoid stage fright, try memorizing an opening line.

Staying within the time frame you are given is critical, so talk through your presentation to see how much time is needed for each slide. Make sure you build time in for discussion and questions. Avoid reading the slides to your audience. The slides should be simple, and your oral presentation will fill in the details to provide further explanation. Provide your audience with notes if you want to make sure that the information is not misinterpreted or if others require the information in order to carry out their tasks.

Wear clothing that is appropriate to your audience. You want them to focus on what you are saying, not what you are wearing. Arrive at the location of the presentation early to set up and test any equipment you plan to use (see Figure 2-23). Remember to turn off your cell phone or any electronic device that could cause unplanned disruptions to the flow of your presentation. Finally, consider asking

FIGURE 2-23: Arriving early to set up and test presentation equipment is ultimately time well spent.

someone you trust to help you. Your assistant can run the presentation remotely or jot down questions and comments from the audience so that you can address them later.

Public Speaking

It is natural to be nervous when you speak in front of an audience, but the more you practice the easier it gets. Figure 2-24 lists some of the do's and don'ts of public speaking.

It is not a big deal if you forget a minor point or two during your presentation. If you get distracted and temporarily lose your train of thought, ask your audience if they have any questions. This will give you time you to recover. Listen carefully to questions and feedback at the end of your presentation. They can help you assess whether you have conveyed your message successfully (see Figure 2-25). When your presentation is over, take some time to reflect on what went well and what can be done better to improve the next presentation opportunity.

FIGURE 2-24: The Do's and Don'ts of Public Speaking.

Do	Don't
▶ speak normally and with enthusiasm. ▶ project your voice; if necessary use a microphone. ▶ speak clearly. ▶ repeat key points and emphasize them with deliberate pauses. ▶ keep your eyes on the audience. ▶ admit when you don't know the answer to a question, and convey a sincere desire to find and return an answer.	▶ speak too fast or too slowly. ▶ turn your back to the audience. ▶ hide behind the lectern. ▶ read directly from the slides or your notes. ▶ apologize if you can't answer a question but admit that you don't know the answer and that you will try to find out and get back to the questioner.

FIGURE 2-25: *Observe your audience's feedback. They will help you assess whether you have conveyed your message successfully.*

Image copyright Yuri Arcurs, 2010. Used under license from Shutterstock.com.

Career Profile

CAREERS IN ENGINEERING

An Engine for Change

Most of us only think about emissions standards when we see a plume of black smoke pouring out of a semitrailer's tailpipe, but Jess Crompton thinks about them all the time (Figure 2-26). In fact, it's her job to think about them. Crompton is a senior engineer with Caterpillar, the Peoria, Illinois, manufacturer of construction and mining equipment. Crompton doesn't build those distinctively yellow trucks we've all seen, but she does make sure they cause as little harm to the environment as possible.

"My current job is in the Aftertreatment Controls Group," Crompton says. "I help design and develop control systems for our products' engine-exhaust emissions technology."

Reducing the pollution caused by a combustion engine can be done in two major ways, according to Crompton. You can reduce the emissions produced by the combustion process or you can treat the exhaust that comes out of the engine. This second option includes "after-treatment controls."

"Our machines are required to meet certain pollutant limits," Crompton says, "so we are designing, testing, and validating the systems that will enable this."

On the Job

Along with the rest of the Aftertreatment Controls Group, Crompton is currently working on the Environmental Protection Agency's (EPA's) Tier IV program, the next round of emissions legislation for off-highway machines. "There's a lot of back and forth between the EPA and the industry I work in," Crompton says. "They do a lot of research to get a sense of what limits are possible, and we share our issues and concerns with them. The upcoming set of standards is incredibly challenging to meet, so it's a pretty important time."

Another important time for Crompton was the three months she spent in Europe as part of Caterpillar's Engineering Rotational Development Program. "I was based in Geneva, Switzerland," she says, "but I got to travel to several other countries to research and document

FIGURE 2-26: *Jess Crompton is a senior engineer with Caterpillar.*

© Cengage Learning 2012

the emissions regulations that might impact Caterpillar's aftertreatment business. In addition to learning how widely the regulations vary, I gained exposure to cultural differences, both personally and professionally."

Inspiration

Crompton didn't always know she wanted to be an engineer because she didn't always know what an engineer was. "As a child, I loved math," she says. "But during high school, when students start considering their college or career paths, I wasn't sure what you could do with math. There didn't seem to be many options. In fact, the only one I knew of was becoming a teacher. Luckily, there were several people in my family and community who suggested engineering. At the time, I didn't really know what that was. So I did some research, got a slight feel for some of the engineering disciplines. Still, it was mostly a leap of faith when I chose engineering as my major in college. And it turned out to be a natural fit."

Education

When she started working toward a physics degree at Augustana College in Rock Island, Illinois, Crompton still wasn't sure where she was headed. A bachelor's and a master's degree in mechanical engineering from the University of Illinois solidified her thoughts. "It's not as if I knew at any one point that I wanted to be an engineer," Crompton says, "but each step led to the next until I'd fully developed the skills and appreciation to be one."

Advice to Students

Having taken her time to commit to engineering as a profession, Crompton hopes tomorrow's engineers will learn from her mistakes. "It was difficult for me at the beginning," she says. "I never had an understanding of what an engineer was. I tried to read about it, but I wish I'd talked to engineers and spent some time with them, job shadowed them. Engineering is a huge challenge, but if you work hard and believe you can do it, it can be incredibly fulfilling."

SUMMARY

In this chapter you learned the following:

- Engineers use design processes to systematically solve problems. One such process involves a 12-step method:

  Step 1—Define the problem
  Step 2—Brainstorm
  Step 3—Research and generate ideas
  Step 4—Identify criteria and constraints
  Step 5—Explore possibilities
  Step 6—Select an approach
  Step 7—Develop a design proposal
  Step 8—Model or prototype
  Step 9—Test and evaluation
  Step 10—Refine
  Step 11—Create the final solution
  Step 12—Communicate the results

The first part of the process involves the creation of a design brief that summarizes the most important information about a design project. Research is then conducted to gain a better understanding of the nature of the problem, as well as to identify any existing solutions that may address similar problems. This helps to stimulate creative ideas that often surface during brainstorming sessions. The experienced designer recognizes the importance of keeping an engineer's notebook throughout a design project. This notebook serves as a record of important information that is found during research, as well as a repository for solution ideas. Once the important information has been shared by and between the members of a design team, the team will begin to generate large numbers of ideas in order to have a greater chance of finding a solution. A decision matrix is used to narrow down the ideas, and it provides a logical means of choosing an idea that is most likely to result in an optimum solution. Once an idea is selected, it is further refined through detailed sketches and computer models. Once the idea

has been developed into a potential solution via a design proposal, a prototype is built, tested, and refined. Because of the iterative nature of the design process, the engineer may need to repeat earlier steps in the process in order move forward with a solution. Once the final solution has been worked out, it is then communicated to the client.

- Brainstorming combines individual brain power with human interaction to generate as many ideas as possible for solving a design problem. Following the brainstorming guidelines helps establish an environment that eliminates fear and encourages team members to voice their ideas freely. Traditional brainstorming involves several people sketching and sharing their ideas in a rapid fashion. The stepladder and reverse-brainstorming techniques are two other types of brainstorming methods that can help generate solutions to problems.

- Sketching is a form of visual communication that designers use to capture and convey ideas. Sketches may take the form of three-dimensional pictorials or two-dimensional multiviews. Dimensions are placed on multiview drawings to communicate an object's height, width, and depth, as well as information about the sizes and locations of features such as holes.

- Engineers use computer-aided design (CAD) programs to develop and document a design solution's geometry. CAD programs are used in conjunction with computer-aided manufacturing (CAM) and rapid prototyping technologies to create models or prototypes. Designers have several different types of rapid-prototyping methods to choose from, including stereolithography (SLA), fused deposition modeling (FDM), selective laser sintering (SLS), and powder-binder printing.

SUMMARY

(continued)

- Engineers utilize a systematic analysis procedure when solving mathematics- and science-based problems. The steps to solve an engineering analysis problem are:

  Step 1—State the problem
  Step 2—Create a diagram
  Step 3—Make assumptions
  Step 4—Choose equations
  Step 5—Substitute and solve
  Step 6—Check the solution
  Step 7—Discuss the results

- The final step in a design process involves communicating the results, usually in the form of a digital presentation. The most popular software program used today to create and deliver digital presentations is Microsoft PowerPoint. A presentation's effectiveness depends on several factors, including:

  - how well the presenters understand their audience,

  - the brevity and the clarity of the message,

  - how the information on each slide is arranged into a visually attractive and readable composition,

  - the degree to which the presenters practice their presentation, and

  - the quality of the presenters' public speaking skills.

BRING IT HOME

1. What is a design process?
2. What makes design an iterative process?
3. What is a design brief, and what information does it contain?
4. What is brainstorming, and what are the different ways in which it is carried out in the 12-step design process that you have studied?
5. Why do engineers conduct research?
6. What is the difference between primary and secondary research?
7. What is the purpose of an engineer's notebook, and what kinds of information are typically recorded in it?
8. What is the difference between criteria and constraints?
9. What methods do engineers use to identify and select the most promising ideas for further development?
10. What is the purpose of a decision matrix, and how is one created?
11. What kind of information is usually contained in a design proposal?
12. What is the difference between a model and a prototype?
13. Give an example of a kind of test that might be performed on a prototype?
14. What are the different ways in which a final solution is communicated?
15. Name a technique used by design teams to help generate ideas.
16. Why are there guidelines for brainstorming?
17. Explain the difference between a pictorial and a multiview sketch.
18. What dimensions (width, depth, or height) are shown in the top, front, and side views of a multiview drawing?
19. Identify two advantages that CAD systems have over traditional methods for producing technical drawings by hand.
20. What is the purpose of finite element analysis?
21. What are CAM programs used for?
22. How do rapid prototyping technologies decrease the development time for a design solution?
23. Identify four rapid prototyping technologies and describe the method by which each creates a model.
24. List the steps of the engineering analysis procedure.
25. Identify five characteristics of an effective digital presentation.

Problem Statement: When students use the hallway drinking fountains, their hands leave bacteria and germs on the control surfaces as a result of direct physical contact. This causes the spread of colds, flu, and other illnesses.

Design Statement: Design, build, and test a hands-free control device that can be retrofitted to your school's existing hallway drinking fountains that will help prevent the spread of germs.

Use the design process that you learned about in this chapter as a guide. Document all of your project work in an engineer's notebook and keep a daily log of your efforts. When you are finished, communicate your design solution via PowerPoint presentation to your local board of education.

CHAPTER 3
The Mechanical Advantage

GPS
DELUXE

| START LOCATION | DISTANCE | END LOCATION |

Menu

Before You Begin

Think about these questions as you study the concepts in this chapter:

1. What is a force, and how is it applied?

2. What is the difference between work and power?

3. What is mechanical advantage, and how is it a trade-off between the effort and load with respect to force and distance?

4. What is the difference between a machine's actual mechanical advantage (AMA) and its ideal mechanical advantage (IMA), and why is one always smaller in magnitude than the other?

5. What are the six simple machines, and how do you calculate their mechanical advantage?

6. What is a compound machine, and how do you calculate its ideal mechanical advantage?

Our human-built world is filled with mechanical devices called *machines* that we seamlessly integrate into our everyday routines. Though we often think of them as being large, complex, and even dangerous, not all machines fit this paradigm. For example, the pop-top that you bend up to open a can of soda is a mechanical device that falls under the category of a *simple machine* (see Figure 3-1).

You can think of a machine as any device, either fixed or moving, that helps you accomplish a task. Machines are used to change the direction of an input force or motion and are capable of multiplying force, distance, torque, or speed. Motion may occur in a straight line (*linear motion*), around a pivot point (*circular motion*), or oscillate back and forth along a straight line (*reciprocal motion*).

This chapter serves to introduce the concepts of force, work, power, mechanical advantage, and simple machines. People who are involved with the design, manufacture, installation, operation, analysis, or maintenance of mechanical systems are expected to have a thorough understanding of these concepts.

> **Machine:**
> any device, either fixed or moving, that helps in accomplishing a task.

Figure 3-1: *The pop-top on a can of soda acts as a simple machine.*

FORCE, WORK, AND POWER

Force

A **force** is a push or pull that acts on an object to initiate or change its motion or to cause deformation. Most forces that we work with in engineering applications come under the category of **contact forces**. These are forces transmitted from one physical object to another object through physical contact. Forces that act on an object without physical contact, such as gravity and magnetism, fall under the category of **long-range forces**. Several units are used to measure force. The most common metric unit in the **International System of Units** (SI) is the newton (N). The unit of force that is most commonly used in the **United States customary system of units** is the pound (lb). One pound of force is equal to approximately 4.4 newtons.

> **Force:**
> a push or pull that acts on an object to initiate or change its motion or to cause deformation.

When we work with machines, we describe a *force* as being either an *effort force* or *load force*. An effort force can be thought of as an input force that is generated by a person, motor, engine, magnetic field, spring, moving water, wind, or any other phenomena that serves to impart energy into a machine or system. If you want to raise a flag to the top of a flagpole, then you must apply an effort force to the pulley rope (see Figure 3-2).

The *output* of a mechanical system is called the *load* or *resistance*. A load can be thought of as the weight of the object that needs to be moved or as the resistance that a mechanical device must overcome to accomplish its task. For example, your weight serves as a load when you step onto an elevator. The spring inside a ballpoint pen also serves as a load because it provides a resistance that must be overcome in order to make the pen work.

Work

Engineers are especially concerned with the topic of *work* and how it applies to mechanical systems. They design machines to make work easier for us. In fact, many engineers spend their entire careers trying to design devices that perform work more efficiently. Work is a measure of the amount of energy that is transferred when a force acts on an object, causing the object to move or deform (see Figure 3-3). In some instances, a force may be exerted on an object but no work is accomplished. For example, if you push on a wall, and the wall does not move, then you have not performed work.

To calculate work (W), one must multiply the magnitude of the force (F) by the amount of *displacement* (d) that results (see Equation 3-1). This assumes that the force is constant and acts in the same direction in which the load travels. Displacement is a measure of how far an object moves as measured along a straight line from start to finish. Displacement is a vector quantity, which means it has both magnitude and direction:

$$W = F\,d \qquad \text{(Equation 3-1)}$$

FIGURE 3-2: The pulling action on the pulley rope serves as the effort force that is needed to raise the flag. The weight of the flag serves as the load force.

VisionsofAmerica/Joe Sohm/Photodisc/Getty Images.

FIGURE 3-3: Work occurs when a force is applied to a load, causing the load to move.

© Cengage Learning 2012

The common SI unit of displacement is the meter (m), and the common U.S. customary unit is the inch (in) or foot (ft), depending on the magnitude of the displacement. It is important to note that *distance*, which is a scalar quantity (magnitude only), should not be confused with displacement. **Distance** is also a measure of length, but it is not confined to a straight-line path.

The greatest amount of work is accomplished when the constant force acts parallel to the direction that the load moves. If the force does not act in the same direction as an object's motion, then the component force that acts parallel to the object's motion must be determined. This is done by multiplying the magnitude of the applied force by the cosine of the angle by which it deviates from the object's direction of motion (see Equation 3-2). When the direction of an applied force and the resulting motion are parallel, then the angle is zero degrees. The cosine of zero degrees is 1. The result of the work formula is a scalar quantity that is measured in foot-pounds (ft · lb) or inch-ounces (in · oz) in the U.S. customary system, and newton-meters (N · m) or joules (J) in the SI system. One joule is equal to one newton-meter.

$$W = F\, d \cos \theta \qquad \text{(Equation 3-2)}$$

Example

Problem: A woman applies a constant 100-lb force to a box at a 20° angle as measured from the horizontal (see Figure 3-4). As a result, the box moves 12 feet across a level surface. How much work did the woman accomplish?

$$W = F\, d \cos \theta$$
$$W = 100\,\text{lb} \times 12\,\text{ft} \times \cos 20°$$
$$W = 100\,\text{lb} \times 12\,\text{ft} \times 0.94$$
$$W = 1{,}128\,\text{ft} \cdot \text{lb}$$

FIGURE 3-4: **Example work problem.**

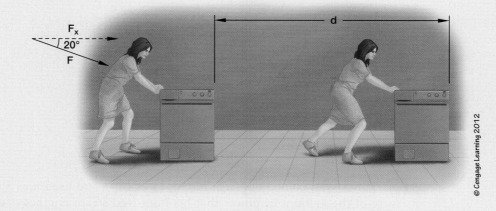

Power

The concept of work is independent of time. Therefore, the amount of work that is accomplished by a 100-lb force moving a box a distance of 12 feet will be the same if the task takes 20 seconds or 20 hours. You may have heard the expression "Time is money!" Business owners tend to use this expression a lot because they want the most

Power:

a measure of the amount of work that is accomplished per unit time.

amount of work accomplished in the shortest possible time. If you consider how much time it takes to accomplish work, then you are dealing with the concept of *power*. Power (P) is a measure of the amount of work (W) that is accomplished per unit time (t) (see Equation 3-3). The SI unit of power is the *watt*; named in honor of James Watt for his significant improvements to the design of the steam engine and his work in defining the concept of power during the late 1700s. The U.S. customary unit of power is the foot-pound per second. One watt is equal to 0.738 ft · lb/sec.

$$P = \frac{W}{t} \qquad \text{(Equation 3-3)}$$

Returning to our previous problem in which a box was moved 12 feet, we learned that the woman moving the box performed 1,128 ft · lb of work. If she accomplished the task in 20 seconds, how much power did she exert in watts?

$$P = \frac{W}{t}$$

$$P = \frac{1,128 \text{ ft} \cdot \text{lb}}{20 \text{ sec}}$$

$$P = 56.4 \text{ ft} \cdot \text{lb/sec}$$

$$1 \text{ watt} = 0.738 \text{ ft} \cdot \text{lb/sec}$$

$$P = \frac{56.4 \text{ ft} \cdot \text{lb/sec}}{0.738 \frac{\text{ft} \cdot \text{lb/sec}}{\text{watt}}}$$

$$P = 76.5 \text{ watt}$$

If the task of moving the box 12 feet took 20 hours to accomplish, how much power did the woman exert in watts?

$$1 \text{ hour} = 3,600 \text{ sec/hour}$$

$$t = 20 \text{ hours} \times 3,600 \text{ sec/hour}$$

$$t = 72,000 \text{ sec}$$

$$P = \frac{W}{t}$$

$$P = \frac{1,128 \text{ ft} \cdot \text{lb}}{72,000 \text{ sec}}$$

$$P = 0.016 \text{ ft} \cdot \text{lb/sec}$$

$$1 \text{ watt} = 0.738 \text{ ft} \cdot \text{lb/sec}$$

$$P = \frac{0.016 \text{ ft} \cdot \text{lb/sec}}{0.738 \frac{\text{ft} \cdot \text{lb/sec}}{\text{watt}}}$$

$$P = 0.022 \text{ watt}$$

James Watt coined the term horsepower (hp), and used it as a standard measure of the amount of power that his improved steam engines could produce in comparison to the amount of power that the average workhorse could generate. Horsepower is a non-SI unit of power that is still used in the United States, especially in reference to the power output of combustion engines (see Figure 3-5).

One horsepower (hp) is equal to approximately 746 watts. Human beings cannot generate large amounts of horsepower, which is why we design automobiles and aircraft to move us from place to place instead of pedaling a bicycle or rowing a boat for hours or even weeks. A highly trained athlete can only output approximately 0.25 hp for a period of several hours. The average, healthy human has the

FIGURE 3-5: (a) A typical chainsaw engine is rated between 2 and 6 hp. (b) A lawn tractor engine produces about 11 hp. (c) The V-8 engine in a 1967 Chevy Camaro SS is rated at 325 hp. (d) The power outputs of all these engines are miniscule when compared to the 37,000,000 hp that is generated by the three main engines on a space shuttle orbiter.

(a)&(b) © iStockphoto.com/DonNichols. (c) Image copyright Patrick James Bryk, 2010. Used under license from Shutterstock.com. (d) © iStockphoto.com/Marje.

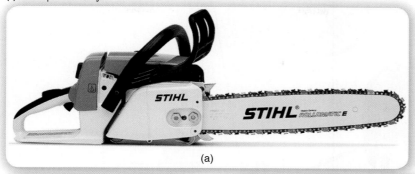
(a)

(b)

(c)

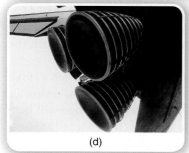
(d)

ability to generate a sustained power of approximately 0.1 hp. This is about how much power was exerted when the woman from our previous problem moved the box 12 feet in 20 seconds.

$$1 \text{ hp} = 746 \text{ watt}$$
$$P = \frac{72.4 \text{ watt}}{746 \text{ watt}}$$
$$P \approx 0.1 \text{ hp}$$

OFF-ROAD EXPLORATION

To learn more about the extraordinary things that can be done with human power, watch the film *The Flight of the Gossamer Condor*.

Most people don't understand what *horsepower* is. The term was created by James Watt, who marketed the steam engine as a replacement for workhorses that were used to power water pumps in mines. He estimated that an average workhorse could move a 550-lb load a distance of 1 foot in 1 second (see Figure 3-6), so he made this the equivalent of one "horsepower." Watt's creation and coining of the horsepower unit allowed him to market his steam engine to mining companies in a way that made the power output of his machines understandable to the customer. In honor of Watt's contributions, the scientific community made the *watt* the standard SI unit of power.

Continued

FIGURE 3-6: One horsepower is equivalent to moving a 550-lb load a distance of 1 foot in 1 second.

550 lb

1 foot

© Cengage Learning 2012

MECHANICAL ADVANTAGE

The term *mechanical advantage* will inevitably come up at some point if you are involved with the design of machines. Mechanical advantage can be thought of as the ratio of the output force (load) produced by a machine to the input force (effort) that is applied to the machine. It is also a ratio between the distance over which the effort force acts to the distance moved by the load.

Some machines are designed to amplify an input force. This occurs when a small effort force is exerted over a large distance to move a larger load across a shorter distance. Under these circumstances, a machine's mechanical advantage will be a value that is greater than 1. Some machines require a large effort force that acts over a short distance to move a smaller load across a larger distance. In this case, the mechanical advantage will be a value between 0 and 1. There are two types of mechanical advantage: *ideal mechanical advantage* (IMA) and *actual mechanical advantage* (AMA).

Mechanical advantage:

the ratio of the output force produced by a machine to the input force that is applied to the machine.

Ideal Mechanical Advantage

Ideal mechanical advantage (IMA) is a ratio between the distance across which an effort force acts and the resulting distance across which the load acts (see Equation 3-4). The term *ideal* means that no energy is lost to *friction*, deformations that occur in structural components under load, or other energy-absorbing phenomena:

$$\text{IMA} = \frac{\text{effort distance}}{\text{load distance}} \qquad \text{(Equation 3-4)}$$

If either distance is unknown, then a machine's ideal mechanical advantage can be calculated by dividing the load force (L) by the *ideal effort force* (E_i) (see Equation 3-5). The ideal effort force is the amount of input force that is needed to move a load under ideal circumstances in which a machine operates at 100% efficiency.

Therefore, the amount of work that is put into the machine equals the amount of work that comes out of the machine.

$$IMA = \frac{\text{load force}}{\text{ideal effort force}} \qquad \text{(Equation 3-5)}$$

$$IMA = \frac{L}{E_I}$$

Actual Mechanical Advantage

Unfortunately, ideal circumstances never exist, and no machine operates at 100% efficiency. Energy is lost to heat between the input and output of a machine as a result of friction. **Friction** is a type of resistance force that results when two objects move against each other. Friction always opposes motion. Mechanical devices such as bearings, as well as lubricants such as graphite, oil, grease, and silicone are used to reduce friction.

Actual mechanical advantage (AMA) is a ratio between the load force (L) and the *actual effort force* (E_A) (see Equation 3-6). The **actual effort force** is the amount of input force that is needed to move a load under circumstances in which a machine operates at less than 100% efficiency. Unlike ideal mechanical advantage, a machine's *actual mechanical advantage* takes into account the losses that occur because of friction and other phenomena, such as the deformation (elongation or bending) of a structural member under load. For this reason, a simple machine's actual mechanical advantage will always be less than its ideal mechanical advantage:

$$AMA = \frac{\text{load force}}{\text{actual effort force}} \qquad \text{(Equation 3-6)}$$

$$AMA = \frac{L}{E_A}$$

> **Friction:**
> a type of resistance force that results when two objects move against each other.

Efficiency

If all machines are incapable of operating at 100% efficiency, how do we know how efficient a machine is? To answer this question, we first have to take a closer look at the concept of *efficiency*. **Efficiency** is a ratio between the amount of work that goes into and comes out of a machine. It is also a ratio between a machine's actual and ideal mechanical advantage (see Equation 3-7). The symbol for efficiency is the Greek letter η (pronounced "eta"). The value of the efficiency is expressed as either a decimal value between 0 and 1 or as a percentage. For example, an efficiency value of 0.85 is equivalent to 85%. A machine's efficiency can never be equal to or greater than 1 (100%), because there will always be losses.

$$\text{Efficiency} = \frac{\text{actual mechanical advantage}}{\text{ideal mechanical advantage}} \qquad \text{(Equation 3-7)}$$

$$\eta = \frac{AMA}{IMA}$$

> **Efficiency:**
> the ratio between a machine's actual and ideal mechanical advantage.

The formulas for ideal mechanical advantage, actual mechanical advantage, and efficiency apply to all simple machines. In the following sections, we will introduce other equations that are used to calculate ideal mechanical advantage, the magnitudes of the effort and load forces, and the distances across which the effort or load acts along with their respective simple machines.

THE SIX SIMPLE MACHINES

Aside from the fact that both are built from stone, what do the Egyptian pyramids and the Roman aqueducts have in common? Both of these colossal achievements of ancient human engineering employed the assistance of *simple machines*.

A **simple machine** is a mechanical device that is used to change the direction or the magnitude of a single applied force. The six classical mechanical devices that are considered to be simple machines include the inclined plane, the wedge, the lever, the wheel and axle, the pulley, and the screw (see Figure 3-7). Though the identity of who invented each simple machine has been lost to history, the operating principles behind many of these devices were explained in the writings of a Greek engineer and mathematician named Archimedes during the 3rd century B.C.E.

FIGURE 3-7: Examples and applications of the six simple machines. (a) Inclined plane. (b) Wedge. (c) Lever. (d) Pulley. (e) Wheel and axle. (f) Screw.

(a) Image copyright nicobatista, 2010. Used under license from Shutterstock.com. (b) © iStockphoto.com/goldenlobby. (c) Image copyright Picsfive, 2010. Used under license from Shutterstock.com. (d) © iStockphoto.com/Digiphoto. (e) © iStockphoto.com/grybaz. (f) © iStockphoto.com/dial-a-view.

Simple machines are used to:

▶ move a heavy load over a short distance using a small effort force,

▶ move a small load over a large distance using a greater effort force, and

▶ change the direction or magnitude of an effort force.

The old adage "You can't have your cake and eat it too" also applies to simple machines. No one simple machine can be used to decrease a necessary effort force while simultaneously increasing the distance that is moved by a load. Either effort or distance must be sacrificed.

The Inclined Plane

The simplest of all the simple machines is the *inclined plane*. An inclined plane is characterized as a stationary, flat surface that is set at an angle and is used to bridge two planes that are offset by some vertical distance. The inclined plane is used for moving an object (load) from a lower to a higher elevation that, for whatever reason, is impractical to traverse directly. A wheelchair ramp is a common application of the inclined plane (see Figure 3-8).

Inclined plane:

a stationary, flat surface that is set at an angle and is used to bridge two planes that are offset by some vertical distance.

FIGURE 3-8: **A common application of the incline plan is a wheelchair ramp.**

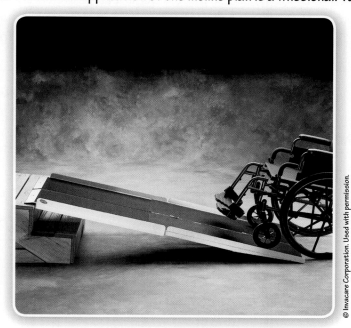

© Invacare Corporation. Used with permission.

Even with friction, it is often easier to roll or slide a heavy object up a ramp than it is to lift the same object vertically using sheer muscle power. For example, most people do not have the strength to lift a motorcycle or 4-wheeler into the back of a truck bed. Instead, wood planks or metal beams are used as a makeshift ramp so the load can be safely pushed onto the truck bed. The same amount of work is accomplished when using a ramp as would be accomplished if the load had been lifted directly onto the truck bed. A relatively small effort force is exerted over a relatively long distance in comparison to a large effort force being exerted over a short distance. The reduced effort force that is required to accomplish the same task displays *mechanical advantage*.

The geometry of an inclined plane is synonymous with that of a right triangle (see Figure 3-9). The hypotenuse of the right triangle is considered the *slope length* (s). The *slope angle* is denoted as theta (θ). The side opposite the slope angle represents

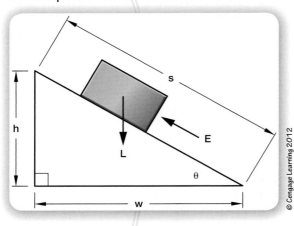

FIGURE 3-9: Variables associated with calculating the mechanical advantage of an inclined plane.

© Cengage Learning 2012

the *overall height of the incline* (h). Though not directly used to calculate mechanical advantage, the *base width* (w) of the inclined plane is synonymous with the *adjacent side* of the slope angle.

Because the effort acts along the length of the slope, causing the load to be raised a distance that is equal to the vertical height of the incline, the ideal mechanical advantage of an inclined plane is a ratio between these two distances (see Equation 3-8):

$$IMA_{Inclined\ Plane} = \frac{slope\ length}{vertical\ height\ of\ the\ incline} \quad \text{(Equation 3-8)}$$

$$IMA_{Inclined\ Plane} = \frac{s}{h}$$

The values of the load force (L), ideal effort force (E_I), and actual effort force (E_A) are also variables that can be used to calculate the ideal and actual mechanical advantage of an inclined plane (see Equations 3-5 and 3-6).

Example

Problem: The inclined plane shown in Figure 3-10 is used to move a 50-lb load a vertical distance of 36 inches. The *actual* amount of effort force that is needed to move the load up the inclined surface is 25 lb. Use this information, along with the dimensions given in the illustration, to answer the following questions.

a. What is the length of the inclined plane's slope?

$$\sin \theta = \frac{opposite}{hypotenuse} \Rightarrow \sin \theta = \frac{h}{s} \Rightarrow s = \frac{h}{\sin \theta}$$

$$s = \frac{36\ in}{\sin 20°}$$

$$s = \frac{36\ in}{0.342}$$

$$s = 105.26\ in$$

FIGURE 3-10: Inclined plane problem.

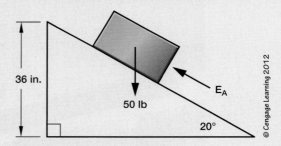

© Cengage Learning 2012

b. What is the *ideal mechanical advantage* of this inclined plane?

$$IMA_{Inclined\ Plane} = \frac{s}{h}$$

$$IMA_{Inclined\ Plane} = \frac{105.26\ in}{36\ in}$$

$$IMA_{Inclined\ Plane} = 2.92$$

c. What is the *actual mechanical advantage* of this inclined plane?

$$AMA = \frac{L}{E_A}$$

$$AMA = \frac{50\ lb}{25\ lb}$$

$$AMA = 2$$

(continued)

d. How *efficient* is this inclined plane system?

$$\eta = \frac{AMA}{IMA}$$

$$\eta = \frac{2}{2.92}$$

$$\eta = 0.68 \text{ or } 68\%$$

e. What is the *ideal effort force* that would be needed to push the load up the inclined plane in the absence of friction?

$$IMA = \frac{L}{E_I} \Rightarrow E_I = \frac{L}{IMA}$$

$$E_I = \frac{50 \text{ lb}}{2.92}$$

$$E_I = 17.12 \text{ lb}$$

The Wedge

To understand the importance of the *wedge*, imagine trying to eat without the ability to use your teeth. A **wedge** is a rigid object with sloping sides that moves against or through a load in order to exert a force. The shape of a wedge is often that of two inclined planes set base to base. The major difference between the inclined plane and the wedge is that the inclined plane functions as a stationary machine. Also, an inclined plane does not move through a load.

Our ancient ancestors discovered that certain types of stone would crack and flake in ways that produce razor-sharp edges. They used these stone wedges as tools to skin animal hides, chop down trees, dig out boat hulls, and tip the ends of spears. The ancient Egyptians used both bronze and wood wedges to quarry the stones that make up the enormous temples and pyramids that we still marvel at today. The wood wedges were pounded into cracks in the rock and then soaked with water. As the wood absorbed the water, the wedge would expand and split the stone.

Most of the cutting tools used today are examples of wedges. Other common applications of the wedge include zippers, plows on farming equipment, and keys (see Figure 3-11). The ridges on a key and the tumblers inside a lock are shaped like wedges. As the key slides into the lock, the tumblers move up and down the ridges on the key until they are oriented in such a way as to allow the key to rotate the locking mechanism.

As it is true for all simple machines, the ideal mechanical advantage of a wedge is a ratio between the distance across which the effort acts and the resulting distance that the load moves. Imagine an axe that is used to split a wood log. As the full length of the axe is driven into the wood, the log will split into two parts. The leading edges on both sides of the split move along the slope length of the wedge until they are separated by a gap that is equal to the thickness of the wedge (see Figure 3-12).

The ideal mechanical advantage of a wedge is the ratio between the *slope length* (s) and the *thickness of the wedge* (t), as shown in Equation 3-9.

$$IMA_{\text{Wedge}} = \frac{\text{slope length}}{\text{thickness of the wedge}} \qquad \text{(Equation 3-9)}$$

$$IMA_{\text{Wedge}} = \frac{s}{t}$$

> **Wedge:**
> a rigid object with sloping sides that moves against or through a load to exert a force.

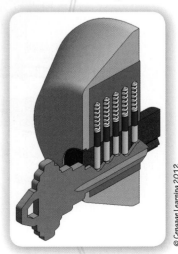

FIGURE 3-11: A wedge can be thought of as a moving inclined plane. The ridges on a key and the tumblers inside a door lock are common applications of the wedge.

© Cengage Learning 2012

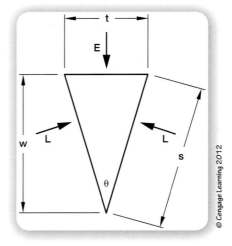

Because of their similar geometry, the wedge and the inclined plane share a common characteristic with respect to mechanical advantage. As the incline angle (θ) decreases, the mechanical advantage increases. In other words, if an inclined plane or wedge becomes longer and thinner, it will provide greater mechanical advantage. This is one of the reasons why chefs insist on using knives that are very sharp. Less effort is needed to cut an object with a sharp knife than with a dull knife.

Example

Problem: The hydraulic-powered log splitter shown in Figure 3-13 is used to split large logs in half. A wedge, located on one end of the device, provides the cutting action. The shape of the wedge is an isosceles triangle that is 4 inches thick by 8 inches wide. An *actual* effort force of 500 lb is required to split the log in the illustration. The cutting action of the wedge is 90% efficient. Use this information to answer the following questions.

FIGURE 3-13: **(a) This hydraulic-powered log splitter uses the principle of the wedge. (b) The geometry of the wedge is an isosceles triangle that is 8 inches wide by 4 inches thick.**

(a)

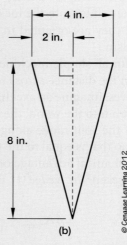

(b)

(continued)

a. What is the *slope length* on either side of the wedge?

$$s = \sqrt{w^2 + (0.5 \times t)^2}$$

$$s = \sqrt{(8\text{in})^2 + (0.5 \times 4 \text{ in})^2}$$

$$s = \sqrt{64 \text{ in}^2 + (2 \text{ in})^2}$$

$$s = \sqrt{64 \text{ in}^2 + 4 \text{ in}^2}$$

$$s = \sqrt{68 \text{ in}^2}$$

$$s = 8.25 \text{ in}$$

b. What is the ideal mechanical advantage of this wedge?

$$IMA_{Wedge} = \frac{s}{t}$$

$$IMA_{Wedge} = \frac{8.25 \text{ in}}{4 \text{ in}}$$

$$IMA_{Wedge} = 2.06$$

c. What is the actual mechanical advantage of this wedge?

$$\eta = \frac{AMA}{IMA} \Rightarrow AMA = \eta \times IMA$$

$$AMA = 0.9 \times 2.06$$

$$AMA = 1.85$$

d. How much *resistance* does the log provide against the cutting action of the wedge?

$$AMA = \frac{L}{E_A} \Rightarrow L = AMA \times E_A$$

$$L = 1.85 \times 500 \text{ lb}$$

$$L = 925 \text{ lb}$$

e. If friction is not a consideration, what would the *ideal effort force* need to be to split the log?

$$IMA = \frac{L}{E_I} \Rightarrow E_I = \frac{L}{IMA}$$

$$E_I = \frac{925 \text{ lb}}{2.06}$$

$$E_I \approx 450 \text{ lb}$$

The Lever

Archimedes once said, "Give me a lever long enough and a fulcrum on which to place it, and I shall move the world." Of all the simple machines, the operating principles of the lever are perhaps the most widely employed in the design of machines. A **lever** is a rigid bar that is allowed to rotate at some angle about a pivot point called the *fulcrum*. It is used to move a load when an effort force is applied to the bar. Depending on where the **fulcrum** is located on the lever, the lever may be used to change the direction of an effort force, multiply an effort force, or increase the distance that is traveled by a load. Levers are divided into three categories: 1st-class, 2nd-class, and 3rd-class levers.

> **Lever:**
>
> a rigid bar that is allowed to rotate at some angle about a pivot point called the *fulcrum*. It is used to move a load when an effort force is applied to the bar.

Fun Facts

As is the case with the inclined plane and the wedge, the invention of the lever predates recorded human history. Historians believe that the inclined plane, wedge, and lever were commonly used during the Stone Age, which lasted from approximately 2.5 million B.C.E. to 4000 B.C.E.

As a lever pivots about its fulcrum, the path that is generated by both the effort and the load will form an arc. As such, the distance from the fulcrum to the effort or the load is the radius of the respective arc (see Figure 3-14).

FIGURE 3-14: **The distance from the fulcrum to either the load or the effort can be thought of as a radial distance.**

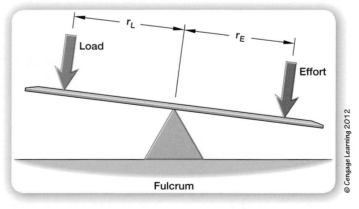

© Cengage Learning 2012

The *ideal mechanical advantage* of *any* lever is a ratio between the *radius of the effort arm* (r_E) and the *radius of the load arm* (r_L) (see Equation 3-10):

$$IMA_{Lever} = \frac{\text{radius of the effort arm}}{\text{radius of the load arm}} \quad \text{(Equation 3-10)}$$

$$IMA_{Lever} = \frac{r_E}{r_L}$$

OFF-ROAD EXPLORATION

To learn about how the ancient Egyptians used the lever and other simple machines to build and raise their giant obelisk monuments, watch the NOVA film *Secrets of Lost Empires: Pyramid/Obelisk*.

FIGURE 3-15: (a) If the fulcrum of a 1st-class lever is located midway between the load and the effort, then the lever will not exhibit mechanical advantage. (b) If the fulcrum is located closer to the load, then the lever will amplify the effort force. (c) If the fulcrum is located closer to the effort, then the lever can be used to move the load across a relatively large distance.

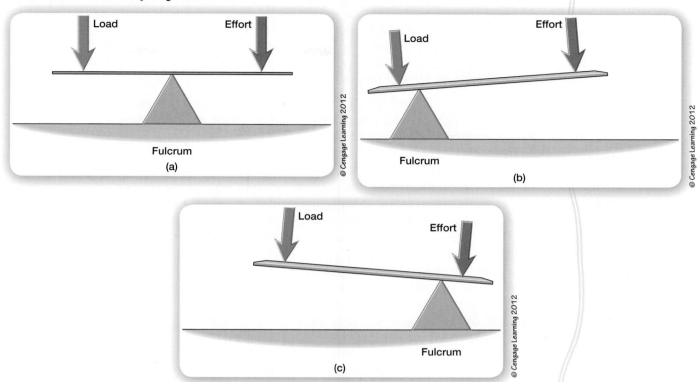

On a **1st-class lever**, the fulcrum is located somewhere between effort and load. Of the three lever classes, the 1st-class lever is the only one in which the effort force and the load force act in the same direction. It is also the only class of lever that can be modified to amplify an effort force or the distance that is moved by the load (see Figure 3-15).

No mechanical advantage exists if the fulcrum is located midway between the effort and the load because the radius of the effort arm will be equal to the radius of the load arm. In this case, the 1st-class lever serves only to change the direction of the applied effort force. Moving the fulcrum toward the load will decrease the effort force needed to lift the load. Moving the fulcrum toward the effort will increase the distance traveled by the load but require an effort force that is greater than the load.

If you have ever opened a can of paint with the end of a screwdriver or used a pry bar to remove nails from a piece of lumber, then you have used a 1st-class lever. You may have used a seesaw at the playground when you were younger. This, too, is an example of a 1st-class lever (see Figure 3-16). If two children of unequal weight were using the see-saw, what did they have to do to balance the lever arm?

FIGURE 3-16: (a) A pry bar and (b) a see-saw are common examples of a 1st-class lever.

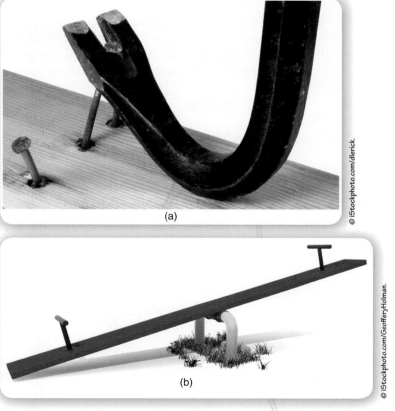

Your Turn

The beam balance, which operates on the principle of the 1st-class lever, was the first weight-measuring device ever invented. In this activity you will be using a playground see-saw as a balance beam to calculate a friend's weight. You will need a writing utensil, your engineer's notebook, a tape measure, a weight scale, a calculator, a playground see-saw, and two of your friends for this activity.

1. Measure your weight using the weight scale and record the value in your engineer's notebook.
2. Ask two of your friends to join you at a playground that is equipped with a see-saw. It is important that at least one of your friends has a weight that is significantly more or less than your weight (10 to 20 lb difference).
3. Locate the see-saw and measure the distance from the see-saw's fulcrum to the center of one of the seats. It is assumed that this distance will be the same on both ends of the see-saw.
4. Make a sketch of the see-saw and record the measured distance in your engineer's notebook.
5. Ask your friend whose weight differs most from yours to sit at one end of the see-saw while you sit at the other end. Give the writing utensil, tape measure, and your engineer's notebook to the friend who is not sitting on the see-saw.
6. Have the person who weighs the most move toward the see-saw's fulcrum until the bar becomes balanced and level. Alternatively, you can move away from the fulcrum until balance occurs. Make sure that no one's feet are touching the ground and are not resting on the see-saw. It is important that one of the two people on the see-saw remain at the initial seat location.
7. Ask your friend with the tape measure to measure the distance from the see-saw's fulcrum to the center of the location where either you or your friend moved in order to balance the see-saw. Ask the person to record the measurement in your engineer's notebook.
8. Use the following equation to estimate your friend's weight. Record your calculations in your engineer's notebook.

$$\text{your friend's estimated weight} = \frac{\text{your weight} \times \text{your distance to the fulcrum}}{\text{distance from your friend to the fulcrum}}$$

9. Ask the friend who joined you on the see-saw to weigh him- or herself (preferably with the same weight scale that you used) and record this value in your engineer's notebook.
10. Use the following equation to calculate the percentage of error between your friend's estimated weight and his or her actual weight. Record your calculations in your engineer's notebook.

$$\% \text{ error} = \frac{\text{your friend's estimated weight} - \text{his/her actual weight}}{\text{your friend's actual weight}} \times 100$$

11. Speculate as to the reasons for the error between your friend's estimated weight and his or her actual weight. Record your ideas in your engineer's notebook.

FIGURE 3-17: The load is loacted between the effort and the fulcrum on a 2nd-class lever.

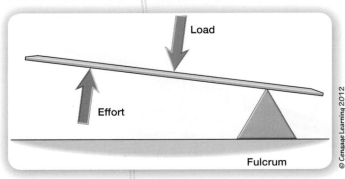

© Cengage Learning 2012

On a **2nd-class lever**, the load force and the effort force must act in opposite directions. Also, the load is always located between the fulcrum and effort, which means the mechanical advantage of any 2nd-class lever will always be greater than 1 (see Figure 3-17). Moving the load closer to fulcrum will increase the mechanical advantage. Moving the load closer to effort decreases the mechanical advantage.

Second-class levers are used in applications where an effort force must be amplified, such as when a person uses a wheelbarrow to lift a load of dirt or when someone tries to crack open the tough shell of a nut (see Figure 3-18).

FIGURE 3-18: (a) A wheelbarrow and (b) a nut cracker are common examples of a 2nd-class lever.

(a)

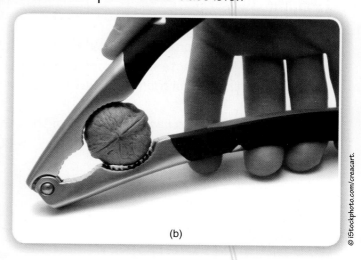

(b)

FIGURE 3-19: The effort is loacted between the load and the fulcrum on a 3rd-class lever.

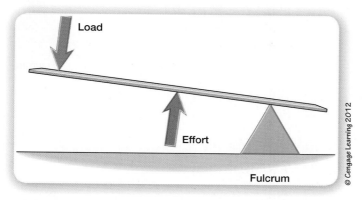

On a **3rd-class lever**, the effort is always located between the load and fulcrum (see Figure 3-19). For this reason, the mechanical advantage of a 3rd-class lever will always be a value between 0 and 1. As was the case with the 2nd-class lever, the load force and the effort force must act in opposite directions.

Common applications of 3rd-class levers include backhoes, shovels, barbecue tongs, and tweezers. Even your forearm functions as a 3rd-class lever (see Figure 3-20). The next time you lift a dumbbell, think about the fact that your bicep muscle is generating a force that is greater than the amount of weight that you are holding in your hand.

FIGURE 3-20: (a) The bicep muscle and your forearm function as a 3rd-class lever, as does a pair of (b) tweezers.

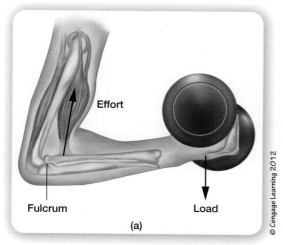

(a)

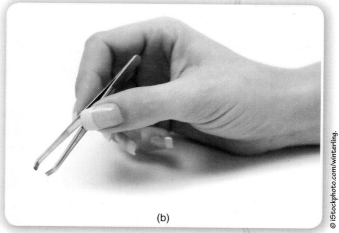

(b)

Problem: The schematic representation of the 2nd-class lever shown in Figure 3-21 depicts a load weighing 150 lb that acts 4 feet from the fulcrum point. An analysis of the lever shows that it is 90% efficient. Use this information to answer the following questions.

a. What is the *ideal mechanical advantage* of the lever?

$$IMA_{Lever} = \frac{r_E}{r_L}$$

$$IMA_{Lever} = \frac{8 \text{ ft}}{4 \text{ ft}}$$

$$IMA_{Lever} = 2$$

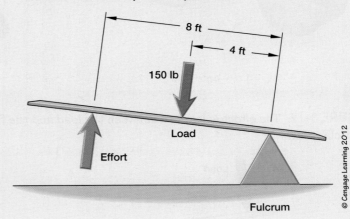

FIGURE 3-21: Example lever problem.

8 ft

4 ft

150 lb

Load

Effort

Fulcrum

© Cengage Learning 2012

b. What is the *ideal effort force* that would be needed to move the lever in the absence of friction?

$$IMA = \frac{L}{E_I} \Rightarrow E_I = \frac{L}{IMA}$$

$$E_I = \frac{150 \text{ lb}}{2}$$

$$E_I = 75 \text{ lb}$$

c. What is the *actual mechanical advantage* of this system?

$$\eta = \frac{AMA}{IMA} \Rightarrow AMA = \eta \times IMA$$

$$AMA = 0.9 \times 2$$

$$AMA = 1.8$$

d. What is the *actual* amount of effort force that must be applied to the lever to move the load?

$$AMA = \frac{L}{E_A} \Rightarrow E_A = \frac{L}{AMA}$$

$$E_A = \frac{150 \text{ lbs}}{1.8}$$

$$E_A = 83.33 \text{ lbs}$$

Third-class levers are used primarily to move a load over a large distance or to reduce the magnitude of an effort force. The effort force will always be greater than load force, which is why a 3rd-class lever can never provide mechanical advantage.

The Wheel and Axle

Many ancient inventions were made out of wood, which makes it difficult to find archeological evidence that has survived the passing of time. Much of the evidence that archeologists find comes in the form of ancient carvings in stone or clay that depict the existence of an invention or technology. Such is the case with the *wheel and axle*. A **wheel and axle** is a rotating device that consists of a wheel or crank that is attached to a smaller diameter axle.

Historians credit the ancient Mesopotamians with the development of the wheel as far back as 5000 B.C.E. Aside from its application to the cart, one of the first important applications of the wheel and axle took the form of the potter's wheel around 3300 B.C.E. This allowed for an increase in the volume and quality of pottery, which was used to carry water and store food. Another major historical application of the wheel and axle occurred during the 2nd century B.C.E. in ancient Greece with the invention of the waterwheel. Waterwheels were used for millennia to grind grain and power saw mills.

The wheel, or in some cases a crank, is locked to and shares its center axis with a smaller diameter axle. Common examples of the wheel and axle include door knobs, the steering wheel on an automobile, and flywheels on manufacturing equipment (see Figure 3-22).

> **Wheel and axle:** a rotating device that consists of a wheel or crank that is attached to a smaller diameter axle.

FIGURE 3-22: Any device that utilizes a hand crank or a flywheel operates on the principle of the wheel and axle. Such is the case with (a) a windlass that is used to raise a boat anchor or (b) the hand feed lever on a drill press.

(a)

(b)

You can think of the wheel and axle as a continuously rotating lever. In fact, continuous rotary motion is the only quality that differentiates a lever from a wheel and axle because levers are designed to operate across finite angles. Therefore, the wheel and axle can be made to exhibit the same qualities as a 2nd- or 3rd-class lever, depending on whether the effort force is applied to the wheel or to the axle. Mechanical advantage is achieved when the wheel and axle behaves like a 2nd-class lever (see Figure 3-23).

The *ideal mechanical advantage* of a wheel and axle is calculated by dividing the *radius of the wheel* (r_W) by the *radius of the axle* (r_A), as shown in Equation 3-11. It should be noted that this equation assumes that the effort force is applied to the wheel rather than to the axle.

FIGURE 3-23: The ideal mechanical advantage of a wheel and axle is a ratio between the wheel radius and the axle radius, assuming the effort force is applied to the wheel.

$$\text{IMA}_{\text{Wheel \& Axle}} = \frac{\text{radius of the wheel}}{\text{radius of the axle}} \quad \text{(Equation 3-11)}$$

$$\text{IMA}_{\text{Wheel \& Axle}} = \frac{r_W}{r_A}$$

When a relatively small effort force is applied to the outside rim of the wheel or crank handle, a large *torque* is applied to the axle. This is the reason why a pipe is used to elongate a wrench handle when trying to remove a nut that is "frozen" to a bolt. The concept of *torque* will be described in greater detail in Chapter 4, Mechanisms. For now, you can think of **torque** as being the rotational equivalent to the concept of force.

When a greater effort force is applied to the axle, the wheel and axle behave like a 3rd-class lever. An example of this occurs on the back wheel of a bicycle where the effort force is applied to a small diameter sprocket that is affixed to a larger diameter wheel (see Figure 3-24). The advantage, in this case, can be seen when you consider the circumferential distance that is covered by the wheel in relation to the distance covered by the axle. It is for this reason that wheeled vehicles, such as bicycles and automobiles, have the effort force applied to the axle in order to move the vehicle a greater distance per revolution.

FIGURE 3-24: This bicycle's power-transfer system applies a large effort force to the back wheel axle, which allows the load (the weight of the bicycle and its rider) to move a greater distance for every revolution of the wheel.

© iStockphoto.com/hamurishi.

Example

Problem: Figure 3-25 shows a bucket that is used to raise water from a deep well via a wheel and axle system that is 75% efficient. When full, the bucket weighs 20 lbs. The axle is 8 inches in diameter, and the distance from the center of the axle to the crank handle is 16 inches. Use this information to answer the following questions.

a. What is the *ideal mechanical advantage* of the wheel and axle?

$$IMA_{Wheel\ \&\ Axle} = \frac{r_W}{r_A}$$

$$IMA_{Wheel\ \&\ Axle} = \frac{16\ in}{4\ in}$$

$$IMA_{Wheel\ \&\ Axle} = 4$$

(continued)

b. What is the *actual mechanical advantage* of the wheel and axle?

$$\eta = \frac{AMA}{IMA} \Rightarrow AMA = \eta \times IMA$$

$$AMA = 0.75 \times 4$$

$$AMA = 3$$

c. Under *ideal* conditions, how much effort must be applied to the crank handle to raise the bucket of water?

$$IMA = \frac{L}{E_I} \Rightarrow E_I = \frac{L}{IMA}$$

$$E_I = \frac{20\ lb}{4}$$

$$E_I = 5\ lb$$

d. What is the *actual* amount of effort force that must be applied to the crank handle to raise the bucket of water?

$$AMA = \frac{L}{E_A} \Rightarrow E_A = \frac{L}{AMA}$$

$$E_A = \frac{20\ lb}{3}$$

$$E_A = 6.67\ lb$$

FIGURE 3-25: This well is an example of how a wheel and axle was used in the everyday task of gathering water.

© iStockphoto.com/alexsol.

The Pulley

If you have ever raised a flag on a pole or used an elevator to move between the floors of a building, then you have used a *pulley*. A **pulley** is a free-spinning wheel (usually grooved) around which a rope, chain, or belt is passed. Cranes that are designed to move cargo containers to and from ships also use pulley systems, as do cranes that are designed to lift materials to the tops of construction sites. Sailors have used pulleys for millennia to raise and lower the sails of their ships (see Figure 3-26).

Some scholars believe that the pulley was first developed by the ancient Mesopotamians as early as 1500 B.C.E., where it was used to move water. Drawings from the 8th century B.C.E. show that the fixed pulley was used by the Assyrians in what is now modern-day Iraq.

The fixed pulley, shown in Figure 3-27, is the simplest of all pulley systems. The pulley is secured, or fixed to an immovable surface such as the ground, a wall, or ceiling. One end of the rope is connected to the load, with the effort force being applied to the other end. The rope, being segmented by the pulley wheel, is divided into two *cables*. The cable that runs between the pulley and the load is used to support the weight of the load and is called a *support cable*.

> **Pulley:**
>
> a free-spinning wheel (usually grooved) around which a rope, chain, or belt is passed.

FIGURE 3-26: Pulleys have been used on (a) construction cranes and as part of (b) the rigging on sailing ships for thousands of years.

(a)

(b)

© IStockphoto.com/matsou.

© IStockphoto.com/Zootzims.

FIGURE 3-27: A single fixed pulley provides no mechanical advantage.

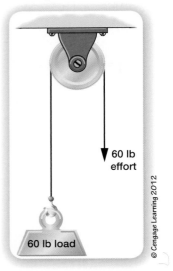

60 lb effort

60 lb load

© Cengage Learning 2012

A **support cable** carries all or part of the load force. The section of the rope to which the effort force is applied is called the **effort cable**. In the case of a single fixed pulley, the effort cable serves only to change the direction of the applied effort force. Therefore, under ideal conditions, a 60-lb effort force would be needed to move a 60-lb load. In this case the effort cable does not divide the load force in any way. If the effort force acts in the same direction that the load would travel if unsupported (which is downward in this case), then the effort cable cannot be considered a support cable. It is important that you be able to recognize how many support cables are present in a pulley system because the number of support cables indicates the ideal mechanical advantage of the pulley system (see Equation 3-12).

$$IMA_{Pulley} = \text{number of support cables} \qquad \text{(Equation 3-12)}$$

A fixed pulley has an ideal mechanical advantage of 1 because there is only one support cable. To prove this is true, you can envision the pulley as a rotating 1st-class lever (see Figure 3-28). The load and effort forces are applied on either end of the pulley wheel where their respective cables are tangent to the wheel. The axle serves as the fulcrum of the lever arm. Note that the fulcrum is centered between the effort and the load locations, which makes the load arm radius (r_L) equal to the effort arm radius (r_E). As you learned earlier, a 1st-class lever that has equal load arm and effort arm lengths will have an ideal mechanical advantage of 1.

Friction exists at the wheel's fulcrum. As a result, some of the energy that is applied at the effort is converted into heat. As it is true with all simple machines, moving the load requires an actual effort force that is greater in magnitude than the ideal effort force.

If the pulley is removed from its fixed location and connected to the load, then the pulley is free to move with the load. One end of the rope is connected

FIGURE 3-28: A fixed pulley with one support cable is analogous to a 1st-class lever.

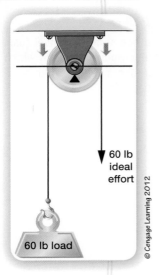

60 lb ideal effort

60 lb load

© Cengage Learning 2012

to a fixed location, and the other end is connected to the effort. In this application, the pulley is known as a *runner*. Both cables are used to support the weight of the load and are thus considered support cables. As stated previously, the way to tell that the effort cable is in fact a support cable is to look at the directions that the load and the effort are acting. In this case, the effort force acts upward, which is opposite the direction that the load wants to move if left unsupported. Because there are two support cables in this pulley system, the ideal mechanical advantage is 2. To prove this, you can envision the runner as a rotating 2nd-class lever (see Figure 3-29). The effort force is applied to the right side of the wheel where the effort cable is tangent to the pulley wheel. Because the load is mechanically connected to the pulley and in line with the axle, the axle becomes the load point. The lever arm fulcrum is therefore located on the left side of the wheel where the support cable is tangent to the pulley wheel. Note that the load point is centered between the fulcrum and the effort locations, which makes the load arm radius half the length of the effort arm radius. As you learned earlier, a 2nd-class lever exhibits this kind of configuration and always has a mechanical advantage that is greater than 1.

Under ideal circumstances, an effort force that is half the magnitude of the load force would be required to move a load that is connected to a single runner. In other words, 1/2 unit of effort force is needed to move 1 unit of load. The disadvantage is that the user must pull upward on the rope and thus work against the force of gravity. Recognize, however, that not all pulley systems are oriented up and down; some work horizontally, where gravity is of less consideration. Another disadvantage is that the amount of effort cable that must pass through the user's hands is twice the distance that the load moves (see Figure 3-30). This is congruent with the idea that mechanical advantage occurs when a smaller effort force is applied over a distance that is greater than the distance that is moved by the load.

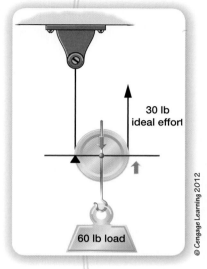

FIGURE 3-29: A single movable pulley (runner) has two support cables and is analogous to a 2nd-class lever.

30 lb
ideal effort

60 lb load

© Cengage Learning 2012

FIGURE 3-30: A runner allows the user to lift a load with half the effort, but the load will only move half the distance of the effort cable.

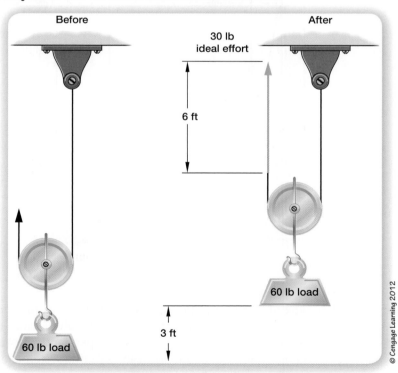

Before

After

30 lb
ideal effort

6 ft

60 lb load

60 lb load

3 ft

© Cengage Learning 2012

Because the ideal mechanical advantage of any simple machine is a ratio between the effort and load distances, the IMA of a pulley system can be found by dividing the length of the displaced effort cable (d_E) by the load displacement (d_L) (see Equation 3-13):

$$IMA_{Pulley} = \frac{\text{length of the displaced effort cable}}{\text{load displacement}}$$ (Equation 3-13)

$$IMA_{Pulley} = \frac{d_E}{d_L}$$

A single pulley alone constitutes a simple machine. Two or more pulleys used together form a *compound machine*. Pulleys can be used in combination to lift especially heavy loads or to perform more complex tasks. A **block and tackle** is an arrangement of two or more pulleys that are strung together to lift or move a load. The block and tackle is employed when mechanical advantage must be achieved, and the effort force must act in the same direction as the load. Figure 3-31 shows two pulley systems that have an ideal mechanical advantage of 2. Each of the two support cables in the block and tackle carries one-half of the load force. The effort force acts in the same direction as the load, so the effort cable is not considered a support cable. Adding more pulleys increases the amount of friction that occurs in the system and reduces the overall efficiency of the system.

FIGURE 3-31: (a) A block and tackle that consists of one fixed pulley and one runner exhibits the same mechanical advantage as (b) a single runner but allows the user to apply an effort force that acts in the same direction as the load.

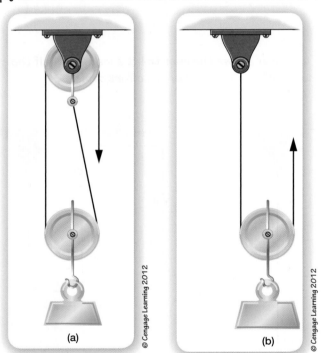

(a)

(b)

© Cengage Learning 2012

© Cengage Learning 2012

There is another type of compound pulley system that serves the same purpose, but its ideal mechanical advantage is slightly more difficult to determine (see Figure 3-32). All of the pulleys in this system act as independently moving runners. If you were to count the number of support cables in this system, you might

FIGURE 3-32: This compound pulley system consists of three independently moving runners.

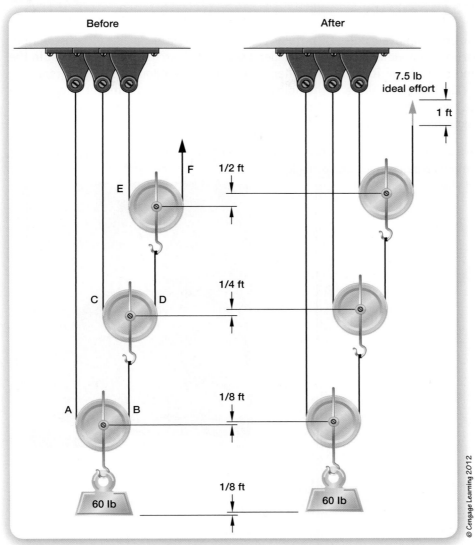

deduce that the ideal mechanical advantage is 6. However, the method of counting support cables only applies when those cables are connected to the same block. A **block** is a pulley that contains multiple wheels that spin independently but move in unison. In this case, each pulley moves independently.

To determine the ideal mechanical advantage of this system, the 60-lb load force must be sequentially divided by its support cables. In the example shown in Figure 3-32, the 60-lb load is directly supported by cables A and B, which will each carry 30 pounds. The 30-lb load on cable B is further divided by cables C and D, each of which carries 15 pounds. The 15-lb load on cable D is further divided by cables E and F, each of which carries 7.5 pounds. Therefore, an ideal effort force of 7.5 pounds would need to be applied to cable F to move the load on the pulley system. To calculate the ideal mechanical advantage of this pulley system, the load force is divided by the ideal effort force, resulting in an ideal mechanical advantage of 8. Remember, the disadvantage comes in the amount of effort cable that must pass through the user's hands. In this case and under ideal circumstances (no friction), the user must pull 8 feet of rope to move the load a distance of 1 foot.

Point of Interest

The ancient Egyptians and Romans employed pulleys in crane systems called *derricks* that were used to move heavy stone blocks and other large items. The Greek historian Plutarch (A.D. 46–120) wrote that Archimedes (287–212 B.C.E.) was challenged by King Hiero II of Syracuse to single-handedly launch a fully loaded ship from its place of construction. The legend states that Archimedes succeeded in the challenge by using a compound pulley system that employs multiple pulleys to significantly reduce the necessary effort force (see Figure 3-33).

FIGURE 3-33: (a) According to legend, Greek mathematician and engineer Archimedes used a compound pulley system to single-handedly move (b) a fully loaded ship.

(a)

(b)

Example

Problem: The pulley system shown in Figure 3-34 is used to move a 60-lb load a vertical distance of 10 ft. The *actual* amount of effort force that is needed to move the load is 15 lb. Use this information to answer the following questions.

a. What is the *ideal mechanical advantage* of this pulley system?

$$\text{IMA}_{\text{Pulley}} = \text{number of support cables}$$

$$\text{IMA}_{\text{Pulley}} = 6$$

b. What is the *actual mechanical advantage* of this pulley system?

$$\text{AMA} = \frac{L}{E_A}$$

$$\text{AMA} = \frac{60 \text{ lb}}{15 \text{ lb}}$$

$$\text{AMA} = 4$$

c. How *efficient* is the pulley system?

$$\eta = \frac{\text{AMA}}{\text{IMA}}$$

$$\eta = \frac{4}{6}$$

$$\eta = 0.67 \text{ or } 67\%$$

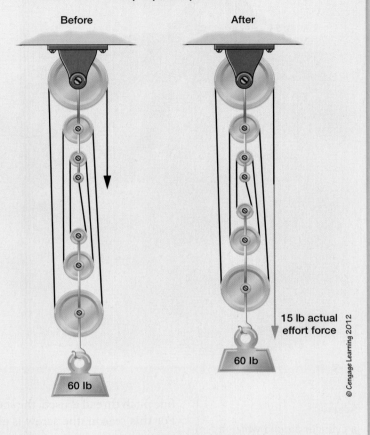

FIGURE 3-34: Example pulley problem.

Before   After

15 lb actual effort force

60 lb

60 lb

© Cengage Learning 2012

d. What is the *ideal effort force* that would be needed to move the load on the pulley system?

$$\text{IMA} = \frac{L}{E_I} \Rightarrow E_I = \frac{L}{\text{IMA}}$$

$$E_I = \frac{60 \text{ lb}}{6}$$

$$E_I = 10 \text{ lb}$$

e. What is the *length of the effort cable* that must pass through the user's hands to raise the load to the desired height?

$$\text{IMA}_{\text{Pulley}} = \frac{d_E}{d_L} \Rightarrow d_E = \text{IMA}_{\text{Pulley}} \times d_L$$

$$d_E = 6 \times 10 \text{ ft}$$

$$d_E = 60 \text{ ft}$$

The Screw

There is debate amongst historians as to when the screw was first developed. Some suggest that it was invented in the 7th century B.C.E. by the Assyrians and employed in the use of irrigation systems that fed the Hanging Gardens of Babylon. Others believe that the screw was invented by Archimedes during the 3rd century B.C.E. In both cases, the screw is said to have been used to raise water (see Figure 3-35). Historians do agree that by the 1st century B.C.E., wooden screws had found an application in the design of mechanical presses that were used in both olive oil and wine production throughout the Mediterranean region.

A **screw** is a cylinder about which a helical groove or thread is wound. It is used to convert rotary motion into linear motion and is capable of generating large output forces. It is for these reasons that the screw is used in countless designs of valves, presses, and jacks, as well as an endless variety of clamping devices (see Figure 3-36).

An important geometric feature of the screw is its **pitch (p)**, which is the linear distance that the screw travels in one revolution. For a single helix screw thread, the pitch can be determined by measuring the linear distance from a point on one screw thread to the corresponding point on the adjacent thread (see Figure 3-37). When a fine pitch thread is used, the screw can be made to achieve very precise linear motion. For this reason, the screw is also employed in precision measuring and calibration

Screw:

a cylinder around which a helical groove or thread is wound. It is used to convert rotary motion into linear motion and is capable of generating extremely large output forces.

Pitch (p):

the linear distance traveled in one revolution of a screw.

FIGURE 3-36: *The screw in this vise converts the rotary motion of the handle into linear motion of the movable jaw. The screw gives the user the ability to generate especially large clamping forces.*

© Cengage Learning 2012

© iStockphoto.com/DNY59.

FIGURE 3-37: (a) For a single-helix screw thread, the pitch can be determined by measuring the linear distance from a point on one screw thread to the corresponding point on the adjacent thread. (b) Screws threads that have very small pitch values are used in precision measuring tools such as the zero- to 1-inch micrometer.

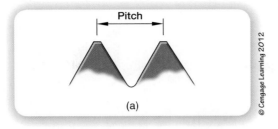

Pitch

(a)

© Cengage Learning 2012

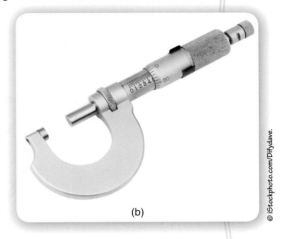

(b)

© iStockphoto.com/Difydave.

instruments, as well as precision manufacturing equipment. Example applications include the micrometer and lead screw–operated lathes and milling machines.

The unified screw thread and the International Organization for Standardization (ISO) metric thread are the two most common thread types used in the United States. The standards that are used to identify these different thread types differ slightly and are identified in Figure 3-38. As such, it is important to be able to distinguish between the two thread types and to properly interpret their values.

FIGURE 3-38: (a) A 1–8 UNC thread callout denotes a screw that has a 1-inch major diameter and 8 threads for every 1 inch of linear distance as measured parallel to the screw axis. (b) An M24 × 3 (metric) thread callout denotes a screw that has a 24-millimeter major diameter and a thread pitch that is 3 millimeters.

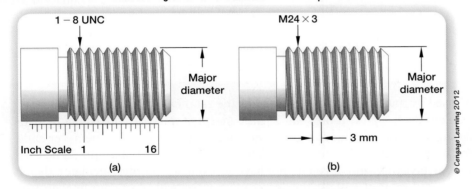

1 – 8 UNC

Major diameter

Inch Scale   1   16

(a)

M24 × 3

Major diameter

3 mm

(b)

© Cengage Learning 2012

To calculate the pitch of a unified screw thread, one must divide the value of 1 inch by the number of screw threads that occur in 1 inch as measured parallel to the screw axis (see Equation 3-14). For example, the pitch of a 1–8 unified national course (UNC) thread is found by dividing 1 inch by the number 8. The resulting pitch value is 0.125 in. The pitch of a screw thread must be known in order to calculate the screw's ideal mechanical advantage.

$$p = \frac{1}{\text{number of threads per inch}} \qquad \text{(Equation 3-14)}$$

As mentioned earlier, a tremendous mechanical advantage can be generated by devices that employ the screw. The reason for this becomes evident when you think about the relatively large circumferential distance around which the effort force must act in comparison to the resulting short linear distance that the load moves

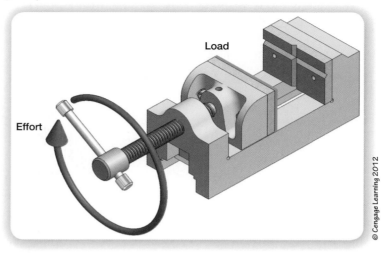

for every revolution of the screw. This linear distance is equal to the pitch of the screw thread (see Figure 3-39).

To calculate the ideal mechanical advantage of a screw, one must know the radial distance from the screw's center axis to the point where the effort force is exerted. Multiplying the value of π by twice this radial distance will produce the circumferential distance that is traveled by the effort in one revolution. Rotating the screw once will advance the screw a linear distance that is equal to the screw pitch. Therefore, dividing the circumferential distance by the length of the screw pitch will reveal the ideal mechanical advantage for the screw (see Equation 3-15):

$$IMA_{Screw} = \frac{\text{circumferetial distance traveled by the effort in one revolution}}{\text{screw pitch}} \qquad \text{(Equation 3-15)}$$

$$IMA_{Screw} = \frac{2\pi\, r_E}{p}$$

The Greek symbol π (pi) is used by mathematicians to represent the ratio between a circle's circumference and its diameter, which is a fixed value, slightly more than 3. The concept of π has fascinated mathematicians around the world for millennia because its value is an irrational number. Its decimal value never ends, and there is no repeating pattern in the numbers. Scientific calculators have a π button that should be used when performing calculations that involve this ratio. If you do not have access to a scientific calculator, then it is often acceptable to use the abbreviated value of 3.14. As the degree of accuracy will affect your final answer, it is best to ask your instructor about the method that he or she prefers.

Example

Problem: The screw mechanism shown in Figure 3-40 is called a *nut splitter*. It is used to remove a stripped, fused, or otherwise damaged nut from a bolt. A wrench is used to turn a 1/2–20 unified national fine (UNF) threaded rod that pushes a cutting tool into the nut. The effort force is applied to the wrench at a radial distance of 10 inches from the screw's center axis. A 2,000-lb shearing force (load) is required to split the nut. The actual mechanical advantage of the screw is 750. Solve for the following:

(continued)

a. What is the *pitch* of the screw thread?

$$p = \frac{1 \text{ inch}}{\text{number of threads in one inch}}$$

$$p = \frac{1 \text{ in.}}{\left(\frac{20}{1 \text{ in}}\right)}$$

$$p = 0.05 \text{ in}$$

b. What is the *ideal mechanical advantage* of the screw?

$$\text{IMA}_{Screw} = \frac{2\pi r_E}{p}$$

$$\text{IMA}_{Screw} = \frac{2 \times 3.14 \times 10 \text{ in}}{0.05 \text{ in}}$$

$$\text{IMA}_{Screw} = \frac{62.8 \text{ in}}{0.05 \text{ in}}$$

$$\text{IMA}_{Screw} = 1,256$$

c. How *efficient* is the screw system?

$$\eta = \frac{\text{AMA}}{\text{IMA}}$$

$$\eta = \frac{750}{1,256}$$

$$\eta = 0.60 \text{ or } 60\%$$

d. What is the magnitude of the *ideal effort force* that must be applied to the wrench handle to split the nut?

$$\text{IMA} = \frac{L}{E_I} \Rightarrow E_I = \frac{L}{\text{IMA}}$$

$$E_I = \frac{2,000 \text{ lb}}{1,256}$$

$$E_I = 1.59 \text{ lb}$$

e. What is the magnitude of the *actual effort force* that must be applied to the wrench handle to split the nut?

$$\text{AMA} = \frac{L}{E_A} \Rightarrow E_A = \frac{L}{\text{AMA}}$$

$$E_A = \frac{2,000 \text{ lb}}{750}$$

$$E_A = 2.67 \text{ lb}$$

FIGURE 3-40: (a) A nut splitter is a combination of the screw and the wedge. (b) It is used to remove a nut that has become locked to a bolt because of corrosion.

Courtesy of Northerntool.com.

(a)

© Cengage Learning 2012

(b)

COMPOUND MACHINES

Thinking back to the part of the chapter that introduced pulleys, you may remember that two or more pulleys working together were referred to as a *compound machine*. The nut splitter in the previous problem is also an example of a compound machine. A **compound machine** is a combination of two or more simple machines that work together to accomplish a task.

Once early engineers had mastered the concept of mechanical advantage and the individual use of the simple machines, they began to combine the simple machines in ways that would serve to further multiply an effort force. The ideal mechanical advantage of a compound machine is the product of all the ideal mechanical advantages of the simple machines of which it is comprised. To illustrate this, let's imagine that you are an ancient Greek engineer who has been given the challenge of designing a crane that incorporates two or more different simple machines. Your challenge is summarized in the design brief shown in Figure 3-41.

> **Compound machine:**
>
> a combination of two or more simple machines that work together to accomplish a task.

FIGURE 3-41: Design brief used to summarize the challenge of designing an ancient crane.

Design Brief	
Client:	Alexander the Great, King of Macedonia
Designer:	Hannah Chapman, Royal Engineer
Problem Statement:	Large timbers and quarried stone that are bound for the Greek empire must be lifted onto and off of sailing ships.
Design Statement:	Design a crane that makes use of more than one type of simple machine to lift heavy objects, and calculate its mechanical advantage.
Constraints:	1. Available materials: timber, rope, and wrought iron 2. The crane must be self-supporting. 3. The maximum load will be 300 lb. 4. The maximum ideal effort must be no greater than 10 lb.
Deliverables:	▶ sketch that summarizes the proposed solution and provides the necessary dimensions for all related calculations ▶ calculations for the ideal mechanical advantage of each simple machine of which the solution is comprised ▶ calculations for the overall ideal mechanical advantage of the compound machine ▶ calculations for the ideal effort needed to lift a 300 lb load

Given the limitations of the materials and the know-how of the time, you might have come up with a design for a crane that looks similar to what historians believe the ancient Greeks invented (see Figure 3-42).

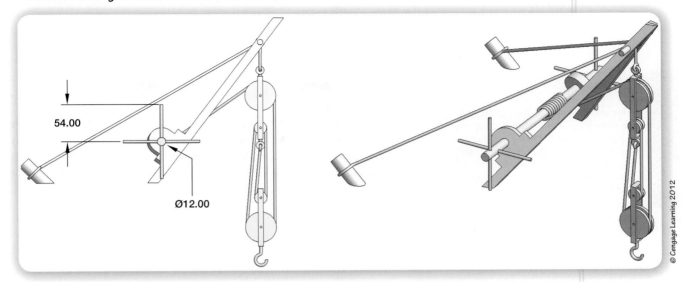

54.00

Ø12.00

© Cengage Learning 2012

Looking closely at the design, we can see that it consists of a wheel and axle as well as a block and tackle. To begin the process of calculating the overall ideal mechanical advantage of this compound machine, we will start with the pulley system. The effort cable attaches to a winch and acts in a downward direction. You may recall that an effort cable that acts in the same general direction as the load will not function as a support cable. The remaining cables between the fixed pulley block and the runner block do serve as support cables. The ideal mechanical advantage for this type of pulley system is equal to the number of *support cables*. Therefore, the ideal mechanical advantage of this pulley system is 4:

$$IMA_{Pulley} = \text{number of support cables}$$
$$IMA_{Pulley} = 4$$

Now we turn our attention to the wheel and axle. The effort cable from the pulley system winds around an axle that is 12 inches in diameter. Each of the four handles extends 54 inches from the center of the axle. Given these dimensions and assuming that the person is applying the effort force at the very end of the handle, the ideal mechanical advantage of the wheel and axle is 9:

$$IMA_{Wheel \& Axle} = \frac{r_W}{r_A}$$
$$IMA_{Wheel \& Axle} = \frac{54 \text{ in}}{6 \text{ in}}$$
$$IMA_{Wheel \& Axle} = 9$$

As mentioned earlier, the ideal mechanical advantage of a compound machine is the *product* of the ideal mechanical advantages of the simple machines of which it is comprised. Therefore, the ideal mechanical advantage of this crane is 36.

$$IMA_{Compound Machine} = IMA_{Pulley} \times IMA_{Wheel \& Axle}$$
$$IMA_{Wheel \& Axle} = 4 \times 9$$
$$IMA_{Wheel \& Axle} = 36$$

As it is true for a simple machine, the ideal mechanical advantage of a compound machine describes the ratio between the load force and the ideal effort force. Under ideal conditions, a person would have to apply 8.33 lb of effort force to the end of one of the handles in order to lift a 300-lb load. This value falls within the required 10-lb effort force limit. Had the ideal effort force exceeded the specified limit, what changes might you make to the crane design in order to meet the criteria?

$$IMA = \frac{L}{E_I} \Rightarrow E_I = \frac{L}{IMA}$$

$$E_I = \frac{300 \text{ lb}}{36}$$

$$E_I = 8.33 \text{ lb}$$

OFF-ROAD EXPLORATION

Having a pulley that is capable of lifting a large stone block is one thing. Holding onto the stone is another problem. Perform an Internet search using the key phrase "Lewis lifting device" to find out how large stone blocks are secured to the pulley during transport.

Given the significant jump in mechanical advantage that can be achieved by a compound machine, it is easy to understand why so many of the machines that are used today are not limited to only one type of simple machine. Figure 3-43 shows an example of how an engineer might document his or her design ideas and calculations using an engineer's notebook.

FIGURE 3-43: Engineer's notebook sketches of the proposed design solution with calculations.

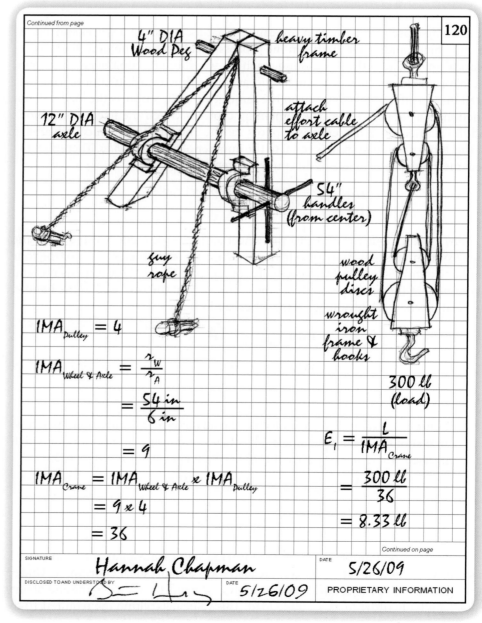

SUMMARY

In this chapter, you learned:

- A *force* is a push or pull that acts on an object to initiate or change its motion or to cause deformation. Forces are divided into two categories: contact forces and long-range forces. Contact forces are generated when two physical objects touch. Long-range forces are generated by phenomena such as gravity and magnetism. The force that is *input* or applied to a machine is called the *effort force*. The *load force* is the weight of the object that a machine is moving or the *resistance* that a machine must overcome in order to accomplish its task. As such, the *output* of a machine is the force that is applied to the load.

- *Work* (W) is achieved when a force is applied to an object, causing the object to be displaced. Common units of measurement for work are the foot-pound (ft · lb) in the U.S. customary system of units, and the newton-meter (N · m) or joule (J) in the International System of Units (SI). *Power* (P) is the time rate at which work is achieved or work per unit time. It is measured in foot-pounds per second (ft · lb/sec) and horsepower (hp) in the U.S. customary system of units. SI units include the newton-meter per second (N · m/sec) and watt.

- *Mechanical advantage* is the ratio of the output force (load) produced by a machine to the input force (effort) that is applied to the machine. It is also a ratio between the distance over which the effort force acts to the distance moved by the load. If the distance across which the effort acts is larger than the distance across which the load moves, then the mechanical advantage will be greater than 1. If the distance across which the load moves is larger than the distance

across which the effort acts, then the mechanical advantage will be a value between 0 and 1.

- *Ideal mechanical advantage* (IMA) is what a machine would exhibit under ideal circumstances, where friction and other energy-absorbing phenomena do not exist. As such, the work performed by the ideal *effort force* (E_I) is transferred, undiminished, to the *load* (L). Unfortunately, the presence of friction and other energy-absorbing phenomena makes this impossible. Therefore, the *actual effort force* (E_A) needed to move a load will always be greater than the ideal effort force. A machine's *actual mechanical advantage* (AMA) reflects the losses that occur between a machine's input and output. Because the actual effort force is always greater than the ideal effort force, a machine's ideal mechanical advantage will always be greater than its actual mechanical advantage. The following equations apply to all simple machines.

$$\text{IMA} = \frac{\text{Load Force}}{\text{Ideal Effort Force}} = \frac{L}{E_I}$$

$$\text{AMA} = \frac{\text{Load Force}}{\text{Actual Effort Force}} = \frac{L}{E_A}$$

- The inclined plane, wedge, lever, wheel and axle, pulley, and screw make up the six *simple machines*. The *ideal mechanical advantage* of the various simple machines can be found by dividing the distance across which the effort acts by the resulting distance across which the load moves. As such, each simple machine has its own specific equation that can be used to calculate its ideal mechanical advantage (see Figure 3-44).

(*continued*)

■ A compound machine, such as a crane or a nut splitter, is a combination of two or more simple machines that work together to accomplish a task. The ideal mechanical advantage of a compound machine is the *product* of the ideal mechanical advantages of the simple machines of which it is comprised.

FIGURE 3-44: Equations for calculating the ideal mechanical advantage of the six simple machines.

Inclined Plane

$$\text{IMA}_{\text{Inclined Plane}} = \frac{\text{slope length}}{\text{vertical height of the incline}} = \frac{s}{h}$$

Wedge

$$\text{IMA}_{\text{Wedge}} = \frac{\text{slope length}}{\text{thickness of the wedge}} = \frac{s}{t}$$

Lever

$$\text{IMA}_{\text{Lever}} = \frac{\text{radius of the effort arm}}{\text{radius of the load arm}} = \frac{r_E}{r_L}$$

Wheel and Axle

$$\text{IMA}_{\text{Wheel \& Axle}} = \frac{\text{radius of the wheel}}{\text{radius of the axle}} = \frac{r_W}{r_A}$$

Note: This equation assumes that the effort force is being applied to the wheel.

Pulley

$$\text{IMA}_{\text{Pulley}} = \frac{\text{length of the displaced effort cable}}{\text{load displacement}} = \frac{d_E}{d_L}$$

Screw

$$\text{IMA}_{\text{Screw}} = \frac{\text{circumferetial distance traveled by the effort}}{\text{screw pitch}} = \frac{2\pi\, r_E}{p}$$

1. What is a force?
2. What is the difference between a contact force and a long-range force?
3. Give two examples of forces that fall under the category of long-range forces.
4. What is the difference between an effort force and a load force?
5. Why might an engineer be more concerned with the power that a machine can generate rather than the amount of work that a machine can accomplish?
6. What is horsepower?
7. How is mechanical advantage a trade-off between force and distance?
8. What is the difference between a simple machine's ideal and actual mechanical advantage?
9. Identify two things that cause a simple machine to be less than 100% efficient.
10. Identify an application of the inclined plane.
11. What is the difference between an inclined plane and a wedge?
12. Identify an application of the wedge.
13. What is the difference between a 1st-, a 2nd-, and a 3rd-class lever?
14. Identify applications of a 1st-, a 2nd-, and a 3rd-class lever.
15. A 2nd-class lever has an effort arm radius of 2 inches and a load arm radius of 1 inch. What is the lever's ideal mechanical advantage?
16. Identify an application of the wheel and axle.
17. If the actual mechanical advantage of a wheel and axle is 10, what does this value represent?
18. Identify an application of the pulley.
19. How does a fixed pulley behave like a 1st-class lever?
20. How does a movable pulley behave like a 2nd-class lever?
21. At a glance, how can one determine the ideal mechanical advantage of a block and tackle?
22. Identify an application of the screw.
23. What is a screw's pitch?
24. How do you determine the screw pitch of a unified screw thread?
25. How is it that the screw can generate a large mechanical advantage?
26. How is a nail clipper an example of a compound machine?
27. How do you calculate the ideal mechanical advantage of a compound machine?

EXTRA MILE

Problem Statement: The director of a local museum wants to incorporate a hands-on, interactive sculpture to enhance the museum's permanent exhibit on simple machines and other ancient technologies.

Design Statement: Design, build, and test a human-powered sculpture that circulates a baseball continuously through a series of obstacles. The sculpture must incorporate the wheel and axle, the screw, and the inclined plane. The minimum amount of time allotted for the baseball to complete one cycle through the sculpture is 20 seconds. The solution must fit within a space that is 4 feet tall, 2 feet wide, and 2 feet deep.

CHAPTER 4
Mechanisms

Menu

START LOCATION	DISTANCE	END LOCATION

Before You Begin

Think about these questions as you study the concepts in this chapter:

1. What is a mechanism, and how does it differ from a simple machine?

2. What are the different types of linkage systems, and what are their applications?

3. What is the purpose of a cam and follower, and what kinds of products contain them?

4. What are different types of bearings that are used in mechanical devices?

5. What is a gear, and what conditions must be satisfied to have two gears mesh properly?

6. What are the most common types of gears, and what are their applications?

7. What is the difference between a simple and a compound gear train?

8. How do you calculate the gear ratio, speed ratio, and torque ratio of a simple and compound gear train?

9. What components make up a chain drive mechanism?

10. What factors must be considered when an engineer designs a chain drive mechanism?

Many engineers design physical objects that solve our problems, fulfill our needs or wants, and make our lives easier. Some of the solutions they create are static (un-moving), such as buildings and bridges. However, think about how many objects you have used today that involve motion or rely on some type of movement to accomplish their task. The linkage system within a reclining chair, the bearings within the wheels of the car or school bus that brought you to school, the gears within a pencil sharpener, and the sprocket and chain on your bicycle are all examples of *mechanisms* that are used routinely throughout our human-built world. More often than not, these mechanical devices are used in combination to achieve a desired result. To generate solutions to problems, engineers must draw from a technology database from which their minds can make connections to create new and innovative ideas. Learning about these devices and how they work is not only an integral part of the education of the design engineer but also an exploration of the creative works of those who have come before us.

When you press a button to eject a DVD from a computer, a *mechanism* attached to the DVD tray changes the rotary motion of an electric motor into the tray's linear motion. When you ride a bicycle, it is a *mechanism* that transfers the energy from your leg muscles to the rear wheel of the bicycle. Whereas a *simple machine* is a single mechanical device that functions alone to perform work, a mechanism is an assemblage of moving mechanical components that are supported by a rigid structure. This chapter is by no means a complete treatise on all of the different mechanisms that exist, but we do pay attention to linkages, cams, bearings, gears, and chain drive mechanisms.

OFF-ROAD EXPLORATION

Cornell University's College of Engineering maintains a vast multimedia resource—the Kinematic Models for Design Digital Library (KMODDL)—that is dedicated to the study of mechanisms. To explore images and videos of the various mechanisms that are talked about in this chapter, visit *http://kmoddl.library.cornell.edu/*.

Mechanism:
an assemblage of moving mechanical components that are supported by a rigid structure.

Linkages

Being one of the simplest and most versatile types of mechanisms, *linkage systems* are incorporated by mechanical engineers into the design of industrial equipment, tools, and the inner workings of commercial products such as reclining chairs and automobiles. The term linkage refers to an assembly of rigid mechanical components within a mechanism that are linked together for the purpose of transmitting force and controlling motion. In machines, linkage systems are often used to change rotary motion into linear motion and vice versa.

Building on the idea of a compound machine, you can think of a linkage as a series of interconnected levers. The linkage system that forms the structure of a desk lamp allows the user to orient the lamp into many different positions (see Figure 4-1). Large clamping forces can also be generated by linkages, which is why they are used in tools such as toggle clamps and locking pliers. The force that you apply to the hand crank on a reclining chair is transmitted to a linkage system, which moves the various surfaces of the chair into different positions along predetermined paths. Not all linkages are movable. Some must remain in a stationary position and serve as fixed connection points for other movable components such as the stationary framework of a computer printer.

The human skeletal system is the biological equivalent of a mechanical linkage system in that bones act as rigid components that are connected for the purpose of controlling motion and transmitting force. Some joints, such as the shoulder, are capable of 360° of motion. Other joints can only move through a fixed angle range, such as the

Linkage:
an assembly of rigid mechanical components within a mechanism that are linked together for the purpose of transmitting force and controlling motion.

FIGURE 4-1: (a) This desk lamp, (b) locking pliers, and (c) reclining chair use linkage systems in different ways to manipulate force and motion.

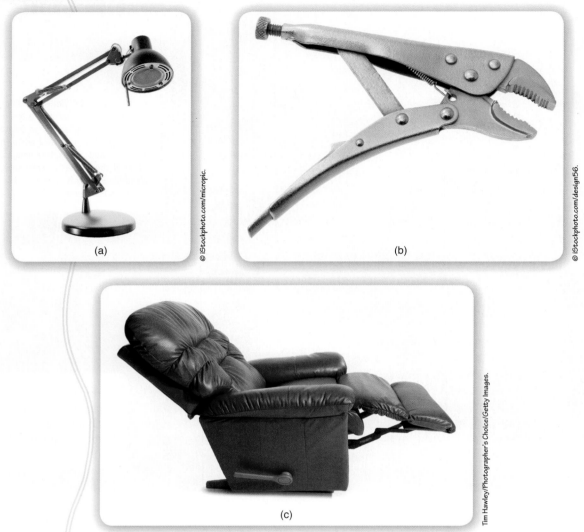

(a)

© iStockphoto.com/micropic.

(b)

© iStockphoto.com/design56.

(c)

Tim Hawley/Photographer's Choice/Getty Images.

FIGURE 4-2: Schematic symbols used to represent linkage elements.

Pin joint   Fixed joint (Pivot)   Slider   Rocker, Crank, or Bar

© Cengage Learning 2012

elbow and the knee. When used together, the coordination of the angular motions of these links allows your hand to move along a straight line as your brain tells your muscles to reach out and grab an object.

Linkage systems are comprised of basic elements such as a rockers, cranks, connecting bars, and sliders that are joined together via movable pin joints. A connection is also made to a fixed pivot joint on a machine's housing or on an immovable surface such as a wall or ceiling. Engineers use schematic symbols to represent these basic elements when sketching, analyzing, and communicating the motions of a linkage system's components (see Figure 4-2).

FOUR-BAR LINKAGE The four-bar linkage is one of the most common types of linkages used in mechanical systems (see Figure 4-3). It consists of three movable bars, two fixed pivot joints, and two pin joints. The fourth bar is a fixed structure, such as a machine's rigid framework, that provides the connection to the fixed pivot joints.

The toggle clamp shown in Figure 4-4 is an example of a four-bar linkage. Joints A and D are part of the machine's rigid frame and serve as the fixed pivot points.

Bars AB and CD act as **rockers** because each has one fixed pivot point and is only allowed to rotate back and forth through a fixed angle range. Bar BC serves as a link or connecting bar; in this case, it provides the input force to the system because it is connected to the clamp handle.

As the toggle clamp handle is moved upward in a counterclockwise direction, the clamp is disengaged and moves in a clockwise direction. When the handle is rotated clockwise to its extreme limit, the clamp rotates counterclockwise into position. Notice that joints B, C, and D appear to be in line with each other when the clamp

FIGURE 4-3: Example of a four-bar linkage.

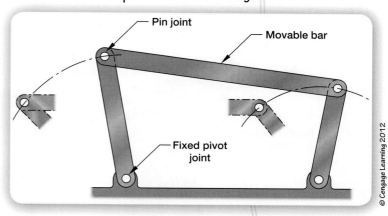

© Cengage Learning 2012

FIGURE 4-4: Toggle clamps are used in fixtures to hold workpieces in place during manufacturing operations. This clamp utilizes a four-bar linkage system.

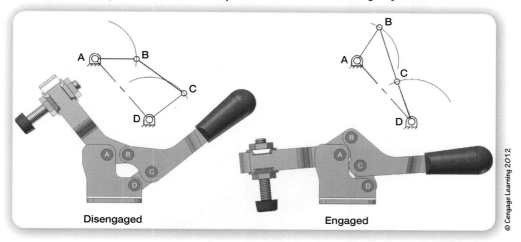

© Cengage Learning 2012

is fully engaged. In fact, pin joint C is slightly below the center line that spans the distance between pin joint B and fixed joint D. When this happens, the clamp is locked in position. This creates a tremendous holding force that counteracts any upward force that may be generated by a workpiece as it is being machined.

A *parallel linkage* is a type of *four-bar linkage* system that follows the geometry of a **parallelogram**. The two rocker bars on the syringe-powered robot shown in Figure 4-5 are of equal length. The horizontal distance between their pivot points are of equal spacing on the base and the gripper platform. The resulting geometry of the linkage system takes the shape of a parallelogram. As the slider moves, it causes the rocker bars to rotate. The rocker bars remain parallel to each other, and the gripper platform remains in a horizontal position.

A variation of the parallel linkage is the *scissor linkage*. The scissor linkage is used in devices that are designed to amplify linear motion such as hydraulic-powered scissor jacks. They are also employed in the design of pantograph machines, which are used by technical illustrators to replicate or resize drawings (see Figure 4-6). Engravers also use pantographs to cut letters, numbers, or decorative patterns into metal or plastic surfaces using templates.

SLIDER-CRANK Another common type of linkage is the *slider-crank*. This linkage system is used to convert reciprocating linear motion into continuous rotary motion or vice versa. A **crank** is a bar that has one fixed pivot point. Unlike the rocker, a crank has the ability to rotate 360°. The *slider* often takes the form of a piston that moves back and forth inside a fixed housing such as a

OFF-ROAD EXPLORATION

To learn more about mechanisms that are used in manufacturing, read *Jig and Fixture Design*, by Edward G. Hoffman, 2004, ISBN 1-4018-1107-8.

FIGURE 4-5: This syringe-powered robot arm incorporates a parallel linkage, which allows the gripper platform to remain horizontal as it moves.

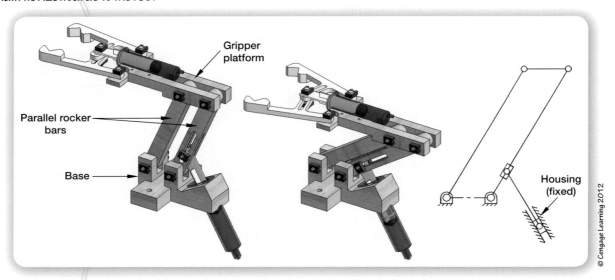

FIGURE 4-6: Scissor linkages are employed in the design of (a) industrial scissor jacks (Skyjack SJ 7127 Rough Terrain Scissor Life) and (b) pantographs.

(a)

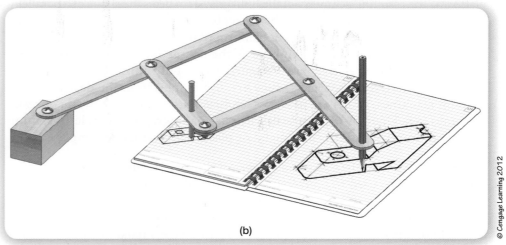

(b)

Points of Interest
James Watt and the Parallel Linkage

James Watt is sometimes mistakenly credited with the invention of the steam engine. In fact, it was Watt's numerous innovations to the steam engine that have made his name more often associated with the technology than that of its actual inventor, Thomas Newcomen. Newcomen invented the steam engine in 1712, which was frequently employed to the task of pumping water out of mine shafts (see Figure 4-7). Perhaps the most recognizable feature on a steam engine is the large beam that acts as a 1st-class lever. The beam pivots back and forth like a see-saw as force is transferred from a steam-powered piston to a water pump or some other device.

As the large beam teeters back-and-forth, the path formed by a point on either end of the beam takes the shape of an arc. This poses a mechanical problem. How do you transfer the force from a piston rod that moves up and down in a linear fashion along a fixed axis to a point on the end of a beam that is moving along the arc of a circle? A direct, rigid connection cannot be made because the end of a rigid steam-engine piston rod cannot move along this arc path and stay aligned with the axis of its fixed piston cylinder. Newcomen solved this problem by adding an arch head to the end of beam, which he connected to the piston rod via chain links. As the beam teetered, the chain would wrap around or unwind from the arch head, thus remaining tangent to the axis of the piston rod. Unfortunately, the chain could only transfer force when placed in tension, which was caused by a vacuum formed on the bottom side of the piston as steam was rapidly cooled. The weight of the pump on the other side of the beam would move the piston back up the cylinder, providing the return stroke.

Of the numerous innovations that Watt made to the steam engine, he was especially proud of his addition of the parallel linkage system between the beam and the piston rod in 1784. Like Newcomen's chain design, Watt's linkage system solved the problem of keeping the piston rod aligned with the cylinder bore as it transmits force to the rotating beam (see Figure 4-8). The parallel linkage and the piston rod mechanism, being comprised of rigid components, can transfer the force in both tension and compression. When used in conjunction with his improved double-acting cylinder, this linkage system allows the steam engine

FIGURE 4-7: **Illustration of a Newcomen steam engine that shows how chains were used to solve the problem of converting an arc motion to a linear motion.**

From Karsnitz, O'Brien, Engineering Design: An Introduction. © 2009 Delmar Learning, a part of Cengage Learning, Inc. Reproduced with permission. www.cengage.com/permissions.

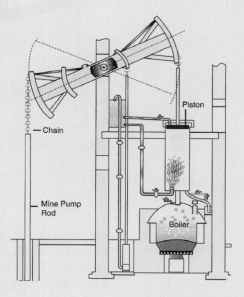

(continued)

piston to provide power during both the up and down stroke. Watt's innovations to the steam engine made it more reliable and more efficient than Newcomen's design. The steam engine replaced the water wheel as the driver of industrial machinery and freed factories of the need to be located next to rivers and creeks.

FIGURE 4-8: (a) Watt's steam engine replaced (b) Newcomen's chain connection between the engine piston rod and the beam with a parallel linkage, thus allowing the piston rod to be rigidly connected to the beam and provide power on both the up and down strokes.

(a) From Karsnitz, O'Brien, *Engineering Design: An Introduction*. © 2009 Delmar Learning, a part of Cengage Learning, Inc. Reproduced with permission. www.cengage.com/permissions. (b) © Cengage Learning 2012

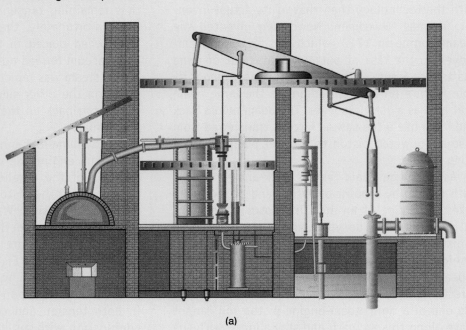

(a)

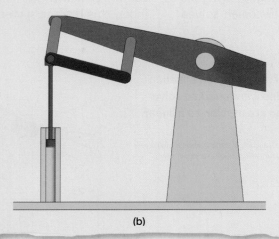

(b)

hollow cylinder (see Figure 4-9). This kind of mechanism is found in piston-type internal-combustion engines, where the linear motion of the slider (piston) is transferred to a crankshaft. It is also found within piston-type air compressors, where the crank is connected to an electric motor or gas-powered engine shaft.

The Cam and Follower

Before the widespread use of computer control systems and electromechanical devices such as servomotors, precision timing and control of linear motion within

FIGURE 4-9: A slider-crank linkage mechanism is used to change the continuous rotary motion of a crank into reciprocating linear motion of a slider and vice versa.

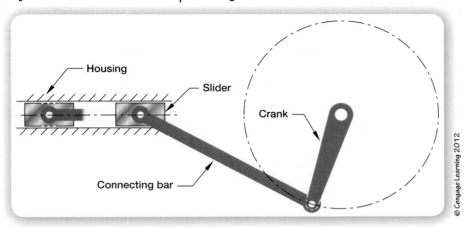

machines was often the product of the *cam and follower*. The **cam and follower** is a mechanism that changes continuous rotary motion into **intermittent** linear motion. The cam serves as the rotating component, which makes contact with the follower. The follower may take the form of a rod that moves back and forth along its axis, or a lever arm that teeters as one end of the lever arm rides along the contours of the cam. There are several types of cam-and-follower mechanisms, but the simplest and most common is the *plate cam* (see Figure 4-10). The plate is made from a flat piece of rigid material that is cut to a specific shape called the *cam profile*.

A cam profile can take on an infinite number of shapes, depending on the application (see Figure 4-11). Unless the plate cam has an integral shaft, a hole is located at a specific point on the cam for interfacing with a *camshaft*. A separate mechanical device, called a *follower*, rides along the surface of the cam profile as the plate rotates with the camshaft. The follower will rise or fall a specific linear distance over a given degree of rotation. The total follower displacement is the difference between the distances of the two points on the cam profile that have the shortest and longest radial distance from the camshaft's center axis. The point with the shortest radial distance will occur along an imaginary circle called the *base circle*.

An engineer will first create a cam-displacement diagram to show the timing and motion of the follower

Cam and follower:
a mechanism that changes continuous rotary motion into intermittent linear motion.

FIGURE 4-10: The simple plate cam is used to convert rotary motion into intermittent linear motion. A follower moves up and down as it rides along the surface of the plate cam profile.

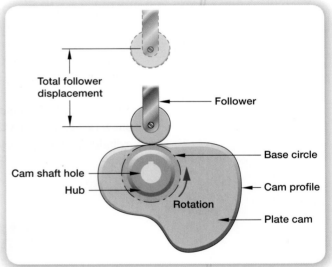

FIGURE 4-11: The cam profile may be symmetrical or irregular, depending on the application. In the case of a simple circular disk, the camshaft will be offset from the center of the disk.

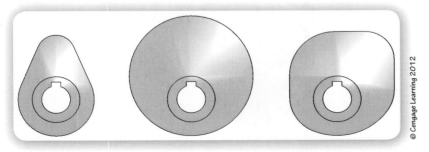

FIGURE 4-12: To understand the relationship between a plate cam's geometry and its displacement diagram, imagine the plate cam divided into (a) equal angular increments. The radial distances from the base circle to the points along the perimeter of the cam profile correspond with the plotted points shown on (b) the displacement diagram.

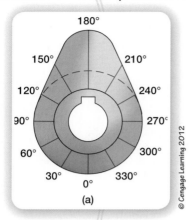

(a)

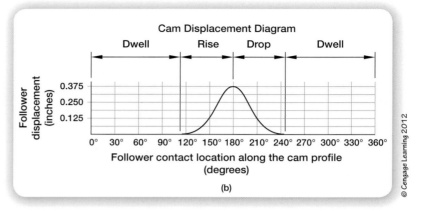

(b)

(see Figure 4-12). The shape of the curve represents the motion of the follower as the cam makes one full revolution. The cam follower may exhibit periods of constant velocity, uniform acceleration, a dwell in motion, or any combination thereof. When the diagram is finished, polar coordinates (angle and distance) taken along the diagram's curve are used to plot points on the actual cam and form the shape of the cam profile.

The most widely recognized and continued use of the cam and follower occurs within the piston-type internal-combustion engine. One or more camshafts are located inside the engine, and are lined with a series of plate cams. Each cam is responsible for the timing and linear motion of a cylinder valve. There are at least two valves per engine cylinder. One valve opens to allow air and fuel to enter the combustion chamber. The second valve opens to vent the exhaust gases after combustion has occurred. Both valves must be closed at the time of combustion (see Figure 4-13).

Cam-type clamping mechanisms are used on manufacturing fixtures to hold parts in place during machining operations. For example, the gold-colored cam

FIGURE 4-13: The inner workings of this piston-type internal-combustion engine show two camshafts that are used to coordinate the opening and closing of cylinder intake and exhaust valves.

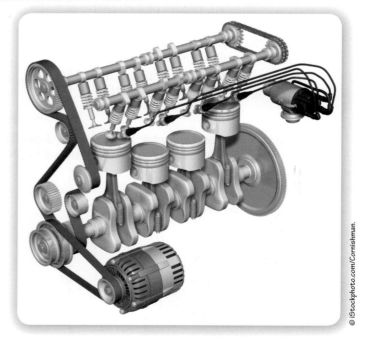

Fun Fact

A music box is an application of the cam and follower (see Figure 4-14). The cylinder within the music box functions as a cam plate and is "programmed" with a series of protrusions or pins. The pins are precisely located around the circumference of the cylinder to control the timing of the musical notes. A musical note is made when a pin strikes a tooth on a metal comb, which acts like a follower. The comb is a flat piece of metal that is cut into a shape that is similar to the teeth on a hair comb, except the teeth are of varying lengths and arranged to form a musical scale. The song is played out in one full revolution of the cylinder, thereby making the timing of each sound a function of the angle of the cylinder in relation to its starting position.

Hand-operated music boxes often employ speed-reducing gear trains to slow down the *rotational speed* (how fast an object rotates around an axis) of the hand crank. This means that the operator may need to turn the hand crank 20 times to rotate the cylinder one time.

FIGURE 4-14: A music box contains two mechanical components called the *cylinder* and *comb* that function as a cam plate and follower, respectively. This music box is driven by a hand crank and employs a speed-reducing gear train.

© iStockphoto.com/ThomasOlsson.

action fixture clamps shown in Figure 4-15 have two components. The head of the cap screw is offset from the center axis of the screw. This allows the screw head to function as a plate cam. As the screw turns, the screw head moves the hexagonal washer toward the workpiece to generate a sideways clamping pressure.

OFF-ROAD EXPLORATION

Use the Internet to research images and video about the 18th-century watchmakers Pierre Jaquet-Droz and Henri Maillardet and their cam-driven automatons.

FIGURE 4-15: Cams are also used in the design of clamping mechanisms.

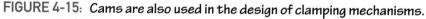

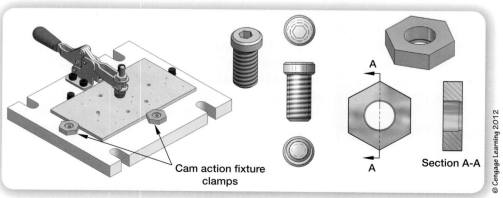

Cam action fixture clamps

A

A

Section A-A

© Cengage Learning 2012

Case Study »→

Timberkits: Designing Wood Mechanism–Driven Toys

Eric Williamson is a designer and craftsman from Llanbrynmair, Mid Wales (in the United Kingdom), who has been creating unique wooden toys, called *automata*, since the 1970s. The toys use simple mechanisms such as linkages and cams to convert the rotary motion of a hand-powered crank into lifelike animation of wooden animal figures, machines, and even people (see Figure 4-16). Eric's source of inspiration is his reverence for human anatomy and nature's ability to create graceful motions that are extremely difficult to replicate mechanically.

always see something quite amazing and improbable in the simplest, most mundane objects, and are fascinated to find that a pack of pieces of wood can become a drummer, a caterpillar, or a dragon."

The design process that Eric uses to "carry an idea through to the feasible" is unique and involves going back and forth between sketching and modeling. "My approach would be best described as that of an artist who is aware that his ideas have to be manufactured; and have to sell! First, I do a bit of 'scribbling.' Then I make a model to see how practical the idea is. After this, I go back to the drawing board (plain, old-fashioned board with mathematical instruments) and 'tighten up' the drawing a bit. The process goes on like this until I'm

Eric Williamson Automata Guru.

(a)

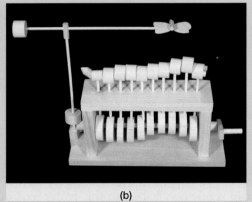
Timberkits US Caterpillar.

(b)

Figure 4-16: (a) Eric Williamson has been designing and building wooden toys since the 1970s. (b) His line of commercial wooden toy kits, called Timberkits, use simple mechanisms to animate models such as this caterpillar and butterfly.

With the help of his wife, Alison, Eric has developed his automata into a commercial line of self-assembled wooden toys called *Timberkits*, which are now sold across the world. Despite the growing presence of high-tech toys, Eric's automata have been selling for almost 20 years. He believes that many modern toys and activities for children leave no room for imagination. "Our imaginations are powerful. Children can

ready to make 'technical drawings,' which specify the dimensions of every aspect of each component."

Eric sends his finished models to other people who "pick holes" in his ideas and provide feedback on what the "real world" might think about them. He also works with retailers to find out what they are interested in, too. A finished prototype is then sent to a manufacturer in China for mass production using

Case Study ⟫⟫→

(continued)

traditional and computer-aided manufacturing methods. Step-by-step assembly instructions are provided with each kit, which are accompanied by hand-sketched isometric and oblique pictorial drawings. One example of Eric's Timberkit models is the T-REX (see Figure 4-17).

crank into a swinging motion that sends the left leg forward and backward. The right arm and bottom jaw are part of a four-bar linkage system that controls the opening and closing of the mouth of the T-REX. As the right arm rotates upward, the jaw appears to close.

Eric and his wife hope to inspire children, both young and old, to make their own automata using

Figure 4-17: *The parts of the T-REX Timberkit are animated through a series of interconnected linkage mechanisms and cams.*

© Cengage Learning 2012

The T-REX assembly is driven by a hand-powered crank mechanism that is attached to a plate cam (see Figure 4-18). The plate cam is in turn connected to the toy's left leg via a linkage rod and pin joint. The rod changes the continuous circular motion of the

whatever resources are available. "We want people to make their own models, and we hope ours will be the starting point for that process." To learn more about Timberkits, visit www.timberkits.com.

Figure 4-18: *(a) A modified version of the slider and crank is used to animate the toy's left leg. (b) The motion of T-REX's jaw is controlled by a four-bar linkage system.*

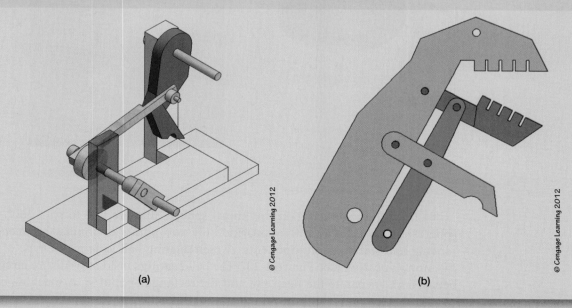

© Cengage Learning 2012

(a)   (b)

Bearings

If you look inside a machine, the chances are good that you will find at least one component that rotates or slides against a stationary component. How do stationary objects like automobile transmission system housings or the trucks on a skateboard remain in a fixed position while still being attached to rotating components? The answer lies in a category of devices called *bearings*. Bearings are used to reduce friction where two surfaces meet and slide against each other as a result of linear or rotary motion. We will look at two categories of bearings—those that are used in linear-motion applications and those that are used in rotary-motion applications.

LINEAR MOTION BEARINGS Linear-motion bearings apply to situations in which one component must slide along another in one direction. Plain bearings have no moving parts and are shaped like cylindrical tubes. They are made from materials such as bronze metal and nylon polymer, which naturally exhibit low friction (see Figure 4-19). A plain bearing may also be called a *sleeve, bushing,* or *journal bearing.* Plain bearings are press fit into holes and used in situations where a tight clearance exists between the outside diameter of a shaft and the bore that it slides through. In the plastics industry, metal dowels are used to align two or more plates in an injection mold. The holes through which these dowels pass are often fitted with plain bearings.

FIGURE 4-19: (a) Plain bearings are made from materials such as bronze that exhibit low friction. (b) The four large alignment holes in the corners on this split metal mold have been outfitted with plain bearings to aid the sliding action of the alignment pins on the matching mold half.

(a)

© Cengage Learning 2012

(b)

Courtesy of Kurt Sutton/Hulton Archive/Getty Images.

Ball-bearing slides contain tiny metal spheres that are free to spin 360°. The spheres are used because they provide only two points of contact between the moving and stationary components. A recirculating ball slide is one type of linear motion bearing (see Figure 4-20). The ball bearings inside this device circulate between and separate the fixed rail and the rectangular shuttle. A machine component that must exhibit linear motion can be attached to the shuttle via standard size machine screws.

FIGURE 4-20: A recirculating ball slide consists of a shuttle that moves back and forth along a fixed guide rail.

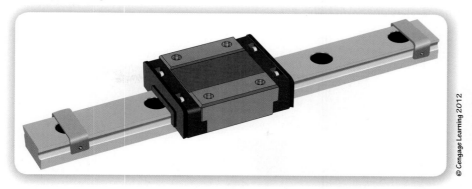

ROTARY MOTION BEARINGS Rotary motion bearings are designed to interface with rotating shafts. Radial load bearings and thrust bearings are two commonly used devices that fall under this category. It should be noted that bearings require lubrication unless their inner components are sealed, in which case permanent lubrication has already been applied and is protected by special seals.

Radial load bearings are used when a rotating shaft must be supported by a fixed structure. This type of bearing is made of several components that move independently. The inner and outer rings of most bearings are separated by a string of spherical metal balls or cylinders that allow one of the rings to spin with the shaft while the other remains stationary (see Figure 4-21).

FIGURE 4-21: (a) A rotating shaft within a machine will exhibit a radial load. (b) A radial load bearing has an outer and inner ring that rotate independently because of a series of metal balls or cylinders that are evenly spaced between them. Low friction results from the small amount of surface contact that occurs between the spinning elements.

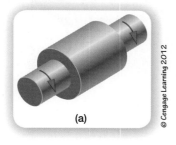

(a)

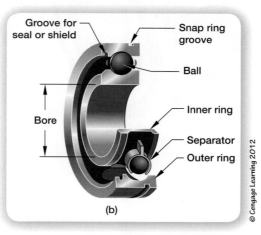

(b)

Either the outer or inner ring of the bearing may be free to spin. When a bearing is press fit onto the end of an axle, the inner ring of the bearing will spin with the axle while the outer ring remains stationary and is supported by a fixed structure. Gear shafts in transmission systems use bearings in this way. The outer ring of a bearing that is press fit into a skateboard wheel will spin with the wheel while the inner ring remains stationary with the truck axle (see Figure 4-22).

Some shafts generate axial forces as well as radial loads. For example, a spinning shaft that is oriented vertically will cause end thrust because of its weight. Special types of gears, like the ones shown in Figure 4-22a also generate end thrust as they rotate. When this occurs, a tapered roller bearing or thrust bearing is used

FIGURE 4-22: (a) Bearings are often press fit onto the ends of gear shafts, wherein the outer ring is stationary and the inner ring turns with the shaft. (b) The outer ring of a bearing that is press fit into a skateboard wheel will turn with the wheel while the inner ring remains stationary with the axle.

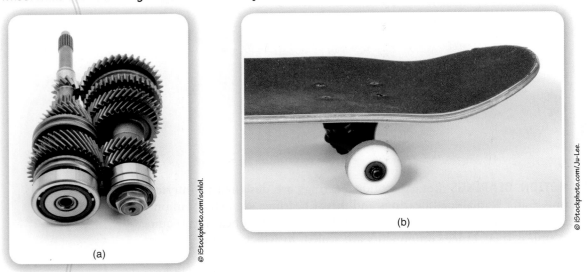

© iStockphoto.com/schlol.

(a)

© iStockphoto.com/Ju-Lee.

(b)

FIGURE 4-23: (a) If a rotating shaft also exhibits end thrust, then special kinds of bearings such as (b) a tapered roller bearing or (c) thrust bearings must be used.

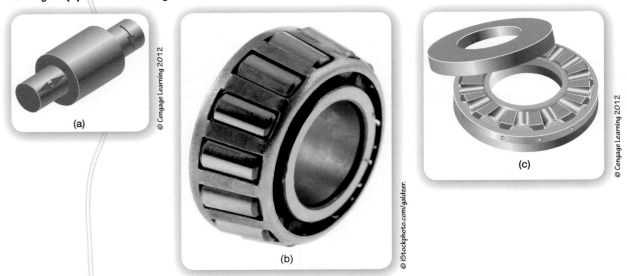

© Cengage Learning 2012

(a)

© iStockphoto.com/galdzer.

(b)

© Cengage Learning 2012

(c)

(see Figure 4-23). Thrust bearings operate in much the same way as radial load bearings, except that the independently moving rings are top and bottom plates that are separated by ball bearings or cylinders.

Gears

Perhaps more so than any other object, the gear is used in graphic images to symbolize a marriage between science, technology, engineering, and mathematics. Why would such an object be so revered as to be offered up time and again as the epitome of applied human knowledge?

Gears are used to control rotational direction, rotational speed, and torque in mechanical systems that involve precisely controlled rotary motion. They operate behind the scenes in many commonly used commercial products. You probably touch some device at least once a day that has gears buried somewhere inside it

FIGURE 4-24: Gears can be found in (a) the mechanism that opens and closes a DVD tray within a computer and (b) a manually operated can opener.

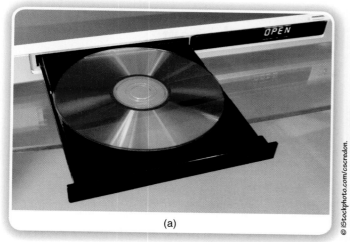

(a)

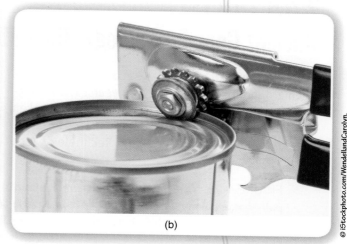

(b)

(see Figure 4-24). For example, if you have used a manually operated can opener to open a can of vegetables, opened and closed a push button–operated DVD tray, or sharpened a pencil with a mechanical pencil sharpener, then you have used a device that incorporates gears.

Gears give us the ability to precisely control the output of a motor or engine by decreasing speed and increasing torque to more appropriate and useful levels. The gears that are found in automobile transmission systems allow the high speed of an engine's crankshaft to be reduced while increasing the torque to the drive axle. When starting from rest, an automobile requires a large amount of torque at a very low speed to move the mass of the vehicle. Once the vehicle is in motion, the gears in the transmission system must "shift" in order to allow the wheels to spin faster. This increase in speed also results in a decrease in torque, which is why a vehicle cannot start out in fifth gear. Conversely, gears are also used in transmission systems to increase speed and decrease torque. An example application of this can be found inside a wind-turbine nacelle (see Figure 4-25).

FIGURE 4-25: This cutaway illustration of a wind turbine nacelle shows how gears are used to increase the speed of the generator shaft.

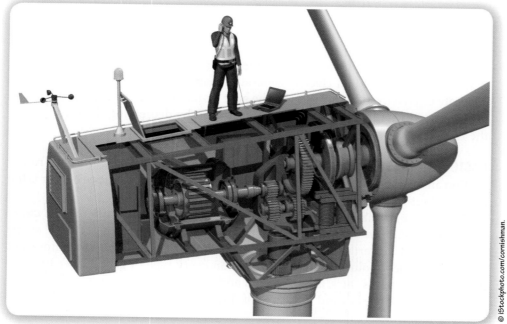

Fun Facts

Ancient Gear Technology: The Antikythera Mechanism

According to archeological records, gears have been around for more than 2,100 years. The Antikythera mechanism, which was discovered in sunken wreckage near the island of Crete in 1901, is the oldest evidence of ancient gear technology (see Figure 4-26). Scientists speculate that it was made around 100 B.C. by Greek engineers and was used to calculate the positions of the sun, moon, and other planets. The device is comprised of more than 30 bronze gears with triangle-shaped teeth.

FIGURE 4-26: (a) The Antikythera mechanism is the earliest known geared device. Through painstaking analysis, scientists have been able to discern enough information to build (b) models that show what this device may have looked like.

(a) Photo: Rien van de Weygaert, Groningen, the Netherlands. For more images, go to www.astro.rug.nl/weygaert/antikytheramechanism.html. (b) John Gleave.

(a)   (b)

A wind-turbine rotor generates a tremendous amount of torque, much more than is needed to spin an electric generator shaft. Because it is more efficient to generate electricity at higher speeds, the gear system found inside a wind-turbine nacelle increases the speed of the generator shaft by decreasing the torque from the rotor shaft.

BASIC GEAR GEOMETRY A gear is a toothed wheel that is used to transmit rotary motion and torque from one shaft to another. The most common type of gear is the external spur gear (see Figure 4-27).

Gears have specially designed teeth that are defined by a mathematical curve called an *involute curve*. An involute curve is similar to the cross-sectional shape that is formed by certain types of snail shells (see Figure 4-28). It is important to note that the involute curve is not used on every type of gear. Other types of mathematical curves, such as the hypoid and cycloid, are used on more complex gear designs that are beyond the scope of this text.

> **Gear:**
>
> a toothed wheel that is used to transmit rotary motion and torque from one shaft to another.

For almost 2,000 years, the triangle and other simple shapes were used as the basis for gear-tooth geometry. Unfortunately, these shapes could not transmit speed and torque in a smooth and constant fashion. As a result, gear-driven mechanical devices suffered from excessive vibration, noise, and wear. In 1767, a Swiss mathematician and physicist named Leonhard Euler proposed the involute curve as a viable gear-tooth profile for solving these problems. Euler developed the mathematics for generating an involute tooth profile and is considered the father of involute-based gears.

FIGURE 4-27: Gears are toothed wheels that provide smooth, continuous transmission of rotary motion and torque from one shaft to another. Note that mated external spur gears spin in opposite directions.

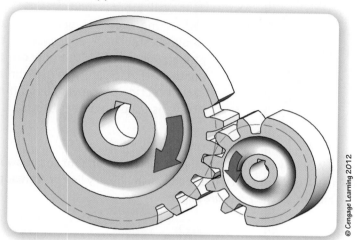

© Cengage Learning 2012

FIGURE 4-28: Spur gears incorporate a mathematical curve, called (a) an involute curve. The shape of this curve is very similar to the cross-sectional shape of (b) certain types of snail shells. (c) Involute curves form the working parts of a gear's teeth.

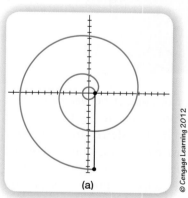

(a)

© Cengage Learning 2012

(b)

© iStockphoto.com/xelf.

(c)

© Cengage Learning 2012

The only part of a spur gear's tooth that comes into contact with another gear is the part that is formed by the involute curve. A simple experiment can help you envision how an involute curve is generated. You will need a long piece of rope, a large bag of popcorn, and a friend to help you.

Step 1: Find a large tree that has a relatively uniform circular trunk. Tie one end of the rope around the tree trunk.

Step 2: Wrap the rope around the tree trunk once, returning to where the knot is located. You have now established a section of the rope that is approximately the same size as the tree's circumference.

Step 3: Move as close to the tree trunk as possible and hold the rope tight to your torso. Start unwinding the rope from around the tree, making sure to keep the rope taught and tangent to the tree's circular trunk.

Step 4: Ask your friend to drop a trail of popcorn along the path that your feet generate as you walk around the tree.

Step 5: When the rope has been completely unwound, step back and observe the curve that is revealed by the popcorn trail. The path is that of an involute curve. Only a small section of the very beginning of this curve is used on a gear tooth.

FIGURE 4-29: When two spur gears mesh, the point of contact between the teeth (shown as a red dot) will move along an imaginary line of contact. The angle of this line is known as the *pressure angle*.

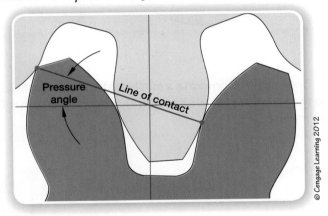

OFF-ROAD EXPLORATION

To learn more about gear terminology and the mathematics that drive a gear's geometry, read *Engineering Drawing and Design*, by David A. Madson et al., 2004, ISBN 0-7668-1634-6

FIGURE 4-30: The shape of a gear's tooth is a function of the pressure angle.

When two spur gears mesh, the point of contact between the teeth will occur on the surface of the gears' involute curves. It is at this point of contact where motion and torque are smoothly transferred from one gear shaft to another. As the gear teeth roll past each other, the point of contact will move along an imaginary *line of contact* (see Figure 4-29). The angle that defines the slope of the line of contact is called the *pressure angle*, and it is an important value that determines the shape of the gear's teeth.

A gear that is based on a shallow pressure angle, such as 14.5°, will have relatively square-shaped teeth (see Figure 4-30). A gear that is based on a steep pressure angle, such as 25°, will have more pronounced triangular-shaped teeth. Only gears that are based on the same pressure angle will be able to mesh properly.

Another important part of a gear's geometry that is not visible to the naked eye is the **pitch circle**, an imaginary circle around which the teeth on a gear or any other

FIGURE 4-31: The pitch circles of two mating gears must be tangent to each other for the gears to mesh properly. However, if the gear teeth do not have the same diametral pitch value (tooth size), they cannot mesh together.

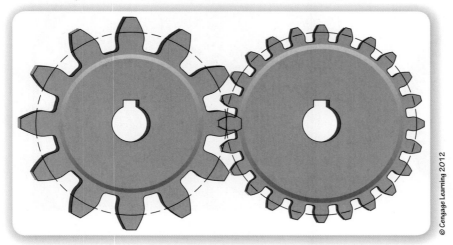

© Cengage Learning 2012

uniformly toothed device are evenly spaced (see Figure 4-31). It is important to know the pitch circle diameters of meshing gears when laying out the hole locations for the gear shafts on a gear plate. The center distance between two mating external spur gears is equal to the sum of the radii of the gears' pitch circles. Only when the pitch circles of two mating gears are tangent to each other will the gears be able to mesh properly.

The size of a gear tooth, called the **diametral pitch**, is determined by the number of teeth that occur in 1 inch of arc length along the pitch circle. A gear that has a large diametral pitch value will have smaller teeth than a gear that has a small diametral pitch value. Only gears of the same tooth size can mesh together. The pressure angle, pitch circle diameter, and diametral pitch are critical values that must be considered when designing gear-drive mechanisms (see Figure 4-32).

FIGURE 4-32: Three Rules for Spur Gears in Mesh

Rule 1   Only gear teeth that are based on the same pressure angle will be able to mesh properly.

Rule 2   Only when the pitch circles of mating gears are tangent to each other will the gears be able to mesh properly.

Rule 3   Only gear teeth that have the same diametral pitch value (tooth size) will be able to mesh properly.

GEAR TYPES   There are dozens of types of gears, most of which are beyond the scope of this text. Aside from the external spur gear, other types of gears that you might work with in the classroom or find within common domestic products include the internal spur gear, helical gear, bevel gear, and the worm and wheel. Gears are classified based on the orientations of their gear shafts, such as parallel shaft gears, nonparallel and intersecting shaft gears, and nonparallel and nonintersecting shaft gears. Gears are not limited to purely rotary motion, and may also be used to convert rotary motion to linear motion (and vice versa). Such is the case with the rack and pinion.

The external spur gear falls under the category of parallel shaft gears. External spur gears are by far the simplest, most commonly used, and easiest to manufacture of all gear types. An external spur gear can be used in conjunction with an *internal spur gear* (also called a *ring gear*). Internal spur gears are used in situations where the input power shaft must be directly in line with the output shaft. Such is the case with most automatic-transmission systems in automobiles, gear-reduction systems in cordless drills, and even the common manually operated pencil sharpener.

FIGURE 4-33: (a) An internal spur gear will only mesh with an external spur gear. Both gears rotate in the same direction. (b) This type of gear is often found in an automobile's automatic transmission system.

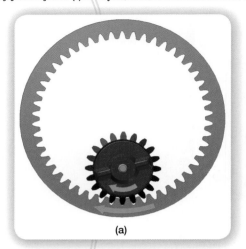

(a)

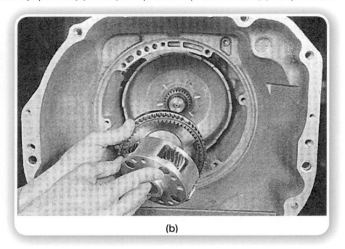

(b)

An internal spur gear can only be used in conjunction with an external spur gear (see Figure 4-33), wherein the three rules for gears in mesh must still be observed. Unlike two external spur gears in mesh, mated internal and external spur gears will turn in the same direction.

FIGURE 4-34: (a) A helical gear runs more quietly and can handle more torque than an external spur gear. The end thrust that is generated by mating helical gears must be compensated for with (b) special types of thrust bearings.

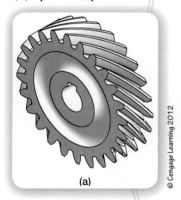

© Cengage Learning 2012

(a)

© iStockphoto.com/galdzer.

(b)

A *helical gear* is similar to an external spur gear, except that the teeth wrap around the gear hub in the form of a helix (see Figure 4-34a). This feature allows more teeth to be in contact at one time, resulting in greater load capacity. Like spur gears, the shafts of mating helical gears must be parallel. Helical gears provide smoother and quieter operation than their spur gear equivalents. Such gears are used in transmission systems because they shift more easily than spur gears. One of the disadvantages of helical gears is the formation of end thrust. The sliding action along the helical teeth results in lateral forces that would make the gears slip past each other in opposite directions if the shafts are not securely held in place. The gear shafts must therefore be equipped with thrust bearings to counteract this tendency (see Figures 4-22a, 4-23, and 4-34b).

The *worm and wheel* falls under the category of gears that have nonparallel and nonintersecting shafts. A worm is a screw thread that shares similar geometric features as the gear (called a *wheel*) to which it meshes. It also serves as the input to the gear system. The wheel serves as the output and may take the form of a standard spur gear or a helical gear (see Figure 4-35). It takes one full revolution of the worm to rotate the wheel only one tooth distance. Therefore, a worm that interfaces with a 40-tooth wheel must spin 40 times to turn the wheel only one time. It is for this reason that the worm and wheel are used in machines that require large speed reductions and proportionally large increases in torque. As a mechanism, the worm and wheel tends to be self-locking in that the wheel is not able to turn the worm.

FIGURE 4-35: (a) A worm gear is actually a screw that shares similar geometric features as the wheel to which it interfaces. (b) The worm shaft and the wheel shaft are nonparallel and nonintersecting.

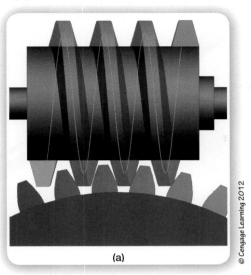
(a)

(b)

FIGURE 4-36: The worm and wheel is found in devices that must exhibit precision rotary motion, such as (a) a machinist's rotary table. It is also employed in (b) electric winch systems that are used in elevators and building cranes.

(a)

(b)

The worm and wheel is frequently used in devices that are capable of precise angle calibration such as a rotary table used in the machining industry (see Figure 4-36). It is also used in the design of electric winch systems that generate enormous pulling forces. Such systems are found in elevators and giant building cranes.

Straight-tooth bevel gears belong to the category of nonparallel and intersecting shaft gears. Bevel gears have a conical form, and their teeth are the same shape as those found on spur gears. Unlike spur gears, bevel gear teeth taper as they draw closer to the apex of the gear cone. Whereas spur gears are interchangeable, bevel gears must be designed in pairs. The most recognizable application of this type of gear can be found between a drill chuck and its mating chuck key (see Figure 4-37).

A lesser known, though extremely important application of straight-tooth bevel gears is found within a device called a *differential* (see Figure 4-38). The differential is part of an automobile's drive system. As a car turns, its wheels on the inside curve will spin slower than the wheels

OFF-ROAD EXPLORATION

To view short films that detail the working principles of a differential, visit *www.youtube.com* and perform the key phrase search, "how differential gear works."

FIGURE 4-37: (a) A drill chuck and (b) chuck key use matched bevel gears.

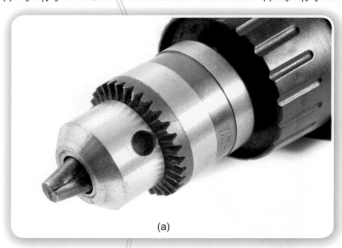

(a)

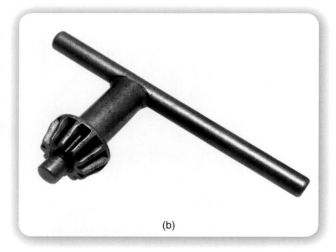

(b)

FIGURE 4-38: Differential gear trains are found in automobile drive systems and incorporate the use of straight-tooth bevel gears. The device shown allows the wheels to still receive power from the engine while turning at different speeds as the car turns.

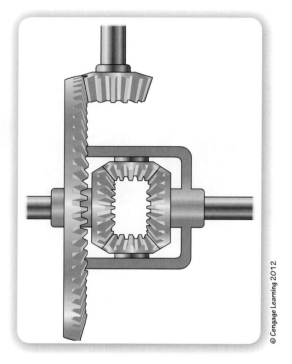

© Cengage Learning 2012

on the outer curve. If the wheels are not powered by the motor and are free to spin independently, then this does not pose a problem. A differential is used to maintain power to two wheels and still allow both to spin at different rotational speeds. All-wheel drive vehicles have two differentials, one for the front axle and one for the rear axle.

A *rack and pinion* consists of a spur gear (called a *pinion*) that interfaces with a flat bar into which straight gear teeth have been cut (called a *rack*). It is used to convert rotary motion into linear motion and vice versa. The rack and pinion is employed in manually operated arbor press machines (see Figure 4-39). This tool is used to force pulleys onto shafts, or pins and bearings into holes where an

FIGURE 4-39: (a) A rack and pinion is used in the design of (b) a manually operated arbor press. (b) Baileigh Industrial.

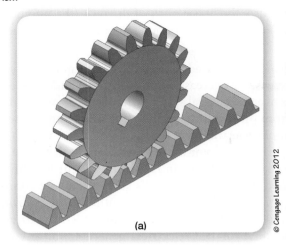
(a)

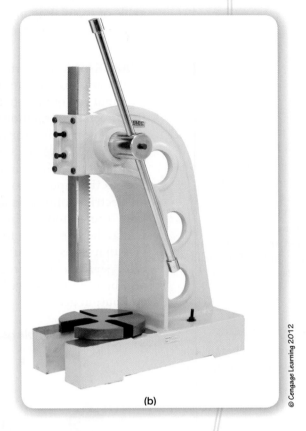
(b)

interference fit must occur. The rack and pinion is also used in automobile steering systems, wherein the rotary motion of the driver's steering wheel is converted to linear motion within the car's steering mechanism.

If the location of the pinion shaft is fixed, then the rack will be free to move back and forth as the pinion rotates. If the rack is held stationary, then the center of the pinion will move along a linear path that is parallel to but offset from the rack (see Figure 4-40). Regardless of which component is fixed, either will travel a distance that is equal to the circumference of the pinion's pitch circle for each revolution of the pinion.

FIGURE 4-40: If the rack is held stationary, then the center of the pinion will move along a linear path that is parallel to but offset from the rack.

Rack and pinion gears must be properly matched if they are to work together. The straight sides of the rack teeth are inclined at an angle that is equal to the pressure angle of the mating pinion. Also, the linear pitch of the rack (the number of teeth that occur in one linear inch) must equal the pinion's diametral pitch of the pinion.

GEAR TRAINS The term **gear train** is used to describe two or more gears in mesh. Gear trains are used to increase or decrease rotational speed and torque between the input and output shafts of a mechanical system. They can also be used to change the direction of rotation from one shaft to another. The following focuses on two types of gear trains: the simple gear train and the compound gear train.

A **simple gear train** consists of two or more gears in mesh that have only one gear per shaft. The simple gear train shown in Figure 4-41 identifies the 10-tooth gear as the input (also called the *driving* gear). As such, it may be connected to an electric motor, a hand crank, wind turbine, or any other source of rotary motion. The 40-tooth gear is identified as the output (also called the *driven* gear). It may be connected to a conveyor belt, electrical generator, or any other device that utilizes rotary motion. The 20- and 30-tooth gears serve as *idlers*. An **idler gear** is used to bridge the gap between the input and output gears. They may also be used to change or maintain the direction of rotation between the input and output gears.

Gear train:

two or more gears in mesh that are used to increase or decrease rotational speed and torque and change or maintain rotational direction between the input and output shafts of a mechanical system.

Simple gear train:

two or more gears in mesh that have only one gear per shaft.

FIGURE 4-41: Simple gear trains have only one gear per shaft.

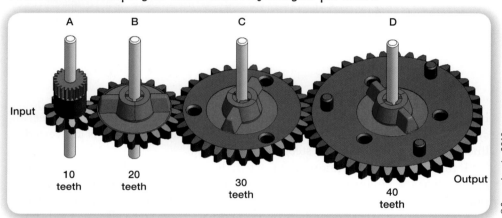

© Cengage Learning 2012

Compound gear train:

a gear train that has at least one idler gear shaft that carries two gears of different size that rotate at the same speed.

If two gears of different size are locked to the same idler gear shaft, as shown in Figure 4-42, then the result is a **compound gear train**. Compound gear trains have a greater ability to increase or decrease the rotational speed and torque between the input and output shafts while keeping the size of the individual gears to reasonable dimensions.

FIGURE 4-42: A compound gear train exists when at least one idler gear shaft carries two or more gears of different size that spin in unison.

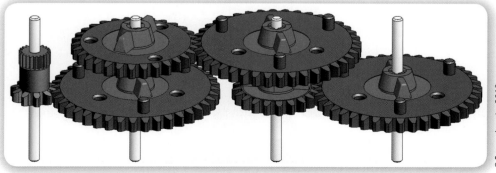

© Cengage Learning 2012

FIGURE 4-43: *Example of a simple gear train.*

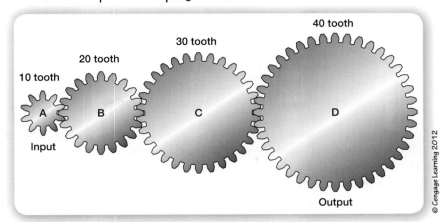

CALCULATING GEAR RATIOS IN GEAR TRAINS The *mechanical advantage* of a gear train is identified by its gear ratio. This is the ratio between the number of revolutions of the input gear and output gear in a gear train. The process of calculating the gear ratio of any gear train begins by counting the number of teeth (*n*) that exist on each gear. Using the simple gear train shown in Figure 4-43, we can see that the input gear is locked to shaft A and has 10 teeth. The output gear is locked to shaft D and has 40 teeth. The 20- and 30-tooth gears are locked to shafts B and C, respectively, and serve as idlers between the input and output gears.

The next step involves identifying the sets of mating gears along the gear train. It is important to treat each set of mating gears as a separate simple gear train, wherein one gear acts as the input gear and the other acts as the output gear. In this case, gears A and B, B and C, and C and D make up the three sets of mating gears along the gear train.

The gear ratio for each set of mating gears is then determined using the following equation.

$$GR = \frac{\# \text{ of teeth on the output gear}}{\# \text{ of teeth on the input gear}} = \frac{n_O}{n_I} \qquad \text{(Equation 4-1)}$$

Gear A serves as the input and gear B serves as the output for the first set of gears in the simple gear train. Therefore,

$$GR_{AB} = \frac{n_B}{n_A}$$

$$GR_{AB} = \frac{20 \text{ teeth}}{10 \text{ teeth}}$$

$$GR_{AB} = \frac{2}{1}$$

This process is repeated for the remaining sets of mating gears.

$$GR_{BC} = \frac{n_C}{n_B} \qquad\qquad GR_{CD} = \frac{n_D}{n_C}$$

$$GR_{BC} = \frac{30 \text{ teeth}}{20 \text{ teeth}} \qquad\qquad GR_{CD} = \frac{40 \text{ teeth}}{30 \text{ teeth}}$$

$$GR_{BC} = \frac{3}{2} \qquad\qquad GR_{CD} = \frac{4}{3}$$

The end-to-end gear ratio of the entire gear train is then calculated by multiplying the gear ratios of each pair of mating gears along the gear train using the following equation:

$$GR = GR_{AB} \times GR_{BC} \times GR_{CD} \qquad \text{(Equation 4-2)}$$

Gear ratio:

the ratio between the number of revolutions of the input (driving) gear and output (driven) gear in a gear train.

Therefore,

$$GR = GR_{AB} \times GR_{BC} \times GR_{CD}$$

$$GR = \frac{2}{1} \times \frac{3}{2} \times \frac{4}{3}$$

$$GR = \frac{4}{1} = 4{:}1$$

The resulting end-to-end gear ratio of the *simple gear train* is 4:1. This means the 10-tooth input gear would have to turn four times for every one turn of the 40-tooth output gear. It is common practice to express gear ratios as 1:n (wherein n represents a number other than 0 or 1) for speed-increasing gear trains, and n:1 for speed reduction gear trains. A 4:1 gear ratio would therefore represent a speed reduction.

The end-to-end gear ratio of a *simple gear train* will be maintained regardless of the number and size of the idler gears that it contains. This is proven by the fact that the values representing the idler gears canceled out in the multiplication, leaving only the values representing the 10-tooth input gear and the 40-tooth output gear. As a result, our first equation (Equation 4-1) may be used to calculate the gear ratio of a simple gear train. It must be understood that this *only* applies to a simple gear train.

$$GR_{AD} = \frac{n_O}{n_I}$$

$$GR_{AD} = \frac{40 \text{ teeth}}{10 \text{ teeth}}$$

$$GR_{AD} = \frac{4}{1} = 4{:}1$$

Point of Interest

To save time and avoid confusion, engineers make schematic drawings of gear trains in which centerlines are used to represent the pitch circles of the mating gears (see Figure 4-44). Such drawings are easier to create and serve as an effective tool for communicating complex information.

FIGURE 4-44: *Schematic drawing of a compound gear train that consists of 10-, 20-, 30-, and 40-tooth gears.*

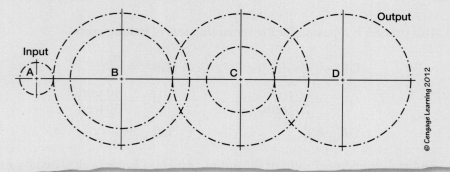

© Cengage Learning 2012

Calculating the gear ratio of a compound gear train, such as the one shown in Figure 4-44, is identical to the process we just used for the simple gear train.

Step 1: Identify the number of teeth that exist on each gear.

We can see that the input gear is locked to shaft A and has 10 teeth. The output gear is locked to shaft D and has 40 teeth. Two gears are locked to shaft B. Following the order of the mating gears along the gear train, we can see that the larger gear (B_1) has 40 teeth, and the smaller gear (B_2) has 30 teeth. Because both gears are locked to the same shaft, they will rotate with the same speed. This is also the case with shaft C, wherein the larger gear (C_1) has 40 teeth, and the smaller gear (B_2) has 30 teeth.

Step 2: Identify the sets of mating gears along the gear train.

Gears A and B_1, gears B_2 and C_1, and gears C_2 and D make up the three sets of mating gears along the gear train. The gear ratio for each set of mating gears is then determined using the next equation.

Step 3: Calculate the gear ratio for each set of mating gears along the gear train.

$$GR_{AB_1} = \frac{n_{B1}}{n_A} \qquad GR_{B_2C_1} = \frac{n_{C1}}{n_{B2}} \qquad GR_{C_2D} = \frac{n_D}{n_{C2}}$$

$$GR_{AB_1} = \frac{40 \text{ teeth}}{10 \text{ teeth}} \qquad GR_{B_2C_1} = \frac{40 \text{ teeth}}{30 \text{ teeth}} \qquad GR_{C_2D} = \frac{40 \text{ teeth}}{20 \text{ teeth}}$$

$$GR_{AB_1} = \frac{4}{1} \qquad GR_{B_2C_1} = \frac{4}{3} \qquad GR_{C_2D} = \frac{2}{1}$$

Step 4: Multiply the gear ratios for each set of mating gears along the gear train to determine the end-to-end gear ratio.

$$GR = GR_{AB_1} \times GR_{B_2C_1} \times GR_{C_2D}$$

$$GR = \frac{4}{1} \times \frac{4}{3} \times \frac{2}{1}$$

$$GR = \frac{32}{3} = \frac{10.67}{1} = 10.67{:}1$$

The resulting end-to-end gear ratio of the *compound gear train* is 10.67:1. This means the 10-tooth input gear would have to turn 10.67 times for every one turn of the 40-tooth output gear. This gear ratio represents a speed reduction.

Note that the input and output gears on the compound gear train have the same number of teeth as the input and output gears in the previous simple gear-train example. It is important to remember that the simplified equation for calculating the gear ratio of a simple gear train will not work for a compound gear train because the idler gears in a compound gear train *directly affect* the end-to-end gear ratio.

One might question, "If the relationship between the driving and driven gears is all that matters, why design compound gear trains? Why not make gear trains simpler by using only two gears and get rid of the idler gears altogether?" One reason for designing and using compound gear trains is illustrated in Figure 4-45.

Both gear trains have the same distance between input and output shafts, as well as the same gear ratio. The proportions between the pitch diameters of the two mating gears in the simple gear must match the 10.67:1 ratio. As a result, the output gear is large. Gear trains are often lubricated with grease or oil, which must be contained by an enclosure called a *gear box*. You can see that a gear box needed to enclose the *simple gear train* would be considerably larger than the housing needed to enclose the *compound gear train*.

SPEED AND TORQUE RATIOS One of the major advantages to using gear trains is the ability of the designer to manipulate *rotational speed*. As mentioned earlier,

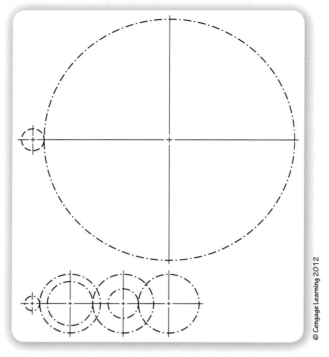

rotational speed (lowercase Greek omega, ω) is a measure of how fast an object rotates about an axis. The speed ratio (SR) is the ratio between the rotational speed of the input gear and output gear in a gear train. Though calculated differently, the speed ratio and gear ratio of a gear train are identical values. When the gear ratio (GR) and the rotational speed of the input shaft (ω_I) are known, the rotational speed of the output shaft (ω_O) can be calculated using the following equation:

$$\text{SR} = \frac{\omega_I}{\omega_O} = \frac{\text{GR}}{1} \qquad \text{(Equation 4-3)}$$

Returning to our previous compound gear train example (see Figure 4-44), imagine that the input gear is attached to an electric motor shaft. If the motor rotates at 600 revolutions per minute (RPM) under load, what would be the rotational speed of the output shaft?

$$\text{SR} = \frac{\omega_I}{\omega_O} = \frac{\text{GR}}{1}$$

$$10.67 = \frac{600 \text{ RPM}}{\omega_O}$$

$$\omega_O = \frac{600 \text{ RPM}}{10.67} = 56.23 \text{ RPM}$$

If the speed ratio of the compound gear train is 10.67:1, and the rotational speed of the input shaft is 600 RPM, then the rotational speed of the output shaft will be 56.23 RPM.

Another major advantage to using gear trains is the ability to manipulate *torque*. In Chapter 3 The Mechanical Advantage, you learned that *torque* is the rotational equivalent of the concept of force. Because we have now reached a point of engineering application where the magnitude of this rotational force can be manipulated through a mechanism such as a gear train, we must recognize how to calculate torque so that its effects can be predicted. Therefore, we will build on our previous definition by recognizing torque as the measure of the tendency of a force to rotate a body on which it acts around an axis.

The symbol for torque is the lowercase Greek letter τ (pronounced "tau"). Torque is generated when a force is applied tangential to a shaft, which is also perpendicular to an imaginary radial arm that extends from the shaft's fulcrum or axis of rotation (see Figure 4-46).

FIGURE 4-46: Example of a torque problem.

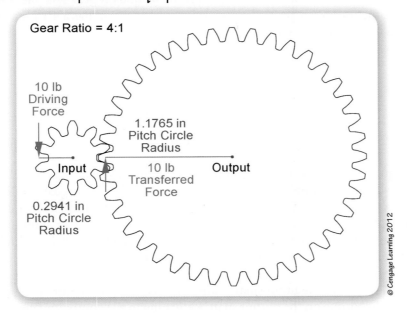

The torque value is the product of the radial arm distance (r) and the magnitude of the perpendicular force ($F_\perp$), as shown in Equation 4-4. In the U.S. customary system of units, torque is commonly measured in ounce-force-inches (ozf · in), pound-force-inches (lbf · in), and pound-force-feet (lbf · ft). Common SI units of torque include the newton-meter (N · m), gram-force-centimeter (gf · cm), and kilogram-force-meter (kgf · m).

$$\tau = rF_\perp \qquad \text{(Equation 4-4)}$$

Torque is increased when a small input gear meshes with a larger output gear. Therefore, there are two torque values that must be considered between two gears in mesh: the torque on the input gear (τ_I) and the torque on the output gear (τ_O). If a 10-lb force is applied perpendicular to the teeth of a 10-tooth gear that is in mesh with a 40-tooth gear (as shown in Figure 4-46), how much torque will result on both gear shafts? To solve this problem, the pitch diameters of both gears must be known so that the radius values can be used. It is standard practice for gear manufacturers and commercial suppliers to provide the pitch diameters of their gears. Under ideal conditions, the 10-lb force will be transmitted, undiminished, to the 40-tooth output gear. Therefore, we can calculate the input torque and output torque using Equation 4-4 and the dimensional information given in Figure 4-46.

$$\tau_I = r_1 F_\perp \qquad\qquad \tau_O = r_2 F_\perp$$

$$\tau_I = 0.2941 \text{ in} \times 10 \text{ lbf} \qquad \tau_O = 1.1765 \text{ in} \times 10 \text{ lbf}$$

$$\tau_I = 2.941 \text{ lbf} \cdot \text{in} \qquad\qquad \tau_O = 11.765 \text{ lbf} \cdot \text{in}$$

The ratio between the torque values on the input and output gears is inversely related to the gear ratio. If the gear ratio and the torque value on one of the gears are known, then the **torque ratio (TR)** can be used to calculate the unknown torque value on the remaining gear (see Equation 4-5):

$$TR = \frac{\tau_I}{\tau_O} = \frac{1}{GR} \qquad \text{(Equation 4-5)}$$

In the simple gear train example from Figure 4-46, gear 1 rotates four times for every one rotation of gear 2. However, the torque on gear 1 is one-quarter of the value the torque on gear 2. Therefore, a gear train that has a 4:1 gear ratio will exhibit a torque ratio of 1:4:

$$TR = \frac{\tau_I}{\tau_O}$$

$$TR = \frac{2.941 \text{ lbf} \cdot \text{in}}{11.765 \text{ lbf} \cdot \text{in}}$$

$$TR = \frac{1}{4} = 1:4$$

Torque will increase along a gear train as rotational speed decreases. This is the reason why an automobile transmission must provide a large gear reduction between the engine crankshaft and the wheel axle to move a vehicle from a stationary position. As the car moves faster, its inertia will require the gear train to provide less torque and greater speed. Unfortunately, you cannot have both high speed and high torque. One must come at the expense of the other because the two are inversely proportional.

If the torque on the input shaft is known, then the torque ratio can be used to calculate the torque on the output shaft. Returning to the compound gear train example given in Figure 4-44, we know that the gear ratio is 10.67:1. If the input gear is attached to the shaft of a motor that generates 120 ozf · in of torque, how much torque will be generated on the output shaft?

$$TR = \frac{\tau_I}{\tau_O} = \frac{1}{GR}$$

$$\frac{120 \text{ ozf} \cdot \text{in}}{\tau_O} = \frac{1}{10.67}$$

$$\tau_O = 120 \text{ ozf} \cdot \text{in} \times 10.67$$

$$\tau_O = 1{,}280.4 \text{ ozf} \cdot \text{in}$$

If the torque ratio of the compound gear train is 1:10.67, and 120 ozf · in of torque is applied to the input shaft, then 1,280.4 ozf · in of torque will be generated on the output shaft.

Sprockets and Chain Drives

Sprocket and chain drive mechanisms are another way of transferring and manipulating rotational speed and torque between rotating shafts in a machine. Though they are less accurate and not as compact as geared systems, power chain mechanisms do have the advantage of being easily assembled. They can also span distances that are either impractical or impossible for gears. Chain mechanisms are also used to change rotary motion into linear motion. Examples of this can be seen in the design of conveyors and the track systems found on military tanks and large construction equipment. The most widely recognized application of the power chain mechanism can be found on the average bicycle (see Figure 4-47).

FIGURE 4-47: **(a) A sprocket and chain drive mechanism on a bicycle is used to transfer and manipulate rotational speed and torque between the pedal shaft (input) and back-wheel axle (output). Sprocket and chain drives are also used on (b) construction equipment track systems to change rotary motion into linear motion.**

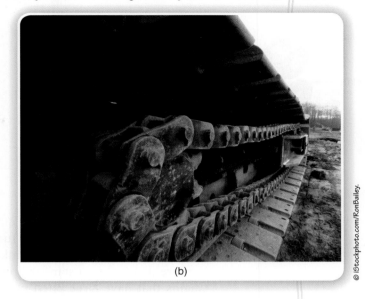

(a)

(b)

© iStockphoto.com/mipan.

© iStockphoto.com/RonBailey.

A **sprocket** is a toothed wheel that is used in conjunction with a continuous chain to transfer rotational speed and torque from an input shaft to an output shaft (see Figure 4-48). Though they are similar to gears, sprockets are not designed to mesh together.

FIGURE 4-48: **Basic components of a power chain system.**

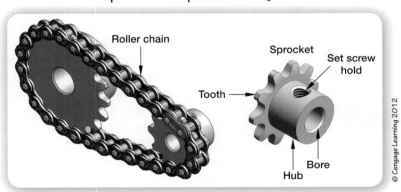

© Cengage Learning 2012

A **chain** is a series of links that are uniform in length and fitted together to form a continuous band. Of the various types of chains that are used, the most common is the *roller chain* (see Figure 4-49). Roller chains consist of two repeating components: the pin link and the roller link. The roller link is the part of the chain that makes contact with the teeth on the sprockets. A connecting link is used to join two roller links at either end of a chain to form a continuous loop. The connecting link is held in place by a link plate and a spring clip.

The center distance between sprockets is important, although there is more room for error in this kind of drive system in comparison to geared mechanisms.

FIGURE 4-49: Components that make up a roller chain assembly.

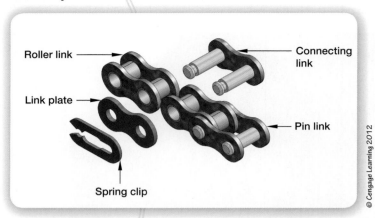

Roller link

Link plate

Spring clip

Connecting link

Pin link

© Cengage Learning 2012

FIGURE 4-50: The chain link pitch is an important distance that is used to calculate the center distance between the input and output shafts, as well as the length of the chain.

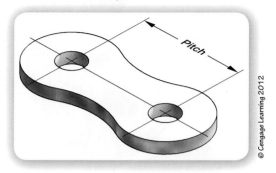

Pitch

© Cengage Learning 2012

The sprocket center distance is a function of the **chain pitch**, which is the distance between centers on one chain link (see Figure 4-50). Roller chains are available in more than a dozen standard pitch sizes that range from 0.250 in. to 3.000 in. Chains that have large pitch values are able to handle greater load capacities. For example, the working load for a size 25 chain (which has a 0.250 in. pitch) is 140 lb, whereas the working load for a size 240 chain (which has a 3.000 in. pitch) is 22,250 lb.

DESIGNING ROLLER CHAIN DRIVE SYSTEMS Several factors must be considered when designing roller chain drive systems. Sprockets that are joined by a chain will spin in the same direction. Because a chain elongates under load, it will eventually need to be retightened by either removing a link or increasing the center distance between the sprockets. An adjustable idler sprocket may also be used to return a loose chain to its proper tension. One side of the chain is expected to sag slightly during operation. It is for this reason that an optimal arrangement of the sprockets will occur along a horizontal line as shown in Figure 4-51. If the system must be inclined, then it should not be offset at an angle that exceeds 45°. Also, the chain must have a minimum 120° of contact around the smaller of the two sprockets.

The design of a roller chain drive system also involves several calculations. To illustrate this process, let's imagine that you are one of several mechanical systems designers for a FIRST Tech Challenge (FTC) robotics team, and that you have been given the task of designing the roller chain drive system that will power the rear wheels of the robot. Your design challenge is summarized in the design brief shown in Figure 4-52.

Research is an early step in almost every engineering design process, and it is usually an activity that engineers repeat throughout a design challenge. Engineers don't know everything, but they are especially good at finding answers to questions that lead to a vast body of knowledge over many years. When an engineer begins a design challenge, he or she may not fully understand the problem or the constraints that must be followed. Therefore, a design challenge often begins with the engineer identifying unknowns. For us, a #25 chain link does not provide enough information about the dimensions of the actual chain components. During your research, you would most likely find an ANSI (American

FIGURE 4-51: The smaller of the two sprockets in a chain drive system must have a minimum of 120° of chain contact.

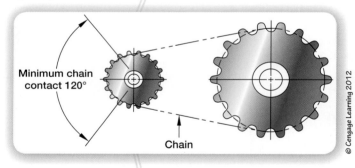

Minimum chain contact 120°

Chain

© Cengage Learning 2012

FIGURE 4-52: Design brief for a robot's roller chain drive mechanism.

Design Brief

Client:	FIRST Tech Challenge Teammates
Designer:	Shawn Biehler, FTC Mechanical Systems Sub-Team
Problem Statement:	A robot must use a standard 12V DC kit motor to power two rear wheels that are locked to the same drive shaft.
Design Statement:	Design and calculate the necessary dimensions for a roller chain drive mechanism that will transfer power from the standard 12V DC kit motor to the robot's rear wheel drive shaft.

Constraints:

1. You must use the #25 roller chain provided in the robot kit of parts.
2. The drive sprocket will be the 10-tooth sprocket provided in kit (0.251 in diameter shaft hole through hub).
3. The 12V DC motor has a 0.236 in diameter shaft and a maximum rotational speed of 152 RPM.
4. The distance between the centers of the drive and driven sprockets can be no greater than 12 in.
5. The maximum rotational speed of the rear wheel drive shaft must be approximately 85 RPM.
6. The chain will be lubricated manually.

Deliverables:

- ▶ calculations for the chain drive ratio
- ▶ calculations for the number of teeth on the driven sprocket
- ▶ calculations for the maximum and minimum distance between the drive and driven sprockets
- ▶ distance between the drive and driven sprockets
- ▶ calculations for the chain length in pitches
- ▶ calculations for the length of the roller chain
- ▶ sketch that summarizes the proposed solution and gives recommendations for the rear wheel axle designer based on the dimensions of the driven sprocket

National Standards Institute) standards table that is used to identify the pitch values for various chain link sizes (see Figure 4-53). Recognizing that you are limited to a #25 chain link that is provided in the robot's kit of parts, you find that the standard *pitch distance* (p) is 0.250 in.

The next step involves calculating the *chain drive ratio* (CDR) between the drive sprocket and the driven sprocket. The **chain drive ratio** is similar to the concept of a gear train's speed ratio; it is a ratio between the rotational speeds of the input

FIGURE 4-53: Table used to identify the standard pitch dimensions of roller chain links.

ANSI B29-1: Standard Dimensions for Roller Chains			
Size (#)	Pitch (inch)	Roller DIA (inch)	Working Load (lb)
25	0.250	0.130	140
35	0.375	0.200	480
40	0.500	0.312	810
50	0.625	0.400	1,430
60	0.750	0.469	1,980
80	1.000	0.625	3,300
100	1.250	0.750	5,072

(driver) and output (driven) sprockets. The following equation is used to calculate the *chain drive ratio*:

$$\text{CDR} = \frac{\omega_I}{\omega_O} \qquad \text{(Equation 4-6)}$$

If the 10-tooth sprocket is attached directly to the DC motor shaft, then the maximum rotational speed of the input shaft (ω_I) is 152 RPM. Your design brief also identified that the maximum rotational speed of the output shaft (ω_O) is approximately 85 RPM. Therefore,

$$\text{CDR} = \frac{\omega_I}{\omega_O}$$

$$\text{CDR} = \frac{152 \text{ RPM}}{85 \text{ RPM}}$$

$$\text{CDR} = \frac{1.79}{1} = 1.79{:}1$$

The chain drive ratio also reflects the ratio between the number of teeth on the driven sprocket (n_O) and the number of teeth on the drive sprocket (n_I). Because the output shaft must spin slower than the input shaft, it is logical to assume that the driven sprocket will be larger than the 10-tooth driver sprocket. We can use the following equation to verify this assumption.

$$\text{CDR} = \frac{n_O}{n_I} \qquad \text{(Equation 4-7)}$$

Therefore,

$$\text{CDR} = \frac{n_O}{n_I}$$

$$1.79 = \frac{n_O}{10 \text{ teeth}}$$

$$n_O = 10 \text{ teeth} \times 1.79 = 17.9 \text{ teeth} \approx 18 \text{ teeth}$$

Commercially available sprockets have total tooth values that are even numbers. It is also impossible to acquire a sprocket that contains a fraction of a tooth. Even if such a component existed, it would not be able to function as part of a chain drive mechanism that is based on a fixed chain pitch. Therefore, the number of teeth on the driven sprocket must be rounded to 18. As a result, the robot's output shaft will not spin at exactly 85 RPM, but it will be close. This trade-off is still within the limits of the constraints as identified in the design brief. After searching through online commercial sources of roller chain sprockets, you decide on an 18-tooth sprocket that matches the chain pitch and make note of the critical dimensions for the rear wheel axle designer (see Figure 4-54).

FIGURE 4-54: The sprocket hub size, axle hole diameter, and method used to affix the sprocket to a shaft are important considerations when designing roller chain drive systems.

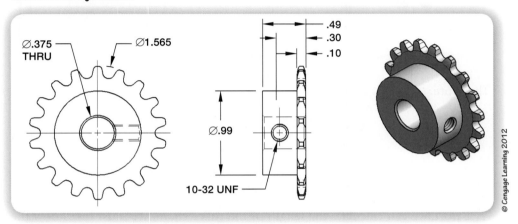

© Cengage Learning 2012

The next step in the process involves establishing a center distance between the input and output shafts. Manufacturers recommend that an ideal center distance be between 30 and 50 times the value of the chain pitch (p), with the distance never exceeding 80 times the chain pitch in extreme cases. Though you have been given a maximum 12-inch center distance between sprockets, this does not tell you what the ideal distance should be. To establish an ideal distance, calculate the maximum and minimum center distances based on the above recommendations.

Minimum center distance = $30p$

Minimum center distance = 30×0.25 in

Minimum center distance = 7.5 in

Maximum center distance = $50\,p$

Maximum center distance = 50×0.25 in

Maximum center distance = 12.5 in

Given the minimum and maximum ideal center distances of 7.5 inches and 12.5 inches, respectively, we can see that one of these values is above the allotted 12-inch constraint. To conserve material and keep the weight of the design down, a sprocket center distance of 8 inches will be used (see Figure 4-55).

The next step involves converting the established center distance to an equivalent *number of chain pitches between centers* (C). To do this, we divide our established center distance between the sprockets by the pitch (p) of the chain link:

$$C = \frac{\text{center distance between sprockets}}{p}$$ (Equation 4-8)

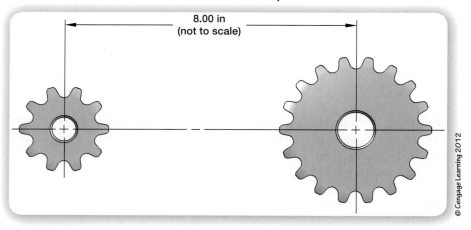

8.00 in
(not to scale)

© Cengage Learning 2012

Therefore,

$$C = \frac{\text{center distance between sprockets}}{p}$$

$$C = \frac{8 \text{ in}}{0.25 \text{ in}}$$

$$C = 32 \text{ chain pitches}$$

Before we can calculate the length of the roller chain, we must first know the total number of teeth (S) of the driver (n_I) and driven (n_O) sprockets, as well as the difference in the number of teeth (F) between the driver and driven sprocket:

$$S = n_I + n_O \qquad\qquad \text{(Equation 4-9)}$$

$$F = n_O - n_I \qquad\qquad \text{(Equation 4-10)}$$

$$S = n_I + n_O$$

$$S = 10 \text{ teeth} + 18 \text{ teeth}$$

$$S = 28 \text{ teeth}$$

$$F = n_O - n_I$$

$$F = 18 \text{ teeth} - 10 \text{ teeth}$$

$$F = 8 \text{ teeth}$$

These values are used in Equation 4-11 to calculate the chain length in terms of the total chain pitches:

Chain length in chain pitches = $2C + S/2 + K/C$ \qquad (Equation 4-11)

The variable K is a constant that is taken from a manufacturer's table (see Figure 4-56), and is based on the difference in the number of teeth between the larger and smaller sprocket (F). The table shows that a difference of 8 teeth is equal to a K constant of 1.62.

Therefore,

Chain length in chain pitches = $2C + S/2 + K/C$

Chain length in chain pitches = $(2 \times 32) + 28/2 + 1.62/32$

Chain length in chain pitches = $64 + 14 + 0.0506$

Chain length in chain pitches = $78.056 \approx 80$

FIGURE 4-56: Manufacturer's table of sprocket teeth *K* constants.

F	K	F	K	F	K
1	0.03	11	3.06	21	11.17
2	0.10	12	3.65	22	12.26
3	0.23	13	4.28	23	13.40
4	0.41	14	4.96	24	14.59
5	0.63	15	5.70	25	15.83
6	0.91	16	6.48	26	17.12
7	1.24	17	7.32	27	18.47
8	1.62	18	8.21	28	19.86
9	2.05	19	9.14	29	21.30
10	2.53	20	10.13	30	22.80

The chain length in chain pitches must be rounded to the next larger even whole number because we cannot have a fraction of a chain pitch. Rounding to an even number avoids the use of an offset link, which adds a cost to the design that is unnecessary (see Figure 4-57). One of the 80 chain links must be a connecting link.

FIGURE 4-57: An **offset link** is needed along with a **connecting link** to join the ends of a roller chain that is made up of an odd number of chain links.

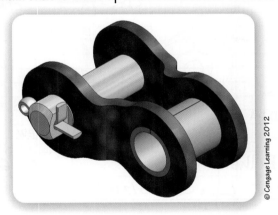

© Cengage Learning 2012

The last step in the process involves converting the chain length in pitch distances to the actual end-to-end center distance of the chain in inches. To do this, we multiply the chain length in chain pitches by the pitch value of one chain link. Therefore,

Chain length = chain length in chain pitches × chain pitch

Chain length = 80 × 0.25 in

Chain length = 20 in

As with any design endeavor, it is a good idea to keep a running record of your ideas and your calculations by using an engineer's notebook (see Figure 4-58).

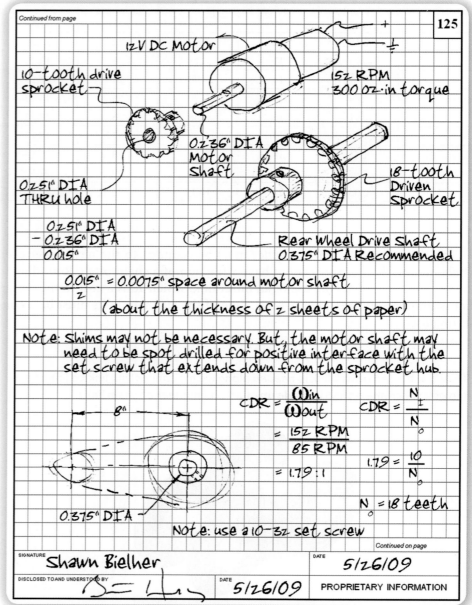

SUMMARY

In this chapter you learned:

- Whereas a simple machine is a single mechanical device that performs work, a mechanism is comprised of two or more moving mechanical components that operate according to the same principles on which simple machines are based. Examples of mechanisms include various types of linkage systems, the cam and follower, bearings, gears, and chain drives.

- The most common types of linkage systems found in industrial and commercial products include the four-bar linkage, parallel linkage, scissor linkage, and the slider-crank mechanism.

- The cam and follower is another mechanism that is found within the inner workings of piston engines and pumps. It is used to convert rotary motion into intermittent linear motion, such as when the camshaft in an engine controls the opening and closing of intake and exhaust valves. A cam plate is a disk that can take on any shape, depending on the application. A follower, which may be a rod or the end of a lever, will follow along the cam's profile as the cam plate spins.

- Bearings are used on machines, tools, pulleys, and wheels to reduce the friction that is generated when two objects move against each other. Most of the bearings that are used fall into one of two categories: linear motion bearings and rotary motion bearings. The simplest type of bearing is the bushing. Rotary motion bearings are used where a rotating shaft meets and must be supported by a fixed structure. Some rotating shafts also generate end thrust, which can be compensated for with tapered roller bearings or thrust bearings.

- A gear is a toothed wheel that transmits rotary motion and torque from one shaft to another. Gears are classified according to the orientation of their shafts. The most common types of parallel shaft gears are external and internal spur gears, and helical gears. Straight-tooth bevel gears are the most common of the nonparallel and intersecting shaft gears. The worm and wheel is a common example of a nonparallel and nonintersecting shaft gear system. Lastly, the rack and pinion is a type of gear that sits in a category of its own because a gear rack does not have a rotating shaft. This type of gear mechanism is used to change linear motion into rotary motion and vice versa.

- Two or more gears that work together form a gear train. If each shaft in the gear train carries only one gear, then it is said to be a simple gear train. A compound gear train is formed when at least one idler gear shaft carries two gears of different size that rotate at the same rotational speed. Three conditions must be satisfied for spur gears to properly mesh together: (1) They must share the same tooth size (diametral pitch), (2) the shape of the teeth on both gears must be based on the same pressure angle, and (3) the distance between the gear shafts must be such that the pitch circles of the mating gears are tangent to each other.

- A gear train's gear ratio is a ratio between the number of revolutions of the input gear and the output gear. The number of teeth on each gear can also be used to calculate the gear ratio of a simple or compound gear train. As such, the ratio between the number of teeth on the output gear to the number of teeth on the input gear is determined for each set of mating gears along the gear train. These individual gear ratios are then multiplied to determine the end-to-end gear ratio of the entire gear train. A gear train's speed ratio is a ratio between the rotational speed of the input gear and the rotational speed of the output gear. Its value is identical to the gear ratio, which can be used to calculate an unknown shaft speed. A gear train's torque ratio is a ratio between the torque on the output gear and the torque on the input gear. Its value is the inverse of the gear ratio, which can be used to calculate an unknown torque value.

(continued)

■ The roller chain and sprocket is one of the most common types of chain drive mechanism used in industrial and commercial products. Like gears, chain drive mechanisms are used for transferring motion and manipulating rotational speed between rotation shafts. Although they are less precise than gears, power chain systems can span distances that would be impractical for gears. A roller chain is a continuous band of two repeating components of uniform length called the *pin link* and the *roller link*. The roller links make contact with the teeth on a sprocket.

■ Several factors must be considered when designing a chain drive mechanism. Unlike mating gears, sprockets that are connected by a chain will spin in the same direction. The magnitude of the load that is carried by the chain decreases as the size of the chain decreases. Chain size is related to the distance between the centers on one chain link, called the *pitch*, which determines the optimum center distance between the driving and driven sprockets. The ratio between the rotational speeds of the driving and driven sprockets is called the *chain drive ratio*.

BRING IT HOME

1. Explain how a mechanism differs from a simple machine.
2. Identify five different types of mechanisms, and identify one application of each.
3. What is the difference between a rocker and a crank?
4. What does a four-bar linkage consist of?
5. Identify one application of a four-bar linkage.
6. Explain what a scissor linkage is used for.
7. What kinds of devices use a slider-crank linkage system?
8. What is the difference between a cam and a follower?
9. What is the purpose of a cam and follower mechanism?
10. Identify three applications of either the cam and follower or cam-action mechanisms.
11. What is a cam-displacement diagram?
12. What is the purpose of a bearing?
13. Identify three different types of bearings and give examples of where they would be used.
14. How does a radial load bearing work?
15. Identify three objects that contain gears.
16. How are gears classified?
17. What is an involute curve, and what does it have to do with gears?
18. Name five types of gears and give an example of where each is used.
19. What are the three conditions that must be satisfied for two gears to mesh properly?

20. What is the difference between a simple and a compound gear train?
21. Why aren't all gear trains made to be simple gear trains?
22. What is a gear ratio?
23. What is an idler gear, and does it affect the end-to-end gear ratio in a simple and compound gear train?
24. If a 20-tooth input gear meshes with a 40-tooth output gear, what is their gear ratio?
25. What is rotational speed?
26. What is the relationship between a gear train's gear ratio and its speed ratio?
27. What is torque, and how is it calculated?
28. What is the relationship between a gear train's gear ratio and its torque ratio?
29. What is a sprocket?
30. Why might an engineer use a chain drive system instead of a gear system?
31. What are the two repeating components that make up a roller chain, and how are they connected to form a continuous loop?
32. What does the chain pitch value represent?
33. What is the ideal distance between the driving and driven sprocket in a chain drive system?
34. What does the chain drive ratio represent?
35. Under what circumstances would an offset link need to be used in a roller chain?

Problem Statement: During the manipulation of sensitive materials (such as low-level nuclear waste), dangerous conditions exist wherein workers cannot risk direct contact with the objects they handle. This brings about a need for specialized devices that can be operated from remote locations.

Design Statement: Design, build, and test a syringe-powered robot arm that incorporates the use of linkage mechanisms to move one ping-pong ball from a given location in a work environment and drop it down a 1½ in. (nominal size) PVC tube in the shortest possible time. The ping-pong ball represents the sensitive material. The work environment represents a space in which dangerous conditions exist that prevent human beings from making direct contact with the material.

FIGURE 4-59: Syringe-powered robot arm test environment. All dimensions are in inches unless otherwise specified.

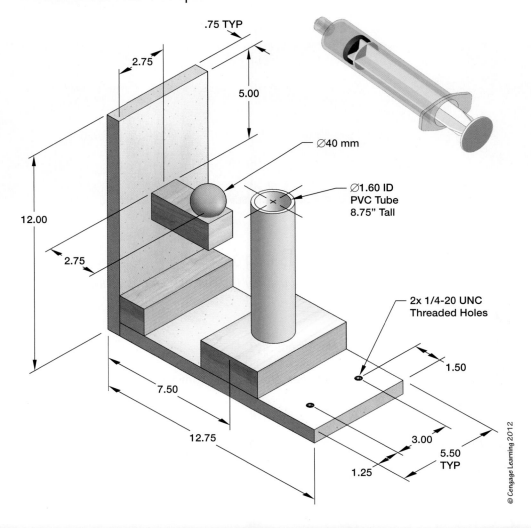

.75 TYP
2.75
5.00
Ø40 mm
Ø1.60 ID
PVC Tube
8.75" Tall
12.00
2.75
2x 1/4-20 UNC
Threaded Holes
1.50
7.50
12.75
3.00
5.50
TYP
1.25

© Cengage Learning 2012

CHAPTER 5
Energy

Before You Begin

Think about these questions as you study the concepts in this chapter:

1. What is energy?

2. What are the different forms of energy?

3. What are the differences between conduction, convection, and radiation?

4. What is heat?

5. What are the different types of temperature scales?

6. How do we convert from one temperature scale to another?

7. What are the units of energy and how is energy calculated?

8. What are the laws of thermodynamics?

9. What are the five most common renewable energy sources, and how are they used?

10. What are four nonrenewable energy sources, and how are they used?

Energy is defined as the ability to cause matter to move or transform. Humans have been using, wasting, selling, conserving, and fighting over energy for thousands of years. Energy is what makes things happen. Understanding what energy is and how it affects your everyday life will help you see the world in an entirely new way.

Before we explore energy here on Earth, we must first look at where most of our energy comes from: the sun. You can think of the sun as a gigantic nuclear furnace that continuously sends its energy through 93 million miles of empty space before reaching Earth's atmosphere. Traveling at the speed of light, the sun's energy takes about eight and one half minutes to get to you.

The sun warms up and evaporates the water in Earth's oceans and lakes, forming clouds. The clouds then move over land and release their moisture in the form of rain and snow. Earth's great rivers and glaciers depend on the sun to keep them flowing. Differences in air and water temperature are responsible for wind and ocean currents. Plants are able to absorb the energy from the sun through the process of photosynthesis. This process allows plants to convert water and carbon dioxide into sugar (glucose), which they use for food.

Animals eat the plants grown using the sun's energy. The plants give animals the energy for breathing, growing, running, jumping, and searching for more food. This process works its way up the food chain as animals eat other animals that ate the plants. The energy that originally came from the sun has been transformed by the plants and animals from one form to another (see Figure 5-1).

We don't know everything about energy, but we know much more than we did a century ago. We can measure it, describe its behavior, and even use mathematical formulas to predict what it's going to do. Scientists use their knowledge of energy to explore energy even further, whereas engineers use their knowledge of energy to solve problems and create new technologies for the benefit of humanity.

1

> **Energy:**
> the ability to cause matter to move or transform.

(a)

Matter is changed into electromagnetic energy in the sun. A small amount of the sun's energy reaches the leaves of a banana tree.	Through photosynthesis, the tree converts the sun's energy into chemical energy stored in the carbohydrate molecules in the tree's cells.	A monkey eats a banana from the tree and converts the banana's chemical energy into mechanical energy for leaping from tree to tree.

(b)

Figure 5-1: *Even monkeys get their energy from the sun.*
Image copyright worldswildlifewonders, 2010. Used under license from Shutterstock.com.

FORMS OF ENERGY

Although energy comes in many forms, we can put them all into two categories: **potential energy**, which is stored energy, and **kinetic energy**, which is the energy of motion.

Potential Energy

The amount of energy stored in a body at rest is called *potential energy*. A famous example of potential energy can be found in the nursery rhyme "Humpty Dumpty" (see Figure 5-2). When Humpty was sitting on the wall, gravity was pulling him toward the ground even before he fell. This provided the potential energy for him to fall to the ground and led to his demise.

There are four types of potential energy:

1. Potential chemical energy
2. Potential mechanical energy
3. Potential nuclear energy
4. Potential gravitational energy

POTENTIAL CHEMICAL ENERGY Potential chemical energy is energy stored in the chemical bonds of molecules. The potential energy comes from the atomic bonds that, like glue, hold the molecules together. Food, gasoline, natural gas, propane, and biomass are all examples of potential chemical energy.

POTENTIAL MECHANICAL ENERGY Potential mechanical energy is energy that is stored in an object because of the application of a force. A catapult is a good example of potential mechanical energy. Catapults store a large amount of energy that can be used to launch an object a certain distance. Whether a spring is compressed or a rubber band is stretched, potential mechanical energy is stored and ready to be used.

POTENTIAL NUCLEAR ENERGY Potential nuclear energy is energy stored in the nucleus of an atom. An immense amount of energy holds the nucleus together. Nuclear power plants harness a tiny portion of this energy by splitting the nucleus of a uranium atom in a process called *nuclear fission*. When the nucleus is split, a tremendous amount of energy is released in the form of heat. *Heat* is a form of energy that flows because of differences in **temperature**.

POTENTIAL GRAVITATIONAL ENERGY Potential gravitational energy is energy stored because of an object's position or place in a gravitational field. A roller coaster takes advantage of gravity. As you are towed up the first hill of the roller coaster, you gain potential gravitational energy. When you reach the top, you have enough energy to travel through the barrel rolls, loops, and heart-pounding hills as you make your way to the end.

Kinetic Energy

Kinetic energy is energy resulting from motion. This energy can be observed in moving objects, waves, electrons, molecules, and substances. There are five types of kinetic energy:

1. Electrical energy
2. Radiant energy
3. Thermal energy
4. Motion energy
5. Sound energy

Potential energy:

the amount of energy stored in a body at rest.

Kinetic energy:

the energy of motion.

FIGURE 5-2: Humpty Dumpty has the potential energy that will eventually lead to a great fall.

© Cengage Learning 2012

Temperature:

a measure of the average kinetic energy of the molecules of a substance.

ELECTRICAL ENERGY Electrical energy is the flow of tiny charged particles we call *electrons*. We commonly refer to this energy as *electricity*. Electricity provides the energy we use every day for lighting, cooking, heating, entertainment, and more. More information on electricity can be found in Chapter 6.

RADIANT ENERGY Radiant energy is electromagnetic energy that travels in waves. Examples of radiant energy are light, radio waves, microwaves, and X-rays. The sun produces massive amounts of electromagnetic energy through nuclear fusion. This energy travels 93 million miles through the vacuum of space as waves before reaching Earth.

FIGURE 5-3: Space shuttle *Discovery* lifts off on April 5, 2010, to begin a 15-day mission to the International Space Station.

Image credit: Scott Andrews, courtesy of NASA.

THERMAL ENERGY Thermal energy is the internal energy of a body or a substance resulting from the vibration and movement of atoms and molecules. You can think of these atoms and molecules as bumper cars. If the cars could only travel at a snail's pace, few collisions or vibrations would occur. However, if the cars could travel at the speed of a race car, the resulting collisions or vibrations would be astounding. The thermal energy of a substance is measured by the intensity of these vibrations. Everything in the universe has some thermal energy because an absence of thermal energy implies a temperature of absolute zero.

MOTION ENERGY Motion energy is the kinetic energy contained in an object or a substance that is moving. The movement of a train, the flight of a rocket (Figure 5-3), and a baseball flying through the air are all examples of motion energy. Wind energy is also an example of motion energy.

SOUND ENERGY Sound energy is energy that travels in longitudinal waves that expand and compress the substances it travels through. Examples of sound energy include ultrasound and audible sound.

WHY ENERGY IS IMPORTANT

Nothing happens without energy. Humans have been using energy to do work for them for thousands of years. Early civilizations used fire for cooking and to provide light and warmth. Centuries of explorers and traders took advantage of wind energy to power their ships around the globe. The conveniences of modern life are made possible by the abundance of energy at our fingertips. Think about the energy you used since you woke up this morning. Did you turn on a light, take a bus to school, use a cell phone, or listen to an iPod? Did you eat anything this morning? If you did, the food you ate provides the energy for your brain to function and the energy for you to get to your next class. Can you think of any other things you did this morning that required energy?

HEAT

Heat (usually denoted by the letter Q) is the flow of, or potential flow of energy from a higher-temperature substance to a lower-temperature substance. Heat is a form of energy and is measured using the same units that are used for work and mechanical energy.

Heat:
the flow of, or potential flow of energy from a higher-temperature substance to a lower temperature substance.

Heat Transfer

Trying to collect and store usable heat is like trying to carry water using a leaky bucket. Both will eventually escape from their containers.

Imagine you ran out of hot water to take a bath and your only option is to fill the bathtub with ice cold water. Your body temperature is roughly 98.6°F. Let's say the temperature of the water is 50°F. As you probably have guessed, when you get into the bath, you're going to become very cold. Do you know why? The answer is heat flow. Heat always flows from hot to cold. Because the temperature of your body is higher than the water temperature, the heat will flow from your body into the water. Over time, the water temperature will increase and your body temperature will decrease.

Why do you wear shorts in the summertime and long pants in the winter? The answer to this question is also heat flow. If it's really cold outside and your body temperature is higher than the air temperature, then the heat your body produces will flow into the air. To prevent this heat from escaping, you can wear pants, a jacket, a hat, and gloves. If it's warm outside, and your body temperature is closer to the air temperature, you want your body to easily transfer its heat to the air. The solution to this problem is to expose more of your skin to the air by wearing shorts and a T-shirt.

If you make a cup of hot cocoa and set it out on the table, why does is get colder? A more technically correct question would be, why does it get less warm? The answer to this, too, is heat flow. Because the liquid is warmer than the air in your kitchen, the heat will flow from your hot cocoa into the air. You won't notice any change in air temperature from one cup of hot cocoa, but imagine if you had 1000 cups of hot cocoa sitting on the table.

Heat transfer can occur in three forms: *conduction*, *convection*, and *radiation*.

CONDUCTION When a cast-iron pan is placed over an open flame, heat slowly moves up the handle toward the cooler end (Figure 5-4). On a molecular level, the molecules in the pan initially have much more kinetic energy than the molecules in the end of the handle. These molecules are vibrating rapidly, colliding with other molecules next to them. A molecule or atom's individual kinetic energy is transferred up the handle of the pan as faster moving atoms and molecules transfer their energy to slower moving atoms and molecules. These in turn vibrate faster, causing adjacent slower molecules farther up the handle to also vibrate faster. This process continues as the kinetic energy of the molecules is transferred all the way to the end of the handle. This process is known as *thermal conduction*. Thermal conduction occurs when heat (thermal energy) is transferred within a substance through particle to particle collisions.

All materials conduct heat to some degree, but some materials are better thermal conductors than others. Materials such as glass and Styrofoam are poor conductors of heat, whereas metals such as steel and aluminum are excellent conductors. Generally, materials that are good conductors of electricity also make good thermal conductors. Copper and aluminum are good thermal and electrical conductors.

CONVECTION The transfer of heat through the movement of warmed matter in a fluid is called convection. The simplified graphic in Figure 5-5 shows warm air rising from the land and moving over the cooler waters of the sea where it begins to cool. At night the process is reversed as the warmer waters of the sea heat the air above, causing the air to rise and move toward the cooler land where the air is cooled. The convection currents will continue until the land and water are the same temperature.

FIGURE 5-4: When a cast iron pan is placed over an open flame, heat slowly moves up the handle toward the cooler end.

© Cengage Learning 2012

Thermal conduction:

occurs when heat (thermal energy) is transferred within a substance through particle-to-particle collisions.

Convection:

the transfer of heat through the movement of warmed matter through a fluid.

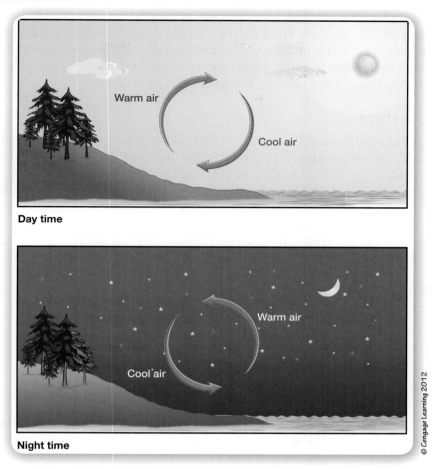

© Cengage Learning 2012

RADIATION You can't escape radiation and its effects. When you warm a meal in a microwave, play outside on a sunny day, get an X-ray, or listen to the radio, you are being bombarded with radiation (see Figure 5-6). Radiation refers to the energy transferred by electromagnetic waves and is commonly referred to as **electromagnetic radiation**. Radiation has the ability to transfer energy through empty space unlike convection and conduction, which both need matter to transfer energy. **Thermal radiation** is the transfer of heat (thermal energy) from one place to another through electromagnetic waves.

FIGURE 5-6: **The sun is constantly radiating, or sending out, an enormous amount of energy.**

Image copyright Yuri Arcurs, 2010. Used under license from Shutterstock.com.

Shaking, Rattling, and Rolling: How Temperature and Kinetic Energy Are Related

Let's take another look at the bumper car example. Each bumper car represents an atom or a molecule inside an object or a substance. If the bumper cars don't move, no work is done and no energy is transferred. When the bumper cars are moving, they contain kinetic energy. If they bump into one another, the kinetic energy of one car is transferred to another. The faster the bumper cars move, the more work that is being done and the more energy that is being transferred.

There are approximately 7,900,000,000,000,000,000,000,000 (7.9×10^{24}) molecules of water (H_2O) in 1 cup of water. Because it would be impossible to keep track of the kinetic energy of each molecule, measuring the temperature of water is actually measuring the *average* kinetic energy of all of those molecules. When the temperature of water increases, those molecules vibrate, spin, and slam against each other more often and with greater force. As the temperature of the water decreases, the average kinetic energy of those molecules decreases.

Temperature

Everyone has their own definition of hot and cold. You might think it is cold outside when the temperature is below 60°F, and someone else might think below 30°F is cold. Imagine if meteorologists only told you "It's going to be hot today" or "It's going to be cold today" instead of telling you the exact temperature. The term "hot" to a person living in Southern California might mean a temperature above 110°F, but it might mean a temperature above 80°F for an Alaskan native. The use of thermometers and temperature scales eliminates the confusion caused by using terms such as *hot* and *cold* to describe temperature.

THERMOMETERS Today, three types of thermometers are used to measure temperature:

1. Mercury and alcohol
2. Digital
3. Infrared

Each type is appropriate for certain applications.

FIGURE 5-7: This outdoor thermometer uses a red liquid that expands when warmed. The liquid rises inside the capillary tube, which is calibrated to show temperatures for easy reading.

Mercury and Alcohol Thermometers Mercury is metal that exists in a liquid state at room temperature. Mercury thermometers—mercury encapsulated in glass—take advantage of the fact that this metal, like many substances, expands when heated and contracts when cooled. A mercury-filled bulb at the bottom of the thermometer is connected to a thin, evacuated tube. When the temperature increases, the mercury expands and is forced upward into the glass tube, which is marked with a calibrated scale that allows the temperature to be read. Because of the health risks posed by mercury, it has been replaced over the last few decades with nontoxic liquids, which are usually colored red. An alcohol thermometer like the one shown in Figure 5-7 is now commonly used to measure outdoor air temperature.

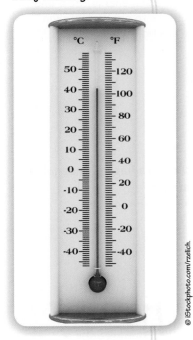

© iStockphoto.com/rzelich

Digital Thermometers If you're running a fever, chances are you would measure your temperature using a digital thermometer (see Figure 5-8). These devices use a sensor called a *thermistor* to measure temperature. As this sensor becomes warmer, its electrical resistance decreases. A microprocessor (a tiny computer) correlates the resistance to a temperature, which is displayed on the screen.

Infrared Thermometers Infrared thermometers are commonly used in high-temperature applications such as furnaces and kilns and in situations where using other types of thermometers is not possible. You are radiating energy in the form of electromagnetic waves right now. These waves are more commonly referred to as *infrared radiation*. You can feel this radiation when you lay out in the sun as the heat is transferred to your skin through infrared radiation. An infrared thermometer has a lens that focuses the infrared radiation onto a thermopile, which is an electronic device that can detect the intensity of infrared radiation. A microprocessor converts

FIGURE 5-8: A digital thermometer.

© iStockphoto.com/Charles Brutlag.

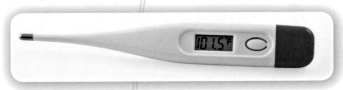

the measured infrared radiation into a temperature value and displays it on an LCD screen. There are also infrared cameras that can detect the radiation emitted from an object and then display it (see Figure 5-9). A few applications of infrared cameras include:

▶ firefighting (to detect hot spots),

▶ search and rescue (to find missing persons),

▶ contracting (to conduct home energy audits), and

▶ law enforcement (for night vision).

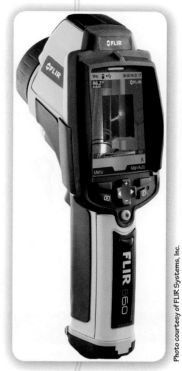

FIGURE 5-9: A thermal imaging camera.

Photo courtesy of FLIR Systems, Inc.

TEMPERATURE SCALES What is your body temperature? If you live in the United States, your answer would most likely be 98.6 degrees Fahrenheit (98.6°F). However, if you live most other places in the world, your answer would most likely be 37 degrees Celsius (37°C). Both numbers describe the same intensity of heat; they just do so using different scales.

Fahrenheit and Celsius Scales The Fahrenheit scale sets the boiling point of water equal to 212° and the freezing point of water equal to 32°, whereas the Celsius scale sets the boiling point of water equal to 100° and the freezing point of water equal to 0° (see Figure 5-10).

FIGURE 5-10: *Celsius versus kelvin temperature scales.*

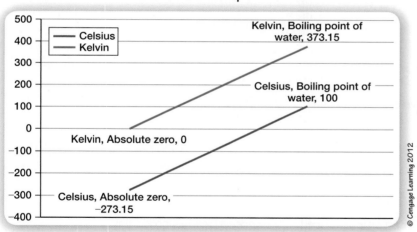

© Cengage Learning 2012

Kelvin Scale Developed by Scottish physicist William Thompson, the kelvin scale is widely used by scientists and engineers. Each degree on the scale is called a kelvin (K). By international agreement, the kelvin is never referred to as degrees kelvin or kelvin degrees. Instead, a temperature of 300 kelvins would be written as 300 K (not 300°K). The magnitude of 1 kelvin is equal to that of 1 Celsius degree. The difference between the boiling and freezing points of water is 100°C or 100 K. Science has proven that there is an absolute lowest temperature, which is called absolute zero. This is the lowest point on the kelvin scale and is defined as zero (0 K). On the kelvin scale, the freezing point of water is 273.15 K, so to convert from Celsius (T_C) to kelvin (T_K), you must add 273.15 to the Celsius temperature as shown in Equation 5-1.

$$T_K = T_C + 273.15 \qquad \text{(Equation 5-1)}$$

FIGURE 5-11: *Temperature-conversion formulas.*

	From Fahrenheit (T_F)	From Celsius (T_C)	From kelvin (T_K)
To Fahrenheit (T_F)		$T_F = \frac{9}{5} T_C + 32$	$T_F = \frac{9}{5} T_K - 459.67$
To Celsius (T_C)	$T_C = \frac{5}{9}(T_F - 32)$		$T_C = T_K - 273.15$
To kelvin (T_K)	$T_K = \frac{5}{9}(T_F - 32) + 273.15$	$T_K = T_C + 273.15$	

CONVERTING TEMPERATURES Converting from one temperature scale to another is very useful. For example, if you were to visit Canada, where the Celsius scale is used to measure temperature, you could use the simple formulas shown in Figure 5-11 to easily convert the temperature to the Fahrenheit scale.

$$T_F = \frac{9}{5} T_C + 32$$

Example

Problem: If the temperature in your classroom is 68°F, what is the temperature in degrees Celsius and in kelvin?

a. To find the temperature in degrees Celsius, use Equation 5-2:

$$T_C = \frac{5}{9}(T_F - 32) \qquad \text{(Equation 5-2)}$$

By substituting 68°F for T_F, we can solve for T_C:

$$T_C = \frac{5}{9}(68°F - 32)$$

$$T_C = 20°C$$

b. Use Equation 5-3 to convert the temperature from degrees Celsius to kelvin:

$$T_K = T_C + 273.15 \qquad \text{(Equation 5-3)}$$

$$T_K = 20°C + 273.15$$

$$T_K = 293.15 \text{ K}$$

FINDING CHANGES IN TEMPERATURES Change in temperature indicates how much heat a system has gained or lost. When the temperature of a system increases, it is a positive change in temperature. When the temperature of a system decreases, it is a negative change in temperature.

UNITS OF ENERGY

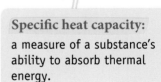

The Btu

In the United States, energy production and consumption is commonly measured in **Btu** or *British thermal units*. The energy in 1 Btu is capable of raising the temperature of 1 pound of water 1°F. Therefore, it would take 10 Btu of energy to raise the temperature of 10 pounds of water by 1°F or to raise the temperature of 1 pound of water 10°F. Another common unit of energy that appears on natural gas utility bills in the U.S. is the *therm*. 100,000 Btu is equal to 1 therm.

CALCULATING THE HEAT NEEDED TO CHANGE TEMPERATURE To calculate how much heat (energy) is required to change the temperature of a substance you must know:

- ▶ the mass of the substance,
- ▶ the desired change in temperature,
- ▶ the specific heat capacity of the substance.

Specific heat capacity (c) is a measure of a substance's ability to absorb thermal energy. See Figure 5-12 for specific heat capacities of common liquids and solids.

As shown in Equation 5-5, the amount of heat (Q) is equal to the mass (m) of the substance times the substance's specific heat capacity (c) times the substance's change in temperature (ΔT):

$$Q = mc\Delta T \qquad \text{(Equation 5-5)}$$

> **Specific heat capacity:**
> a measure of a substance's ability to absorb thermal energy.

The common units for specific heat capacity are J/kg·°C, J/kg·K, and kcal/kg·°C in the metric system and Btu/lb·°F in the U.S. customary system. For example, the specific heat capacity of water is 1 kcal/kg·°C, 4,190 J/kg·°C, and 1 Btu/lb·°F.

The Joule

The SI unit for work and energy is the **joule (J)**; named in honor of the English physicist James Prescott Joule. During his studies into the nature of heat in the 1840s, Joule discovered that the temperature of a system can be raised by adding energy in the form of heat or by performing work on the system. In terms of

> **Joule (J):**
> SI unit for work and energy.

FIGURE 5-12: Specific heat capacities of common liquids and solids.

From Serway, *Physics for Scientists and Engineers, 8e.* 2010 Brooks/Cole, a part of Cengage Learning, Inc. Reproduced with permission. www.cengage.com/permissions.

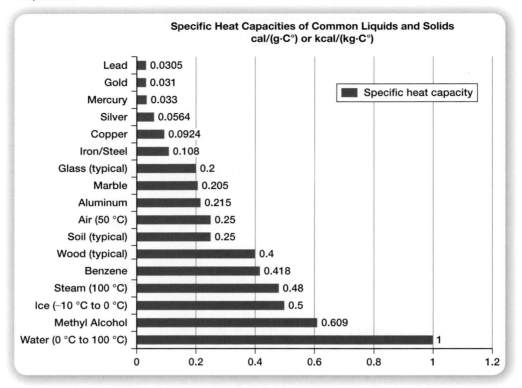

Specific Heat Capacities of Common Liquids and Solids
cal/(g·C°) or kcal/(kg·C°)

Material	Specific heat capacity
Lead	0.0305
Gold	0.031
Mercury	0.033
Silver	0.0564
Copper	0.0924
Iron/Steel	0.108
Glass (typical)	0.2
Marble	0.205
Aluminum	0.215
Air (50 °C)	0.25
Soil (typical)	0.25
Wood (typical)	0.4
Benzene	0.418
Steam (100 °C)	0.48
Ice (−10 °C to 0 °C)	0.5
Methyl Alcohol	0.609
Water (0 °C to 100 °C)	1

mechanical work, one joule is equal to the amount of work that is accomplished when a force of 1 newton acts over a distance of 1 meter.

The Calorie

calorie (cal):

the amount of heat needed to raise the temperature of one gram of water by one degree Celsius.

Another common unit for measuring heat is the **calorie (cal)**, which is equivalent to 4.19 J. One calorie is equal to the amount of amount of heat needed to raise the temperature of 1 gram of water by 1 degree Celsius. Using Equation 5-5, you can see that if the mass of 1 gram of water (0.001 kg) is multiplied by the specific heat capacity of water (1.00 J/kg·°C) times the change in temperature (1°C), the amount of heat needed to raise the temperature by 1 degree Celsius is 1 calorie:

$$Q = mc\Delta T$$

$$Q = (0.001 \text{ kg})(4{,}186 \text{ J/kg} \cdot °\text{C})(1°\text{C})$$

$$Q = 4.19 \text{ J or 1 cal}$$

Point of Interest

Food energy is expressed in Calories. The food Calorie (capitalized to distinguish it from the gram calorie) is a measurement of the food's energy content made available through respiration. The food Calorie (Cal) is equal to 1 kilocalorie (kcal) or 1,000 cal. The United States Government requires food manufacturers to label the energy content of their products in food Calories, thus allowing customers to control their energy intake.

Problem: The swimming pool at Brockport High School holds roughly 198,000 gallons of water and is kept at a temperature of 80°F year round using a natural gas heater. If you were to completely drain and then refill the pool with 50°F water, how much energy would be required to heat the water back to 80°F, and how much would that energy cost?

Step 1. Knowing that 1 British thermal unit (Btu) is the amount of energy required to raise the temperature of 1 pound of water 1°F, we must first find the mass of the water in the pool using Equation 5-6. One gallon of water has a mass of 8.35 lb.

$$\text{Mass of water in pool} = \text{number of gallons} \times \text{mass per gallon} \qquad \text{(Equation 5-6)}$$
$$= 198,000 \text{ gallons} \times 8.35 \text{ lb/gallon}$$
$$= 1,653,300 \text{ lb}$$

Step 2. Next, we must calculate the amount of energy needed to heat the pool using Equation 5-5. Remember, 1 therm is equal to 100,000 Btu:

$$\text{Energy to heat pool (Btu)} = mc\Delta T$$
$$= 1,653,300 \text{ lb} \times 1 \text{ Btu/lb} \cdot {}°\text{F} \times (80°\text{F} - 50°\text{F})$$
$$= 49,599,000 \text{ Btu or } 495.99 \text{ therms}$$

Step 3. Natural gas is purchased in therms. Assuming natural gas costs $0.74892* per therm, find the cost of heating the pool water using Equation 5-7.

$$\text{Ideal cost of heating the water} = \text{Therms} \times \text{Cost per therm} \qquad \text{(Equation 5-7)}$$
$$= 495.99 \text{ therms} \times \$0.74892 \text{ per therm}$$
$$= \$371.46$$

NOTE: Water heaters cannot convert all of the chemical energy contained in natural gas to thermal energy. Therefore, the actual cost of heating the pool would be greater than $371.46 because of the inefficiencies in the system. Newer water heaters have higher efficiency ratings than older models, meaning they can convert more of the chemical energy in the natural gas to thermal energy.

---

*The cost of natural gas varies because of supply and demand among other factors. The cost of natural gas can usually be found on your local utility's website.

COMPARING INCANDESCENT AND FLUORESCENT BULBS

Americans have waged war on the high cost of energy by making improvements to their homes that will reduce their energy consumption. An easy and inexpensive way to reduce energy costs is to replace energy-hogging incandescent light bulbs with efficient compact fluorescent light bulbs (CFLs). The graph shown in Figure 5-13 compares the power consumption of incandescent and compact fluorescent light bulbs. When you purchase a light bulb, what you are really interested in is how much light it will produce. The intensity of the light produced is measured in *lumens*. The higher the lumen rating, the more light the bulb will produce.

FIGURE 5-13: Incandescent versus compact fluorescent lightbulbs.

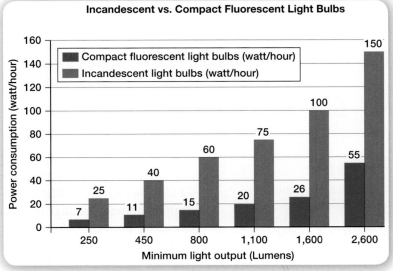

© Cengage Learning 2012

© iStockphoto.com/Jill Fromer.

Watt (W):

the SI unit of power.

For example, if you go the store and find the 100-watt incandescent lightbulb shown in Figure 5-14, you will see it produces 1,600 lumens. In comparison, a compact fluorescent bulb that also produces 1,600 lumens only consumes 26 watts.

Power (P) is the amount of energy conversion or transfer (Q) that occurs per unit time (t) as shown in Equation 5-8. The **watt (W)** is the SI unit of power that measures the rate of energy conversion. One watt is equal to 1 joule of energy per second (see Equation 5-9).

$$P = Q/t \qquad \text{(Equation 5-8)}$$

$$1 \text{ watt} = 1 \text{ joule}/1 \text{ second} \qquad \text{(Equation 5-9)}$$

Real-World Examples

In the U.S. customary system, 1 watt is equal to 3.414 Btu/hr. To find the amount of energy in British thermal units (Btu) needed to light a bulb for a period of 1 hour, you must multiply the light bulb's wattage rating by 3.414 Btu/hr and the amount of time (in hours) that the bulb will be illuminated.

Example

Problem: According to the graph in Figure 5-13, a 75-watt incandescent light bulb and a 20-watt compact fluorescent bulb give off 1,100 lumens. How much more thermal energy (in Btu) is given off by the 75-watt bulb per hour compared to the 20-watt CFL?

a. To find out how much thermal energy is given off by the 75-watt bulb in one hour:

$$75 \text{ W} \times \frac{3.414 \text{ Btu/hr}}{1 \text{ W}} \times 1 \text{ hr} = 256.05 \text{ Btu}$$

b. To find out how much thermal energy is given off by the 20-watt compact fluorescent light bulb in one hour:

$$20 \text{ W} \times \frac{3.414 \text{ Btu/hr}}{1 \text{ W}} \times 1 \text{ hr} = 68.28 \text{ Btu}$$

c. To determine how much more thermal energy is given off by the incandescent light bulb:

$$256.05 \text{ Btu} - 68.28 \text{ Btu} = 187.77 \text{ Btu}$$

Power (P):

the amount of energy conversion or transfer that occurs per unit time.

In the previous example, you learned that switching from a 75-watt incandescent bulb to a 20-watt CFL amounts to a savings of 187.77 Btu for every hour the CFL is used. Over the lifetime of this bulb (10,000 hours), it would save 1,877,700 Btu. This energy savings is equivalent to the amount of heat energy that is produced by burning 185 pounds of coal. Think about it—that's an average-sized adult's weight in coal over the lifetime of one CFL.

Example

Problem: Switching from a 75-watt incandescent bulb to a 20-watt CFL saves 187.77 Btu for every hour it is used. If the average life expectancy of a CFL is 10,000 hours, calculate the energy savings in pounds of coal. The thermal energy generated by burning coal is approximately 10,084.5 Btu per pound.

Step 1. To determine energy savings over the life of the compact fluorescent bulb:

$$187.77 \text{ Btu/hr} \times 10,000 \text{ hr} = 1,877,700 \text{ Btu}$$

Step 2. To convert energy savings from Btu to pounds of coal:

$$\text{Energy savings} = \frac{1,877,700 \text{ Btu}}{10,084.5 \text{ Btu/lb}}$$

$$= 186.2 \text{ pounds of coal}$$

Example

Problem 1: If cold water enters your house at a constant temperature of 52°F (11.1°C) and an average bathtub holds roughly 50 gallons (189.3 liters) of water, how much energy (in joules) is needed to raise the temperature of the bathwater to 100°F (37.8°C)? Remember, water has a specific heat capacity of 1 Btu/lb · °F (4,186 J/kg · °C) and 1 gallon of water has a mass of 8.35 lb (3.79 kg).

$$Q = mc\Delta T$$

$$Q = (50 \text{ gal} \times 3.79 \text{ kg/gal}) \times 4,186 \text{ J/kg} \cdot °C \times (37.8°C - 11.1°C)$$

$$Q = 2.12 \times 10^7 \text{ J}$$

Problem 2: A cubic foot of natural gas contains 1.071×10^6 joules of energy. It will take 19.79 cubic feet of natural gas to heat 189.3 liters of water. If the cost of natural gas is $0.74892 per therm, what is the ideal cost of heating the bathwater?

Step 1. Using Equation 5-5, calculate the amount of energy needed to heat the bathwater in therms:

$$\text{Energy to heat bathwater (Btu)} = mc\Delta T$$

$$= (50 \text{ gal} \times 8.35 \text{ lb/gal}) \times 1 \text{ Btu/lb} \cdot °F \times (100°F - 52°F)$$

$$= 20,040 \text{ Btu or } 0.2004 \text{ therm}$$

Step 2. Using Equation 5-7, calculate the cost of heating the bathwater:

$$\text{Ideal cost of heating bathwater} = \text{Therms} \times \text{Cost per therm}$$

$$= 0.2004 \text{ therm} \times \$0.74892 \text{ per therm}$$

$$= \$0.15$$

THE LAWS AND FUNDAMENTALS OF THERMODYNAMICS

Thermodynamics:

a branch of physics that is based on the fundamental laws of nature concerning energy and mechanical work.

Thermodynamics is a branch of physics that is based on the fundamental laws of nature concerning energy and mechanical work. The history of the science of thermodynamics is closely tied to the invention and development of the steam engine. *Thermo* means heat, and *dynamic* means movement. However, thermodynamics encompasses much more than just the study of heat transfer.

Zeroth Law of Thermodynamics: Thermal Equilibrium

When no heat flows between two thermodynamic systems that are in contact with each other, the systems are in *thermal equilibrium*. The *zeroth law of thermodynamics* states that if two bodies (A and B) are in thermal equilibrium with a third body (C), then all three bodies are in thermal equilibrium with each other (see Figure 5-15). This law seems rather obvious, but its purpose is to validate the use of temperature measurements to verify a state of thermal equilibrium between two systems that are not in direct contact.

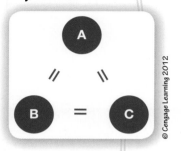

FIGURE 5-15: **Thermal equilibrium.**

© Cengage Learning 2012

First Law of Thermodynamics: Conservation of Energy

Conservation of energy:

the scientific law that states that energy can be neither created nor destroyed.

We've spent a lot of time talking about energy flow and transformation. The *first law of thermodynamics*—the **conservation of energy**—states that the total amount of energy never changes. Energy can be neither created nor destroyed. This means that the total amount of energy entering a system must equal the total amount of energy exiting a system (see Figure 5-16). Energy can flow and change its form, but the total amount of energy never changes. If this is true, then why are we so concerned about using less energy? The answer to this question lies in the second law of thermodynamics.

FIGURE 5-16: **Energy can be neither created nor destroyed.**

© Cengage Learning 2012

Second Law of Thermodynamics: The Usefulness of Energy

The *second law of thermodynamics* explains why energy cannot be reused again and again. It states that as energy changes forms, its "usefulness" decreases.

Point of Interest

Let's take a look at how the chemical potential energy contained in an energy bar (Figure 5-17a) is transformed into kinetic energy by your body (Figure 5-17b). Energy bars are packed with carbohydrates. The carbohydrate is perhaps the most important source of energy for athletes. No matter what sport you play, carbohydrates provide the energy you need to operate your muscles. After you eat carbohydrates, your body breaks them down into simpler sugars (glucose, fructose, and galactose). These sugars are absorbed into your bloodstream. The cells in your muscles convert the chemical energy in these sugars to mechanical energy (motion) and thermal energy (heat). If you don't need the glucose right away, it gets stored in your muscles and liver in the form of glycogen. Once these stores are filled up, any extra glycogen is stored as fat.

(Continued)

FIGURE 5-17: **(a) Energy bars contain carbohydrates that (b) athletes use to boost their mechanical energy.**

(a)

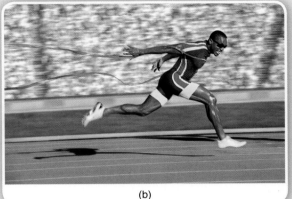

© iStockphoto.com/caziopela.

© iStockphoto.com/Pete Saloutos.

(b)

Entropy is the measure of the amount of unusable energy in a system. The second law of thermodynamics also states that the efficiency of any system must always be less than 100%.

For example, one gallon of jet fuel contains roughly 130,000 Btu of energy. The engines of a jet aircraft like the one shown in Figure 5-18 are capable of using the fuel's highly concentrated energy to propel the aircraft. Some of the energy is converted directly into waste heat that is dissipated into the atmosphere. Even the energy used to propel the aircraft is converted back into heat as the aircraft encounters the friction of the air molecules slightly heating the air as it passes through it. Eventually, every bit of energy is transformed into "low-grade" energy in the form of heat. The energy does not disappear—it is transformed into a less useful form. The heat produced by the aircraft will eventually be radiated away from Earth into outer space.

Entropy:

a measurement of the amount of unusable energy in a system.

Third Law of Thermodynamics: Heat Flow

The *third law of thermodynamics* states that it is impossible to reach a temperature of absolute zero. It is possible to come close to absolute zero, but it is not possible to actually reach it by any process or series of processes. A loophole in the second law of thermodynamics would allow a system to be 100% efficient if absolute zero could be reached, but the third law of thermodynamics closes this loophole.

Fundamentals of Thermodynamics

A **thermodynamic system** is a quantity of matter or a region in space that is separated by a boundary from its surroundings. For example, the coffee inside a thermos

FIGURE 5-18: **An F-15E Strike Eagle U.S. Air Force photo/Staff Sgt. Aaron Allmon.**

© Cengage Learning 2012

FIGURE 5-19: A thermos is a great example of a thermodynamic system.

© iStockphoto.com/YouraPechkin.

(see Figure 5-19) is a thermodynamic system. Everything outside the thermos is the *surroundings*. The *boundary* is the thermos, which is located between the system and the surroundings.

Thermodynamic systems can be open or closed. An **open thermodynamic system** allows mass and energy to flow freely through the boundary. Your lungs are an example of an open thermodynamic system. Air that enters your lungs is warmed by your body and is eventually pushed out. A **closed thermodynamic system** does not allow mass to cross the boundary from the system to its surroundings (or vice versa), but it does allow energy to be transferred across the boundary in the form of work or heat. An example of a closed thermodynamic system would be a light bulb. Electrical energy, light, and heat pass through the boundary, but the mass contained within the bulb cannot pass through the boundary.

RENEWABLE ENERGY SOURCES

Renewable energy sources are those that can be replenished in short order. As shown in Figure 5-20, the five most common renewable energy sources are biomass, hydroelectric power, geothermal, wind, and solar.

Renewable energy source: any source of energy that can be replenished in short order.

Biomass

Living or recently living plant material is the source of all biomass energy or *bioenergy*. It has been used for as long as humans have been burning wood to keep warm and cook food. Although wood is the largest biomass energy resource today, other plants and plant-derived materials can also be used. Biomass fuel sources include:

▶ living or recently living plant material,

▶ dedicated energy crops,

▶ fast-growing plant matter,

▶ forest, mill, and agricultural waste,

▶ organic consumer waste, and

▶ organic industrial waste.

BIOMASS POWER

Biomass power, or **biopower**, uses the chemical energy stored within biomass to generate electricity. Biopower plants use technologies including direct firing, co-firing, gasification, pyrolysis, landfill gases, and anaerobic digestion.

Direct Firing The majority of biopower plants use *direct-firing* systems that work by burning biomass to produce steam that is used to turn the blades of turbines. The turbines are connected to generators that produce electricity. To increase the overall energy efficiency of the system, some direct-firing systems use the steam from their power plants to aid in manufacturing processes or to heat buildings.

Co-Firing *Co-firing* refers to blending biomass with fossil fuels in existing power plants. Coal-fired power plants can use co-firing systems to significantly reduce emissions.

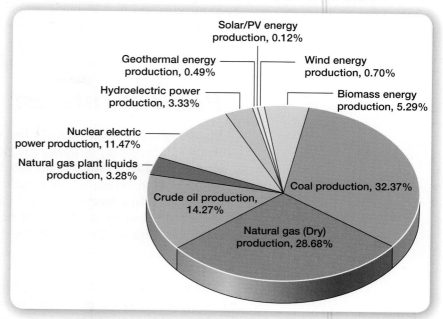

FIGURE 5-20: U.S. energy production by fuel source.
From Energy Information Administration, Annual Energy Review 2008, www.eia.goe.gov, accessed May 26, 2010.

Solar/PV energy production, 0.12%
Geothermal energy production, 0.49%
Wind energy production, 0.70%
Hydroelectric power production, 3.33%
Biomass energy production, 5.29%
Nuclear electric power production, 11.47%
Natural gas plant liquids production, 3.28%
Coal production, 32.37%
Crude oil production, 14.27%
Natural gas (Dry) production, 28.68%

Gasification *Gasification* systems expose biomass to high temperatures in an oven with limited oxygen to convert biomass into synthesis gas, a mixture of hydrogen and carbon monoxide. The synthesis gas, or *syngas,* can then be burned in a conventional boiler, chemically converted into other fuels or products, or used to replace natural gas in a gas turbine to generate electricity.

Pyrolysis *Pyrolysis* uses a thermochemical process that is similar to gasification. In pyrolysis, oxygen must be completely excluded while the biomass is exposed to high temperatures. This will pyrolyze biomass to a liquid rather than gasify it. As with syngas, pyrolysis oil can be used to generate electricity or as a chemical source for making plastics, adhesives, and other bioproducts.

Landfill Gases A landfill might seem like an unlikely energy source, but the natural decay of biomass produces *methane gas* that can be captured and used for power production. Wells can be drilled in landfills to release the methane from decaying organic matter. The pipes from each well transport the methane gas to a central processing facility where it is filtered and cleaned before burning. This process not only produces electricity but also reduces the release of methane (a damaging greenhouse gas) into the atmosphere.

Anaerobic Digestion *Anaerobic digestion* can also be used to produce methane from biomass. Naturally occurring bacteria are used to decompose biomass in the absence of oxygen in closed vessels. This decomposition produces methane gas that is suitable for power production. The process can also be used to convert possibly troublesome wastes, such as those found in sewage-treatment plants, into usable compost.

ETHANOL

As a nation, our concerns about our dependence on foreign oil, global warming, offshore drilling, smog, and the environmental and political costs of our nation's 20,000-barrel-a-day petroleum addiction have never been higher. What if a solution to all of those problems is already here and ready to fuel the cars we drive? By using our abundant, renewable resources, we can produce alternative fuels to fill up our gas tanks rather than continuing to use fossil fuels. These alternative fuels add no net greenhouse gases to the atmosphere.

Ethanol is an alternative fuel produced from renewable and sustainable resources, and it has already proved to be realistic, efficient, and economical. Ethanol is made by fermenting and then distilling simple sugars from corn, sugarcane, or cellulosic feedstock such as agricultural waste, paper pulp, and switchgrass.

Point of Interest

The thought of using ethanol to fuel cars is as old as the automobile itself. Henry Ford originally designed the Model T to run on a mix of gasoline and alcohol (ethanol), calling it the fuel of the future. Thomas Midgley, the inventor of high-octane leaded gasoline, drove a car powered by an ethanol-gasoline mix to a meeting of the Society of Automotive Engineers in 1921. Since then, various political and economical factors have prevented the widespread use of ethanol as our fuel of choice.

There are two ways to convert biomass into ethanol: sugar-platform and thermochemical conversion processes. However, some hybrid processes are also under development to combine parts of each method.

Sugar-Platform Processes All sugar-platform processes use living organisms, such as yeasts, to convert simple sugars into ethanol. Like all living organisms, they have a favorite "food" and a favorite environment that keeps them healthy. The yeasts that convert simple sugars to ethanol have been used for centuries to make beer, wine, and liquor.

Scientists have been searching for and trying to create organisms that can break down the components of biomass containing cellulose (the fibrous substance that makes up the cell walls of plants), hemicellulose (found in wood), and lignin (also found in wood) into simple sugars that can then be fermented into ethanol. Although these "designer" enzymes have been manufactured to break down cellulose and hemicellulose into simple sugars, breaking down the lignin has been more difficult. Most sugar-platform processes assume that the lignin will be used to fuel the ethanol-making process.

OFF-ROAD EXPLORATION

Conduct a study of the energy efficiency of gasoline, ethanol, and ethanol-blended fuels for each of the following factors:

▶ Miles per gallon

▶ Energy content per gallon

▶ Octane ratings

Thermochemical Processes The thermochemical process is a new one. It uses enzymes to break down the cellulose in woody fibers so that more of the plant can be used to make cellulosic ethanol. These processes transform all of the nonmineral components of the biomass into synthesis gas or syngas. The syngas is passed over a catalyst that accelerates the conversion of syngas into ethanol and other alcohols. Besides producing ethanol and other alcohols, the syngas may be used to produce synthetic petroleum products, fertilizer, plastics, and many other products. As stated earlier, it can also be used as a biofuel to generate electricity.

Hydropower

Hydroelectric power plants take advantage of a naturally occurring, continuous process called the *water cycle*. Energy from the sun evaporates water from Earth's rivers, lakes, and oceans. When the water vapor reaches cooler air in the atmosphere, it condenses and forms clusters of tiny water droplets that we call *clouds*. This moisture replenishes the water in lakes and rivers as it falls back to Earth in the form of rain or snow. Gravity does the rest of the work as the water moves from high ground to low ground, eventually making its way back into the oceans. The force of this moving water can be extremely powerful.

Hydroelectric power plants need falling water in order to operate. There is no better source of falling water than from a naturally occurring waterfall such as Niagara Falls, where the first hydroelectric power plant was constructed. If there is no natural waterfall, a human-made dam is used instead.

Dams are commonly built in areas where the walls on either side of a river can produce an artificial lake or reservoir. There are roughly 80,000 dams in the United States, but less than 3% of all dams are currently used to produce power. The rest of the dams are used solely for flood control and irrigation.

Dams can increase pressure by raising the height of the water behind the dam (also known as the *head*), and they can regulate the amount of water flowing through them. Specially designed gates called *spillway gates* restrict the amount of water flowing from the reservoir. During times of heavy rain, the spillway gates can be opened wide to prevent water from spilling over the top of the dam.

Centuries ago, people harnessed the energy of flowing water to grind grain or to run sawmills. Today, the energy of flowing water is used to generate electricity. Hydroelectric power plants use water to spin the turbines that power electric generators.

A common hydroelectric power plant has three major components (see Figure 5-21):

1. a powerhouse in which turbines spin generators to produce electricity,
2. a dam that can control the flow of water, and
3. a reservoir where water can be stored.

FIGURE 5-21: **Hydroelectric power plant.**

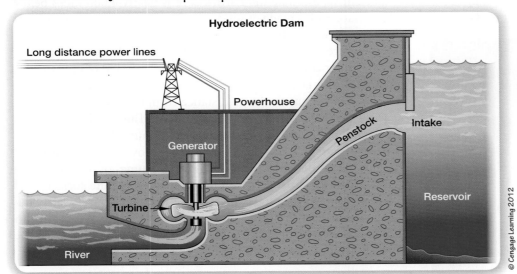

© Cengage Learning 2012

To generate electricity, a dam opens its gates to allow water from the reservoir to flow down through gigantic tubes called **penstocks**. A turbine is positioned at the end of a penstock, where the fast-moving water spins the blades of the turbine. The turbine is connected to a generator, which produces electricity. The electricity is transported over vast distances by way of power lines to local utility companies where it is sold and transported to customers.

Geothermal

Geothermal energy makes use of the heat from within Earth. The word *geothermal* comes from a combination of the greek words *geo,* meaning Earth, and *therme,* meaning heat. People use geothermal energy to heat buildings, produce electricity, and provide hot water for multiple uses.

Penstock:
a pipe or sluice that controls the flow of water from behind a dam.

OFF-ROAD EXPLORATION

▶ Research the chronology of hydroelectric power at Niagara Falls.
▶ Design a system to convert tidal energy into electricity.
▶ Design a system to convert wave energy into electricity.

FIGURE 5-22: A section view of Earth's core.

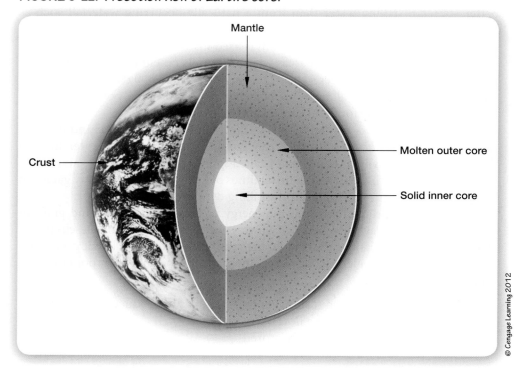

FIGURE 5-23: The Strokkur geyser in Iceland erupts about every 20 minutes.

Scientists estimate the Earth's core is nearly 4,000 miles below the surface and is made of solid iron. This inner core is surrounded by a molten iron outer core (see Figure 5-22). It is estimated that the temperature of the core is between 5,000°F and 11,000°F. Surrounding the outer core is a layer thought to be composed of molten rock (magma). This layer, called the *mantle*, is approximately 1,800 miles thick. That's about the same distance from Chicago to Las Vegas. Earth's outermost shell is called the *crust*. This insulating shell is broken in pieces called *plates* that are continuously in motion. The theory of *plate tectonics* describes how the movement of entire continents causes the crust to crack, allowing plumes of magma to rise up through the crust. Where the magma reaches the surface, volcanoes form. In locations where there is underground water, the magma can heat the water and circulate it back to the surface to create wonders such as hot springs, mud pots, and geysers like the one shown in Figure 5-23.

HIGH-TEMPERATURE GEOTHERMAL ENERGY When geothermal reservoirs (made of magma and a constant supply of water) are located near the surface, we can get to them by drilling wells. Some of the deepest wells extend more than 2 miles into Earth's crust. These reservoirs are found by drilling exploratory wells. If a reservoir is found, a larger production well is drilled to allow the hot water and steam to escape to the surface. The water and steam are used to generate electricity at power plants close to the production wells. There are four different types of geothermal power plants: flash steam, dry steam, binary, and hybrid.

Flash Steam Plants A *flash steam plant* works by pumping water into geothermal reservoirs where the water is superheated and then pumped back to the surface. When the water reaches the surface, where the pressure is significantly less, it rapidly boils and turns into steam. The steam is used to turn a turbine, which spins a generator and makes electricity. Unlike coal, oil, and nuclear power plants, flash steam plants produce no pollution.

Dry Steam Plants *Dry steam plants* use the steam escaping from the reservoir to turn a turbine, which spins a generator to produce electricity. The Geysers is a dry steam power plant in northern California that has been producing electricity since 1960. It is the largest known dry steam field in the world and supplies electricity to more than 1.8 million people.

Binary Power Plants *Binary power plants* produce electricity using hot but not boiling water. The heat from the geothermal hot water is passed through a heat exchanger that transfers the heat to another liquid to produce electricity. The water from the geothermal reservoir is pumped through the heat exchanger and then is returned back to the reservoir to be reheated and recycled. On the other side of the heat exchanger is a working fluid (usually hydrocarbons or an ammonia–water mixture) that boils at a lower temperature than water. The vapor from the boiling liquid drives turbines connected to generators to produce electricity. Because both systems are kept isolated from one another, binary power plants have the advantage of being pollution free.

Heat exchanger:
a device that efficiently transfers heat from one substance to another.

Hybrid Plants Some power plants combine flash and binary systems to make use of the steam and hot water. These power plants are referred to as *hybrid plants*. Both Hawaii and Iceland use hybrid power plants to provide electricity to thousands of homes.

GEOTHERMAL HEATING AND COOLING If you were to travel about 25 feet underground, you would find the temperature there to be fairly constant year round. If you live in a temperate region, the ground temperature remains approximately 52°F year round. This steady ground temperature will usually be cooler than the air during the summer months and warmer than the air during the winter months. The temperature difference makes geothermal heating and cooling possible.

Geothermal heating and cooling systems work by pumping water through long pipes buried underground. The water is cooled (or warmed, depending on the season) by Earth's crust and returned back to the building. The water then enters a heat exchanger, which transfers heat to the building during the winter months and transfers heat away from the building during the summer months. The water is then sent back to the pipes buried below ground to begin the process over again. Geothermal heating and cooling systems can reduce heating and cooling costs by 50% to 70%, greatly reducing the dependence on fossil fuels for heating.

Wind

Wind is an abundant and largely untapped renewable energy source. Caused by Earth's uneven heating, wind is the flow of a warmer air mass toward a cooler air mass. For example, the air over land heats up much faster than the air over the sea. Because heat always flows from hot to cold, this "sea breeze" is the reason most coastlines are ideal locations for wind turbines.

Wind turbines like the one shown in Figure 5-24 work like electric fans but in reverse. Instead of the fan making the wind, wind turbines use the wind to produce

FIGURE 5-24: *Components of a wind turbine.*

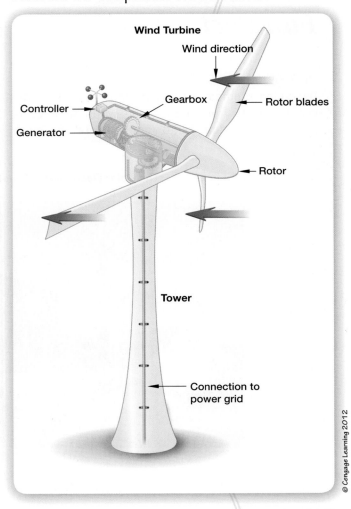

Wind Turbine
Wind direction
Controller
Gearbox
Rotor blades
Generator
Rotor
Tower
Connection to power grid

© Cengage Learning 2012

electricity. They operate in much the same way as hydroelectric turbines. Instead of water, however, wind pushes on the blades of the turbine, causing them to spin. The blades are connected to a gearbox that increases the revolutions per minute (RPMs) of a generator shaft. The generator produces electricity that can be fed into the power grid for residential and commercial customers to use.

Solar

The sun is constantly radiating, or sending out, enormous amounts of energy. This radiant energy is commonly known as *solar energy*. Only a miniscule amount (about one-billionth) of the total energy emitted by the sun strikes Earth. Even though seemingly insignificant, the sun's energy is an enormous energy source for humans.

When a car is left in the hot sun in the middle of summer, the temperature in the car will quickly rise higher than the temperature outside. This is because the sun's radiant energy passes through the windows of the car and is absorbed by the seats, walls, and floor. The glass allows the sun's radiant energy to pass through but it allows only a small amount of heat to escape through conduction. This is how a *solar collector* works. A greenhouse is an example of a solar collector. Many homeowners around the country use solar energy for space heating. Some are able to save upward of 50% off of their home-heating costs.

There are three types of solar space heating systems: passive, active, and hybrid systems that combine both passive and active elements.

Fun Fact

In 2009, the U.S. Department of Energy's Solar Decathlon program assembled 20 college and university teams from around the world in a competition to design, build, and operate the most attractive and energy-efficient solar-powered house. This winning design was created by the Technische Universität Darmstadt, in Darmstadt, Germany (see Figure 5-25).

FIGURE 5-25: The Solar Decathlon in Washington, D.C.

Jim Tetro, U.S. Department of Energy Solar Decathlon.

PASSIVE SOLAR HEATING Buildings designed using *passive solar* techniques use the structure as a solar collector. Just like a greenhouse utilizes the energy of the sun to provide heat year round, a passive solar building can do the same. Architects designing these structures are able to maximize the amount of solar energy absorbed by changing the size and placement of windows and altering the shape of the structure itself. For example, windows typically face south because the sun shines from the south in North America.

ACTIVE SOLAR HEATING *Active solar* heating systems use mechanical equipment, such as blowers and pumps, and an outside source of energy to maintain a constant temperature within a system. Most active solar systems use custom solar collectors that look like black boxes covered with glass. Black metal plates inside the boxes convert the radiant energy from the sun into usable heat. A working fluid such as air or water is pumped through the collectors and is warmed by this heat. The working fluid is used to heat the building using an ordinary forced-air system. These solar collectors are usually put high up on the south side of roofs where tall trees and buildings cannot shade them. Similar systems can be used to supplement water heating. This can save the average homeowner hundreds of dollars every year.

Photovoltaic Cells Photovoltaic (PV) cells are used in some active solar heating systems to generate electricity. PV cells are made of semiconductor materials such as silicon. When light strikes the PV cell, its energy is absorbed by the semiconductor. This generates a voltage across the cell, which allows free electrons to move through a circuit to power fans, heat pumps, and other devices in homes.

SOLAR POWER PLANTS Much like solar cells, solar power plants use solar energy to generate electricity. Because the solar radiation that reaches any one spot on Earth is so spread out, in order for a solar power plant to be viable, the light must be focused to produce the high temperatures required to generate electricity. There are currently three types of solar power plants that use mirrors or other reflective surfaces to magnify the sun's energy as much as 5,000 times its normal intensity: solar parabolic troughs, solar parabolic dish or engine systems, and solar power towers.

Solar Parabolic Troughs Solar parabolic troughs use miles of reflective troughs that focus sunlight onto a pipe located at the parabola's focal point (see Figure 5-26). A working fluid such as oil is circulated inside the pipes to collect the heat energy and transfer it to a heat exchanger. The heat exchanger heats water to produce steam, which drives a turbine connected to a generator to produce

FIGURE 5-26: Reflectors focus solar energy on the absorber tube in which a working fluid such as oil is superheated and pumped into a heat exchanger to produce steam. The steam turns a turbine connected to a generator to produce electricity.

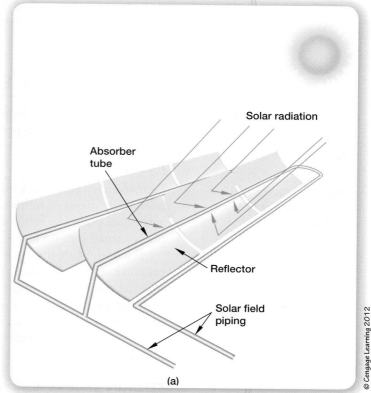

(a)

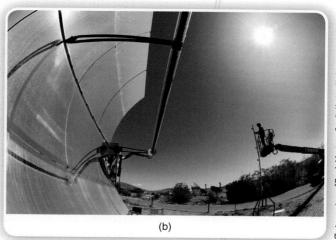

(b)

FIGURE 5-27: A solar power tower.

Image courtesy of BrightSource Energy.

electricity. The largest parabolic trough power plant in the United States is located in the Mojave Desert in California.

Solar Parabolic Dish and Engine Systems *Solar parabolic dish and engine systems* use mirrors in the shape of a dish to collect and concentrate the sun's radiant energy into a small area where a receiver is located. The receiver transfers the sun's energy to a heat engine, commonly a Stirling cycle engine that converts heat into electricity. The solar dish has the highest efficiency (30%) of all the solar technologies that have been demonstrated on a large scale.

Solar Power Towers *Solar power towers* use hundreds of large sun-tracking mirrors (heliostats) to focus the sun's energy on the receiver at the top of a tower (see Figure 5-27). A working fluid is heated in the receiver to generate steam. The steam is used in a conventional turbine generator to produce electricity.

NONRENEWABLE ENERGY SOURCES

Nonrenewable energy sources are those that cannot be replenished in our lifetimes. Examples include petroleum, natural gas, coal and nuclear energy.

Petroleum

Petroleum is a fossil fuel. Fossil fuels are appropriately named because they formed from the remains of billions of tiny plants and animals that died millions of years ago. These plants were buried by thousands of feet of sediment and sand that eventually turned into rock.

The weight of the rock above subjected this organic mixture to enormous pressure and heat, causing the mixture to undergo chemical changes. Compounds made of hydrogen and carbon atoms transformed into hydrocarbons. The end product is an oil-saturated rock that is much like a wet sponge. It's important to note that all organic material buried underground does not turn into oil; certain geological conditions must exist for the transformation to take place.

REFINING CRUDE OIL Crude oil is the oil that is taken directly from the ground. This thick, black oil must be refined into usable products such as gasoline, diesel fuel, and asphalt.

Crude oil's first stop is a refinery where it is processed. Some refineries are located near oil wells, but usually the crude oil must be transported to the refinery by ship, pipeline, barge, truck, or train.

Oil refineries clean and separate the crude oil into gasoline and hundreds of other useful products using a process called *fractional distillation*. Every refinery performs three basic steps: separation, conversion, and treatment. Figure 5-28 shows the initial separation process. Inside the distillation tower, liquids and vapors are separated into components (fractions) according to their weight and boiling point. The fractions with the lowest density, including gasoline and liquid petroleum gas (LPG), are vaporized and rise to the top of the tower, where they are drawn off and condensed back to liquid form. Medium density fractions such as kerosene and diesel stay in the middle of the distillation tower. Fractions with the highest density, called *gas oils*, have the highest boiling points and settle toward the

Nonrenewable energy source:

an energy source that cannot be replenished in our lifetime.

Fossil fuel:

fuel formed from the remains of tiny plants and animals that died millions of years ago.

bottom. The conversion process uses heat and chemicals to break apart the molecules of the heavier fractions to produce more gasoline and other petroleum fuels. Treatment is the final stage of the refining process in which technicians carefully mix selected fractions together to form various fuels and by-products.

Natural Gas

Natural gas is often considered a nonrenewable fuel source. However, methane, the main ingredient in natural gas, can be produced using renewable sources. Methane is colorless and odorless, and, most importantly, it gives off a great deal of thermal energy when burned. Like oil, natural gas is thought to have formed after the remains of tiny sea animals and plants that died millions years ago were exposed to extreme pressure and heat. After the natural gas was formed, it migrated upward through tiny pores in the surrounding rock. Some of the gas seeped to the surface, where it dissipated into the atmosphere. Layers of impermeable rock, shale, or clay trapped large amounts of natural gas, forming vast underground reservoirs. The natural gas can be extracted from these reservoirs by drilling a hole through the rock layers.

Natural gas is a flammable combination of mostly **hydrocarbon** gases. Though methane is the main component of natural gas, it also contains ethane, propane, butane, and carbon dioxide (see Figure 5-29). These other chemicals are separated during the refining process and sold individually for use in devices such as gas-powered grills.

FIGURE 5-28: A diagram of an oil-fractioning column.

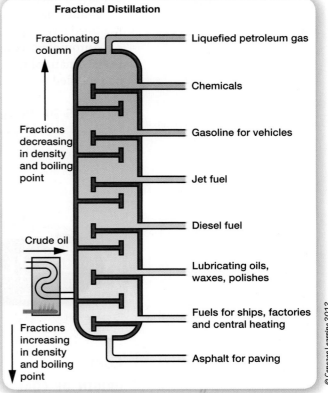

© Cengage Learning 2012

Hydrocarbon:
an organic compound that contains only carbon and hydrogen.

FIGURE 5-29: The composition of natural gas.

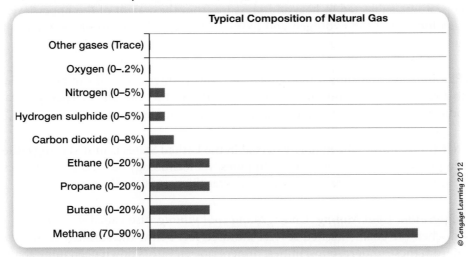

© Cengage Learning 2012

CONSUMERS OF NATURAL GAS Natural gas represents about 29% of the energy we use in the United States every year. The largest consumer of natural gas is industry, which commonly uses it as a heat source to manufacture products. It is used in the production of fertilizers, pharmaceuticals, photographic film, ink, glue, paint, plastics, laundry detergents, and insect repellent.

Residential and commercial customers are the second largest consumers of natural gas. Many homes and businesses use natural gas to run water heaters, stoves, furnaces, and clothes dryers. Natural gas is an ideal fuel to use in the home because it burns cleanly.

Natural gas can also be used to produce electricity. After coal and uranium, it is the third largest producer of electricity in the United States.

Automobile companies are now selling vehicles that are powered by compressed natural gas (see Figure 5-30). Gasoline- and diesel-powered vehicles can be modified to run on natural gas.

FIGURE 5-30: **The Honda Civic GX runs on compressed natural gas.**

Courtesy of American Honda.

Coal

ORIGIN OF COAL Coal comes from plants and other organic matter that lived during the Pennsylvanian period millions of years ago. Throughout this period, low-lying coastal swamps and deltas covered much of what is now the eastern and midwestern Unites States, creating conditions ideal for coal formation. As these plants died, they fell into the swamp and began to decay. This process continued for more than 30 million years, creating many layers of partially decayed organic matter. The weight of the upper layers compressed the lower layers to form peat. The peat was subjected to heat and pressure from the weight of the upper layers, which forced out the hydrogen and oxygen and left behind carbon rich deposits that we know today as coal.

TYPES OF COAL As shown in Figure 5-31, North American coal can be classified into four categories based on carbon content and the amount of energy it can produce: anthracite, bituminous, sub-bituminous, and lignite.

MINING THE COAL The majority of coal in the United States is mined using either underground or surface mining techniques. As technology advances, the process of extracting coal has drastically changed over the past 30 to 35 years.

Underground Coal Mining *Underground coal mining*, also called *deep mining*, is used when the coal is located several hundred feet below the surface. To remove coal from these underground mines, miners ride elevators down deep mine shafts, where they extract the coal using continuous mining machines. These machines have gigantic spinning drums with sharp teeth that cut into the walls and pulverize the coal. The coal is then loaded on a conveyor belt and transported out of the mine.

Surface Coal Mining *Surface coal mining* is less expensive than underground mining and produces the majority of the coal in the United States. It is commonly used when the coal is located less than 200 feet below the surface. The surface-mining

FIGURE 5-31: *Types of coal in North America.*

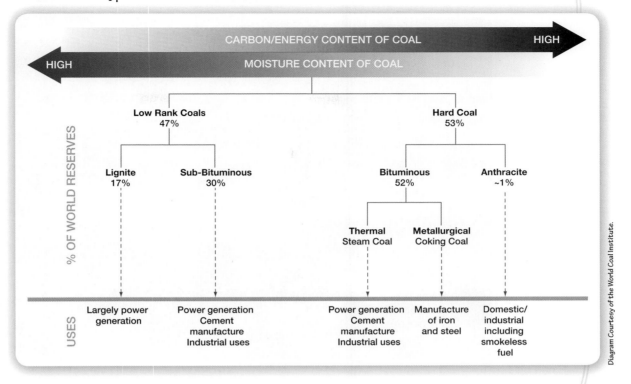

Diagram Courtesy of the World Coal Institute.

FIGURE 5-32: *A surface mining machine.*

Image copyright nng. Used under license from Shutterstock.com.

technique uses giant machines to remove the topsoil and layers of rock to expose large beds of coal (see Figures 5-32 and 5-33). When the coal is removed and the mining is finished, the dirt and rock are returned to the pit, the topsoil is replaced, and the area is replanted. The land can then be used for farming, wildlife refuge, recreation, or industry.

USES OF COAL

Generating Electricity Approximately 92% of the coal mined or imported in the United States is used to generate electricity. Coal-burning plants use the heat generated from the burning coal to boil water and produce steam, which rapidly expands to push its way through the giant turbine blades such as those shown in

FIGURE 5-34: A steam turbine during repair. When this turbine is fully operational, steam will push through the blades, causing the turbine to rotate in much the same way a pinwheel rotates when you blow on it.

Figure 5-34. This causes the blades to turn in much the same way a pinwheel turns as you blow on it. The turbine blades are connected to generators that generate electricity as they turn. Coal-burning power plants produce more than 50 percent of the total electricity generated in the United States.

FIGURE 5-35: U.S. coal usage by sector (in tons).

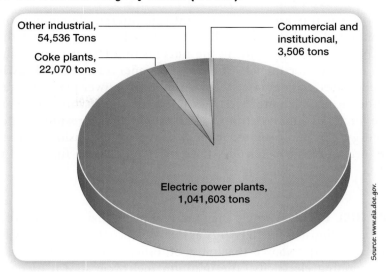

Other industrial, 54,536 Tons

Coke plants, 22,070 tons

Commercial and institutional, 3,506 tons

Electric power plants, 1,041,603 tons

Source: www.eia.doe.gov.

Industrial Applications Though the vast majority of coal is used to generate electricity, coal also has industrial applications (see Figure 5-35). Coal contains large amounts of methanol and ethylene, as well as other ingredients that are used to make fertilizers, plastics, tar, synthetic fibers, and medicines. The paper and concrete industries also burn large amounts of coal.

Steel Production *Coke* is a form of coal that is used in the production of steel. It is produced by heating coal to more than 2,000°F in an oxygen-free oven. The coke is then removed from the oven, rapidly cooled, and shipped to a steel plant. Iron is smelted as a result of the extremely high temperatures that are generated from the burning of coke. The carbon from the coke combines with the iron to form steel.

For Export During 2008, more than 81.5 million tons of coal was exported to other countries from the United States (see Figure 5-36). The majority of U.S. coal exports go to Canada, Brazil, the Netherlands, and the United Kingdom.

OFF-ROAD EXPLORATION

Track the development of nuclear fusion energy at *www.llnl.gov/*.

FIGURE 5-36: Total U.S. coal exports in 2008 (in tons).

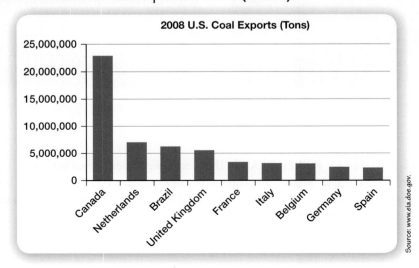

Source: www.eia.doe.gov.

Nuclear Energy (Uranium)

Nuclear power plants are used to tap into the incredible amount of energy that is contained within the smallest building block of matter: the atom. **Nuclear fission** involves splitting the nucleus of a heavy atom such as uranium 235 into smaller nuclei. The process generates tremendous heat energy, which is used to generate steam. All nuclear power plants currently use this method to produce electricity. **Nuclear fusion** involves combining or fusing the nuclei of smaller atoms together, which also results in the release of heat energy. The sun produces energy using nuclear fusion. Scientists and engineers are working to harness the power of nuclear fusion to produce electricity.

Nuclear fission:

a nuclear reaction in which a nucleus is split into smaller nuclei, resulting in the release of energy.

Nuclear fusion:

a nuclear reaction in which smaller nuclei are fused together, resulting in the release of energy.

URANIUM FUEL CYCLE A single uranium fuel pellet that is the size of a pencil eraser contains almost as much energy as one ton of coal. Like other fuel sources, uranium is an energy source that must be processed to produce an efficient fuel for generating electricity. The uranium fuel cycle shown in Figure 5-37 demonstrates the process of mining, refining, converting, enriching, using, and storing uranium fuel.

FIGURE 5-37: Uranium enrichment process.

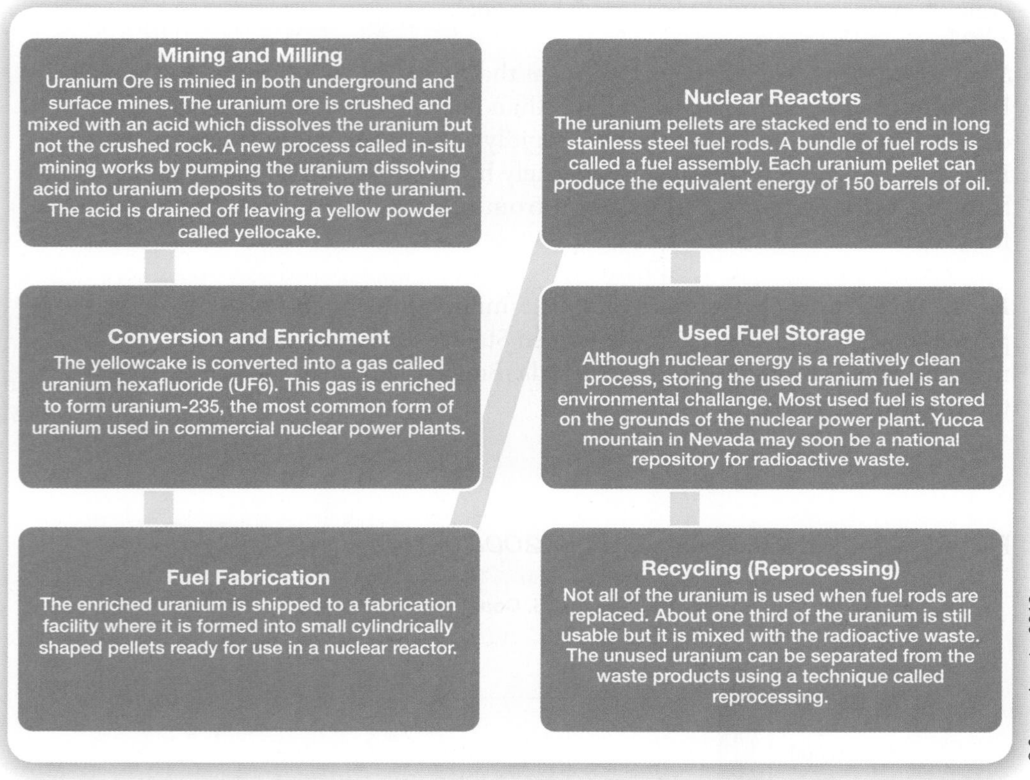

Mining and Milling
Uranium Ore is minied in both underground and surface mines. The uranium ore is crushed and mixed with an acid which dissolves the uranium but not the crushed rock. A new process called in-situ mining works by pumping the uranium dissolving acid into uranium deposits to retreive the uranium. The acid is drained off leaving a yellow powder called yellocake.

Conversion and Enrichment
The yellowcake is converted into a gas called uranium hexafluoride (UF6). This gas is enriched to form uranium-235, the most common form of uranium used in commercial nuclear power plants.

Fuel Fabrication
The enriched uranium is shipped to a fabrication facility where it is formed into small cylindrically shaped pellets ready for use in a nuclear reactor.

Nuclear Reactors
The uranium pellets are stacked end to end in long stainless steel fuel rods. A bundle of fuel rods is called a fuel assembly. Each uranium pellet can produce the equivalent energy of 150 barrels of oil.

Used Fuel Storage
Although nuclear energy is a relatively clean process, storing the used uranium fuel is an environmental challange. Most used fuel is stored on the grounds of the nuclear power plant. Yucca mountain in Nevada may soon be a national repository for radioactive waste.

Recycling (Reprocessing)
Not all of the uranium is used when fuel rods are replaced. About one third of the uranium is still usable but it is mixed with the radioactive waste. The unused uranium can be separated from the waste products using a technique called reprocessing.

INSIDE A NUCLEAR POWER PLANT There are two types of nuclear power plants used in the United States: pressurized-water reactors (PWRs) and boiling-water reactors (BWRs). More than 70% of the reactors in the United States are pressurized-water reactors. A diagram of a PWR is shown in Figure 5-38.

FIGURE 5-38: A pressurized-water reactor.

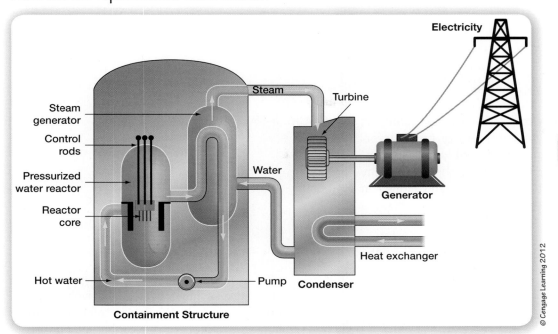

How a Pressurized-Water Reactor Works

1. Inside the reactor containment structure, water is kept under high pressure to prevent it from boiling as it flows through the reactor (see Figure 5-38).
2. The nuclear fission chain reaction in the fuel rods superheats the water to approximately 600°F (315.5°C).
3. This superheated water is then pumped into the heat exchanger, where the cooler water flowing from the condenser becomes steam. The steam drives a conventional steam turbine and generator that produces electricity. The pressurized water from the primary reactor loop never comes into contact with the water in the secondary loop where the steam is generated.
4. The condenser converts the steam back into water after it passes over the blades of the turbine. The condenser loop uses either a cooling tower or a large body of water to regulate the temperature of this loop.

ENERGY EFFICIENCY AND CONSERVATION

The concept of *energy efficiency* involves using technology to reduce the energy consumed to perform the same function. *Energy conservation* simply means using less energy. We all use energy every day. Every one of us can make a difference by reducing the amount of energy we consume. Reducing the amount of energy we consume will benefit our quality of life and the environment.

The United States uses nearly 189 billion Btu of energy every minute of every day every year. This accounts for approximately 20% of the energy produced worldwide. China, the second largest consumer of energy, consumes about 15% of the world's total energy production. The U.S. Department of Energy divides energy usage into three categories: residential and commercial, transportation, and industrial. You can help by making choices and taking actions that will reduce the total amount of energy used in all three categories.

The average American family spends hundreds of dollars annually on home utility bills. Unfortunately, a large amount of that purchased energy goes to waste. Fortunately, there are many things you can do to save energy and money in your home. Here are some ways *you* can make a difference:

▶ Find out which parts of your house use the most energy by performing a home energy audit. A home energy audit will locate the areas which need improvement and suggest the most effective ways to cut your energy costs. A free online home audit calculator is located at www.energysavers.gov.

▶ Always check for the Energy Star logo when purchasing new electronics and appliances (see Figure 5-39). Energy Star is a cooperative program of the U.S. Environmental Protection Agency (EPA) and the U.S. Department of Energy that helps everyone save money and protect the environment by promoting energy efficiency and conservation. Visit www.ENERGYSTAR.gov for more information.

▶ Replace those old incandescent light bulbs with new compact fluorescent light bulbs. Many CFLs use a quarter of the amount of energy that is consumed by incandescent bulbs and they last as much as 10 times longer. CFLs cost more, but they pay for themselves

FIGURE 5-39: *The Energy Star logo.*

(in energy savings) over a relatively short period of time. Visit www.earth911.com to learn how to properly recycle burned out CFL bulbs.

▶ Your electronics use energy even when you are not using them. Turning off the power strip connected to your computer, television, cell phone charger, and so on when you are done using them will reduce the amount of energy you use.

▶ Are you looking for your first car? Consider an energy-efficient hybrid or clean diesel vehicle. These efficient vehicles can save the average owner as much as $1,500 in fuel costs each year. That means you would be using much less energy to get around town.

▶ If you own an automobile, make sure your tires are properly inflated. Tires that have been properly inflated will not only save you money on fuel but will also make them last longer.

Career Profile

ALL-AMERICAN ENGINEER

A college volleyball player who wound up in the oil industry, Gabrielle Guerre likes to keep her options open. And, thankfully, engineering is a profession that speaks to her sense of life's possibilities. "My degree is in mechanical engineering," says the Kansas State University grad, who currently works as a petroleum engineer for Ryder Scott Company, a consulting firm based in Houston, Texas. "But the nice thing about engineering is that you can study one discipline and then have a career in a totally different one. There's always a place for an engineer."

Gabrielle Guerre: Petroleum Engineer, Ryder Scott Company

© Cengage Learning 2012

On the Job

At Ryder Scott Company, Guerre does reserves analysis and auditing for her clients. "A company comes to us because they need to fill a requirement with the SEC (Securities & Exchange Commission) or they're pulling together an annual report or an internal audit," she says. "You're making decisions that have a global influence. The impact that one person can have is remarkable."

Fresh out of college, Guerre took a job with ExxonMobil. "One of the coolest things I got to do was work with steam. One asset I was responsible for had very thick oil, so to decrease its viscosity we would inject steam underground, which would help the oil flow more easily and come to the surface."

Inspirations

Guerre credits her college professors with pointing her in the right direction. "I didn't know what I wanted to do with the rest of my life," she says. "I was still playing volleyball, but I needed a degree that wouldn't limit me in any way when I began to choose a career path. My teachers were especially helpful in working around my volleyball schedule. If I had a question, they were always available."

Sometimes, they asked Guerre to return the favor. "In my dynamics class, my professor once brought me to the front of the lecture hall to demonstrate a kinematics principle with a volleyball," she says. "That was fun."

Education

Having received the benefit of a good education, Guerre is eager to share her knowledge with others. She's been actively involved in several volunteer organizations and likes to introduce youngsters to engineering through classroom demonstrations.

"We'll build a bridge out of paper or show different physics principles," she says.

One demonstration specific to the oil and gas industry even involved Reese's Peanut Butter Cups. "It's a way of helping students understand well design and stimulation techniques," Guerre says.

Fracture techniques have been used for decades to help get more oil and gas out of the ground. And to demonstrate a fracture, Guerre says, "you poke a hole in the middle of a straw and stick one end of the straw in a peanut-butter cup. Then you stand the whole thing in a plastic, see-through cup that you pour Jell-O into. You let the Jell-O cool, but make sure the Jell-O covers the holes in the straw 'well' and the open end of the straw sticks out of the cup some. Then you inject milk (or any colored liquid not the color of the Jell-O) into the open end of the straw with a syringe. The milk has nowhere to go but into the Jell-O, so it creates a fracture."

Advice to Students

Guerre has always believed that using visual learning tools, such as the Jell-O/peanut butter cup demonstration, best allows a student to grasp all the possibilities a person has with an engineering degree. "That person can work anywhere, on anything, in any industry."

SUMMARY

- Energy is defined as the ability to cause matter to move or transform.

- Although energy comes in many forms, we can put them all into two categories: potential energy, which is stored energy, and kinetic energy, which is energy from motion.

- Heat (usually denoted by the letter Q) is the flow of, or potential flow of, energy from a higher-temperature substance to a lower-temperature substance. Heat is a form of energy and is measured using the same units that are used for work and mechanical energy.

- Heat transfer can take three forms: conduction, convection, and radiation.

  - Thermal *conduction* occurs when heat (thermal energy) is transferred within a substance through particle-to-particle collisions.

  - The transfer of heat through the movement of warmed matter in a fluid is called *convection*.

  - Thermal *radiation* is the transfer of heat (thermal energy) from one place to another through electromagnetic waves.

- Temperature is the measure of the average kinetic energy of the molecules of a substance.

- In the United States, energy production and consumption is commonly measured in Btu (British thermal units) and therms; 100,000 Btu equals 1 therm.

- The SI unit for heat energy is the joule. One joule is equal to the amount of work accomplished when a force of 1 newton acts over a distance of 1 meter.

- Another common unit for measuring heat energy is the calorie. One calorie is equal to the amount of heat needed to raise the temperature of 1 gram of water by 1 degree Celsius.

- Power is the amount of energy conversion or transfer that occurs per unit time. The SI unit of power is the watt.

- Thermodynamics is a branch of physics that is based on the fundamental laws of nature concerning energy and mechanical work.

- The scientific law known as the conservation of energy states that energy can be neither created nor destroyed.

- Renewable energy sources are those that can be replenished in short order.

- Nonrenewable energy sources are those that cannot be replenished in our lifetimes.

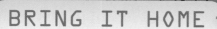

BRING IT HOME

1. When you hold a book above your head, does it have potential or kinetic energy?
2. How are sound waves and electromagnetic waves different?
3. Describe how a solar collector works.
4. How do photovoltaic cells produce electricity?
5. How can the kinetic energy of wind be converted into electricity?
6. What is potential energy?
7. What is kinetic energy?
8. A growing plant turns light energy into what form of energy?
9. When you turn on a television, the electricity is converted into which form(s) of energy?
10. Your body uses the chemical energy in food to produce which form(s) of energy?
11. Does the United States use more renewable or nonrenewable energy sources?
12. List three fossil fuels.
13. What is the difference between passive and active solar heating?
14. Describe how hydroelectric power plants work.
15. How is natural gas used?
16. Where does propane come from?
17. Describe how gasoline is extracted from crude oil.
18. What energy source is a result of the uneven heating of Earth?
19. What is the cleanest fossil fuel used today?
20. What is the difference between nuclear fusion and nuclear fission?

EXTRA MILE

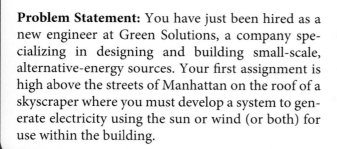

Problem Statement: You have just been hired as a new engineer at Green Solutions, a company specializing in designing and building small-scale, alternative-energy sources. Your first assignment is high above the streets of Manhattan on the roof of a skyscraper where you must develop a system to generate electricity using the sun or wind (or both) for use within the building.

Design Statement: Spend some time brainstorming solutions to the problem and research what solutions are already out there. Choose the best design and create a model of it using recyclable materials. Present your design to your class.

CHAPTER 6
Electrical Systems

GPS DELUXE

	START LOCATION	DISTANCE	END LOCATION

Menu

Before You Begin

Think about these questions as you study the concepts in this chapter:

1 What are the subatomic particles that make up an atom, and how do they relate to electricity?

2 How does the law of charges affect the flow of electricity?

3 What are the different sources of voltage, and how do they generate voltage?

4 What are the differences between voltage, current, and resistance?

5 What electrical properties can a digital multimeter measure, and how are these measurements taken?

6 What are the primary functions of conductors and resistors?

7 How is the resistor color code used to determine the nominal resistance value and tolerance range of a through-hole style fixed resistor?

8 What does Ohm's law reveal about the relationships between voltage, current, and resistance in a conductor?

9 What are the characteristics of series circuits?

10 How do parallel circuits differ from series circuits?

Imagine a world with no iPods, graphing calculators, or cell phones, a world without the conveniences of modern electronics. If you were to travel back in time to 1947, you would probably listen to the evening news on the radio. Black and white television was only beginning to find its way into the homes of the American public (see Figure 6-1). Making a phone call wouldn't be as easy as reaching for your pocket. Most phone lines in those days were party lines: Multiple homes were connected to the same phone line. You would have to wait for your neighbors to finish their conversations before you could make a call. If you wanted to listen to music, you would either tune in to a radio station or put a record on the record player.

Figure 6-1: *In 1947, televisions similar to this one would be found in some homes.*

Radios, televisions, telephones, record players and other electronic devices of that time relied on an electrical component called a *vacuum tube*, which was used to amplify, switch, and control the movement of electricity within a vacuum-sealed glass enclosure (see Figure 6-2). In 1945, the first electronic computer, called ENIAC (for Electronic Numerical Integrator and Computer), occupied an entire room at Penn State University. It used vacuum tubes to control electrical signals. The drawback to using vacuum tubes was their fragility, limited life, heavy power consumption, and bulky size.

Although people in the 1940s were amazed by the capabilities of the vacuum tube, one of the most significant advances of the 20th century was unfolding at Bell Laboratories in New Jersey. In late December of 1947, three scientists—William Shockley, Walter Brattain, and John Bardeen—invented the *transistor*, which replaced vacuum tubes in most applications. Over the years, transistors have become smaller and smaller. Modern computer chips have millions of transistors packed in an area no larger than your fingernail (see Figure 6-3). They are so small that if you were to lay 2,000 transistors end-to-end, they would just about equal the diameter of a human hair.

We are living in the information age. Computers can be found just about everywhere. Computers and modern electronic devices have revolutionized the way we live, communicate, and certainly the way in which we do business. At the core of all of these technological marvels are electrical circuits made of thousands, if not millions, of components, all working together to serve a specific function.

Figure 6-2: *A vacuum tube.*

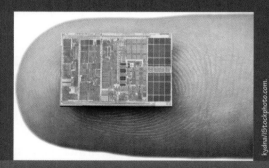

Figure 6-3: *A technician holds a microchip that is made up of millions of transistors.*

Amber

Electricity, electrons, electronic, and other words that begin with *electr* can all be traced to the Greek word *elektor*, meaning "beaming sun." In the Greek language, *elektron* is the word for amber.

Amber is a beautiful bronze-colored stonelike material that glistens with orange and yellow in the sunlight (see Figure 6-4). Amber is actually petri-fied tree sap. The ancient Greeks discovered some weird things about amber. For example, when it was rubbed by fur, it would attract feathers, hair, and other lightweight objects. They had no idea what caused this phenomenon, but the Greeks actually stumbled on one of the first examples of static electricity. The origin of the word *electricity* can be traced to the Latin word *electricus*, which means to "produce from amber by friction."

FIGURE 6-4: These pieces of amber are millions of years old.

© iStockphoto/Vladimir Sazonov.

THE ATOM

Atom:

the smallest indivisible unit of matter and the most basic building block of an element that still re-tains the properties of that element.

All matter—whether solid, liquid, or gas—is composed of atoms (see Figure 6-5). An **atom** is the smallest indivisible unit of matter and the most basic building block of an *element* that still retains the properties of that element. Atoms are usually grouped together with other atoms to form **molecules**, such as H_2O (water).

FIGURE 6-5: Diagram of a carbon atom.

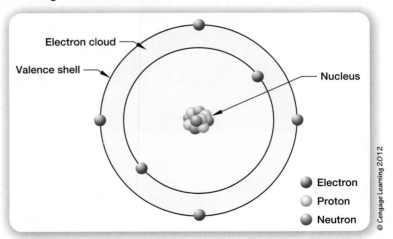

Electron cloud

Valence shell

Nucleus

● Electron
○ Proton
● Neutron

© Cengage Learning 2012

At the center of an atom is its **nucleus**, which consists of tiny subatomic particles called protons and neutrons. A **proton** has a positive electrical charge, and a **neutron** has no charge. Orbiting the nucleus are much smaller negatively-charged subatomic particles called **electrons**. Unlike neutrons and protons, electrons can be shared with other atoms to form chemical bonds. They can also be directed to jump from one atom to another to generate *electricity*. Atoms are mostly empty space. If an atom's nucleus were the size of a tennis ball, then the electrons that surround it would occupy a shell that is as tall as the Empire State Building.

Electron:

an extremely small negatively charged subatomic particle that orbits the nucleus of an atom.

Point of Interest

People from New York City and elsewhere came to Menlo Park, New Jersey, in 1879 to witness something that had never been done before. Thomas Alva Edison and his employees had invited the public to their workshop to see the grounds illuminated with approximately 100 electric incandescent light bulbs.

Much had already been discovered about electricity by the time of Edison's Menlo Park demonstration. For example, Alessandro Volta, an Italian scientist, had invented the electric battery, and English scientist Michael Faraday had created a generator that could produce electricity using magnetism.

Although Edison is credited with inventing the incandescent light bulb, the idea of electric lighting

was not new. Numerous inventors and scientists around the world were busily working to develop various forms of electric lighting. However, at the time of Edison's breakthrough, there were no other forms of electric lighting that were practical for home use. Street lights were generally powered by gas and carbon-arc systems that generated exceptionally bright light by sending electricity across a space between two carbon terminals.

Incandescent bulbs produce light by passing electricity through a fine wire called a *filament*, which then glows. The filament is contained in a vacuum inside a glass bulb. Edison and his employees spent months searching for the perfect filament. Among the first materials to be tested was platinum, but this metal created a filament that was

FIGURE 6-6: **Today we take lighting for granted. (a) Boston's Fenway Park illuminated for a night game with energy-efficient fluorescent and LED light bulbs. (b) An incandescent light bulb similar to the one invented by Thomas Edison.**

(a) Image copyright Kellie L. Folkerts, 2010. Used under license from Shutterstock.com. (b) © Cengage Learning 2012

(a)

(b)

(Continued)

quick to overhead and burn out. After trying hundreds of other materials, Edison's team watched in amazement when they ran electricity through a carbon filament made by burning a small piece of sewing thread. The carbonized sewing thread produced light for 13½ hours. This was the filament that was used during the public demonstration at Menlo Park. Workers later discovered that bamboo filaments could extend the life of the bulbs even more.

Edison's invention came before many homes and businesses had electricity. Electricity had to be readily available to customers to make his invention practical for everyday use. Consequently, he spent the next several years of his life developing a system to produce and distribute electricity using central power plants and a maze of wires running to homes and businesses. It wasn't long before electrical power and the incandescent light bulb spread around the world.

Elements

> **Element:**
> a substance that is composed of only one type of atom, and cannot be broken down by chemical means into a simpler form.

Any material that is composed of only one type of atom is called an **element**. At this time, scientists know of 118 elements. Ninety-four of these elements occur naturally on Earth, and the remaining 24 elements have been created by humans.

The *periodic table of elements* groups the known elements according to their properties and the number of subatomic particles they contain. A number that is associated with an element's chemical symbol identifies that element's **atomic number**. This number tells us how many protons are in one atom of that element. Every element has its own unique atomic number. Therefore, no two elements have the same number of protons. Materials that are composed of two or more elements are called **compounds**.

Ions

> **Atomic number:**
> the number of protons that exist within an atom's nucleus.

Normally, atoms have an equal number of electrons and protons. Because protons have a positive charge and electrons have a negative charge, the charges cancel one another, resulting in an electrically neutral atom.

When an atom gives up or gains an electron from another atom it becomes an **ion**. An atom that gains an electron is called an *anion*. An anion is a negatively-charged ion because it has more electrons in its shell than protons in its nucleus. On the other hand, an atom that gives up an electron becomes a positively charged ion, or *cation,* because it has fewer electrons than protons.

STATIC ELECTRICITY AND THE LAW OF CHARGES

> **Law of charges:**
> opposite charges attract and like charges repel.

The **law of charges** states that opposite charges attract and like charges repel. Consequently, positive charges are attracted to negative charges. In the atom, the electrons are drawn toward the protons within the nucleus. What prevents the electrons from slamming into the nucleus? Centripetal force. Because electrons orbit the nucleus at extremely high speeds, they are pulled away from the protons in much the same way that Earth is pulled away from the sun.

The law of charges can be seen by hanging two Styrofoam balls from the ceiling as shown in diagram A in Figure 6-7. If both balls have no electrical charge (i.e., they are neither positive nor negative), nothing will happen.

In the next experiment, a hard rubber rod is rubbed with fur, making it negatively charged. When the rod is held against both balls, the excess number of electrons the rod gained from the fur is transferred to the Styrofoam balls. Because both balls now have a negative charge, they will repel each other as shown in diagram B in Figure 6-7.

FIGURE 6-7: Styrofoam balls used to demonstrate the laws of charges.

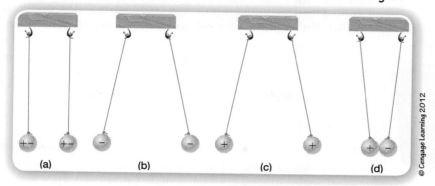

(a)   (b)   (c)   (d)

The same concept holds true for the next experiment in which both balls were given a positive charge. Because like charges repel, the Styrofoam balls separated again (see diagram C in Figure 6-7).

In the last experiment, one of the balls was given a positive charge while the other ball was given a negative charge. Because unlike charges attract, the Styrofoam balls cling together as shown in diagram D in Figure 6-7.

Why does a balloon stick to a wall or ceiling after being rubbed through someone's hair?
Answer: The friction between the two different materials causes a transfer of negative charges from your hair to the balloon. When placed against a wall or ceiling, the balloon's highly-negative charge repels the electrons within the atoms in that area of the wall, causing them to migrate further into the wall. This results in a localized section of wall that has an overall positive charge. The attraction between the negatively-charged balloon and the positively-charged wall area makes the balloon appear to "stick" to the wall.

CURRENT

Current is the movement of charge carriers, such as free electrons, holes, and ions. *Electron current* is the result of free electrons that are directed to move from one atom to another along a conductive path. Because electrons have a negative charge, they are repelled by areas of negative *polarity* and are attracted to areas of positive polarity.

Electricity was first thought of as the flow of positive charges in a conductor from a point of positive polarity to a point of negative polarity; this is commonly referred to as *conventional current*. This theory was formulated long before our present understanding of atoms and their subatomic particles. In fact, many of the theories of electricity, including the development of schematic symbols for polarity-sensitive electronic components, are based on the idea of conventional current. The idea is still valid in that conventional current is the flow of positively-charged holes from one atom to another along a conductive path. As an electron jumps from one atom to another, it leaves behind a positively-charged hole. This positively-charged hole attracts another negatively-charged electron. In essence, the two charge carriers switch positions with electrons moving in one direction (from negative to positive) and positively-charged holes moving in the opposite direction (from positive to negative).

Polarity

Polarity is the condition of being electrically positive or negative. The most common way to represent polarity is by assigning a minus sign (−) for negative charges and positive signs (+) for positive charges. Polarity is a relative concept in that the electrical charge at one point in a *circuit* is identified as positive or negative in relation to another point within the same circuit.

Current:
the movement of charge carriers, such as free electrons, holes, and ions.

Polarity:
the condition of being electrically positive or negative.

FIGURE 6-8: *Every battery has a positive and a negative terminal.*

© iStockphoto/Gewoldi.

Conductor:

a material through which electrons easily flow.

The terminals of the 9-volt battery in Figure 6-8 are labeled with plus and minus signs. The terminal with the minus sign is called the *negative terminal* and is more negatively-charged than the positive terminal. Conversely, the terminal with the plus sign is called the *positive terminal* and is more positively-charged than the negative terminal.

Electron Orbits

Materials that allow electrons to flow easily through them are called **conductors**. Silver, copper, gold, and aluminum are all excellent conductors. So what makes these materials better than others at conducting electricity? The answer lies within the atom's atomic structure. Electrons form orbits, called *shells*, around an atom's nucleus. Each electron shell can contain a set number of electrons. The maximum number of electrons that any one shell can hold may be calculated using the formula $2n^2$, where n represents the shell number. In any atom, the first shell can hold no more than two electrons ($2 \times 1^2 = 2$). The second shell can hold a maximum of eight electrons ($2 \times 2^2 = 8$), and so on. Figure 6-9 illustrates this concept.

FIGURE 6-9: **Maximum number of electrons in each orbit.**

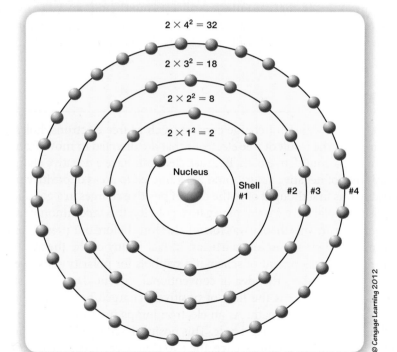

© Cengage Learning 2012

Valence shell:

the outermost shell of an atom.

THE VALENCE SHELL Materials that are excellent conductors have only one or two electrons that occupy the outermost electron shell, called the **valence shell**. Consequently, electrons located in this shell are known as *valence electrons*. The atoms of such materials can easily give up these electrons to neighboring atoms.

SOURCES OF VOLTAGE

If electron current is the result of free electrons that are directed to move from one atom to another along a conductive path, what directs the electrons to move in first place? The answer to this question is *voltage*. Voltage can be thought of as a kind of electrical pressure. Several different sources of voltage are used in electrical circuits including batteries, generators, photovoltaic cells, thermocouples, and piezoelectric devices.

Batteries

Most portable electronic devices such as iPods® and cell phones are powered by an electrochemical device called a *voltaic cell*, which is more commonly known as a battery. A battery generates a voltage by converting chemcial energy to electrical energy. This involves a chemical reaction between two different electrodes that are separated by a chemical solution called an *electrolyte* (see Figure 6-10). The chemical reaction frees more electrons in one electrode than in the other, making one of the electrodes more negatively-charged than the other.

FIGURE 6-10: *Section view of a battery.*

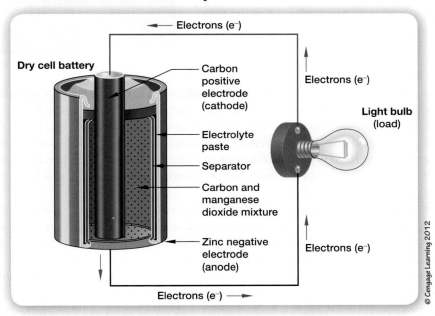

Each electrode is connected to a terminal, which provides a point of connection to an electrical circuit. Because the negatively-charged electrode (anode) is separated from the positively-charged electrode (cathode), the only way for the electrons to travel from the negative terminal to the positive terminal is through an external circuit. The battery's voltage represents the difference in charge between the two terminals. After extended use, the voltage between a battery's electrodes decreases to zero potential difference. It is at this point that a battery is considered "dead."

Generators

A generator is a mechanical device that converts mechanical energy into electrical energy. In the 1830s, the famous English scientist Michael Faraday discovered that a momentary voltage is generated across a wire when the wire is passed through a permanent magnet's magnetic field. His discovery is known today as the principle of **electromagnetic induction**.

Faraday is also credited with the invention of the first electric *dynamo*, the predecessor of today's modern electrical power generators. Faraday's simple dynamo, short for

> **Battery:**
> a device that produces electricity through a chemical reaction between two different metal electrodes separated by a chemical solution called an *electrolyte*.

> **Generator:**
> a device that converts mechanical energy into electrical energy.

> **Electromagnetic induction:**
> the generation of voltage within a conductor as a result of passing the conductor through a magnetic field.

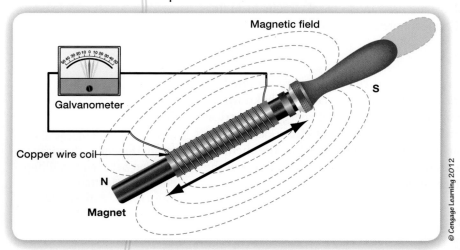

FIGURE 6-12: Turbine generators inside the McNary Dam on the Columbia River.

dynamo-electric machine, was capable of transforming mechanical energy into electrical current based on the relationship between electricity and magnetism. When electrical current passes through a conductor, it creates a magnetic field around the conductor. This also works in reverse: When a magnetic field passes by a conductor, a momentary voltage is produced across that conductor. When connected to a circuit, the voltage will direct electrons to flow. Nearly all electric power that is produced today is based on this principle.

In Figure 6-11, a magnetic rod is passed through a coil of wire. When the rod is moved, a current is produced in the wire that is measured by a *galvanometer*, which detects the presence of electrical current. The direction of the current produced in the circuit is related to the direction in which the magnet was moved.

Repeatedly moving the magnet back and forth will result in an *alternating current*. Modern power plants use the energy contained in coal, oil, uranium, wind, and so on to drive generators to generate electrical current. Generators, like the one shown in Figure 6-12, convert the mechanical energy from a rotating input shaft into electrical energy by spinning large coils of conductive wire within a magnetic field.

Make your own generator using a coil of wire and a strong magnet. Connect a multimeter to both ends of the coil to measure voltage. You should see the voltage increase as the magnet is passed over the coil.

Photovoltaic (PV) Cells

Photovoltaic (PV) cell:

a semiconductor device that generates a voltage as it absorbs photos of light.

Semiconductor materials such as silicon and germanium are used in the creation of **photovoltaic (PV) cells**, which generate voltage as they absorb the energy contained in photons of light. When connected to a circuit, the voltage generated by a PV cell will direct free electrons to move through the circuit's conductive path to power an electrical device such as a lamp (see Figure 6-13). This technology is often used in combination with rechargeable batteries to store charge so that an electrical device can continue to function even when sufficient amounts of light are not available.

FIGURE 6-13: A photovoltaic (PV) cell releases electrons while absorbing photons of light. These electrons are harnessed to produce electric current in a circuit.

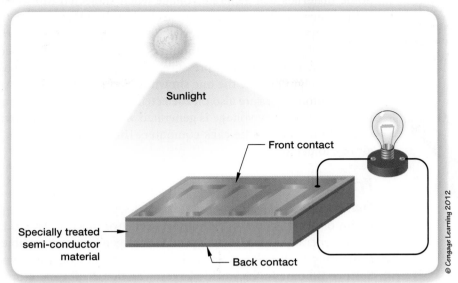

Thermocouples

A **thermocouple** is a device that consists of two dissimilar metals that are bonded together. When heated, a small voltage is generated across the device (see Figure 6-14). The amount of voltage produced is proportional to the temperature. Thermocouples are often used as temperature sensors in electronic temperature control systems.

> **Thermocouple:**
> a device that consists of two dissimilar metals that, when heated, will generate a small voltage across the device.

FIGURE 6-14: A simple thermocouple can be made by twisting the ends of short lengths of copper and iron wire together. A small voltage will be produced when the twisted end of the thermocouple is placed over the flame of a candle. A similar experiment can be done by placing the twisted end on an ice cube.

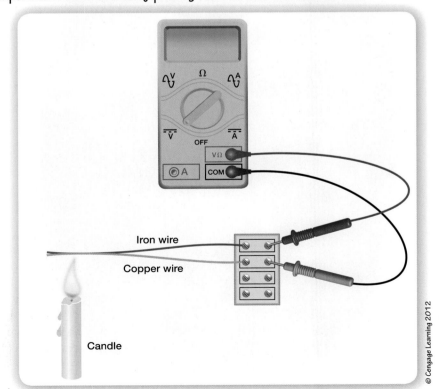

FIGURE 6-15: Piezoelectric igniters are commonly found on gas grills. When the red button is pressed, it comes into contact with a piezoelectric device that produces a quick burst of electricity and sends a spark across a small gap inside the grill. This spark ignites the gas in the grill.

© Cengage Learning 2012

Piezoelectric effect:

a phenomenon in which a material produces a voltage when subjected to mechanical pressure.

Piezoelectric Devices

Some materials will generate voltage when subjected to mechanical pressure, such as squeezing, stretching, bending, or twisting. Quartz crystal is one such material. This phenomenon is known as the **piezoelectric effect**. The amount of voltage produced is directly proportional to the applied pressure. Piezoelectric devices are used in gas grill igniters like the one shown in Figure 6-15. Air-bag sensors in automobiles are also piezoelectric devices. When the sensor is hit, a momentary voltage is generated which causes an electrical signal to be sent to the car's computer. This, in turn, triggers the air bag to deploy.

Friction

Static electricity results from the physical separation and transportation of charges due to friction between two objects. An electrostatic generator, such as a Van de Graaff generator, is capable of generating high voltage static electricity. The Boston Museum of Science is home to the world's largest Van de Graaff generator (see Figure 6-16).

FIGURE 6-16: The Boston Museum of Science is home to the world's largest Van de Graaff generator. Originally designed by professor Robert J. Van de Graaff at the Massachusetts Institute of Technology for atomic research, this giant machine is now used to educate visitors about lightning, electric charge, and storm safety.

© Cengage Learning 2012

COMMON ELECTRICAL UNITS

Electrical units are standard across both the SI and English systems. Figure 6-17 lists the definition, unit, and abbreviation of common electrical quantities. The following sections provide detailed information on current, voltage, resistance, conductance, power, and charge.

FIGURE 6-17: Common electrical quantities.

Quantity	Definition	Unit of Measurement	Abbreviation or Symbol for Unit
Current (I)	The rate of electrical flow	ampere	A
Voltage (V)	Electrical pressure	volt	V
Resistance (R)	Opposition to electrical flow	ohm	Ω
Conductance (G)	The ease of electrical flow	siemens	S
Power (P)	The rate of doing work	watt	W
Charge (Q)	The surplus or shortage of electons	coulomb	C

Current

As we learned earlier, the law of charges states that like charges repel and unlike charges attract. If you were to place a negative charge at one end of a conductor and a positive charge at the other, electrons will flow from the negative side to the positive side (see Figure 6-18). This flow will continue until both sides of the conductor have equal charge.

FIGURE 6-18: Electricity flows from negative to positive.

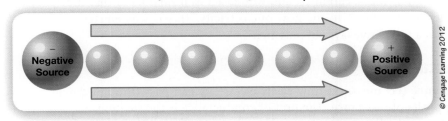

© Cengage Learning 2012

The flow of electricity is similar to water flowing through a pipe. The amount of water flowing through a pipe can be measured by calculating the volume of water passing by a given point over a period of time. In Figure 6-19, a large vessel is being filled with water flowing from a 3-gallon bucket at a rate of 3 gallons per minute.

The *intensity* or quantity of electrical current (I) is the number of electrons flowing past a given point in one second. Current is measured in *amperes,* a unit named for André-Marie Ampère, a French mathematician and physicist who is credited with the discovery of *electromagnetism.* One ampere (or amp) is equal to 1 coulomb per second. One coulomb, which is a basic unit of electrical charge, is approximately 6.24 quintillion, (6.24×10^{18}) electrons. To give you an idea of how big this number is, if each electron were the size of a penny and the pennies were laid out like a carpet, the pennies would cover the surface of the Earth four times over.

Just as water can travel through a pipe in one of two ways, forward or backward, electrical current can flow in one of two ways as well. The direction of current flow depends on the polarity of the power source. Current that flows in only one direction is called *direct current* (DC). A battery is a type of power supply that provides a source of direct current. Current that changes direction is called *alternating current* (AC). The

FIGURE 6-19: Electrical current can be compared to the flow of water. In this case, 3 gallons of water flowed into the bucket over a period of 1 minute, making the rate of water flow 3 gallons per minute.

1 minute

© Cengage Learning 2012

electricity that is supplied to your home and available for use through wall outlets is an example of alternating current. In the United States, AC switches direction 60 times every second.

Voltage

Have you ever wondered what makes the water flow out of your faucets at home? The water in the pipes in your home is under pressure that is generated by pumping water into elevated water towers or by pumps at water treatment plants and repeater pumping stations. In the case of a water tower, the higher the water is pumped, the greater the water pressure will be.

As you've learned, electrons are directed to move from one atom to another as a result of an electrical pressure called *voltage* or *potential difference*. Voltage (V) is the difference in charge between two points. The greater the difference in charge, the greater the voltage will be. The unit of measurement for voltage is the volt (V), named in honor of Alessandro Volta, the Italian scientist who invented the direct current battery. One volt is the potential difference required to move 1 amp of current across a conductor that provides 1 ohm of resistance.

Voltage (V):
the difference in charge between two points; measured in volts.

Resistance

The pipes that water flows through can offer significant resistance. This opposition to the flow of water is caused by friction between the water and the inside surface of the pipe. The greater the friction, the greater the resistance.

Electricity also experiences resistance as it flows. Electrical resistance (R) is the opposition to the flow of electricity, which is measured in ohms (Ω). As we mentioned earlier, some materials conduct electricity better than others. Materials such as copper and aluminum are good conductors and offer little resistance to the flow of electricity. Polymers, ceramics, and woods act as *insulators* in that they provide a great deal of resistance to the flow of electricity. This is why electrical cable are wrapped in a protective layer of polymer.

Resistance (R):
the opposition to the flow of electricity; measured in ohms.

Conductance

Conductance (G) measures how easily electricity can pass through a material. It is the opposite of resistance. The unit of conductance is the siemens (S), named in honor of scientist Ernst von Siemens. As shown in Equation 6-1, the formula for conductance is the reciprocal of the resistance (R):

Conductance (G):
a measure of the ease with which electricity can pass through a material; measured in siemens.

$$G = \frac{1}{R} \qquad \text{(Equation 6-1)}$$

Power

The topic of power has been addressed in Chapter 3 as it relates to mechanical systems and in Chapter 5 as it relates to energy. Here, again, we touch upon the topic of power as it relates to electrical systems. Power (P) is the the rate at which electrical energy is transferred or transformed. As you've learned, the metric unit for power is the watt. In a direct current circuit, power is the product of voltage (V) and current (I) (see Equation 6-2).

Power (P):
the rate at which electrical energy is transferred or transformed; measured in watts.

$$\text{Power} = \text{Voltage} \times \text{Current} \qquad \text{(Equation 6-2)}$$
$$P = V \times I$$

Charge

The basic unit for *electrical charge* (Q) is the coulomb, represented by the capital letter C. The coulomb was named in honor of French physicist Charles Coulomb. As you learned earlier, one coulomb of charge (C) is equal to 6.24×10^{18} electrons.

FIGURE 6-20: A digital multimeter.   FIGURE 6-21: Multimeter probes.

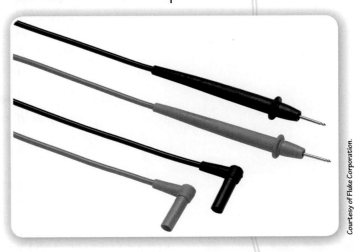

DIGITAL MULTIMETERS

A **multimeter** is a measuring instrument that combines an *ammeter* to measure current, a *voltmeter* to measure voltage, and an *ohmmeter* to measure resistance. Some multimeters have more advanced measuring capabilities such as capacitance and frequency. A digital multimeter (DMM) consists of an interface box (Figure 6-20) and test probes (Figure 6-21). The interface has a digital display to indicate measurements, a dial to change the mode of the meter, and connections for the probes. The test probes in Figure 6-21 have plastic handles that the user holds while placing the tips on the metal probes on the component to be tested. The black probe is always considered the negative probe and is connected to the common port on the multimeter. See Figure 6-22 for common DMM symbols.

FIGURE 6-22: Common multimeter functions and symbols.

Symbol	Function
$\tilde{V}$	AC Voltage
$\overline{V}$	DC Voltage
$m\overline{V}$	DC Millivolt
Ω	Resistance (Ohms)
)))) ▶⊢	Diode/Audible Continuity Test
$\overline{\cdots}\sim$ mA	AC/DC Milliamp
$\overline{\cdots}$ $\sim$A	AC/DC Amp

© Cengage Learning 2012

Safety Note

▶ Always select the proper meter for the job.
▶ Before using a multimeter, always read the safety information enclosed with the unit.
▶ Never use the meter if the test leads look damaged.
▶ Never measure resistance with the power on.
▶ Be sure to adjust all test leads and the rotary switch to the proper position for the measurement you're taking.
▶ When measuring current, never touch the test leads directly to a voltage source.
▶ Keep your fingers behind the finger guards when using the test probes.
▶ Never exceed the maximum voltage or current ratings of the meter.

Using a Digital Multimeter

Many digital multimeters have terminals similar to those shown in Figure 6-23. Carefully observe the following steps when operating a digital multimeter.

FIGURE 6-23: **Test probe connections.**

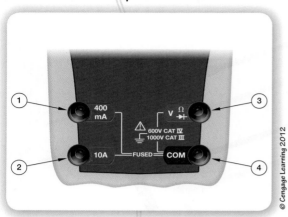

Step 1: Disconnect the power from the circuit being tested.

Step 2: Connect the black test probe to the common terminal 4 (see Figure 6-23).

Step 3: *If you are measuring voltage, resistance, or checking for continuity,* connect the red test probe to terminal 3.

Or

Step 3: *If you are measuring current up to 400 mA,* connect the red test probe to terminal 1.

Or

Step 3: *If you are measuring current up to 10 A,* connect the red test probe to terminal 2.

MEASURING VOLTAGE, RESISTANCE, AND CURRENT

Measuring Voltage

Step 1: Before taking a voltage measurement using your multimeter, you must know whether the voltage is AC or DC. If you are measuring voltage from a battery, you will use the DC setting. However, if you are taking a measurement from an outlet in your home, you will use the AC setting.

Step 2: When measuring the voltage across an electrical component within a circuit, place the multimeter in parallel with the electrical component. Figure 6-24 shows how a voltage reading can be taken across a lamp.

Measuring Resistance

Step 1: Resistance is always measured when the power is disconnected.

Step 2: Isolate the component or components to be measured and then select the proper resistance setting on the multimeter.

Step 3: Touch the probes to both sides of the component and take the reading from the digital display.

Measuring Current

Step 1: Before measuring current in a circuit, disconnect the power from the circuit.

Step 2: Place the multimeter in series with the circuit or components. Figure 6-25 shows how to connect the multimeter to a circuit to measure current. Be sure to connect the red probe to the proper terminal on the multimeter.

Step 3: Select the highest current setting on the multimeter.

Step 4: Turn the power back on the circuit.

FIGURE 6-24: A simple circuit with a multimeter measuring the voltage across a lamp.

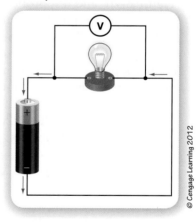

FIGURE 6-25: A simple circuit with an ammeter connected in series to measure current.

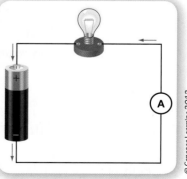

METRIC PREFIXES

Using metric prefixes is very helpful when dealing with very large or small quantities. Figure 6-26 shows common metric prefixes used when communicating electrical measurements. Note that the prefix *pico* is 1,000 times smaller than the prefix *nano*, and the prefix *mega* is 1,000 times larger than the prefix *kilo*. The prefixes shown in Figure 6-26 are those used in *engineering notation*. This is similar to scientific notation; however, the power of 10 in engineering notation is always a multiple of 3.

FIGURE 6-26: *Common electrical prefixes.*

Metric Term	Symbol	Number	Power of 10	Meaning	Common Electronics Abbreviations
pico	p	0.000000000001	1.0×10^{-12}	One millionth of one millionth of one unit	Picoampere (pA)
nano	n	0.000000001	1.0×10^{-9}	One thousandth of one millionth of one unit	Nanoampere (nA) Nanosecond (ns)
micro	μ	0.000001	1.0×10^{-6}	One millionth of one unit	Microampere (μA) Microvolt (μV)
milli	m	0.001	1.0×10^{-3}	One thousandth of one unit	Milliampere (mA) Millivolt (mV)
kilo	k	1000	1.0×10^{3}	One thousand times one unit	Kiloohms (kΩ) Kilovolts (kV)
mega	M	1000000	1.0×10^{6}	One million times one unit	Megaohms (MΩ)

If you found the resistance of a certain component to be 0.0000082 Ω, it is customary to simplify this long number by using a prefix. Starting at the decimal place, count three places to the right; now you are at 0.0082 mΩ (milliohm). If you count three more decimal places to the right, you are now at 8.2 μΩ (micro-ohm). Expressing this same number in engineering notation is easy. Because 1 micro-ohm is one-millionth of 1 ohm, it would be expressed as 8.2×10^{-6} Ω.

COMMON ELECTRICAL COMPONENTS

Resistors

One of the most common electrical components found in circuits is the resistor. As its name implies, a resistor is a simple electronic component that provides resistance to the flow of electricity. They are used intentionally to condition electricity by reducing voltage and limiting current within a circuit so that it conforms to the operational parameters of one or more other electrical components within the circuit. When used in combination, resistors are capable of dividing voltage, which we will discuss later.

RESISTOR COLOR CODE As you learned earlier, resistance is measured in ohms. Instead of printing the actual ohm value on a resistor, the Electronics Industries Association (EIA) established a color coding system to identify the value of a through-hole style fixed resistor. Some surface mount resistors have their resistance labeled on them, but most carbon and wire wound resistors use the EIA

FIGURE 6-27: *A four-band resistor color code.*

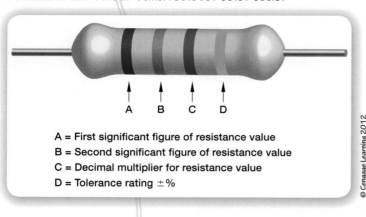

A = First significant figure of resistance value
B = Second significant figure of resistance value
C = Decimal multiplier for resistance value
D = Tolerance rating ±%

color code. The color code consists of three, four, or five bands of color wrapped around cylindrically shaped resistors.

Four-band color codes use the first two bands to indicate the significant figures of resistance value, the third band as the multiplier, and the last band to indicate the tolerance of the resistor (see Figure 6-27). The tolerance is the amount of deviation the rated value may vary from the actual resistance. Resistors with only three bands have no tolerance band; the tolerance is assumed to be ±20%. Resistors with five bands also use the first three bands to indicate the numerical value of the resistor, but the fourth band is a multiplier. The fifth band is the tolerance band (see Figure 6-28).

The placement of the bands on the resistor is critical. The first band is always placed closest to the end of the resistor. In Figure 6-27, the red band is closest to the end of the resistor and is therefore the first resistor-value band.

FIGURE 6-28: **A five-band resistor color code.**

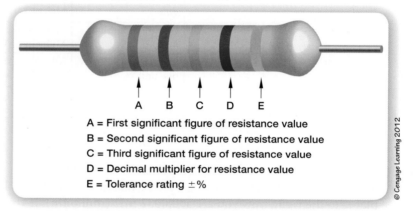

A = First significant figure of resistance value
B = Second significant figure of resistance value
C = Third significant figure of resistance value
D = Decimal multiplier for resistance value
E = Tolerance rating ±%

Tolerance Bands The fourth band on a four-band resistor and the fifth band on a five-band resistor are the tolerance bands. Gold and silver are commonly used for the tolerance band. Gold indicates a tolerance of ±5%, and silver indicates a tolerance of ±10%. A resistor with a value of 100 Ω with a gold tolerance band could have a resistance as high as 105 Ω or as low as 95 Ω and still be within the manufacturer's specifications. A resistor with a value of 100 Ω with a silver tolerance band could have a resistance as high as 110 Ω and as low as 90 Ω and still be within the manufacturer's specifications. If there is no tolerance band, the resistor is said to have a ±20% tolerance, which means that a 100-Ω resistor could be as high as 120 Ω or as low as 80 Ω.

Gold and Silver Colors as Multipliers For resistors with values *lower than 10 ohms,* a special multiplier value is used on the third band (see Figure 6-29). When gold is present in the third band, the first two digits should be multiplied by 0.1. For example, if the first two bands were 27 and the third band was gold, the resistance value would be 2.7 ohms. When silver is present in the third band, the first two digits should be multiplied by 0.01. For example, if the first two bands were 78 and the third band was silver, the resistance value would be 0.78 ohms.

FIGURE 6-29: Resistor color code chart.

Color	Significant Figure in 1st and 2nd Bands	Decimal Multiplier in 3rd Band	Tolerance Color in 4th Band	Special Precision Resistor Tolerance (in 5th Band)
Black	0	1	—	—
Brown	1	10	—	± 1%
Red	2	100	—	± 2%
Orange	3	1,000	—	—
Yellow	4	10,000	—	—
Green	5	100,000	—	± 0.5%
Blue	6	1,000,000	—	± 0.25%
Violet	7	10,000,000	—	± 0.1%
Gray	8	100,000,000	—	—
White	9	1,000,000,000	—	—
Gold	—	0.1	± 5%	—
Silver	—	0.01	± 10%	—
No Band	—	—	± 20%	—

Example

Problem 1: Use the chart in Figure 6-29 to determine the values and tolerance of the resistor shown in (a) Figure 6-30.

FIGURE 6-30

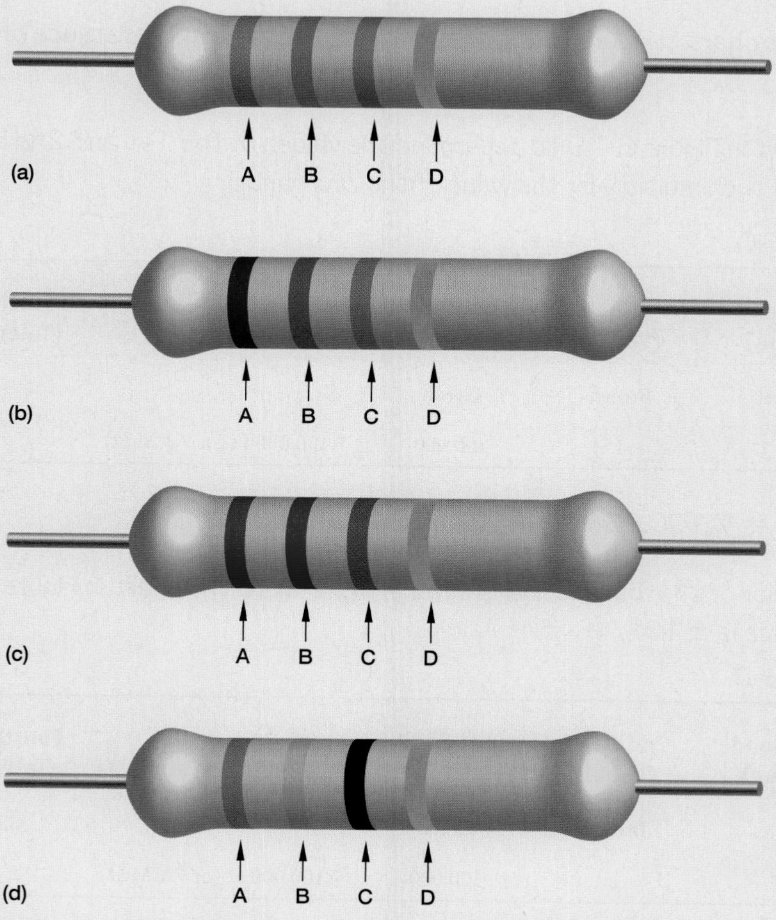

(a)   A B C D

(b)   A B C D

(c)   A B C D

(d)   A B C D

© Cengage Learning 2012

(continued)

1. Refer to the chart in Figure 6-29 to determine the values of the 1st and 2nd bands. Combine these values and then multiply by the value of the 3rd band:

TABLE 6-A:

1st Band (Value)	2nd Band (Value)	3rd Band (Multiplier)	Resistor Value	Fourth Band (Tolerance)
Orange	Green	Blue		
3	5	1,000,000	35,000,000 Ω (or 35 MΩ)	

$$35 \times 1,000,000 = 35,000,000 \ \Omega \ (or \ 35 \ M\Omega)$$

2. The tolerance color of the fourth band is silver. If we refer to the chart in Figure 6-29, we can see that the tolerance is ±10%.

TABLE 6-B:

1st Band (Value)	2nd Band (Value)	3rd Band (Multiplier)	Resistor Value	Fourth Band (Tolerance)
Orange	Green	Blue		Silver
3	5	1,000,000	35,000,000 Ω (or 35 MΩ)	±10%

Problem 2: Use the chart in Figure 6-29 to determine the value and tolerance of the resistor shown in (b) Figure 6-30.

1. Refer to the chart in Figure 6-29 to determine the values of the 1st and 2nd bands. Combine these values and then multiply by the value of the 3rd band:

TABLE 6-C:

1st Band (Value)	2nd Band (Value)	3rd Band (Multiplier)	Resistor Value	Fourth Band (Tolerance)
Violet	Brown	Green		
7	1	100,000	7,100,000 Ω (or 7.1 M Ω)	

$$71 \times 100,000 = 7,100,000 \ \Omega \ (or \ 7.1 \ M\Omega)$$

2. The tolerance color of the fourth band is silver. If we refer to the chart in Figure 6-29, we can see that the tolerance is ±10%.

TABLE 6-D:

1st Band (Value)	2nd Band (Value)	3rd Band (Multiplier)	Resistor Value	Fourth Band (Tolerance)
Violet	Brown	Green		Silver
7	1	100,000	7,100,000 Ω (or 7.1 M Ω)	±10%

Your Turn

Use the chart in Figure 6-29 to determine the values and tolerances of the resistors shown in (c) and (d) in Figure 6-30.

The 10-color resistor code shown in Figure 6-31 is used for resistors with a value *greater than 10 ohms*. Each color is assigned a number from 0 to 9. The resistor color code can be remembered by the following saying:

FIGURE 6-31: *A resistor color code sequence.*

Big Boys Race Our Young Girls But Violet Generally Wins.

By taking the first letter of each of those words, you can rewrite the resistor color code "Black, Brown, Red, Orange, Yellow, Green, Blue, Violet, Gray, White."

There are many different families of parts and components intended to block, control, conduct, switch, store, amplify, and reduce electrical current. The components contained in Figure 6-32 are commonly found in electrical circuits.

ELECTRICAL CIRCUITS

Basic Circuits

An **electrical circuit** is a closed path through which electricity can flow between the terminals of a power source and through one or more electrical components. Some electrical components are used to control the flow of electricity. Other components are used to convert electrical energy into some other form of energy for the purpose of performing a task. The power source serves as a source of electrons and generates the potential difference that causes the electrons to flow. Wires or metal traces provide the pathway through which electricity will flow. A *load* device, such as a buzzer or lamp, is used to transform the electrical energy into another form of energy. Figure 6-33 shows a simple electrical circuit in which a battery provides the voltage to allow electrons to flow from the negative terminal through the wire to a lamp. The lamp converts some of the electrical energy to light energy. From there the electricity flows through another wire to the positive terminal of the battery.

When current is able to flow through a circuit, it is said to be closed. A **closed circuit** provides a path for electrical flow if voltage is applied to it. If the circuit has a break in it and does not allow the flow of electricity through it, it is called an **open circuit**. Sometimes this break is intentional. A control device such as a switch may be inserted between the power source and the load to turn the electricity on and off. Most circuits have some type of switch incorporated into them that allows the circuit to be easily opened or closed. Sometimes breaks are unintentional, such as a broken wire.

> **Electrical circuit:**
> a closed path through which electricity can flow between the terminals of a power source and through one or more electrical components.

FIGURE 6-32: Common electrical components and their schematic symbols.

COMPONENT TYPES	PICTORIAL REPRESENTATIONS	SCHEMATIC SYMBOLS
Conductors (connected)		
Conductors (not connected)		
Cell		
Battery		
Switch (SPST) Single-pole Single-throw		
Switch (SPDT) Single-pole Double-throw		
Switch (DPST) Double-pole Single-throw		
Switch (DPDT) Double-pole Double-throw		
Switch Rotary type		
Switch (NOPB) Push-button type (normally open)		NOPB
Switch (NCPB) Push-button type (normally closed)		NCPB

(a)

COMPONENT TYPES	PICTORIAL REPRESENTATIONS	SCHEMATIC SYMBOLS
Fixed resistor		
Variable resistor		
Voltmeter		
Ammeter milliammeter and microammeter		
Ohmmeter		
Fuses		or
Circuit breaker		
Fixed inductors		AIR CORE IRON CORE
Variable inductor		
Fixed capacitor		
Variable capacitor		

(b)

From Meade, Foundations of Electronics, 5e. 2007, Delmar Learning, a part of Cengage Learning, Inc. Reproduced with permission. www.cengage.com/permissions.

Ohm's Law

Named in honor of Georg Simon Ohm, a German physicist, Ohm's law explains the mathematical relationships between voltage (V), current (I), and resistance (R) in a direct current circuit (see Figure 6-34). **Ohm's law** states that the direct current in a conductor is directly proportional to the voltage applied across the conductor and inversely proportional to the conductor's resistance (see Equation 6-3).

FIGURE 6-33: A simple electrical circuit.

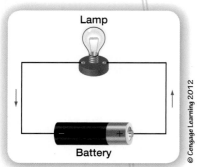

FIGURE 6-34

Voltage	▶ Symbol: **V** ▶ Unit of Measurement: **volt** ▶ Unit Abbreviation: **V**
Current	▶ Symbol: **I** ▶ Unit of Measurement: **ampere "Amp"** ▶ Unit Abbreviation: **A**
Resistance	▶ Symbol: **R** ▶ Unit of Measurement: **Ohm** ▶ Unit Abbreviation: **Ω**

Assuming the resistance remains constant, voltage and current will exhibit a proportional relationship. This means when one quantity increases, the other will increase. Assuming the voltage remains constant, current and resistance will exhibit an inverse relationship. This means when the resistance goes down, the current will increase.

$$\text{Current } (I) = \frac{\text{Voltage } (V)}{\text{Resistance } (R)} \qquad \text{(Equation 6-3)}$$

Figure 6-35 shows the Ohm's law circle, which provides an easy way to identify the 3 mathematical relationships that are defined by Ohm's law. To use it, just cover the variable you are trying to find and whatever's left is your formula. For example, if you are trying to find voltage, cover up the V; you're left with I and R. Because they are next to one another, you will multiply them together to get the equation $V = IR$. If you're solving for current, your equation would be $I = \frac{V}{R}$; and if you're solving for resistance, your equation would be $R = \frac{V}{I}$.

Ohm's law:

the direct current in a conductor is directly proportional to the voltage applied across the conductor and inversely proportional to the conductor's resistance.

FIGURE 6-35: The Ohm's law circle.

© Cengage Learning 2012

Example

Problem: Use Ohm's law to solve for current, voltage, and resistance.

a. To solve for *current* given voltage and resistance (see Figure 6-36):

$$I = \frac{V}{R}$$

$$I = \frac{12\,V}{150\,\Omega}$$

$$I = 0.08\,A \text{ or } 80\,mA$$

FIGURE 6-36

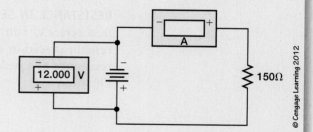

© Cengage Learning 2012

b. To solve for *voltage* given current and resistance (see Figure 6-37):

$$V = I \times R$$

$$V = 4\,A \times 6\,\Omega$$

$$V = 24\,V$$

FIGURE 6-37

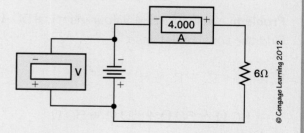

© Cengage Learning 2012

c. To solve for *resistance* given voltage and current (see Figure 6-38):

$$R = \frac{V}{I}$$

$$= \frac{15\,V}{5\,A}$$

$$= 3\,\Omega$$

FIGURE 6-38

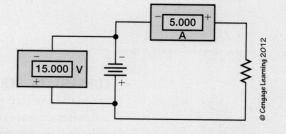

© Cengage Learning 2012

Series Circuits

A **series circuit** provides only one path for electricity to flow through two or more electrical components. If there is a break anywhere in the circuit, the circuit will be opened and there will be no electrical flow. In Figure 6-39, a series circuit is made of one battery and four lights. The lights are connected in series with one another. If one of the lights were to burn out, all of the lights would go out because it would create an open circuit and prevent electrical flow.

FIGURE 6-39: A series circuit.

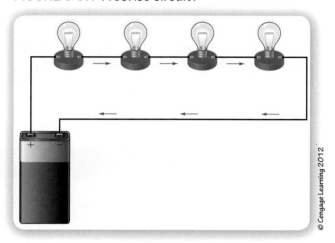

© Cengage Learning 2012

RESISTANCE IN SERIES CIRCUITS Current must pass through all of the resistors in a series circuit because there is only one path for current. Each additional resistor placed in a series circuit adds more opposition to the current. The total resistance of a series circuit is equal to the sum of all resistors in the circuit (see Equation 6-4).

$$R_T = R_1 + R_2 \cdots R_n \qquad \text{(Equation 6-4)}$$

Example

Problem: In Figure 6-40, four identical 50-Ω resistors are connected in series with a 9-V battery. Find the total resistance:

FIGURE 6-40

$R_T = R_1 + R_2 + R_3 + R_4$

$= 50\,\Omega + 50\,\Omega + 50\,\Omega + 50\,\Omega$

$= 200\,\Omega$

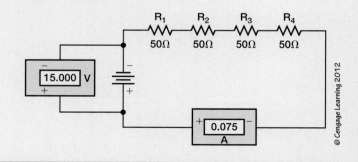

© Cengage Learning 2012

CURRENT IN SERIES CIRCUITS Current in a series circuit remains the same no matter where you take the reading. As you can see in Figure 6-41, all three ammeters measured 0.150 A despite their different locations. Given a constant voltage, current and resistance are inversely proportional. If the total resistance in the circuit shown in Figure 6-41 was decreased, the current would increase. If the voltage was increased and the resistance remained the same, current would increase. This is because voltage and current are directly proportional given a constant resistance.

VOLTAGE IN SERIES CIRCUITS Voltage does not remain constant throughout a series circuit. There are five voltmeters connected to the series circuit shown in Figure 6-42. The voltmeter connected to the power source is measuring the total voltage (V_T). The four remaining voltmeters are positioned over the resistors. Notice that each of those four voltmeters has a different reading. The largest resistor (60 Ω) has the largest voltage (drop) reading (−12 V).

FIGURE 6-42

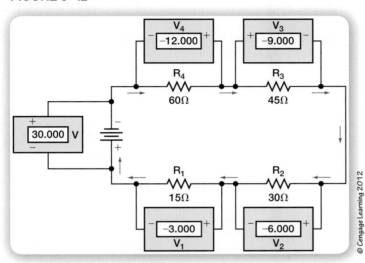

© Cengage Learning 2012

The largest resistor in a series circuit drops the voltage the most, and the smallest resistor drops the voltage the least. This is because each resistor's drop in voltage is equal to current (I) times resistance (R). This concept is known as *voltage drop*. For example, the voltage drop over R1 is −5 V, and the voltage drop over R4 is also −5 V. **Kirchhoff's voltage law** states that the sum of the individual voltage drops in a single loop and the applied voltage must equal zero. In Figure 6-42, the voltage drops are negative because they are an indication of the difference in voltage from one side of the resistor to the other. Because the voltage drops are negative and the applied voltage is positive, they cancel each other out:

$$V_A + V_1 + V_2 + V_3 + V_4 = 0$$

$$30 \text{ V} + (-3 \text{ V}) + (-6 \text{ V}) + (-9 \text{ V}) + (-12 \text{ V}) = 0$$

$$30 \text{ V} - 30 \text{ V} = 0$$

$$0 = 0$$

Voltage Divider Rule Because the same current passes through every component in a series circuit, the voltage drop of each component is equal to the same percentage of the circuit's applied voltage as the resistor is of the total resistance. The voltage drop of each resistor in a series circuit is directly proportional to its resistance value compared to the total resistance of the circuit (see Equation 6-5). The voltage divider rule can be used to calculate the voltage drop of any resistor in a series circuit. If you know the applied voltage, the individual resistor values and the total resistance, you have enough information to solve for each of the voltage drops.

$$V_X = \frac{R_X}{R_T} \times V_T$$

(Equation 6-5)

V_X = Voltage drop across the selected resistor

V_T = The applied voltage to the circuit

R_X = Resistance value of the selected resistor

R_T = Total resistance of the circuit

Example

Problem: Using the circuit shown in Figure 6-42, calculate the voltage drop over R4.

$$V_4 = \frac{R_4}{R_T} \times V_T$$

$$V_4 = \frac{60\,\Omega}{150\,\Omega} \times 30\,V$$

$$V_4 = 12\,V \text{ drop over } R_4$$

Point of Interest

You've probably heard the term *horsepower* (see Figure 6-43). Whether you're purchasing a new car, a lawnmower, or a blender, chances are each one has a horsepower rating. The term *horsepower* was coined by Scottish inventor James Watt (1736–1819). You might recognize his last name because it is stamped on lightbulbs as an indication of the power it uses, but Watt's greatest achievement was the improvement of steam engines.

The invention of the steam engine created a need to equate animal power to the mechanical power produced by the steam engine. While observing ponies that were being used to lift coal from an underground mine, Watt estimated that an average mine pony could lift 220 pounds over a distance of 100 feet in 1 minute. Using the equation W = F · d, he determined that each pony was able to produce 22,000 ft · lb/min of power. Because a pony is smaller than a draft horse, Watt estimated that a draft horse would be able to produce 50% more power, bringing the total to 33,000 ft · lb/min of power. Horsepower is most commonly used in terms of seconds (550 ft · lb/sec).

FIGURE 6-43

POWER IN A SERIES CIRCUIT Electrical energy (Q) is the ability to do work by harnessing the flow of electricity. As you learned in Chapter 5, the joule (J) is the basic unit of energy. It was named after English physicist James Prescott Joule (1818–1889). In electrical systems, 1 joule is the energy required to move 1 coulomb of electrical charge between two points with a potential difference of 1 volt. The watt-hour (Wh) and kilowatt-hour (kWh) are also common units of measurement of electrical energy.

As stated earlier, power is the rate at which electrical energy is transferred or transformed. The watt (W) is the unit of power in electrical systems. One watt is equal to 1 joule of energy being transformed over a period of 1 second (see Equation 6-6).

Watt (W): the unit of power in electrical systems.

$$\text{Power (watt)} = \frac{\text{Energy (joule)}}{\text{Time (second)}} \qquad \text{(Equation 6-6)}$$

Example

Problem: If 4,000 joules of energy are consumed over a period of 50 seconds, what is the power used?

$$P = \frac{Q}{t}$$

$$= \frac{4,000 \text{ J}}{50 \text{ sec}}$$

$$= 80 \text{ W}$$

The Ohm's law and power formula wheel shown in Figure 6-44 provides a graphical representation of the relationships between voltage, current, resistance, and power in a direct current electrical circuit.

FIGURE 6-44: The Ohm's law and power formula wheel.

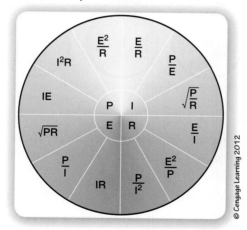

© Cengage Learning 2012

OFF-ROAD EXPLORATION

How to Read Your Gas and Electric Meters

Determining how much electrical power your family consumes is the first step to saving energy. Fortunately, reading your electric meter is easy. The meter consists of a series of clock-style dials with rotating hands to indicate usage (see Figure 6-45). Your meter may have four or five dials. Each hand rotates in the opposite direction of the preceding dial, either clockwise or counterclockwise. Take a reading each day at the same time and record the values in a journal. Compare the changes in value against the days of the week and changes in weather to reveal patterns in energy use.

CALCULATING TOTAL ENERGY CONSUMPTION Your local electrical utility company charges by how much power (watts) your home consumes over a period of time (hours). For example, if you were to leave one 75-watt lightbulb on for a period of 1 hour, the utility company would charge for 75 watt-hour (Wh) of energy (see Equation 6-7). Because the typical home consumes thousands of watts of energy each day, utility companies deal in kilowatt-hours (kWh).

$$\text{Energy (watt-hours)} = \text{Power (watts)} \times \text{Time (hours)} \qquad \text{(Equation 6-7)}$$

Example

Problem 1: If you were to turn on five 2.5-watt light bulbs for a period of 4 hours, what is the total electrical energy consumption in watt-hours?

Q (energy) = P (power) × T (time)

= (5 bulbs × 2.5 watts each) × 4 hours

= 50 Wh (0.05 kWh)

Problem 2: Calculate the power consumption (in watts) if two incandescent light bulbs that are powered by a 9 V power source consume a total of 0.025 A (25 mA).

P (power) = V (voltage) × I (current)

= 9 V × 0.025 A

= 0.23 W

Knowing how to read an electrical utility meter is a necessary skill for keeping track of how much electrical power your home consumes in kilowatt-hours. Stand directly in front of your meter and record the position of the hand starting with the dial on the left. If the hand is between two numbers, always record the lower number. When the hand is between 9 and 0, 0 is considered a 10; therefore, 9 would be the lower number. If the hands are close to or directly over a number, check the dial to the right. If the hand on the dial to the right is between 9 and 0, then record the next lowest number; otherwise, use the exact number. For example, the second dial in Figure 6-45 appears to be a 4; however, it is still a 3 because the third dial is between the 9 and 0.

FIGURE 6-45: The correct reading for this meter is 13,905.

© Cengage Learning 2012

Parallel Circuits

Unlike series circuits, **parallel circuits** provide two or more pathways for electricity to flow. Other major differences between series and parallel circuits include the following:

> ▶ Voltage remains constant throughout all branches of a parallel circuit.

> ▶ Current varies in each branch depending on the resistance value.

> ▶ The total resistance of a parallel circuit is always less than the smallest resistance value of any branch.

VOLTAGE IN PARALLEL CIRCUITS The voltage remains constant across all components in a parallel circuit. Figure 6-46 illustrates that measuring the source voltage or the voltage across any component will yield the same voltage value.

KIRCHHOFF'S CURRENT LAW (CURRENT IN PARALLEL CIRCUITS) **Kirchhoff's current law** states that the current entering a point must equal the current exiting that

Parallel circuit:

provides two or more paths for current to flow.

point. This is a variation of the conservation of energy that states that energy cannot be created or destroyed. Current always seeks the path of least resistance. In a parallel circuit, the resistor with the lowest resistance value will also carry the most current.

Figure 6-47 shows a circuit with three resistors connected in parallel to one another. Think of R1, R2, and R3 as three different-sized pipes that water is being pumped through. Each pipe offers a different resistance to the flow of water. The largest pipe, R1, has the lowest resistance and will allow the most amount of water through. The smaller pipes, R2 and R3, have larger resistance values and limit the flow of water more than R1.

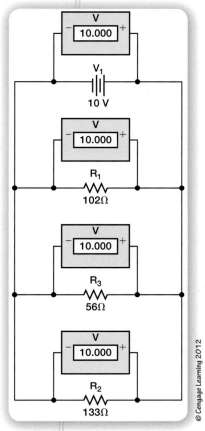

FIGURE 6-47

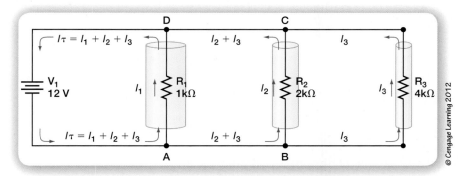

© Cengage Learning 2012

Remember, Kirchhoff's current law states that the current entering a given point must equal the current leaving that point. Therefore, the total current leaving the battery (V_1) must equal the current returning to the battery. The current splits between the three parallel paths. The amount of current entering each path is inversely proportional to the resistance of the individual branches.

When the current leaves the negative terminal of the battery, it heads down the conductor toward point A. At point A, a portion of the current will flow through R1 toward point D. The remaining current at point A will head for point B, where it will split again. A portion of the current at point B will head through R2 toward point C. The remaining current will head for R3 toward point C. This current recombines with the current at point C and then again at point D where it heads toward the positive terminal of the battery.

Finding the amount of current flowing through each branch is easily done by using Ohm's law. Remember, voltage remains the same across all branches of a parallel circuit. Because the voltage across each branch is equal to the voltage across the battery and the resistance values of each branch are given, the only unknown is current.

Example

Problem: Calculate the current for each branch of the parallel circuit shown in Figure 6-47. Remember that $1\text{ k}\Omega = 1{,}000\ \Omega$.

a. To calculate the current through *branch 1*:

$$I_1 = \frac{V_1}{R_1}$$

$$= \frac{12\text{ V}}{1{,}000\ \Omega}$$

$$= 0.012\text{ A }(12\text{ mA})$$

(continued)

b. To calculate the current through *branch 2*:

$$I_2 = \frac{V_2}{R_2}$$

$$= \frac{12\,V}{2{,}000\,\Omega}$$

$$= 0.006\,A\,(6\,mA)$$

c. To calculate the current through *branch 3*:

$$I_3 = \frac{V_3}{R_3}$$

$$= \frac{12\,V}{4{,}000\,\Omega}$$

$$= 0.003\,A\,(3\,mA)$$

d. The *total current* can be calculated by finding the sum of all branch currents:

$$I_T = I_1 + I_2 + I_3$$

$$= 0.012\,A + 0.006\,A + 0.003\,A$$

$$= 0.021\,A\,(21\,mA)$$

RESISTANCE IN PARALLEL CIRCUITS The total resistance of a parallel circuit is referred to as *equivalent resistance*. If you could replace the three resistors in Figure 6-47 with one resistor that would offer the same resistance, that would be the equivalent resistance. The three methods for calculating equivalent resistance are the Ohm's law method, identical resistors method, and the reciprocal method. The number of resistors and their resistance values determine which method(s) can be used.

Ohm's Law Method The Ohm's law method can be used when you know the total current (I_T) and the source voltage (V_T) of the parallel circuit. Using the total current and the source voltage from the previous example, the total resistance (R_T) of the circuit shown in Figure 6-47 can be easily calculated.

Example

Problem: Using the Ohm's law method, find the total resistance of the circuit shown in Figure 6-47.

$$R_T = \frac{V_1}{I_T}$$

$$= \frac{12\,V}{0.021\,A}$$

$$= 571.43\,\Omega$$

Identical Resistors Method When all resistors in a parallel circuit have the same resistance value, the identical resistors method can be used to find the equivalent resistance. The parallel circuit shown in Figure 6-48 has three identical resistors connected in parallel. The equivalent resistance can be found by dividing the value of each resistor (R) by the number of resistors (*n*) in the circuit (see Equation 6-8). The value of each of the resistors is 600 ohms, and there are three resistors in the circuit.

$$R_T = \frac{R}{n}$$
(Equation 6-8)

FIGURE 6-48: Calculating equivalent resistance of a parallel circuit using the identical resistor method.

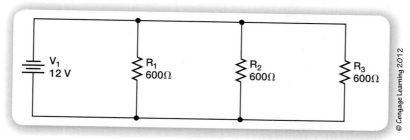

Example

Problem: Using the identical resistors method, find the equivalent resistance of the circuit shown in Figure 6-48.

$$R_T = \frac{R}{n}$$

$$= \frac{600\ \Omega}{3}$$

$$= 200\ \Omega$$

Reciprocal Method If you don't know the total voltage or total current and the resistors do not have the same value, then you can use the reciprocal method to find the equivalent resistance (see Equation 6-9). The total resistance of any parallel circuit is always less than the lowest resistance value. For example, in Figure 6-49 the total resistance of the circuit is going to be less than 450 Ω because that is the lowest resistance value of all branches.

$$R_T = \frac{1}{\left(\frac{1}{R_1}\right) + \left(\frac{1}{R_2}\right) + \left(\frac{1}{R_3}\right) \cdots + \left(\frac{1}{R_n}\right)}$$
(Equation 6-9)

FIGURE 6-49: Calculating total resistance of a parallel circuit using the reciprocal method.

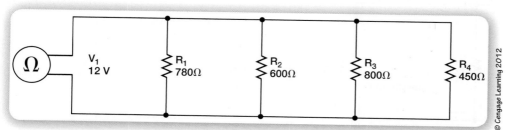

Example

Problem: Using the reciprocal method, calculate the total resistance of the parallel circuit shown in Figure 6-49.

$$R_T = \cfrac{1}{\left(\cfrac{1}{R_1}\right) + \left(\cfrac{1}{R_2}\right) + \left(\cfrac{1}{R_3}\right) + \left(\cfrac{1}{R_4}\right)}$$

$$R_T = \cfrac{1}{\left(\cfrac{1}{750\,\Omega}\right) + \left(\cfrac{1}{600\,\Omega}\right) + \left(\cfrac{1}{800\,\Omega}\right) + \left(\cfrac{1}{450\,\Omega}\right)}$$

$$= 154.5\,\Omega$$

Series–Parallel Circuits

A series–parallel circuit contains components that are connected in series and components that are connected in parallel. Most electronic devices have a combination of series and parallel circuits. From calculators to video-game consoles, these devices are loaded with series–parallel circuitry. These circuits are designed to take advantage of the benefits of both series and parallel circuits. Series–parallel circuits are commonly called *combinational circuits*. They can be easily understood based on what you have already learned about series and parallel circuits.

RESISTANCE IN SERIES-PARALLEL CIRCUITS Figure 6-50 shows a series–parallel circuit with four resistors. The following steps will help you to analyze series–parallel circuits:

Step 1: Draw arrows on the circuit showing the direction of current flow throughout the circuit (see Figure 6-50).

Step 2: Locate and draw a box around each of the subcircuits throughout the series–parallel circuit. For example, the circuit shown in Figure 6-50 consists of one series subcircuit (A) and one parallel subcircuit (B).

Step 3: Solve for the series and parallel subcircuit equivalent resistances. Figure 6-51 shows the equivalent resistance diagram of the series–parallel circuit from Figure 6-50. The resistors in subcircuit A (R3 and R4) are connected in series

FIGURE 6-50: A series–parallel circuit with four resistors.

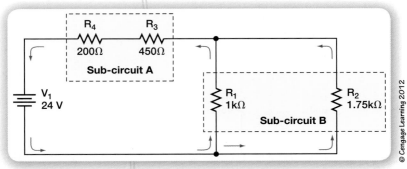

© Cengage Learning 2012

FIGURE 6-51: The equivalent resistance diagram of the series–parallel circuit shown in Figure 6-50.

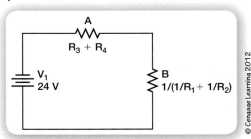

© Cengage Learning 2012

and can be summed. The total resistance of subcircuit A (RA) is equal to R3 + R4.

$$R_A = R_3 + R_4$$
$$= 450 \, \Omega + 200 \, \Omega$$
$$= 650 \, \Omega$$

Step 4: The resistors in subcircuit B are connected in parallel to each other. Using the reciprocal method for calculating equivalent resistance, the formula is:

$$R_B = \frac{1}{\left(\frac{1}{R_1}\right) + \left(\frac{1}{R_2}\right)}$$

$$= \frac{1}{\left(\frac{1}{1,000 \, \Omega}\right) + \left(\frac{1}{1,750 \, \Omega}\right)}$$

$$= 636.36 \, \Omega$$

Step 5: Solve for the total resistance of the circuit. In this case, the circuit has been simplified to a series circuit, allowing the two equivalent resistances to be summed.

$$R_T = R_A + R_B$$
$$= 650 \, \Omega + 636.36 \, \Omega$$
$$= 1,286.36 \, \Omega$$

CURRENT AND VOLTAGE IN SERIES–PARALLEL CIRCUITS Ohm's law is the simplest way to calculate current in a series–parallel circuit. In order to find the total current for the entire circuit, we must calculate the total resistance. The following steps can be used to calculate current in a series–parallel circuit.

Step 1: Simplify the series–parallel circuit shown in Figure 6-52 using equivalent resistances. A simplified version of the circuit is shown in Figure 6-53.

Step 2: Solve for each subcircuit individually. In Figure 6-52, all resistors in series within a subcircuit can be added together. For example, resistors R1 and R2 are connected in series and can be summed (see Figure 6-53).

$$R_1 + R_2 = 200 \, \Omega$$
$$R_3 + R_4 = 450 \, \Omega$$
$$R_5 + R_6 + R_7 = 450 \, \Omega$$

FIGURE 6-52: *Current in a series–parallel circuit.*

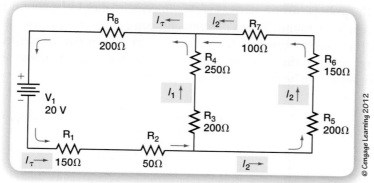

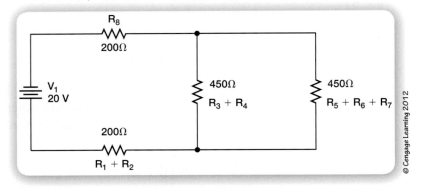

Step 3: Because the resistance values in each of the branches of the parallel sub-circuit are the same, the equivalent resistance will be exactly one-half of the resistance found in both branches. The total resistance of the parallel subcircuit is 225 Ω:

$$R_T = \cfrac{1}{\left(\cfrac{1}{R_3 + R_4}\right) + \left(\cfrac{1}{R_5 + R_6 + R_7}\right)}$$

$$R_T = \cfrac{1}{\left(\cfrac{1}{450\ \Omega}\right) + \left(\cfrac{1}{450\ \Omega}\right)}$$

$$= 225\ \Omega$$

FIGURE 6-54: Equivalent resistance values for the circuit shown in Figure 6-53.

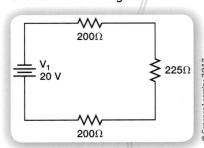

Step 4: The circuit shown in Figure 6-54 is a simplified version of the already simplified circuit shown in Figure 6-53. Because the equivalent resistances have been found, this circuit was simplified to a series circuit where the individual resistance values can be summed to find the total resistance. The total resistance for this series–parallel circuit is 625 Ω:

$$R_T = 200\ \Omega + 225\ \Omega + 200\ \Omega = 625\ \Omega$$

Step 5: The total current can be found using Ohm's law:

$$I = \frac{V}{R}$$

$$= \frac{20\ V}{625\ \Omega}$$

$$= 0.032\ A\ (32\ mA)$$

Step 6: The voltage drops across every component should be found before the current is found. When components are connected in series, the voltage drops can be summed. If components are connected in parallel, the voltage remains the same. In Figure 6-55, the voltage drop across any resistor in series can be found by using Equation 6-10

$$V_X = \frac{R_X}{R_T} \times V_T \qquad \text{(Equation 6-10)}$$

FIGURE 6-55

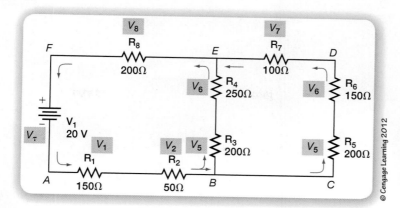

Example

Calculate the voltage drop across resistor R8 in Figure 6-55:

$$V_X = \frac{R_8}{R_T} \times V_1$$

$$= \frac{200\,\Omega}{625\,\Omega} \times 20\,V$$

$$= 6.4\,V$$

Step 7: After the individual voltage drops have been found, the current can be calculated using Ohm's law. See Figure 6-56 for a list of all voltage drops and current for every component in the circuit.

FIGURE 6-56: Voltage drops and current for every component in the circuit.

Resistor	Resistance	Voltage Drop	Current
R_1	150 Ω	4.8 V	32 mA
R_2	50 Ω	1.6 V	32 mA
$R_1 + R_2$	**200 Ω**	**6.4 V**	**32 mA**
R_3	200 Ω	3.2 V	16 mA
R_4	250 Ω	4 V	16 mA
$R_3 + R_4$	**450 Ω**	**7.2 Ω**	**16 mA**
R_5	200 Ω	3.2 V	16 mA
R_6	150 Ω	2.4 V	16 mA
R_7	100 Ω	1.6 V	16 mA
$R_5 + R_6 + R_7$	**450 Ω**	**7.2 V**	**16 mA**
R_8	200 Ω	6.4 V	32 mA

SUMMARY

- Negatively-charged subatomic particles called *electrons* orbit the nucleus of an atom within a series of shells. For conductive materials, the electrons in the outmost shell can be directed to jump from one atom to another, which constitutes electricity.

- An *ion* is a single atom or molecule that has gained or lost one or more electrons, making it positively- or negatively-charged.

- The *law of charges* states that opposite charges attract and like charges repel.

- *Voltage* (*V*) is the electrical pressure that causes electrons to move through a conductor. It is a measure of the difference in charge between two points. Sources of voltage include batteries, generators, thermocouples, and piezoelectric devices.

- *Current* is the movement of charge carriers, such as free electrons, holes, and ions. Electron current involves the movement of electrons from one atom to another along a conductive path.

- The *intensity* of current (*I*) is a measure of the amount of electrons that move past a given point in one second.

- Electrical *resistance* (*R*) is the opposition to the flow of electricity.

- The nominal resistance value and tolerance range for through-hole style fixed resistors is determined using a color code system.

- *Polarity* is the condition of being electrically positive or negative. A plus sign (+) indicates positive charge and a minus sign (-) indicates negative charge.

- The Electronics Industries Association (EIA) set up a color coding system to identify the value of the resistor.

- An *electrical circuit* consists of a source of power, a closed path through which electricity can flow, and one or more electrical components use electricity to perform a task by transforming electrical energy into some other form of energy.

- *Ohm's law* states that the electrical current in a conductor is directly proportional to the voltage across that conductor and inversely proportional to the conductor's resistance.

- A *series circuit* provides only one path for electricity through two or more electrical components. Current is constant in a series circuit.

- *Kirchhoff's voltage law* states that the sum of the individual voltage drops in a single loop and the applied voltage must equal zero.

- In an electrical circuit, *power* (*P*) is a measure of the rate of transmitting or transforming electrical energy. The watt (W) is the unit of power in electrical systems. Power is the rate at which energy flows. One watt is equal to 1 joule of energy being transformed over a period of 1 second.

- Each component in a *parallel circuit* has a direct connection to the power source. Voltage is constant in a parallel circuit.

- *Kirchhoff's current law* states that the current entering a point must equal the current exiting that point.

- A *series–parallel circuit* contains some components that are connected in series and other components that are connected in parallel.

BRING IT HOME

1. Draw and label the parts of an atom.
2. If two objects are both positively charged, will they attract or repel? Explain why.
3. What makes one material a better conductor than another?
4. Define polarity in your own words.
5. Draw and label a diagram of a battery and explain how it works.
6. What is the difference between an anode and a cathode?
7. What is the name of the device that generates a voltage when exposed to light?
8. How many KΩ are there in a MΩ?
9. Convert the following units:
   a. 0.015 ohms to milliohms
   b. 20,000 volts to kilovolts
   c. 35 milliamperes to amperes
   d. 0.75 megaohms to kiloohms
   e. 17,500,000 ohms to megaohms
10. Find the resistance value and the tolerance value for the following color codes. (*Note:* The first color represents the first band, the second color represents the second band, and so on.)
    a. yellow, violet, yellow, and silver
    b. red, red, green, and silver
    c. orange, orange, black, and gold
    d. white, brown, brown, and gold
    e. brown, red, gold, and gold
11. If a precision resistor has a fifth band, how does that affect the tolerance?
12. In a series circuit, if the voltage increases and the resistance remains constant, what should happen to the current?
13. In a series circuit, if the resistance is decreased and the voltage remains the same, what should happen to the current?
14. Draw a schematic diagram showing a voltage source, an ammeter, and a resistor.
15. Explain the similarities and differences between series and parallel circuits.
16. Explain how to find the total resistance of a:
    a. series circuit.
    b. parallel circuit.
17. What law states "the mathematical sum of the voltage drops around a complete circuit must equal the applied voltage"?
18. Create a sketch of a parallel circuit with the following components:
    a. a voltage source
    b. two resistors connected in parallel to one another
    c. a voltmeter positioned to measure total voltage
    d. an ammeter positioned to measure total current
19. Create a sketch of a series–parallel circuit with the following components:
    a. a voltage source
    b. two resistors connected in parallel to one another
    c. two resistors connected in series to one another
    d. a voltmeter positioned to measure total voltage
    e. an ammeter positioned to measure total current

EXTRA MILE

1. Design a simple burglar alarm circuit that will sound a buzzer when your bedroom door is opened.
2. Build a circuit capable of controlling two 9-volt motors. Each motor should be controlled separately.
3. Design a generator that can convert some of the mechanical energy from a bicycle into electrical energy.

CHAPTER 7
Fluid Power Systems

Menu	START LOCATION	DISTANCE	END LOCATION

Before You Begin

Think about these questions as you study the concepts in this chapter:

1 What makes fluid power different from other power transmission methods?

2 How is a hydraulic system different from a pneumatic system, and what advantages does one have over the other?

3 What do hydraulic and pneumatic systems have in common?

4 What is the difference between hydrostatics and hydrodynamics?

5 What does a hydrostatic system amplify in an input force?

6 What is the difference between a fluid's flow rate and its flow velocity?

7 What is Bernoulli's principle, and what are its applications?

8 How is atmospheric pressure different from gauge pressure and absolute pressure?

9 Why are the standard Fahrenheit and Celsius temperature scales inappropriate for calculating the physical properties of gases in confined pneumatic systems?

10 What is the difference between Boyle's law, Charles' law, and Gay-Lussac's law if all three are associated with the temperature, volume, and pressure of a confined gas?

Since the Industrial Revolution, machines have served as the workhorses for our modern, human-built world. Machines give us the ability to mine large quantities of raw materials, harvest crops, build roads and structures, manipulate materials to form countless consumer products, and move people and goods across vast distances. All of these machines have one thing in common: They must all transmit power to accomplish their specific tasks. The three primary power transmission methods used in machines are mechanical systems, electrical systems, and fluid power systems. Figure 7-1 gives examples of components from each of these methods.

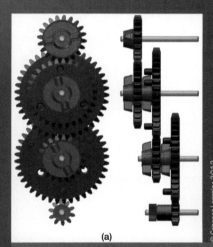

(a)

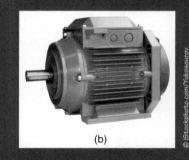

(b)

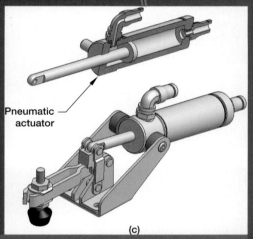

Pneumatic actuator

(c)

Figure 7-1: *Mechanical, electromechanical, and fluid power systems are the three most common methods for transmitting power in industrial applications. Example components from these systems include (a) gears, (b) electric motors, and (c) pneumatic actuators, respectively.*

Fluid power is the use of a confined fluid flowing under pressure to transmit power from one location to another. A fluid can be a gas or a liquid; both media share similar characteristics with respect to *pressure* and the ability to take on the shapes of their containers. Therefore, fluid power systems are divided into two categories: *pneumatics* and *hydraulics*, dealing with gases and liquids, respectively.

Note that many machines use all three forms of power transfer. For example, an automobile uses mechanical components such as gears to transmit torque and speed from the engine to the wheel axles. A starter motor, which is an electromechanical component, is used to turn an automobile engine crankshaft in order to start the internal-combustion process. The fluid power components found within the power-steering and power-brake systems allow a driver to easily control the motion of an automobile that may be more than 20 times the driver's weight.

Almost everyone in the modern world has encountered a hydraulic-powered device, even though they may not have been aware of it. **Hydraulics** is the physical science and technologies that are associated with liquids that are at rest or flowing under pressure. This technology has been used in the power-steering and power-braking systems of virtually every automobile since the 1940s. For the past

Fluid power:

the use of a confined fluid flowing under pressure to transmit power from one location to another.

Hydraulics:

the physical science and technologies associated with liquids that are at rest or flowing under pressure.

215

50 years, owners of so-called low-rider vehicles have outfitted their cars with custom hydraulic-powered adjustable-height suspension systems (see Figure 7-2), that allow a vehicle to execute moves worthy of music videos. However, the most visible application of raw hydraulic power can be seen on heavy construction equipment. If an extremely heavy object needs to be moved, then a hydraulic-powered piece of equipment is often the technology for the job.

(a)

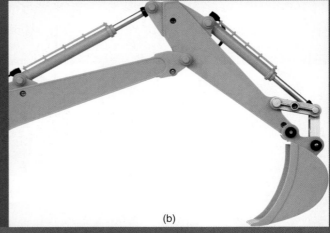
(b)

Figure 7-2: *(a) This lowrider can perform moves that the average car cannot because of its custom hydraulic-powered, adjustable-height suspension system. (b) The same hydraulic components found in the lowrider car are also used to power heavy construction equipment.*

Pneumatics:

the physical science and technologies associated with the mechanics of pressurized gases.

Pneumatics is the physical science and technologies that are associated with the mechanics of pressurized gases. The power of moving air has been used since the beginning of civilization when the first sail ships were used for exploration and trade. Today, we can see examples of pneumatic technology at work in a contractor's nail gun, a dentist's drill, a mechanic's impact wrench, and the air pistons found on an automobile's rear hatchback door (see Figure 7-3). Most

Figure 7-3: *Construction workers use pneumatic nail guns to assemble wall frames, apply flooring, and shingle the rooftops of residential homes.*

consumers don't realize that pneumatic devices are used throughout automated manufacturing work cells to move and workhold goods that must be assembled, inspected, and packaged.

> S T E M
>
> Fluid power technologies are used in most of the major industries, including medical, agriculture, transportation, entertainment, manufacturing, and construction.

Like mechanical systems, fluid power systems give the user the ability to significantly amplify an input force. They are also capable of transmitting power across distances and along pathways that are impractical for mechanical components like gears, chains, and pulleys. Though electrical wires can also be routed along complex paths to supply electricity to motors, a hydraulic pump has a power density (power per unit volume) that is approximately 10 times greater than that of an electric motor. This means that an electric motor that is approximately 10 times the size of a hydraulic pump would be needed to generate an equivalent power output.

Basic Principles of Fluid Power

A closed fluid power system is one in which the fluid is confined to a container or series of containers that are linked together. The fundamental concepts that are common to all *closed* fluid power systems include *area, force,* and *pressure*.

As you have probably learned in your mathematics classes, area (A) is a measure of the two-dimensional space that is enclosed by a shape. Most pneumatic and hydraulic linear actuators are cylindrical. The first step in calculating the volume of a cylinder is to calculate its circular cross-sectional area (see Equation 7-1). The area of a circle may be calculated using either its diameter or its radius value.

$$A_{\text{circle}} = \frac{\pi d^2}{4} = \pi r^2 \qquad \text{(Equation 7-1)}$$

You learned in Chapters 3 and 4 that a force (F) is a push or pull that acts on an object, causing that object to move, change its motion, or deform in some way. The term *pressure* is often used when discussing fluid power systems, though it should not be confused with the concept of force. Pressure (p) is a type of load that occurs when a force is distributed across an area and acts perpendicular to the surface. For example, when you stand on level ground, the weight of your body is spread out across the surface of your feet, with the resulting pressure applied perpendicular to the ground. Though you cannot easily change your weight, you can manipulate the pressure that you exert on the ground by the type of shoes that you wear. For example, high heel shoes will concentrate a person's weight into a small area; especially where the heels meet the floor. This results in a high pressure. A much smaller pressure is exerted on the ground if the same person were to wear a pair of snowshoes, which would distribute his or her weight across a much greater area (see Figure 7-4).

> **Pressure (p):**
> a type of load that occurs when a force is distributed perpendicular to the surface of an object.

FIGURE 7-4: Why would a person wearing (a) a high-heeled shoe generate more pressure against the ground in comparison to the same person wearing (b) a snowshoe?

(a)

© iStockphoto.com/YvanDube.

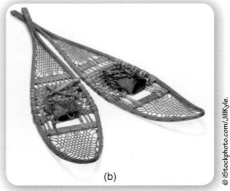

(b)

© iStockphoto.com/JillKyle.

As shown in Equation 7-2, pressure is mathematically defined as the force per unit area. The U.S. customary unit of measure is pounds per square inch (lb/in^2) or psi. The metric unit of pressure is the pascal (Pa), which is equivalent to 1 newton per square meter (N/m^2). One pascal is a very small unit of pressure. In fact, 1 lb/in^2 is equivalent to 6,895 Pa. As such, pressure is often expressed in kilopascals (kPa) or megapascals (MPa).

$$p = \frac{F}{A}$$
(Equation 7-2)

Hydraulic fluid power systems can generate tremendous pressures. The standard operating hydraulic fluid pressure for a commercial aircraft is 3,000 psi, which is comparable to the weight of a midsized vehicle applied to 1 square inch of surface. The development of the Airbus A380 airliner in 2000 (see Figure 7-5) pushed this standard to 5,000 psi.

FIGURE 7-5: Hydraulic fluid pressurized to 5,000 psi is required to move the control surfaces on the giant Airbus A380 commercial airliner, which can seat approximately 850 people.

Image copyright Nalyyer, 2010. Used under license from Shutterstock.com.

Hydraulic versus Pneumatic Systems

How does an engineer determine if a hydraulic system is more appropriate to use than a pneumatic system, and vice versa? Though there are several considerations, the most common deciding factor is based on the compressibility of gases versus the relatively incompressible nature of liquids. Hydraulic fluids do experience compression under pressure, but the amount of compression is extremely small. For example, an oil-based hydraulic fluid will experience a decrease in volume of approximately 0.4% per 1,000 lb/in^2 of pressure. Because fluids are relatively incompressible, hydraulics systems are more appropriate for situations that require precise positioning of objects. Hydraulic systems are also able to generate tremendous power, which allows work to be done on exceptionally heavy loads. For these reasons, almost all heavy construction machines use hydraulic systems. Hydraulic systems are also used in situations that call for "fluid" movements because the output motion of a hydraulic actuator is much smoother than that of a pneumatic actuator. Robots that are designed for both industrial and entertainment purposes take advantage of these qualities.

If an application calls for the quick movement of relatively light objects across short distances, then a pneumatic system may be the most appropriate choice. Because most pneumatic systems use the same air that we breathe, they have a major advantage over hydraulics: an inexhaustible supply of fluid. Also, compressed air is not returned to a holding tank but is exhausted back to the atmosphere once it has run its course through a pneumatic circuit. Pneumatic systems also operate at much lower pressure levels than hydraulic systems (usually around 90 psig).

Neither system is without its disadvantages, which must also be considered. Hydraulic systems can generate extremely high pressures, which causes a hydraulic fluid to heat up and lose its *viscosity*. If the fluid temperature is not taken into consideration, then the system will develop leaks. Generally speaking, hydraulic systems incur higher maintenance costs than pneumatic systems. Leaking fluids also pose safety risks. Many hydraulic oils are petroleum based and inappropriate for use in environments that must remain clean. Petroleum-based oils will also ignite if exposed to an open flame or extremely hot surfaces. Engineers who design aircraft must contend with these drawbacks.

Pneumatic systems have their own unique drawbacks. The compressibility of gas makes pneumatic actuators inappropriate for situations that require precise intermittent motion. Also, the noise that is generated by the compressed gases as they are exhausted to the atmosphere can be significant. Even with large accumulator tanks, an *air compressor* will use a large amount of energy when generating and maintaining a supply of compressed air at constant pressure. This presents significant operating costs. The internal workings of pneumatic components must be lubricated to prevent seizure. This requires the use of a lubricator device within the pneumatic circuit. Water will build up in a pneumatic system as a result of the compression process. This water must be removed on a regular basis to prevent corrosion of the metal components within the system.

MILESTONES IN THE HISTORY OF FLUID POWER

Though important developments and discoveries related to the technology and science of fluid power have occurred over the last 6,000 years (see Figure 7-6), fluid power engineering is still a relatively new discipline. For example, hydraulic-powered machines were a new concept to the industrial world at the beginning of the 19th century. Common domestic uses of fluid power technology were limited to simple devices such as the hand-operated water pump and air bellows that were used to stoke fires. This began to change with the development of oil-based hydraulic fluids and the incorporation of hydraulic-powered automobile brakes in the early 1920s.

One of the most fundamental scientific laws of fluid power was discovered by a French mathematician named Blaise Pascal, who lived during the 17th century. In 1647, Pascal published what is known today as **Pascal's law**. This law describes the behavior of confined fluids (gas or liquid) under pressure. It states that pressure exerted on a confined fluid is transmitted equally and perpendicular to all of the interior surfaces of the fluid's container (see Figure 7-7). Pascal's discovery served as the foundation for the science of hydrostatics, which is used to calculate forces and pressures in closed fluid power systems. It is also the underlying principle behind the operation of hydraulic devices that are capable of amplifying an input force.

Pascal's law:

pressure exerted on a confined fluid is transmitted equally and perpendicular to all of the interior surfaces of the fluid's container.

FIGURE 7-6: Timeline of inventions and scientific discoveries related to fluid power.

Time Period	Invention or Scientific Discovery
200 B.C.E.	Horizontal waterwheel is believed to have been invented in Armenia.
2nd century B.C.E.	Archimedes discovers the principle of buoyancy.
1st century A.D.	Hero of Alexandria invents the Aeolipile, which is considered to be the first steam engine.
5th to 9th centuries A.D.	The first windmill (vertical axis) is developed in Persia for the purpose of pumping water.
1647	French mathematician Blaise Pascal discovers that pressure applied to a confined fluid will be transmitted equally and perpendicular to the surfaces of its container.
1650	German physicist and engineer Otto von Guericke invents the first air pump, which he uses to generate vacuums.
1662	Robert Boyle discovers that a gas kept at a fixed temperature will experience an increase in volume as pressure decreases (and vice versa).
1712	Thomas Newcomen invents the first practical steam-powered engine, which would eventually replace the waterwheel as the primary source of power for industrial equipment.
1738	Daniel Bernoulli discovers that an increase in the speed of a moving fluid is accompanied by a decrease in static pressure.
1795	Joseph Bramah invents the hydraulic press, which replaced the screw press in most industrial applications.
1799	George Medhurst invents the first motorized air compressor, which was used in mining operations.
1918	Aviation pioneer Malcolm Lockheed develops the hydraulic brake system for automobiles.

FIGURE 7-7: (a) Blaise Pascal was the first person to define one of the fundamental scientific principles of hydrostatics. (b) Pascal's law states that pressure exerted on a confined fluid is transmitted equally and perpendicular to all of the interior surfaces of the fluid's container.

(a)

Hulton Archive/Handout/Hulton Archive/Getty Images.

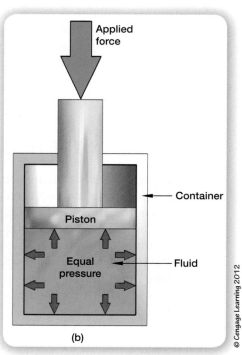

(b)

© Cengage Learning 2012

More than 130 years would pass before a practical technology was developed that employed Pascal's law. That invention took the form of the hydraulic press, which was created by a famous locksmith, Joseph Bramah, in 1795 (see Figure 7-8). Hydraulic presses are still used today to shape metal components such as car body panels and to create interference fits between objects such as gears and gear shafts. If you have ever used a hydraulic jack to raise an automobile for the

FIGURE 7-8: Replica of Joseph Bramah's hydraulic press circa 1801.

Science and Society Picture Library/Contributor/SSPL/Getty Images.

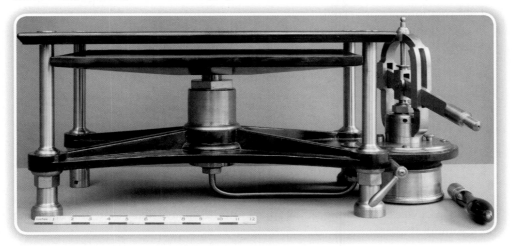

purpose of changing a tire, then you have some idea of the importance of Bramah's invention. Such devices give the user the ability to generate output forces that are well beyond the capability of a single human's muscle power. The science of hydrostatics and the mathematics behind Bramah's invention will be explored later in this chapter.

OFF-ROAD EXPLORATION

To learn more about the history of fluid power, watch the film *Modern Marvels: Hydraulics*.

Pascal's law is an example of a scientific concept that had to wait for advancements in materials and manufacturing before it could be applied to widespread industrial applications. However, the purpose of science is not to further technological advancement. Science is driven by the human desire to explain how the natural world works.

COMMON FLUID POWER SYSTEM COMPONENTS AND SCHEMATIC SYMBOLS

History has shown us that engineering disciplines do not come into being overnight. Each one develops from an "art" to a science over a long period of time. As this happens, the engineers within the discipline will recognize the need to standardize their methods of communication so that technical information can be shared and expanded on. As with every other engineering discipline, professional engineering organizations such as the National Fluid Power Association (NFPA) are responsible for developing and maintaining standards that are associated with the design and representation of pneumatic and hydraulic systems. A standard is a reference that is developed by an authority or through general consent and used as a basis for comparison and verification.

Imagine how much time it would take to communicate an idea for a fluid power system if you had to accurately sketch images of each component and the various ways that they are connected together to form a circuit. Such a task would be a daunting one for even the simplest of designs. Over the years, engineers have developed schematic symbols, which are simple-to-draw, two-dimensional graphics, for each of the different fluid power components found in hydraulic and pneumatic systems. These symbols are carefully arranged and joined together with different types of lines to form schematic circuit diagrams. Engineers use these diagrams to communicate how the components in a fluid power system should connect

Standard:

a reference developed by an authority or through general consent and used as a basis for comparison and verification.

Schematic symbol:

a simplified graphic representation of an electrical, a mechanical, or a fluid power system component.

together. Note that the schematic symbols that appear in this text are not limited to the orientations shown. They may be drawn in any orientation, though increments of 90° rotations are preferred.

Although there are differences between pneumatic and hydraulic system components, both systems will contain at least one of the following:

▶ a device that serves to pressurize the fluid,

▶ a pathway through which the fluid can flow,

▶ a device to control pressure, flow rate, and the direction that a fluid will flow, and

▶ a device that serves as the point of application (where work is performed).

A fluid power system might also include devices that indicate pressure or flow rate. Such devices are monitored by an operator to ensure that a system is behaving as it should.

FIGURE 7-9: (a) Pneumatic hose fittings are easy to connect and disconnect, but they are not designed to withstand the pressures that (b) brass hydraulic hose fittings must contend with.

(a)

(b)

Engineers have scores of pneumatic and hydraulic components to choose from. Addressing all of these components, their schematic symbols, and the myriad ways they may be joined to form circuits is beyond the scope of this text. This chapter will focus on the most common components that are found in hydraulic and pneumatic circuits.

Many hydraulic and pneumatic components have identical functions and schematic symbols. Such components include transmission lines, filters, actuators, and directional control valves. The major differences between these components exist in their connections and the design considerations that account for variations in operating fluid pressures. For example, pneumatic systems often use tubing and quick-disconnect fittings that are made from plastics, which are appropriate for relatively low pressures. Hydraulic components use more rigid connection methods because they may be subject to pressures as high as several thousand pounds per square inch. Hydraulic hoses are made from more durable materials, and in some cases they are formed from extruded metal tubing. Threaded brass fittings are used on the ends of hydraulic lines and are able to withstand extremely high pressures (see Figure 7-9).

Pneumatic and hydraulic hose lines are represented by the same symbols, which also indicate their purpose (see Figure 7-10). A **working line** represents fluid transport to and from an actuator or any other device that performs work in a fluid power system. A **pilot line** represents fluid pressure transmission for the purpose of controlling a **valve**, such as an air-piloted directional control valve in a pneumatic system or a pressure relief valve in a hydraulic system. A double line is used to indicate some type of *mechanical connection* between components such as a shaft that connects an electric motor to a hydraulic gear pump.

Schematic diagrams often show intersecting hose lines, though this does not always mean that the actual lines are connected together (see Figure 7-11). A dot that is approximately five times the width of the schematic hose line should be used to indicate where a physical connection between two or more lines is intended. In such cases, a device such as a **T-connector** is represented.

FIGURE 7-10: A single continuous line shown in a schematic fluid power diagram denotes a working line. A dashed line, representing a pilot line, will appear in a schematic diagram where fluid pressure is used to actuate a control valve. A double line represents a mechanical connection between two components such as a rotating shaft.

————————— Working line
— — — — — — — Pilot line
═══════════ Mechanical connection

FIGURE 7-11: (a) It is important to be able to interpret whether or not intersecting lines in a schematic diagram represent an actual connection between the fluid lines in a fluid power system. (b) T-connectors and similar devices would be represented by connected lines.

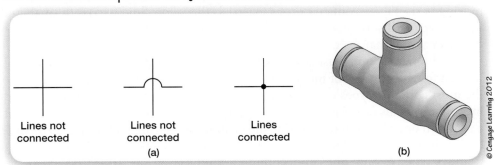

Lines not connected   Lines not connected   Lines connected
(a)

(b)

Both pneumatic and hydraulic systems use special **filters** to remove particulate matter that can damage the inner workings of moving components within a fluid power circuit (see Figure 7-12). Sources of contamination include dirt and small pieces of metal that are released into the fluid as a result of normal wear between

FIGURE 7-12: (a) A compressed air filter or (b) a hydraulic oil filter is one of the first components that a fluid will come in contact with when it enters a fluid power circuit. (c) The schematic symbol for a filter is a diamond with a dashed line oriented perpendicular to the direction of the flow of the fluid.

(a)

(b)

(c)

FIGURE 7-13: (a) Most single-acting cylinders use a spring to automatically retract the piston rod when pressure is cut off. (b) The schematic symbol for a single-acting cylinder is recognizable by the jagged line that indicates the spring within.

(a) Johnson, *Introduction to Fluid Power*, © 2002 Delmar Learning, a part of Cengage Learning, Inc. Reproduced with permission. www.cengage.com/permissions. (b) © Cengage Learning 2012

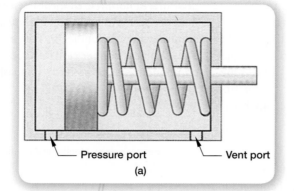

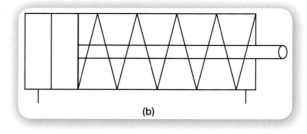

(b)

Pressure port   Vent port

(a)

Actuator:

any device that converts fluid pressure into mechanical motion for the purpose of moving a load.

moving parts. Filters contain porous materials that allow fluid to pass through but which trap (or absorb) and hold minute particles that may be as small as 1/15th the diameter of a human hair.

Fluid power systems use **actuators** to convert fluid pressure into mechanical motion for the purpose of moving a load. Single- and double-acting cylinders are the most common types of linear motion actuators found in both pneumatic and hydraulic systems (see Figure 7-13). A **single-acting cylinder** allows for control of a piston in one direction only. Inside the cylinder is a spring that moves the **piston** back to its nonactuated location once the fluid pressure is removed. A single-acting cylinder requires only one connection to a fluid power line, which occurs at the pressure port. A vent port allows air to enter or escape as the piston moves. This prevents an unwanted vacuum or accumulated air pressure from forming on the side of the cylinder that contains the spring.

Double-acting cylinders are the most common linear actuators found on heavy construction equipment because they allow for control of a piston in both directions (see Figure 7-14). Unlike the single-acting cylinder, this requires connections to fluid power lines on both sides of the piston. As the pressurized fluid

FIGURE 7-14: (a) The double-acting cylinder is controlled in two directions. (b) Its schematic symbol is similar to that of a single-acting cylinder but without the spring.

(a) From Johnson, *Introduction to Fluid Power*, © 2002 Delmar Learning, a part of Cengage Learning, Inc. Reproduced with permission. www.cengage.com/permissions. (b) © Cengage Learning 2012

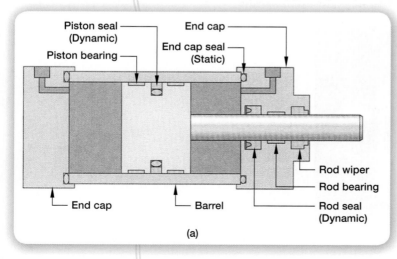

Piston seal (Dynamic)

Piston bearing

End cap

End cap seal (Static)

Rod wiper

Rod bearing

End cap   Barrel   Rod seal (Dynamic)

(a)

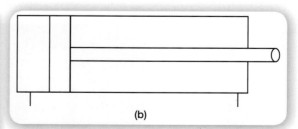

(b)

pushes the piston in one direction, the fluid on the other side of the piston is exhausted to the tank in the case of hydraulics or to the atmosphere in the case of pneumatics.

All fluid power systems incorporate one or more types of control valves. Two of the simplest control valves are the shutoff valve and the check valve (see Figure 7-15). The **shutoff valve** is used to turn all or part of a fluid power system on or off. When open, a fluid may flow through a shutoff valve in either direction, depending on the location of the source of pressure. A **check valve**, however, is a type of one-way valve that allows fluid to flow in one direction only. This type of valve is often incorporated into other types of control valves.

FIGURE 7-15: Two common control valves found in hydraulic and pneumatic systems are (a) the shutoff valve and (b) the check valve.

(a) and (b) From Johnson, *Introduction to Fluid Power*, © 2002 Delmar Learning, a part of Cengage Learning, Inc. Reproduced with permission. www.cengage.com/permissions.

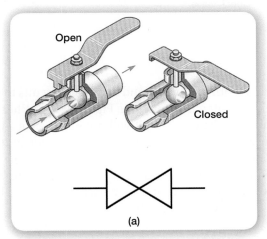
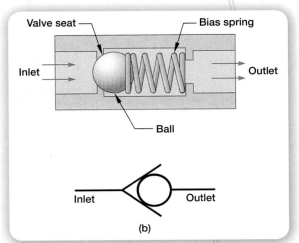

A **shuttle valve** combines the functions of a T-connector and a check valve. It has two inlet ports and one outlet port. Inside the valve is a movable obstruction that shifts to close off one inlet port when the other inlet port experiences a higher pressure (see Figure 7-16). As such, a shuttle valve operates on the principle of a check valve. This type of valve is often used when the control of an actuator must be shared between two independently operated directional control valves. This means that either of the two control valves can be used to move an actuator. The concept is similar to having two electrical switches that can turn a single light on or off.

FIGURE 7-16: A shuttle valve is often used when the control of an actuator must be shared between two independently operated directional control valves.

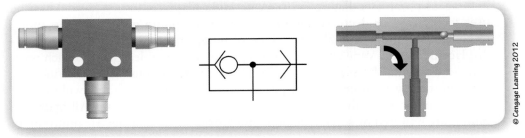

© Cengage Learning 2012

FIGURE 7-17: (a) A flow control valve is used to control the volume of fluid moving in one direction only. (b) The amount of fluid that is allowed to pass is controlled by an adjustable screw. (c) The schematic symbol indicates a passageway constriction, which the arrow shows to be variable. The symbol also shows that a check valve allows the fluid to pass, unrestricted, in the opposite direction.

(a) © Cengage Learning 2012 (b) Adapted from Johnson, *Introduction to Fluid Power*, © 2002 Delmar Learning, a part of Cengage Learning, Inc. Reproduced with permission. www.cengage.com/permissions. (c) © Cengage Learning 2012

(a)

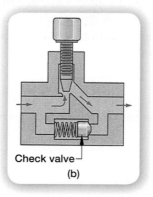
Check valve
(b)

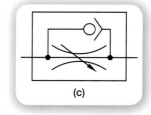
(c)

A **flow control valve** is used in both pneumatic and hydraulic circuits to control the volume of fluid as it flows in one direction only (see Figure 7-17). When a fluid moves through the valve in the opposite direction it will not experience a flow restriction. This type of valve is useful for controlling the speed of a piston as it travels either in or out of a cylinder. Because flow control valves only control the flow of fluid in one direction, a single flow control valve can only control a cylinder's piston speed in one direction. Two flow control valves arranged back-to-back can be used to control a piston's extension and necessary retraction speeds.

Hydraulic System Components

The example of the hydraulic-powered backhoe shown in Figure 7-18 identifies common components that are found throughout many hydraulic systems. These components include a reservoir, filter, pump, pressure relief valve, directional control valve, and an actuator.

A **reservoir** is used in a hydraulic system to store and protect hydraulic oil from outside contamination (Figure 7-19). Unlike a pneumatic system, the fluid used in a hydraulic circuit will return to the reservoir so that it can be recirculated. Hydraulic oil experiences friction as it flows through a circuit, which causes the temperature of the oil to increase. The reservoir acts as a heat exchanger by giving the returning fluid a chance to mix with the rest of the oil before being drawn back into the system. Unlike a pneumatic air compressor tank, the fluid contained in a reservoir is not pressurized.

A hydraulic **pump** is a mechanical device used to generate the hydrostatic pressure that is transmitted to the actuators in a hydraulic circuit. A **fixed-displacement pump** provides a constant pressure, and is the least expensive type because of its simple construction. A common type of fixed-displacement pump is the gear pump (see Figure 7-20). A **variable-displacement pump** allows the user to increase or decrease the pressure that the pump generates. The ability to control the pressure makes this type of pump more expensive than a fixed-displacement type. Vane and axial piston pumps are often designed for such pressure adjustment. The schematic symbol for a hydraulic pump will not indicate whether it is a gear, vane, or axial-piston type. However, the symbol will indicate whether the pressure that the pump generates is fixed or variable. All hydraulic pumps are powered by some type of **prime mover** such as an electric motor or an

FIGURE 7-18: (a) A simple hydraulic circuit is used to power a linear actuator on a backhoe. (b) The schematic diagram for this circuit uses symbols to show how the different hydraulic components are connected together. Note the symbols for the filter, T-connector, and double-acting cylinder.

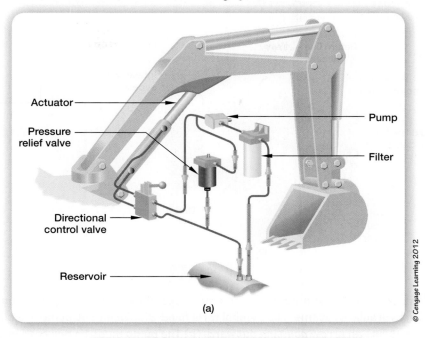

(a)

© Cengage Learning 2012

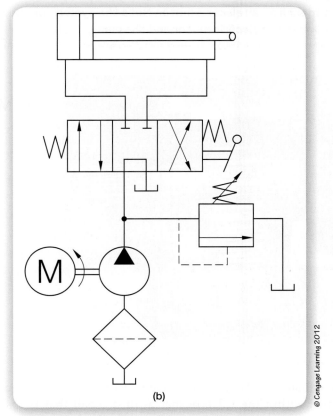

(b)

© Cengage Learning 2012

internal-combustion engine. As such, a pump symbol is often accompanied by the symbol for its prime mover.

A **pressure relief valve** is a type of safety mechanism that is required in even the simplest hydraulic circuits because it will protect the pump and other components from damage caused by excess pressure. Such pressure may be generated

FIGURE 7-19: (a) A hydraulic fluid reservoir is the place from which hydraulic fluid is drawn and to which the hydraulic fluid will return. (b) The schematic symbol looks like a pan connected to a flow line.

(a) Adapted from Reeves, *Technology of Fluid Power*, © 1997 Delmar Learning, a part of Cengage Learning, Inc. Reproduced with permission. www.cengage.com/permissions. (b) © Cengage Learning 2012

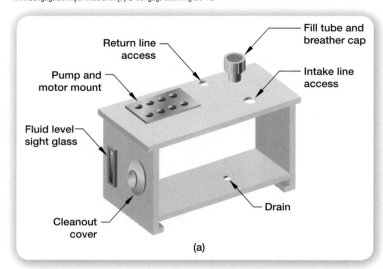

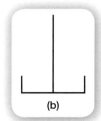

(a)

(b)

FIGURE 7-20: (a) A gear pump is a common type of fixed-displacement pump used in hydraulic systems. The symbol for a fixed-displacement pump is a circle that contains an arrow pointing outward. (b) The pump must be powered by a prime mover such as an electric motor. Note the symbol for the mechanical connection between the electric motor and the pump.

(a) From Reeves, *Technology of Fluid Power*, © 1997 Delmar Learning, a part of Cengage Learning, Inc. Reproduced with permission. www.cengage.com/permissions. (b) © Cengage Learning 2012

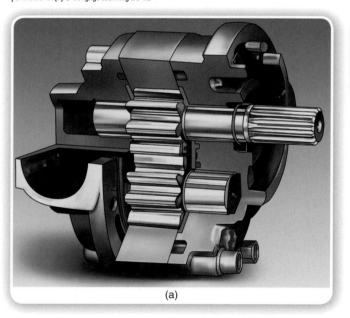

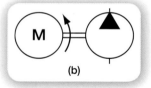

(a)

(b)

when an actuator encounters significant resistance to movement. A spring that is contained within the valve pushes against a ball that keeps the valve closed for normal pressures (see Figure 7-21). A pilot line, represented as a dashed line in the pressure relief valve schematic symbol, allows the fluid in the circuit to push against the ball. If an excess amount of pressure is encountered, the ball will compress the spring and divert the fluid to the reservoir. When the fluid pressure returns to a safe level, the valve will close again. The maximum circuit pressure is set by adjusting a thumbscrew and locknut on the pressure relief valve.

FIGURE 7-21: (a) A closed hydraulic system must be outfitted with a pressure relief valve, which protects the pump from excess pressure. (b) The schematic symbol for a pressure relief valve is connected to the symbol for the reservoir.

(a) From Vockroth, *Industrial Hydraulics*, © 1994, Delmar Learning, a part of Cengage Learning, Inc. Reproduced with permission. www.cengage.com/permissions. (b) © Cengage Learning 2012

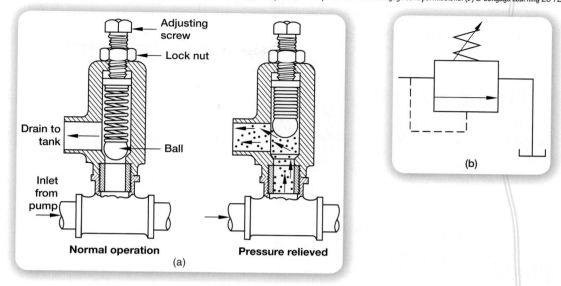

FIGURE 7-22: (a) Manually operated DCVs are outfitted with lever arms or push buttons. (b) On construction equipment, a series of DCVs are positioned next to each other to give the skilled operator the ability to control a machine's complex movements from a single location.

(a) Image copyright Pawel Strykowski, 2010. Used under license from Shutterstock.com. (b) Image copyright Zygimantas Cepaitis, 2010. Used under license from Shutterstock.com.

A **directional control valve (DCV)** is an extremely important device that serves as the control interface between a fluid power system and its operator—which could be a person, a computer, or another component within the system. DCVs are used to control the path that a fluid takes through a circuit (see Figure 7-22).

Schematic symbols for DCVs are represented by two or more boxes. Each box indicates a position that the valve may occupy during operation and contains symbols that show how or if a fluid is able to move through the valve from one port to another. As a general rule, the fluid lines in a schematic diagram are connected to

FIGURE 7-23: The red areas within these cutaway images of a four-way DCV represent pressurized fluid. The blue areas represent exhaust fluid that is returning from an actuator. (a) The springs within this lever-operated four-way DCV will center the spool to a location that locks an actuator in position while redirecting fluid back to the reservoir when the control lever is not being used. (b) This is known as a *tandem center condition*, which is identified by the center box of the DCV's schematic symbol. The symbol for a spring appears on both sides of the valve symbol, which indicates that the spool is spring-centered.

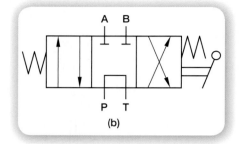

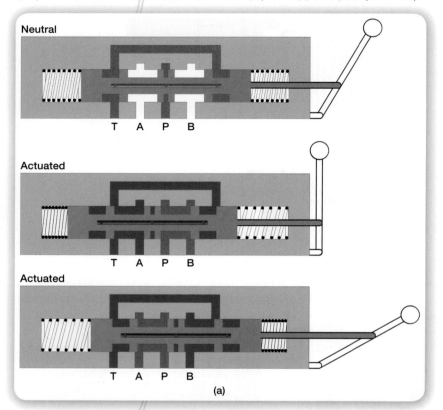

the ports on a DCV symbol that represents the valve's neutral or unactuated state. Figure 7-23 shows a four-way DCV that is used to control the motion of a bidirectional linear or rotary actuator. A four-way DCV has four ports—P, T, A, and B—through which the hydraulic fluid is channeled. Port P is connected to a hydraulic line that provides the "pressure." This line is connected directly or indirectly to the pump. The T port is connected to a return line, which sends the hydraulic fluid back to the "tank." Ports A and B are the working ports, which are connected to the inlets of a bidirectional actuator.

The motion of a spool inside the four-way DCV is controlled by a hand-operated lever. When the spool is in the neutral position, it makes a direct connection between the pressure and tank ports. The neutral position also blocks the flow of fluid to or from the working ports. This causes the actuator that is connected to the DCV to lock in its current position. When the lever is moved to one extreme position, the spool inside the DCV moves to make a direct connection between the pressure port and working port A. The fluid that returns from the actuator moves from working port B to the tank port. When the spool is directed to the other extreme position, pressurized fluid from the pressure port is directed to working port B, with the returning fluid moving from working port A to the tank port.

Four-way valves are designed with different center conditions that determine how the pressurized fluid is redirected when the DCV is in an unactuated state. The example of a four-way DCV shown in Figure 7-23 uses a *tandem center condition*

FIGURE 7-24: (a) The workpiece fixture device found in this manufacturing workcell uses common pneumatic components such as single- and double-acting cylinders. (b) The schematic diagram of the pneumatic circuit shows that the system uses other components as well, such as flow control valves and directional control valves.

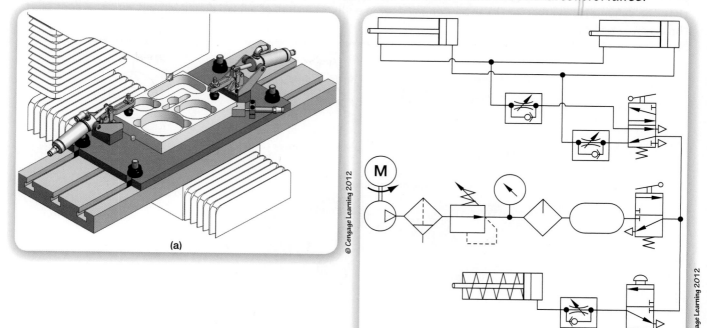

that sends the pressurized fluid directly to the tank when the actuator must remain idle and locked in its position.

Pneumatic System Components

The example of the workpiece fixture system and its schematic diagram shown in Figure 7-24 shows several components that are found throughout pneumatic systems. Many of these components such as filters, flow control valves, and linear actuators are also found in hydraulic systems and have already been addressed. Components that are specific to pneumatic systems often include those that are used to condition and hold the compressed air before it is put to work in the system.

The air that is used in a pneumatic system must pass through a series of conditioning devices before it can be put to use (see Figure 7-25). This process begins when air is drawn into a compressor. The resulting compressed air is

FIGURE 7-25: A standard commercial air-compressor unit contains many components, which are represented by this series of schematic symbols. (a) An electric motor is mechanically connected to (b) an air compressor, which sends pressurized air through (c) a filter. The filter symbol shows that it is outfitted with a manual drain. The filter is connected to (d) a pressure regulator that is, in turn, connected to (e) a pressure gauge and (f) a lubricator. The compressed air is then stored in (g) the receiver tank until it is needed.

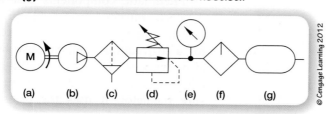

FIGURE 7-26: (a) This two-stage piston compressor functions in a manner that is similar to an internal-combustion engine but without the combustion. Ambient air is drawn into the first stage, compressed, and passed on to the second stage, where it is further compressed. The crankshaft is powered by a prime mover. Note that the schematic symbol for (b) an air compressor is similar to that of a fixed displacement pump, except that the arrow is not solid.

(a) Adapted from Johnson, *Introduction to Fluid Power*, © 2002 Delmar Learning, a part of Cengage Learning, Inc. Reproduced with permission. www.cengage.com/permissions. (b) © Cengage Learning 2012

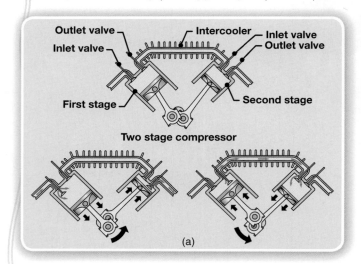

then filtered to remove particulate matter and moisture in the form of water vapor. It then passes through a device that regulates the air pressure to a safe operating level. Oil is then added to the compressed air via a lubricator. The compressed air is then stored inside a pressure vessel for on-demand use.

Pneumatic pressure is generated by an **air compressor**. Like the hydraulic pump, an air compressor does not provide its own power and must be connected to a prime mover. Most compressors use an electric motor or internal-combustion engine to generate the power needed to compress air. Common types of air compressors include reciprocating or piston compressors (see Figure 7-26), rotary-vane compressors, rotary screw compressors, and centrifugal compressors.

A pneumatic **pressure regulator** is used to manually adjust and control the pressure of the compressed air source in a pneumatic system. Such devices usually have a pressure gauge built in so that the user can monitor the pressure and make accurate adjustments whenever needed (see Figure 7-27). Standard operating air pressure for the types of pneumatic tools found in hardware stores is approximately 90 psig.

Pneumatic components that contain moving parts must be lubricated to provide continuous smooth operation. This is not a requirement for a hydraulic system because hydraulic fluid also serves as a lubricant. The **lubricator** is usually placed in series with the filter and the pressure regulator. It uses the pressure from the

FIGURE 7-27: (a) A pressure regulator is used to adjust the air pressure to the desired level. Many come with built-in pressure gauges so users can make accurate adjustments. (b) The pressure regulator schematic symbol shows that an internal air pilot is used to control the air pressure. The symbol for the pressure gage has been included.

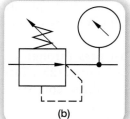

© Cengage Learning 2012

Courtesy of Campbell Hausfeld.

FIGURE 7-28: (a) A lubricator adds a small amount of oil to the compressed air in a pneumatic circuit once it has been filtered. (b) Its schematic symbol is similar to that of a filter.

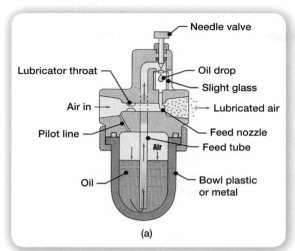

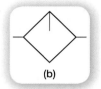

(b)

OFF-ROAD EXPLORATION

To view animations that show how a pneumatic air filter and lubricator work, visit *www.ifps.org/Education/ PneumaticDevices/.*

compressed air to aspirate tiny drops of oil into a fine mist, which mixes with the air and travels through the circuit (see Figure 7-28).

A **receiver tank** acts as a holding device for compressed air before it is drawn into a pneumatic circuit (see Figure 7-29). Receiver tanks are made from welded steel and can be quite large, depending on the system's demand for compressed air.

FIGURE 7-29: The prime mover and two-stage piston compressor on this air-compressor unit are affixed to the top of a large pressure vessel, called (a) the receiver. (b) The schematic symbol for an air receiver looks like a simple capsule.

(a)

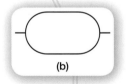

(b)

Not all pneumatic circuit diagrams show the conditioning devices previously described, and some engineers assume that these devices are already present in preexisting pneumatic systems. In such cases, the engineer will represent a conditioned and compressed air source with a simple schematic symbol that is accompanied by a note identifying the required air pressure (see Figure 7-30).

FIGURE 7-30: (a) This schematic diagram of a pneumatic circuit does not identify an air compressor or any of the conditioning devices found in a pneumatic system. (b) Instead, a simple air supply symbol is shown connected to a *directional control valve* symbol. Note that the required air pressure is also identified.

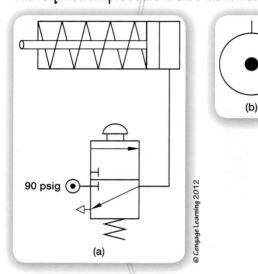

(a)

(b)

DCVs found in pneumatic systems differ from those used in hydraulic systems by the way in which the fluid is "exhausted." Pneumatic DCVs have the freedom to vent exhaust gases back to the atmosphere. This is not an option in a hydraulic system, where the fluid must be returned to the reservoir.

One of the simplest pneumatic DCVs is the three-port, two-way valve (sometimes called 3/2 valve), which acts like an on–off switch. Figure 7-31 shows three variations of a 3/2 that incorporates three common mechanical methods for redirecting airflow within a DCV. Each method incorporates a spring that returns the valve to its neutral state.

FIGURE 7-31: Mechanically operated DCVs are usually actuated by (a) a push button, (b) a roller, or (c) a lever. Each method incorporates a spring that returns the valve to a neutral state.

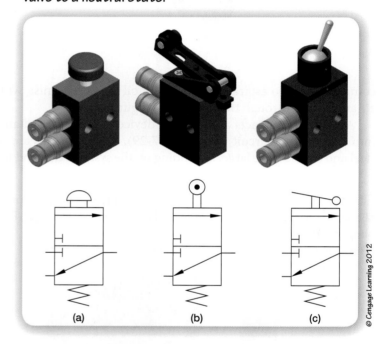

(a)   (b)   (c)

A five-port, two-way valve (sometimes called a 5/2 valve) is often used in pneumatic systems to control a double-acting cylinder (see Figure 7-32). Note that the air supply port on the 5/2 valve is constantly connected to one of the cylinder ports. This makes it impossible for a 5/2 valve to shut off its own air supply. To

FIGURE 7-32: Lever-operated five-port valve, schematic symbol, and airflow illustration. Note the schematic symbol includes the symbols for an air supply. The triangle shapes indicate that the ports will exhaust air to the atmosphere.

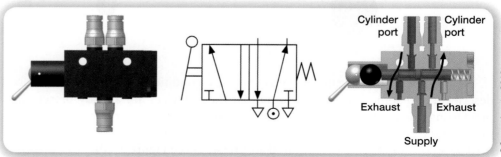

Cylinder port   Cylinder port

Exhaust   Exhaust

Supply

solve this problem, a 3/2 valve is usually placed in series with a 5/2 valve to allow an operator to shut off the flow of compressed air.

Not all DCVs are controlled manually. In fact, most automated manufacturing systems that utilize fluid power circuits are operated by a computer and programmable logic controller (refer to Chapter 8, Control Systems). This is made possible through the use of solenoid-operated DCVs. A solenoid is an electromechanical actuation device that uses the principles of electromagnetism to control the spool within a DCV. A magnetic field is generated when electricity flows through a solenoid coil. The magnetic field causes a plunger to move, which in turn moves the spool within the DCV. The three-port valve shown in Figure 7-33 is controlled with a solenoid.

FIGURE 7-33: (a) This three-port DCV is actuated by a solenoid. (b) The single angle line within a rectangle on top of the DCV symbol identifies a single winding solenoid coil. (c) The solenoid plunger is connected to the spool inside the DCV. When the coil is electrified, the plunger will move to change the direction of the fluid through the DCV.

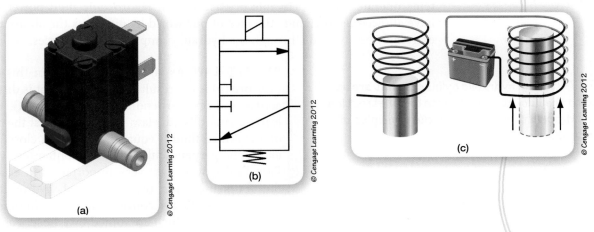

Some DCVs are pilot operated, which means that the spool inside the valve is moved by positive or negative air pressure instead of the mechanical motion of a lever or push button. Figure 7-34 shows an air-operated five-port valve. This type of control valve is often used when the pressure generated by the exhausted fluid from a double-acting cylinder must activate the motion of another double-acting cylinder.

FIGURE 7-34: (a) An air-operated five-port valve with (b) schematic symbol.

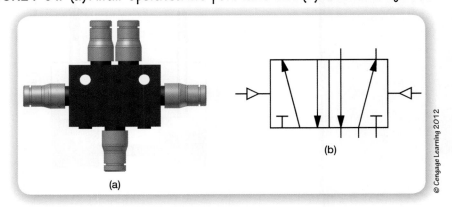

BASIC SCIENTIFIC CONCEPTS OF FLUID POWER

Fluid Mechanics:
the study of the properties of gases and liquids that are at rest or in motion.

A large part of many engineers' formal education includes the study of a branch of applied physics called *mechanics*. Mechanics is the study of forces and their effects on the bodies on which they act. The fields of mechanical engineering, manufacturing engineering, civil engineering, and aerospace engineering involve the application of an important branch of mechanics called *fluid mechanics*. Fluid mechanics is concerned with the study of properties of gases and liquids that are at rest (hydrostatic) or in motion (hydrodynamic).

Fundamentals of Hydrostatics

Hydrostatics:
the study of the properties of fluids that are in a state of static equilibrium (at rest).

Some of the most impressive applications of fluid power technology are based on the principles of hydrostatics, which is the study of the properties of fluids that are in a state of static equilibrium (at rest). Hydrostatic systems, such as those used on the large piece of construction equipment shown in Figure 7-35, transmit energy through fluid pressure and convert it to an output force. The giant linear actuators that power the buckets, spades, and other devices on these systems operate on the principle of Pascal's law. Hydrostatic fluid power systems have allowed us to change the face of our world by giving us the ability to move extremely heavy loads through the hydraulic amplification of forces.

FIGURE 7-35: The hydraulic cylinders that are attached to the bucket on this piece of construction equipment operate according to the principles of hydrostatics.

PASCAL'S LAW As you learned earlier in this chapter, pressure (p) is the result of a force (F) that is distributed across, and acts perpendicular to, an area (A). You also learned about Pascal's law, which states that the pressure of a confined fluid is equal throughout a closed system and acts perpendicular to the containing surfaces. Pascal's discovery was very important and established the foundation for the science of hydrostatics. To illustrate Pascal's law, consider the two cylinders shown in Figure 7-36, which are joined by a hose. Pascal's law tells us that the pressure (p_I) of the fluid in input cylinder A is equal to the pressure (p_O) of the fluid in output cylinder B.

Image copyright Gilles Lougassi, 2009. Used under license from Shutterstock.com.

FIGURE 7-36: As the two connected cylinders represent a closed hydraulic system, Pascal's law tells us that the pressure of the fluid within cylinder A (p_I) must be equal to the pressure of the fluid within cylinder B (p_O).

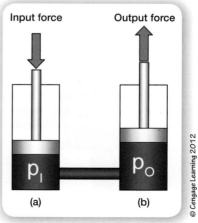

© Cengage Learning 2012

Equation 7-3 is used to express the mathematical relationship between the pressures of the input and output cylinders of a hydrostatic model according to Pascal's law.

$$p_I = p_O \qquad \text{(Equation 7-3)}$$

The pressure of the fluid in input cylinder A (p_I) is a function of the input force (F_I) divided by the area of the input cylinder's piston head (A_I) where it contacts the hydraulic fluid. The same relationship exists within output cylinder B, where its fluid pressure (p_O) is a function of the output force (F_O) divided by the area of the output cylinder's piston head (A_O) where it contacts the hydraulic fluid. Knowing this, we can substitute the input and output forces and areas for the input and output pressures to express Pascal's law in a different way:

$$\frac{F_I}{A_I} = \frac{F_O}{A_O} \qquad \text{(Equation 7-4)}$$

Figure 7-37 shows two hydraulic cylinders of identical geometry that are connected by a hose. Note that the area of both piston heads is 3 in². If a 600-lb force is applied to the top of cylinder C's piston rod, how much fluid pressure will be generated in the system? Also, how much output force will be generated by cylinder D?

To find the answer, we must first calculate the pressure that is generated in the input cylinder (p_I) by dividing the input force (F_I) by the area of the input cylinder's piston head (A_I) where it contacts the hydraulic fluid.

$$p_I = \frac{F_I}{A_I}$$

$$= \frac{600 \text{ lb}}{3 \text{ in}^2}$$

$$= 200 \text{ lb/in}^2 \text{ or } 200 \text{ psi}$$

According to Pascal's law, the 200-psi fluid pressure generated in cylinder C is transmitted equally to cylinder D. We also know that cylinder D has the same cross-sectional area as cylinder C. By algebraically manipulating the variables in Equation 7-4, we can calculate the magnitude of the unknown output force:

$$\frac{F_I}{A_I} = \frac{F_O}{A_O} \Rightarrow F_O = A_O\left(\frac{F_I}{A_I}\right)$$

$$F_O = 3 \text{ in}^2 \left(\frac{600 \text{ lb}}{3 \text{ in}^2}\right)$$

$$= 600 \text{ lb}$$

HYDRAULIC AMPLIFICATION OF FORCE In the previous example, the input and output forces were the same because the geometries of the input and output cylinders were identical. A different output force can be generated by varying the sizes of the two connected cylinders. To prove this, consider the hydrostatic system shown in Figure 7-38. A 100-lb force is applied to the piston rod on cylinder E, which has an area of 2 in² where it contacts the hydraulic fluid. Knowing this, how much fluid pressure will be generated in the system? Also, how much output force will be generated by cylinder F if it has a cross-sectional area of 4 in²?

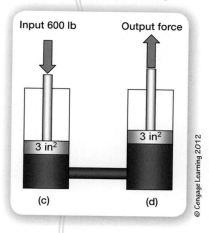

FIGURE 7-37: Example of a closed hydraulic system consisting of two hydraulic cylinders of identical geometry.

FIGURE 7-38: A different output force can be generated by varying the diameters of the input and output cylinders in a hydrostatic system.

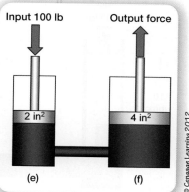

© Cengage Learning 2012

Again, the first step involves calculating the resulting fluid pressure that is generated in the input cylinder (p_I) by dividing the input force (F_I) by the area of the input cylinder's piston head (A_I) where it contacts the hydraulic fluid:

$$p_I = \frac{F_I}{A_I}$$

$$= \frac{100 \text{ lb}}{2 \text{ in}^2}$$

$$= 50 \text{ lb/in}^2 \text{ or } 50 \text{ psi}$$

Pascal's law tells us that the 50-psi fluid pressure generated in cylinder E is transmitted equally to cylinder F. Knowing the cross-sectional area of cylinder F, we can once again algebraically manipulate the variables in Equation 7-4 to calculate the magnitude of the unknown output force:

$$\frac{F_I}{A_I} = \frac{F_O}{A_O} \Rightarrow F_O = A_O \left(\frac{F_I}{A_I} \right)$$

$$F_O = 4 \text{ in}^2 \left(\frac{100 \text{ lb}}{2 \text{ in}^2} \right)$$

$$= 200 \text{ lb}$$

Having a 100-lb input force generate a 200-lb output force constitutes a **hydraulic amplification of force**, which is similar to the concept of mechanical advantage. As you learned in Chapter 3, The Mechanical Advantage, a price must be paid for such an advantage. In this case, the trade-off is seen by the distance that the input cylinder's piston must move, which in this example is double the distance that is moved by the output cylinder's piston.

The ratio between the output and input forces of the pistons in a hydrostatic system is inversely proportional to the distances that their pistons will travel:

$$\frac{\text{Output force}}{\text{Input force}} = \frac{\text{Distance moved by the input force}}{\text{Distance moved by the output force}} \quad \text{(Equation 7-5)}$$

FIGURE 7-39: Hydraulic systems achieve mechanical advantage through the hydraulic amplification of force. In this case, a 100-lb input force, acting over a distance of 10 units, will generate an output force of 1,000 lb that acts over a distance of 1 unit.

$$\frac{F_O}{F_I} = \frac{y_I}{y_O}$$

To illustrate this inverse relationship between force and distance, consider the closed hydraulic system shown in Figure 7-39. The area of piston A, where it contacts the hydraulic fluid, is 1 in². The area of piston B where it contacts the hydraulic fluid is 10 in². A 100-lb force that is applied to piston A will cause the fluid in the hydrostatic system to become pressurized. The equal pressure throughout the fluid acts perpendicular to the surface of piston B, with a resulting output force of 1,000 lb. How far will piston B move?

Before we can solve the problem, we must first recognize the concept of *volume*. **Volume** is the amount of space occupied by a three-dimensional object or enclosed within a container. As the liquid contained within a hydraulic cylinder will take on the form of its container, the volume of that liquid

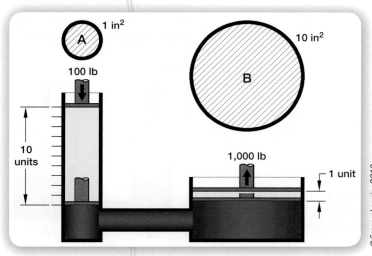

© Cengage Learning 2012

can be determined using the volume formula for a cylinder (see Equation 7-6). The volume (V) of a cylinder is the product of its cross-sectional area (A) and height (h):

$$V_{cylinder} = Ah \qquad \text{(Equation 7-6)}$$

Because the liquid is practically incompressible, the volume of liquid that moves out of input cylinder A (ΔV_I) will be transferred directly to output cylinder B (ΔV_O). The relationship between the transfer of liquid between input and output is given by Equation 7-7:

$$\Delta V_I = \Delta V_O \qquad \text{(Equation 7-7)}$$

The transfer of liquid that occurs between the cylinders can also be expressed by substituting the input and output cylinder areas and heights for the input and output volumes:

$$A_I h_I = A_O h_O \qquad \text{(Equation 7-8)}$$

Recognizing that the height of the cylindrical column of liquid that is removed from cylinder A will not be equal to the height of the column of liquid that is gained by cylinder B, we can algebraically manipulate the variables in Equation 7-8 to calculate the height of the column of liquid that is transferred to cylinder B:

$$A_I h_I = A_O h_O \Rightarrow h_O = \frac{A_I h_I}{A_O}$$

$$h_O = 1 \text{ in}^2 \times \frac{10 \text{ in}}{10 \text{ in}^2}$$

$$= 1 \text{ in}$$

The height of the column of liquid that is gained in cylinder B is equal to the distance that cylinder B's piston will travel.

Point of Interest

The forces that are generated by a typical double-acting cylinder as it extends and retracts are not the same. The reason for this is illustrated in Figure 7-40, which shows how the area of the piston head that comes in contact with the fluid is not equal on both sides of the piston. The difference in area causes a difference in the resulting pushing and pulling forces that are generated by the fluid pressure.

FIGURE 7-40: The areas that come in contact with the fluid on either side of the piston head are not the same size, which results in unequal extension and retraction forces given a constant fluid pressure.

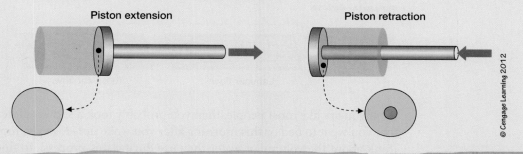

Piston extension

Piston retraction

(continued)

Though manufacturers of linear actuators identify the inside diameters of their cylinders, they do not identify the piston areas. The engineer must calculate the area on both sides of the piston head in order to know how much force will be generated by a given fluid pressure. In some cases, the area that is taken up by the piston rod on one side of the piston head must be taken into consideration.

Fundamentals of Hydrodynamics

Now that we have a basic understanding of the properties of liquids at rest, we turn our attention to a second branch in the science of fluid power called *hydrodynamics*. **Hydrodynamics** is the study of fluids that are in a state of motion. Some of the basic concepts of hydrodynamics include viscosity, the difference between a fluid's flow rate and its flow velocity, and the effect that a change in a fluid conduit's cross-sectional area has on the pressure and velocity of the fluid that flows through it.

Hydrodynamics:
the study of fluids that are in a state of motion.

VISCOSITY An important property of liquids that are used in hydraulic systems is their *viscosity*. **Viscosity** is a measure of a fluid's resistance to flow. The liquid that you are undoubtedly most familiar with is water because it is a necessity of life. Water is characterized as a "thin" liquid that flows easily and quickly. The viscosity of water is so low that it is not used in industrial hydraulic applications because it does not have the proper viscosity to prevent leaking past the seals between hydraulic system components. Fluids such as petroleum-based hydraulic oils are characterized as "thick" and viscous, and provide greater resistance to flow in comparison to water.

Remember: Hydraulic oil serves two purposes within a fluid power system. It must transfer energy through the flow of pressurized fluid, and it must lubricate the moving components within the system. Therefore, hydraulic oil should have enough viscosity to serve as a proper lubricant and to prevent leaking past seals. On the other hand, the viscosity must be low enough to ensure proper flow through **transmission lines** without the need of excessive amounts of pressure.

LIQUID FLOW RATE VERSUS FLOW VELOCITY Under ideal operating conditions, hydraulic fluid should flow through a pipe in a smooth, steady stream. This is referred to as **laminar flow**. If fluid is forced through a transmission line at a velocity that is too high, it will experience **turbulent flow** (see Figure 7-41). To avoid turbulent flow in a fluid system, the engineer must control the fluid *flow rate* and *flow velocity*.

FIGURE 7-41: Laminar flow occurs when a fluid moves in a smooth, steady stream through a system. If the fluid velocity is too great, turbulent flow will result.

© Cengage Learning 2012

Laminar flow   Turbulent flow

If you are like most people, then you probably took a shower either last night before you went to bed or this morning after you woke up. However, most people don't think about the volume of water they use during this process. To find out approximately how much water you use each time you shower, you need to know the average amount of time you spend in the shower along with the flow rate of the showerhead.

Flow rate (*Q*) is the volume (*V*) of fluid that moves past a given point in a system per unit time (*t*) (see Equation 7-9). Flow rates in hydraulic systems are measured in gallons per minute (gpm) or cubic inches per minute (in³/min) in the U.S. customary system. One gallon per minute is equal to 231in³/min. In the metric system, flow rates are measured in liters per minute (lpm), or cubic meters per second (m³/s).

$$Q = \frac{V}{t} \qquad \text{(Equation 7-9)}$$

Let us assume that your showerhead nozzle is rated at 2.5 gpm and that your average shower time is 10 minutes. How many gallons of water do you use on average?

$$Q = \frac{V}{t} \Rightarrow V = Qt$$

$$V = \left(\frac{2.5 \text{ gallons}}{1 \text{ minute}}\right) \times 10 \text{ minutes}$$

$$= 25 \text{ gallons}$$

Example

Problem: A garden hose is used to fill a five gallon bucket in 2 minutes. What is the flow rate of the water through the hose?

$$Q = \frac{V}{t}$$

$$= \frac{5 \text{ gallons}}{2 \text{ minutes}}$$

$$= 2.5 \text{ gpm}$$

Special metering devices called **flow meters** are used to measure a liquid's flow rate. One such device is called a *variable area flow meter* (see Figure 7-42). This device uses a float to indicate a liquid's flow rate in gallons per minute. The meter

FIGURE 7-42: (a) A variable area flow meter is used to measure liquid flow rate through a hydraulic system. (b) A flow meter schematic symbol may appear at the discharge location of a pump in a schematic diagram.

(a) Adapted from Johnson, *Introduction to Fluid Power*, © 2002 Delmar Learning, a part of Cengage Learning, Inc. Reproduced with permission. www.cengage.com/permissions. (b) © Cengage Learning 2012

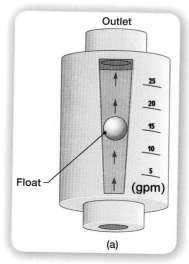

Outlet

25
20
15
10
5
(gpm)

Float

(a)

(b)

must remain in a vertical orientation to work because it relies on gravity to perform its task. As liquid moves upward through a tapered tube within the meter, it pushes against a free-floating obstruction (called a *float*). The float is in turn pulled downward by the force of gravity. The float will appear to be suspended when the force generated by the flowing liquid is balanced by the force of gravity acting on the float. The marks on the side of the meter indicate the flow rate in gallons per minute.

The speed of a fluid moving through a system is referred to as the **flow velocity** (v, pronounced "nu"). Equation 7-10 shows that a fluid's flow velocity is a ratio between the distance (x) that a drop of fluid travels and the amount of time (t) that it takes to travel that distance. Flow velocity is measured in feet per minute (ft/min) or inches per second (in/s) in the U.S. customary system and meters per second (m/s) in the metric system.

$$v = \frac{x}{t}$$ (Equation 7-10)

Example

Problem: If the garden hose from the previous problem is 20 feet long and the water moves from the spigot to the end of the hose in 12 seconds, what is the water's flow velocity in inches per minute?

Step 1: Convert the length of the hose from feet to inches.

$$\frac{12 \text{ in}}{1 \text{ ft}} = \frac{x_{inches}}{20 \text{ ft}}$$

$$x_{inches} = \frac{\left(12 \text{ in} \times 20 \text{ ft}\right)}{1 \text{ ft}}$$

$$= 240 \text{ in}$$

Step 2: Convert the time that it takes for the water to move from the spigot to the end of the hose from seconds to minutes.

$$\frac{60 \text{ seconds}}{1 \text{ min}} = \frac{12 \text{ seconds}}{t_{min}}$$

$$t_{min} = \frac{\left(12 \text{ seconds} \times 1 \text{ min}\right)}{60 \text{ seconds}}$$

$$= 0.2 \text{ min}$$

Step 3: Calculate the water's flow velocity in inches per minute.

$$v = \frac{x}{t}$$

$$= \frac{240 \text{ in}}{0.2 \text{ min}}$$

$$= 1,200 \text{ in/min}$$

Assuming the flow rate is constant, reducing the diameter of a fluid transmission line will cause an increase in *flow velocity*. To better understand the difference between flow rate and flow velocity, imagine that you have been given the task of drinking 12 ounces of cold water in 12 seconds from a container that has a 1-inch diameter opening. This is not hard to do if you are thirsty enough. Now imagine that you have to accomplish this same task using a ¼-inch diameter straw. It is doubtful that a person could generate the necessary flow velocity to move all 12 ounces of liquid in 12 seconds. If a fluid's passageway decreases in cross-sectional area (A), then the fluid's flow velocity (v) must increase in order to maintain a constant flow rate (Q). The relationship between flow velocity, flow rate, and cross-sectional area is identified by Equation 7-11:

$$v = \frac{Q}{A}$$
(Equation 7-11)

Example

Problem: A hydraulic fluid travels through a pipe at 1,200 in/min. The pipe has an inner diameter of 1.25 inches. What is the hydraulic fluid's flow rate in gallons per minute?

Step 1: Calculate the cross-sectional area of the pipe.

$$A = \frac{\pi d^2}{4}$$

$$= \frac{[\pi \times (1.25 \text{ in})^2]}{4}$$

$$= \frac{(\pi \times 1.563 \text{ in}^2)}{4}$$

$$= 1.227 \text{ in}^2$$

Step 2: Calculate the fluid's flow rate in cubic inches per minute.

$$v = \frac{Q}{A} \Rightarrow Q = vA$$

$$Q = 1,200 \text{ in/min} \times 1.227 \text{ in}^2$$

$$= 1,472.4 \text{ in}^3/\text{min}$$

Step 3: Convert the flow rate to gallons per minute.

$$\frac{231 \text{ in}^3/\text{min}}{1 \text{ gpm}} = 1,472.4 \text{ in}^3/\text{min}$$

$$Q_{gpm} = \frac{1,472.4 \text{ in}^3/\text{min}}{231 \text{ in}^3/\text{min}}$$

$$= 6.37 \text{ gpm}$$

Career Profile

CAREERS IN ENGINEERING

Anything's Possible

"Engineering touches pretty much every aspect of our daily lives," says Jamesia Hobbs, a process engineer who's employed by a global consulting firm headquartered in Denver, Colorado. Hobbs, who works out of the company's Atlanta office, was drawn to engineering in part because of the discipline's variety. "There are so many ways for your career to go," she says. "A lot of possibilities. I really like that. I didn't want to stagnate."

Stagnation hasn't been a problem so far. After earning her engineering degree, Hobbs took a job with a national leader in the paper industry. There she got a feel for the kind of engineer she wanted to be. "There are two kinds of engineer," she says, "the cowboy, who's right in the thick of it, and what I call the accounting engineer, who's behind the scenes, making it possible for the cowboys to do their jobs. I realized pretty early on that I am more of the latter. I really enjoy doing design work, putting a package together for a client."

Jamesia Hobbs: Chemical Process Engineer

On the Job

Hobbs' current company has far-reaching interests, and she works for its energy and chemicals division. "Right now, I'm involved with the polysilicon industry, helping design some of the stuff that goes into solar technology," she says. "A current focus is helping companies that want to 'go green.'"

As a chemical engineer, Hobbs is mainly concerned with the raw materials that flow through a plant or facility. She watches such things as temperature ranges, the pressure chemicals are under, and the pipelines they flow through. As a consultant, Hobbs rarely sets foot on site. But her most memorable experience as an engineer almost literally involved getting her feet wet. "When I was working for the paper company, I was the engineer in charge of several projects for a shutdown," Hobbs says. "During a shutdown, a facility is completely turned off for maintenance and cleaning.

On that occasion, there was a huge boiler in the powerhouse that had to be cleaned and inspected, and I had the opportunity to go inside and look around. You hear about this stuff in school, but it's such a revelation to actually be there."

Inspirations

Hobbs didn't always know she wanted to be an engineer. "When I was a child, I changed my mind every week about what I wanted to be," she says, "but I always liked math and science, which were like puzzles to me. And when I learned that a family friend's daughter had gone into chemical engineering, I thought, 'Hmm.'"

Hobbs' parents were supportive but not pushy. "Neither of them had gone to college before I was born," she says, "but they emphasized how important an education was for my future. From a very early age, I knew the importance of getting good grades to prepare for college. And to reinforce the importance of a college education, my mother began pursuing a college degree during my sophomore year in high school."

Education

Hobbs graduated with high honors from North Carolina A&T State University. "It was a really small department," she says. "There were only a dozen or so people in my class, so it was kind of like a family."

Hobbs' favorite teachers administered tough love. "I like teachers who are pretty hard on you," Hobbs says. "And I learn best when there is very little slack."

Advice for Students

Hobbs advises students to develop a solid foundation in math and science to prepare for an engineering education, but she also encourages them to keep their options open. "Think long-term," she says. "Get in touch with an engineer through a firm or an organization and find out about the different ways you can use your degree."

BERNOULLI'S PRINCIPLE Different size hoses, fittings, and valves serve as the pathway for fluid within a fluid power system. As a fluid moves from the source of pressure (pump or compressor), it will usually experience changes in the cross-sectional area of its pathway. These changes in cross-sectional area affect the velocity and pressure of the moving fluid because of a fundamental law of hydrodynamics called *Bernoulli's principle*, named in honor of its discoverer, Dutch-Swiss mathematician Daniel Bernoulli (1700–1782).

Bernoulli's principle states that *the velocity of a fluid increases as the pressure exerted by that fluid decreases*. Likewise, a decrease in a fluid's velocity will cause its pressure to increase. Bernoulli's principle is put to use in devices that incorporate Venturi tubes, which are named in honor of their creator, Italian physicist Giovanni Battista Venturi (1746–1822). A Venturi tube is a constriction that is placed in a pipe that causes a drop in pressure as fluid flows through it. The decrease in fluid pressure is accompanied by an increase in fluid velocity, and vice versa. Perhaps the most recognizable example of a Venturi tube is a giant water-cooling tower found at an electrical power-generation plant (see Figure 7-43). Openings in the base of a cooling tower allow air to be drawn in. Hot water from condensed steam is sprayed into the tower, which heats the air and causes it to move upward. As the air moves up through the tower, its pressure decreases because of the gradual reduction of the tower's cross-sectional area. This causes the air to move faster, which cools the water. The water collects at the bottom of the tower and is recycled back into the power plant's cooling system. The reduction in fluid pressure that occurs as a fluid flows through the constricted section of a pipe is referred to as the Venturi effect.

A vacuum generator is a pneumatic device that incorporates a Venturi tube to generate suction by accelerating the flow of compressed air (see Figure 7-44). Vacuum generators are frequently used in assembly, packaging, and manufacturing operations to lift and clamp objects. A robot that is outfitted with a vacuum generator and rubber suction cup can move an object as delicate as an egg without breaking it. Vacuum generators are also incorporated into fixture plates, which are able to safely and securely hold a workpiece without the use of a vice or strap clamps.

> **Bernoulli's principle:** the velocity of a fluid increases as the pressure exerted by that fluid decreases.

FIGURE 7-43: (a) The giant water-cooling towers found at electrical generation plants are examples of Venturi tubes. (b) A Venturi tube operates according to Bernoulli's principle. The green arrows represent the direction of fluid flow through the tube. The fluid velocity will increase as it moves from the large diameter, high-pressure section of the tube to the small diameter, low-pressure section.

FIGURE 7-44: A vacuum generator uses the Venturi effect to generate suction by accelerating the flow of compressed air.

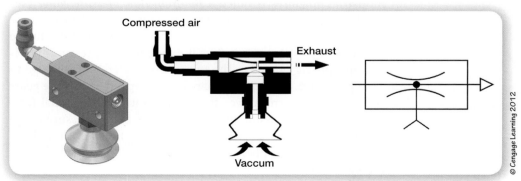

A vacuum generator operates with no moving parts and uses the pressure of the surrounding atmosphere to generate a vacuum. There are three ports: one for compressed air, one for exhaust, and one acting as a vacuum. As the compressed air flows through the vacuum generator to the exhaust port, it experiences a reduction in the pathway's cross-sectional area. As the air velocity increases, the pressure that it exerts perpendicular to its direction of flow will decrease. Note that the vacuum port on the vacuum generator is perpendicular to the pathway of the compressed air. The atmosphere outside the vacuum generator has a pressure that is greater than the pressure exerted by the compressed air perpendicular to its direction of flow. In keeping with the principle that high pressure seeks low pressure, the high pressure atmosphere is drawn into the vacuum generator, resulting in a vacuum (negative air pressure).

Types of Air Pressure

Anyone who has watched a weather report on television has been exposed to the concept of air pressure. Weather forecasters frequently use the terms *high-pressure system* and *low-pressure system* when predicting conditions of sunny or stormy weather, respectively. These terms refer to the differences in atmospheric pressure with respect to the mass of air that is contained within a section of Earth's atmosphere. Engineers who design pneumatic systems are specifically concerned with the concept of air pressure, which they measure using special gauges. The measured pressure in a pneumatic system is referred to as *gauge pressure*. Controlling the gauge pressure in a pneumatic system is critical for efficiency as well as safety. Engineers are also concerned with another type of pressure called *absolute pressure*. This type of pressure is used primarily when performing calculations that also involve the temperature and volume of a confined pneumatic system.

ATMOSPHERIC PRESSURE You probably didn't realize that the air that surrounds you is constantly applying pressure against your body. This is known as **atmospheric pressure** (p_{atm}), and it is generated by the weight of the air above you. In fact, a column of air over a 1-in^2 base area that extends from sea level to the outer extents of the atmosphere weighs approximately 14.7 pounds. Therefore, air pressure at sea level is equal to 14.7 lb/in^2 (psia), or 101 kPa. Sea level pressure is also identified as 1 atmosphere.

GAUGE PRESSURE As discussed earlier, a pressure gauge is used to monitor the air pressure in a pneumatic system. If you have ever checked the air pressure of an automobile or bicycle tire, then you have used a pressure gauge. A common type of pressure gauge used for industrial purposes is called a *Bourdon tube pressure gauge* (see Figure 7-45). A Bourdon tube pressure gauge is used to measure air pressure in a pneumatic system. A curved hollow tube inside the gauge tries to straighten out as it experiences a pressure that is above atmospheric pressure. The tube is connected via a linkage to a simple gear train. As the tube changes its shape, it causes the gears to rotate. This, in turn, moves the pointer on the gauge dial face, which indicates the pressure in psig. The value that a pressure gauge identifies is referred to as **gauge pressure** (p_{gauge}), which should not be confused with atmospheric pressure. A typical pneumatic system operates at 90 psig.

FIGURE 7-45: (a) A Bourdon tube pressure gauge is used to measure air pressure in a pneumatic system. (b) A curved metal tube is located inside the gauge. This tube changes shape as it experiences pressure, and its movement is translated to the pointer on the gauge dial face.

(a) © iStockphoto.com/Difydave. (b) Adapted from Johnson, *Introduction to Fluid Power*, © 2002 Delmar Learning, a part of Cengage Learning, Inc. Reproduced with permission. www.cengage.com/permissions.

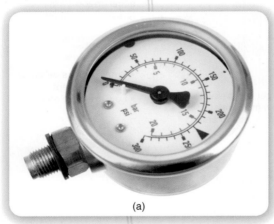

(a)

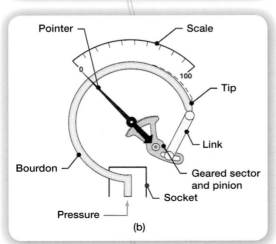

Pointer

Scale

Tip

Link

Geared sector and pinion

Socket

Bourdon

Pressure

(b)

A pressure gauge will not register atmospheric pressure because the weight of the atmosphere also acts on the gauge. Therefore, zero gauge pressure starts at atmospheric pressure. A gauge pressure value that is above atmospheric pressure is considered positive, whereas a gauge pressure value that is below atmospheric pressure is considered negative. The term *vacuum* is also used to describe any pressure below atmospheric pressure (see Figure 7-46).

Absolute Pressure

Figure 7-46 also shows another pressure scale known as **absolute pressure** (p_{abs}). Absolute pressure is based on the concept of a perfect vacuum or absolute zero pressure point, and it is defined as the total pressure exerted on a system, including atmospheric pressure. Because absolute pressure is measured from the lowest possible pressure, the readings will always be positive. Therefore, the absolute pressure is the sum of the pressure measured from a gauge and the atmospheric pressure, as shown by Equation 7-12. It is this pressure scale that scientists and engineers use to perform calculations that relate to the perfect gas laws.

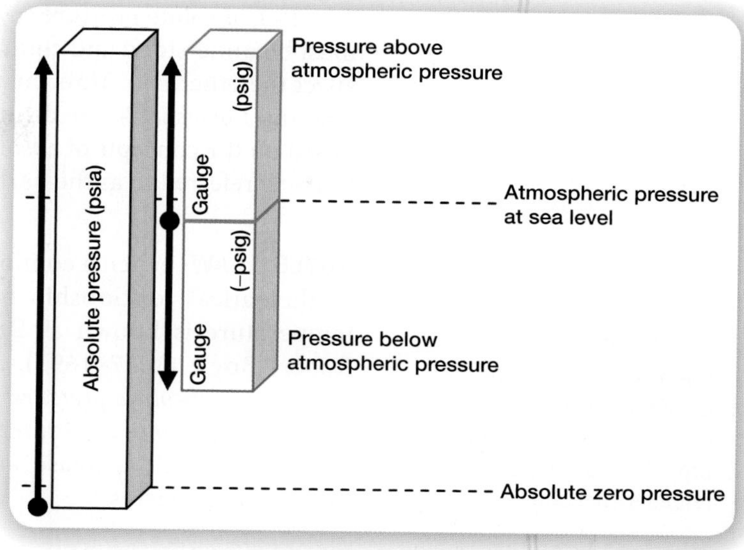

© Cengage Learning 2012

FIGURE 7-46: *Gauge pressure that is above atmospheric pressure is considered positive, whereas gauge pressure below atmospheric pressure is considered negative.*

$$p_{abs} = p_{gauge} = p_{atm} \qquad \text{(Equation 7-12)}$$

Absolute Temperature

Before we can begin our investigation of the perfect gas laws, we must first recognize that the Celsius (°C) and Fahrenheit (°F) temperature scales that we use every day are not directly used in conjunction with the perfect gas laws. Though they are the most widely used temperature scales in both domestic and scientific applications, the zero points on the Celsius and Fahrenheit temperature scales are based on arbitrary physical reactions of matter. For example, 0°C is defined by the temperature at which ice melts under standard atmospheric pressure.

Scientists and engineers use an **absolute temperature** scale, the Kelvin scale, to predict changes in a gas's physical properties. As we saw in Chapter 5, the kelvin (K) is a unit increment of temperature that is based on the point at which matter is devoid of all thermal energy. The unit is named in honor of British physicist and engineer Lord Kelvin (1824–1907), who formulated the scale.

Because temperature is a measure of the movement of molecules, absolute zero (0 K) is the temperature at which all molecular movement stops. It is important to note that the degree symbol (°) is not used to identify a temperature value on the Kelvin scale. Also, unlike the Fahrenheit and Celsius temperature scales, there are no negative values on the Kelvin scale. When studying the properties of gases, it is often necessary to convert temperature values from degrees Celsius or degrees Fahrenheit to Kelvin. Equation 7-13 is used to convert degrees Fahrenheit to Kelvin. Equation 7-14 is used to convert degrees Celsius to Kelvin.

$$K = \frac{(°F + 459.67) \cdot 5}{9} \qquad \text{(Equation 7-13)}$$

$$K = °C + 273.15 \qquad \text{(Equation 7-14)}$$

Perfect Gas Laws

Engineers put gases to work in the design of air pistons, automobile and bicycle tires, shock absorbers, and even lighter-than-air balloons. In each case, a gas's volume (*V*), absolute pressure (p), and absolute temperature (T) are all variables that must be understood and controlled because a change in one of these variables will affect the other two. Three mathematical relationships—*Charles' law*, *Gay-Lussac's law*, and *Boyle's law*—are defined when one of the three variables is kept constant. Based on the concepts of absolute temperature and absolute pressure, they are collectively referred to as the **perfect gas laws**.

BOYLE'S LAW When a confined gas's temperature is kept constant, the resulting mathematical relationship between volume, absolute pressure, and absolute temperature is known as **Boyle's law**, so named in honor of its discoverer, Robert Boyle (1627–1691), an Irish scientist and inventor. Boyle's law states that *a gas's absolute pressure is inversely proportional to its volume, provided its temperature remains constant*. This means as pressure increases, the volume of a confined gas will decrease, and vice versa. Equation 7-15 defines Boyle's law in mathematical terms.

$$p_1 V_1 = p_2 V_2 \qquad \text{(Equation 7-15)}$$

An example application of Boyle's law can be seen in the operation of sealed gas pistons, which are used as shock absorbers on office chairs (see Figure 7-47). When you sit on an office chair, the piston rod retracts into the cylinder, causing the volume of the gas within to decrease. The resulting increase in gas pressure is inversely proportional to the decrease in volume that occurs inside the gas piston cylinder. The piston rod stops moving when the force that is generated by your weight is balanced by the force exerted upward by the piston rod as a result of the gas pressure within the cylinder.

CHARLES' LAW When a gas's pressure is kept constant, the resulting mathematical relationship between volume, pressure, and temperature is known as **Charles' law**. It is named in honor of its discoverer, Jacques Charles (1746–1823), a French inventor

> **Boyle's law:**
>
> the absolute pressure of a confined gas is inversely proportional to its volume, provided its temperature remains constant.

> **Charles' law:**
>
> the volume of a confined gas is proportional to its absolute temperature, provided its pressure remains constant.

FIGURE 7-47: **(a) The gas-piston shock absorber on this office chair is an example application of Boyle's law. (b) As the volume in the gas piston decreases, the pressure that it exerts through its piston rod will increase.**

(a)

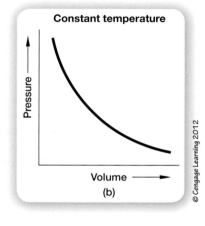

(b)

Problem: A closed flexible container of gas has a volume of 333 cm³ (V_1). The gas exerts an absolute pressure of 113 kPa (p_1) on the inside surfaces of the container. Assuming that the temperature remains constant, what is the volume of the container if the pressure changes to 100 kPa (p_2)?

$$p_1 V_1 = p_2 V_2 \qquad V_2 = \frac{p_1 V_1}{p_2}$$

$$V_2 = \frac{113\ \text{kPa} \cdot 333\ \text{cm}^3}{100\ \text{kPa}}$$

$$= \frac{37{,}629\ \text{kPa} \cdot \text{cm}^3}{100\ \text{kPa}}$$

$$= (376.29\ \text{cm}^3)$$

and scientist. Charles' law states that *the volume of a confined gas is proportional to its absolute temperature, provided its pressure remains constant.* Equation 7-16 defines Charles' law in mathematical terms:

$$\frac{V_1}{T_1} = \frac{V_2}{T_2} \qquad \text{(Equation 7-16)}$$

If you have ever noticed that a basketball takes on a softer quality and has a less responsive bounce during winter, then you have witnessed an example of Charles' law (see Figure 7-48). When a basketball is inflated, its pressure is set to a fixed value. As the temperature of the air within and around the basketball decreases as summer transitions to fall and winter, the volume of the gas inside the ball also decreases proportionally. This assumes that the air pressure inside the basketball remains constant. This relationship between air temperature, pressure, and volume is also why it is important to check the tire pressure of your car regularly to maximize gas mileage and reduce tire wear.

FIGURE 7-48: (a) Charles' law explains why a basketball displays a less responsive bounce in winter than summer. (b) As the temperature decreases, the volume of the air within basketball will decrease. This causes the ball to appear as though it has deflated.

(a) Image copyright Phase4Photography, 2010. Used under license from Shutterstock.com. (b) © Cengage Learning 2012

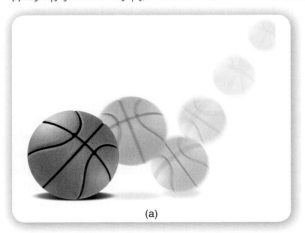

(a)

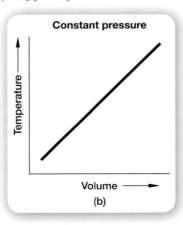

(b)

Problem: A sample of nitrogen gas has a volume of 74.5 cm³ (V_1) at 13°C (T_1). At what temperature (°C) would the gas have a volume of 33.3 cm³ (V_2)?

Step 1: Convert the temperature from degrees Celsius to Kelvin.

$$K = °C + 273.15$$

$$= 13°C + 273.15$$

$$T_1 = 286.15 \text{ K}$$

Step 2: Use Charles' law to solve for the unknown temperature in Kelvin.

$$\frac{V_1}{T_1} = \frac{V_2}{T_2} \quad T_2 = \frac{(T_1 V_2)}{V_1}$$

$$T_2 = \frac{286.15 \text{ K} \times 33.3 \text{ cm}^3}{74.5 \text{ cm}^3}$$

$$= \frac{9,528.8 \text{ K} \cdot \text{cm}^3}{74.5 \text{ cm}^3}$$

$$= 127.90 \text{ K}$$

Step 3: Convert the calculated temperature (T_2) from Kelvin to degrees Celsius.

$$K = °C + 273.15 \Rightarrow °C = K - 273.15$$

$$T_2 = 127.9 \text{ K} - 273.15$$

$$T_2 = -145.25 °C$$

GAY-LUSSAC'S LAW When a gas's volume is kept constant, the resulting mathematical relationship between volume, pressure, and temperature is known as Gay-Lussac's law, named in honor of its discoverer, Joseph Louis Gay-Lussac (1778–1850), a French chemist and physicist. Gay-Lussac's law states *the absolute pressure of a confined gas is proportional to its absolute temperature, provided the gas's volume remains constant.* Equation 7-17 defines Gay-Lussac's law in mathematical terms:

$$\frac{P_1}{T_1} = \frac{P_2}{T_2} \qquad \text{(Equation 7-17)}$$

You may have noticed that there are more remnants of tire blowouts on the highway in summer than in winter. One reason for this can be explained by Gay-Lussac's law (see Figure 7-49). An automobile tire serves as a relatively rigid container for compressed air. The material from which the tire is made has structural limitations, which is why you can find a maximum air pressure value printed on the tire's sidewall. The temperature of the compressed air within the tire increases not only as a result of higher summer temperatures but also as a

Gay-Lussac's law:

the absolute pressure of a confined gas is proportional to its absolute temperature, provided its volume remains constant.

FIGURE 7-49: Gay-Lussac's law explains why (a) more automobile tires experience blowouts during summer months than during winter months. (b) As the temperature increases, the pressure of the air within the tire will increase. If the pressure exceeds the structural limits of the tire material, it will cause the tire to rupture.

(a)

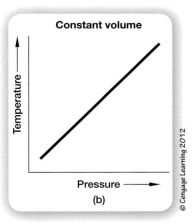

Constant volume

Temperature →

Pressure →

(b)

© iStockphoto.com/MotoEd

© Cengage Learning 2012

result of friction that is generated between the road surface and the tire. This causes an increase in the pressure of the compressed air within the tire. A blowout will occur if the air pressure within the tire exceeds the structural limitations of the tire material.

Example

Problem: A sealed rigid container is used to store a gas. A pressure gauge that is built into the container displayed an initial pressure of 0 psig (p_1) at standard atmospheric pressure and 65°F (T_1). If the temperature changes to 145°F (T_2), what is the new pressure gauge reading if the container remains at standard atmospheric pressure?

Step 1: Convert the gauge pressure to absolute pressure. Note that standard atmospheric pressure is equal to 14.7 psia.

$$p_{1\,abs} = p_{gauge} + p_{atm}$$
$$= 0\ psig + 14.7\ psia$$
$$= 14.7\ psia$$

Step 2: Convert the initial temperature (T_1) from degrees Fahrenheit to Kelvin.

$$K = (°F + 459.67) \times \frac{5}{9}$$

$$T_1 = (65°F + 459.67) \times \frac{5}{9}$$

$$= 291.48\ K$$

(continued)

Step 3: Convert the final temperature (T_2) from degrees Fahrenheit to Kelvin.

$$K = (°F + 459.67) \times \frac{5}{9}$$

$$T_2 = (145°F + 459.67) \times \frac{5}{9}$$

$$= 335.93 \, K$$

Step 4: Use Gay-Lussac's law to solve for the unknown absolute pressure.

$$\frac{p_1}{T_1} = \frac{p_2}{T_2} \Rightarrow p_2 = \frac{p_1 T_2}{T_1}$$

$$p_2 = \frac{14.7 \, psia \times 335.93 \, K}{291.48 \, K}$$

$$= \frac{4938.17 \, psia \cdot K}{291.48 \, K}$$

$$= 16.94 \, psia$$

Step 5: Convert the final pressure (p_2) from absolute pressure to gauge pressure.

$$p_{abs} = p_{2 \, gauge} + p_{atm} \Rightarrow p_{2 \, gauge} = p_{abs} - p_{atm}$$

$$p_{2 \, gauge} = 16.94 \, psia - 14.7 \, psia$$

$$= 2.24 \, psia$$

SUMMARY

In this chapter, you learned the following.

- Fluid power systems use the energy contained in pressurized liquids and gases to transmit forces across distances and in situations that are impractical for mechanical and electromechanical power-transfer methods. In fact, hydraulic systems have a power density that is 10 times greater than that of an electric motor, which makes them an excellent choice for moving control surfaces on aircraft.

- Hydraulic systems use pressurized liquid, whereas pneumatic systems use compressed gases. Hydraulics are self-lubricating and well suited for applications that call for relatively slow and steady movements of large loads that may involve intermittent motions. They generate large amounts of heat, operate at relatively high pressures, and are almost certain to leak hydraulic fluid at some point in time. As such, they are not well suited for application in clean environments or in situations where excessive heat or open flames could ignite the hydraulic fluid. Pneumatic systems operate at relatively low pressures and are well suited for quick, uninterrupted movements of small loads. They can also be used to generate a vacuum for the purpose of lifting, transporting, or work-holding objects. Many pneumatic systems use ambient air, which is a free source of fluid. Unlike hydraulics, they are able to exhaust their fluid back to the atmosphere. Maintaining a constant source of compressed air can be expensive, however, depending on the demand. Unfortunately, pneumatic devices also generate excessive amounts of noise.

- Hydraulic and pneumatic systems do have common elements. Both systems incorporate devices that serve to pressurize a fluid such as a pump or an air compressor. Fluids are often placed through some type of conditioning process to ensure that they will not damage the inner workings of the components in their respective systems. Both pneumatic and hydraulic systems provide pathways through which fluids can flow. Different types of valves are used in both systems to perform identical functions with respect to controlling fluid pressure, flow rate, and the directions that fluids will be allowed to flow within a circuit. Finally, both systems incorporate some sort of output device, such as a linear actuator, where the energy of the pressurized fluid is put to work.

- Hydrostatic systems involve fluids that are in a state of static equilibrium, meaning they do not move. As such, the pressure of the fluid within these systems is constant and acts equally on the inner walls of the containing surfaces. Hydraulic and gas pistons, automobile tires, and bicycle tires are applications of the principles of hydrostatics. Hydrodynamic systems involve fluids that are in motion, and the fluids will experience changes in velocity and pressure depending on the geometry of the passageways through which they flow. An example application of a hydrodynamic system is a pneumatic vacuum generator.

- Pascal's law is a fundamental principle of the science of hydrostatics. It states that pressure exerted on a confined fluid is transmitted equally and perpendicular to all of the interior surfaces of the fluid's container. A hydraulic press, a bottle jack, and any other type of fluid power linear actuator operate according to Pascal's law.

■ Hydraulic amplification of force is achieved when a small input force generates a large output force. This occurs when two linear actuators of different cross-sectional areas are joined together in a closed hydrostatic circuit. Because the pressure within both cylinders is constant, an increased area on the output cylinder's piston head will generate a proportional increase in the output force. However, the input cylinder will have to move a greater distance than the output cylinder. The ratio between the input and output distances that are traveled by the pistons is equal to the ratio between the output and input forces that are applied to those pistons.

■ A fluid's flow rate is associated with the amount of time it takes for a specific volume of fluid to move through a system. Common units of measurement include gallons per minute (gpm), cubic inches per minute (in³/min), liters per minute (lpm), and cubic meters per second (m³/s). A fluid's flow velocity is a function of the amount of time it takes for a fluid particle (such as a drop of liquid) to move a given distance. Flow velocity is measured in feet per minute (ft/min), inches per second (in/s), and meters per second (m/s).

■ Bernoulli's principle states that the velocity of a fluid will increase as the pressure exerted by that fluid decreases, and vice versa. This fundamental law of hydrodynamics is put to work in devices that incorporate a Venturi tube, such as a pneumatic vacuum generator and an electrical power plant's water-cooling towers. As a fluid flows through either of these examples, it will experience a decrease in the cross-sectional area of the channel. This causes an increase in fluid velocity, which results in a decrease in fluid pressure.

■ Atmospheric pressure (p_{atm}) is the weight of a column of air that stretches from sea level to the uppermost portion of Earth's atmosphere. If this column acts on a 1-in² area at sea level, then the pressure of the air is equal to 14.7 psi. Gauge pressure(p_{gauge}) is a measure of the pressure of a fluid in a confined system. Zero gauge pressure begins at atmospheric pressure. Pressure readings from pressure gauges are denoted as psig. Positive gauge pressure is greater than atmospheric pressure. Negative gauge pressure is less than atmospheric pressure and is denoted as a vacuum. Unlike atmospheric or gauge pressure, the absolute pressure scale is based on the concept of zero pressure or a perfect vacuum. Therefore, the absolute pressure scale has no negative values. This type of pressure scale is used in conjunction with the perfect gas laws to calculate the properties of gases in closed pneumatic systems.

■ The zero points of the widely used temperature scales, the Fahrenheit and Celsius scales, are based on arbitrary physical reactions of matter. The Kelvin temperature scale, however, is based on the concept of absolute zero, which is the temperature at which all molecular movement stops. It is this temperature scale that engineers and scientists use to calculate the physical properties of gases in confined pneumatic systems.

■ Boyle's law, Charles' law, and Gay-Lussac's law are collectively referred to as the *perfect gas laws*. Each law expresses the relationship between absolute temperature, absolute pressure, and volume as they relate to confined gases in pneumatic systems. What differentiates one law from another is the fact that one of the variables (temperature, pressure, or volume) must remain constant. Boyle's law is based on a constant gas temperature. Charles' law is based on constant gas pressure. And Gay-Lussac's law is based on a constant volume of gas. The resulting relationships between the changing variables explain the behaviors of devices such as gas pistons (Boyle's law), basketballs that go soft during winter (Charles' law), and the unusual number of automobile tire blowouts that occur during hot summer months (Gay-Lussac's law).

1. What are the three primary power transmission methods used by machines?
2. What is fluid power?
3. Why are gases and liquids considered to be fluids?
4. What is hydraulics?
5. Identify two common applications of hydraulics.
6. What is pneumatics?
7. Identify two common applications of pneumatics.
8. Why are hydraulic systems preferred over electromechanical devices for applications such as aircraft controls?
9. How is the area of a circle calculated, and why would this mathematical formula be useful in the engineering of fluid power systems?
10. What is pressure, and what are its common U.S. customary and metric units of measure?
11. Define the mathematical relationship between pressure, force, and area.
12. Explain why a person would prefer to wear snow shoes versus regular shoes when walking (ill advisedly) across a frozen body of water.
13. How do liquids and gases differ with respect to compressibility?
14. What kinds of applications are hydraulic and pneumatic systems best suited for?
15. Identify three disadvantages of hydraulic systems and three disadvantages of pneumatic systems.
16. Who was Blaise Pascal, and what did he contribute to the science of fluid power?
17. What was the first hydraulic-powered machine to utilize Pascal's law, and who was responsible for its invention?
18. What are standards, and why do engineers develop them?
19. What are schematic symbols, and why are they used to communicate the design of fluid power systems?
20. What common components are found within both hydraulic and pneumatic systems?
21. What is the difference between a working line and a pilot line?
22. How would an engineer tell the difference between connected and nonconnected fluid power lines where the lines cross on a schematic diagram?
23. What is the purpose of a filter in a fluid power circuit?
24. What are the two most common types of linear actuators in fluid power systems, and how do they differ from each other?
25. What is the difference between a shutoff valve and a check valve, and what are these valves used for?
26. Under what conditions might an engineer use a shuttle valve in a fluid power circuit?
27. What is the purpose of a flow control valve?
28. Sketch the schematic symbols for a filter, single-acting cylinder, double-acting cylinder, shutoff valve, check valve, shuttle valve, and flow control valve.
29. What is the purpose of a reservoir in a hydraulic circuit?
30. What kind of mechanical device is used to generate pressure in a hydraulic system?
31. What is a prime mover, and is it needed in a fluid power system?
32. What is the purpose of a pressure relief valve in a hydraulic circuit?
33. What is the function of a directional control valve (DCV), and how are the ports on a hydraulic DCV identified?
34. Sketch the schematic symbols for a hydraulic reservoir, fixed-displacement pump, pressure relief valve, and a four-way tandem center condition DCV.
35. Why is ambient atmosphere put through a conditioning process before it is used in a pneumatic system?
36. What kind of mechanical device is used to generate the air pressure in a pneumatic system?
37. What is the function of a pneumatic pressure regulator?
38. Why is a lubricator used in a pneumatic system?
39. Where is compressed air stored before it is put to use in a pneumatic system?
40. Why might an engineer substitute a simple schematic symbol for a compressed air source in place of the schematic symbols used to identify all the conditioning devices that air must travel through before it is used in a pneumatic system?
41. Sketch the schematic symbols for an air compressor, an air filter outfitted with a manual drain, a pressure regulator, a flow meter, a lubricator, an air receiver tank, and a generic air supply.

(*continued*)

42. Why is a simple pneumatic on–off DCV referred to as a three-port, two-way valve?

43. What are the three most common mechanical actuation methods found on pneumatic DCVs?

44. What is the purpose of a pneumatic five-port, two-way valve?

45. Sketch the schematic symbol for a pneumatic exhaust.

46. What is a solenoid, and how is it used to control a DCV?

47. What is hydrostatics?

48. What is hydraulic amplification of force, and what kind of trade-off does it involve?

49. How is Pascal's law used to determine the hydraulic amplification of force that occurs between two cylinders of dissimilar geometry in a closed hydrostatic system?

50. How do you calculate the volume of liquid contained in a cylindrical actuator?

51. If two cylinders of dissimilar geometry are connected together to form a closed hydrostatic system, why would one cylinder piston travel a greater distance than the other?

52. Why would a typical double-acting cylinder generate different forces between its extension and retraction strokes?

53. What is hydrodynamics?

54. Why is a liquid's viscosity an important consideration in the design of a hydraulic system?

55. What is the difference between turbulent and laminar fluid flow?

56. What is the difference between a fluid's flow rate and its flow velocity?

57. Identify the formula for calculating a fluid's flow rate, along with its common U.S. customary and metric units of measure.

58. What is the purpose of a flow meter?

59. Identify the formula for calculating a fluid's velocity, along with its common U.S. customary and metric units of measure.

60. What is the mathematical relationship between a liquid's flow rate and its flow velocity?

61. What is Bernoulli's principle?

62. Identify two devices that operate according to Bernoulli's principle, and explain what they are used for.

63. What is atmospheric pressure, and what is its value at sea level?

64. What is gauge pressure, and how is it distinguished from other types of pressure?

65. What is absolute pressure, and why would an engineer or scientist be concerned with it?

66. What is the mathematical relationship between atmospheric, gauge, and absolute pressure?

67. When would a scientist or engineer use the absolute temperature scale versus the conventional Fahrenheit or Celsius temperature scales?

68. What are the common variables that are associated with the perfect gas laws?

69. What is Boyle's law? Identify the formula for and one application of Boyle's law.

70. What is Charles' law? Identify the formula for and one application of Charles' law.

71. What is Gay-Lussac's law? Identify the formula for and one application of Gay-Lussac's law.

EXTRA MILE

Problem Statement: Rescue personnel who use hydraulic bottle jacks as part of their standard search-and-rescue equipment are frequently asked questions about these devices during school presentations. Students want to know how such devices operate and how they are able to multiply a person's strength to generate large output forces. The concepts are difficult to explain because the fluid and inner workings of the bottle jack are hidden from view.

(continued)

Design Statement: Design, build, and test a functional concept model of a bottle jack that rescue personnel can use at school presentations to explain how the device operates (see Figure 7-50). Calculate the direct amplification of force that occurs between the input and output cylinders. The solution must:

- be easy to assemble and disassemble;
- utilize water as its fluid source;
- incorporate check valves and a shutoff valve;
- use a 3-mL syringe and a 60-mL syringe as the input and output cylinders, respectively; and
- use a commercially available plastic container as the fluid reservoir.

FIGURE 7-50: A student's bottle-jack concept-model solution ideas recorded in her engineer's notebook.

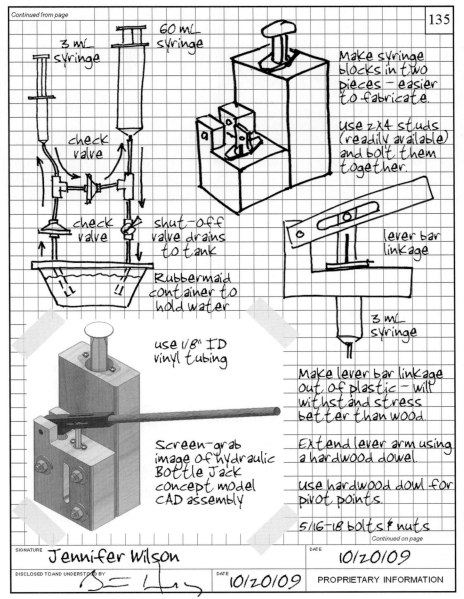

CHAPTER 8
Control Systems

GPS DELUXE

| START LOCATION | DISTANCE | END LOCATION |

Menu

Before You Begin

Think about these questions as you study the concepts in this chapter:

1. What is a system?

2. Identify the components of a system.

3. What is the difference between an open-loop system and a closed-loop system?

4. How do analog and digital signals differ?

5. Are thermistors and potentiometers digital or analog devices?

6. What are some examples of digital input devices?

7. What function do output devices have in a control card system?

8. What roles do processors and controllers play in control systems?

9. What are the characteristics of microcontrollers, computer-based controllers, and programmable logic controllers?

10. What are two common programming languages used in control systems?

11. How is the heating and cooling system in your house a control system?

12. How are process control systems used in industry?

13. How do servomechanisms and numerical control systems control motion?

We can all relate to the expression "out of control." A certain level of anxiety is achieved when things are out of control. Productivity decreases, and mistakes are easily made. Humans have the capability to get themselves in control, but machines do not. That is why we need control systems when we use machines to do complex or highly detailed tasks.

A control system is a set of components working together to perform a given task under the direction of a processor or computer. Control systems for manufacturing equipment were developed between the First and the Second World Wars. As a result, automated manufacturing equipment began to replace manually operated machines. Automated manufacturing equipment minimizes or replaces the need for human labor by giving control of mundane and repetitive tasks to a program-driven computer. The program tells the computer when and how to control output devices such as tool spindles and motion-control motors. The program might also tell the computer to look for feedback information from sensors located on the equipment. If problems arise, this feedback may trigger a response which can dictate necessary steps to correct the process on the fly. A control system would be in place, for example, in an assembly line for filling beverage bottles (see Figure 8-1).

Today's cars can be considered automated equipment because there is a lot more to driving than pressing on the gas pedal and turning the steering wheel. A car's control system is constantly making adjustments according to changing drive conditions. It can adjust the fuel mixture, modify the timing for different RPM ranges, and control the engine's temperature. When the engine light flashes on the dashboard, the car's on-board diagnostic system is telling you that one or more of the car's sensors has sent a message to the car's computer. The engine light also sends a message to the driver that a problem has occurred that requires the attention of a professional.

In this chapter, we will learn about the various components of a basic control system, such as input and output devices, processors, and controllers. We will explore some of the different types of control systems currently available and take a look at how computer programming is used within a control system. Finally, we will explain how control systems make machines more reliable, increasing the overall quality of the products they produce.

> **Control system:**
> a set of components working together to perform a given task under the direction of a processor or computer.

Figure 8-1: *Control systems are used to monitor the fluid levels in a beverage line.*

© iStockphoto.com/fadhlikamarudin.

Input:

information fed into a data-processing system or computer.

Process:

a systemic sequence of actions that is designed to control an output.

Output:

the information produced by a computer, or the actions that result from machines that are controlled by a computer.

Feedback:

information returned to the input of a system in order to provide self-corrective action.

OVERVIEW OF A SYSTEM

Generally speaking, a *system* is a group of interacting or interdependent parts that function together as a whole to accomplish a task. For example, a digital camera is a system that consists of a lens, circuit board, camera body, flash, firmware, light meter, and memory card. This system's task is to capture a moment in time and preserve it as a digital image. As it is true with all systems, the various operations of the digital camera system follow a sequential pattern of **input**, **process**, and **output**. As you point the digital camera and push the shutter button halfway down, the camera's input devices will gather information about lighting and your distance from the subject. This information serves as the input. The camera's computer system processes this information through a software control program which determines the correct focus, aperture, and shutter speed needed to capture the best possible image. The necessary adjustments are made when the shutter button is pressed all the way down. The process results in a captured image, which serves as the system's output. A memory stores the output image so that it may be printed, e-mailed to a friend, or uploaded to a social networking site.

Systems are divided into two categories: *open-loop systems* and *closed-loop systems*.

Open-Loop System

Open-loop systems operate on the assumption that no errors will occur. There is no **feedback** to allow for adjustments once an operation has started. Once an input device is triggered and the process begins, the output will either operate as planned or it will fail.

The motor and shuttle system shown in Figur 8-2 is designed to move along a fixed gear rack, and is controlled via an open-loop system. This system is programmed to continuously repeat a sequence of timed actions that involves moving the shuttle back and forth along the gear rack for 3 seconds in each direction. Ideally, the shuttle should move between the 3-inch and 6-inch marks as identified by the ruler below the gear rack. However, a motor spins faster in one direction. Therefore, the distance the motor travels in one direction will be greater than the distance it travels in the opposite direction, regardless of the uniform time delays contained in the process control program. As a result, the shuttle will drift in one direction over time. Without a feedback component to tell the system that something is going wrong, the system cannot self-correct. Eventually, the shuttle will fall off the gear rack. This doesn't mean that all open loop systems are doomed to fail.

FIGURE 8-2: This figure shows a rack and pinion gear train shuttle that is powered by a 5-volt motor that is intended to move back and forth, continuously, between the 3-inch and 6-inch marks of a ruler using time delays.

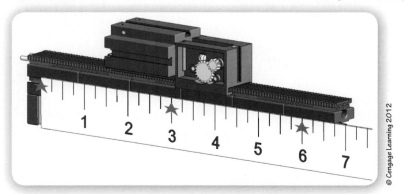

© Cengage Learning 2012

Timing loops work well for short runs and they have their place in some industrial and domestic applications. For example, the sequential operations of a washing machine are controlled with timing cycles.

Closed-Loop System

A closed-loop system also consists of an input, process, and output, but it also incorporates one or more feedback devices that allow a control program to monitor and adjust a system's output (see Figure 8-3). Feedback is generated by devices called sensors. Examples of sensors include limit switches, thermistors, potentiometers, reed switches, and phototransistors. A closed-loop control system receives information from a sensor in the form of an electrical signal. The controller processes the input signals and makes decisions that keep the system operating as planned.

Figure 8-4 shows a different shuttle system that is also designed to move back and forth, continuously, along a fixed gear rack. However, this system incorporates two position sensors located at either end of the fixed gear rack. Instead of relying on time delays, the shuttle moves in one direction until it makes physical contact with a limit switch. When contacted, the switch sends a signal to the controller. The controller processes the information and commands the motor to reverse its direction. The feedback from the limit switches eliminates the need for time delays and keeps the shuttle from falling off the track.

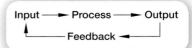

FIGURE 8-3: *Graphical representation of a closed-loop system.*

© Cengage Learning 2012

> **Sensor:**
> a device that responds to a physical stimulus (such as heat, light, sound, pressure, magnetism, or a particular motion) and transmits a resulting signal (as for measurement or operating a control).

FIGURE 8-4: This shuttle device is controlled by a closed-loop system. The shuttle moves back and forth between two limit switches that provide feedback information to the control program.

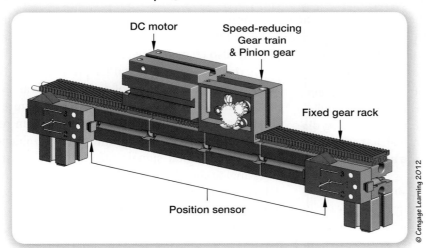

© Cengage Learning 2012

INPUT DEVICES

Input devices, such as sensors, are used to collect information from an environment and feed the information back to a controller in the form of an electronic signal within a closed-loop system. Depending on the sensor, the electrical signal will take the form of an *analog signal* or a *digital signal*, which are distinguished by their different waveforms. Electrical signals are transmitted through wires. However, these signals can be converted to electromagnetic radiation and transmitted wirelessly over great distances and through objects.

Analog Signals

Analog signals are continuous electrical signals that vary in intensity over time and are capable of attaining an infinite number of values or levels within a given range (see Figure 8-5). For example, the voltage across an analog light sensor will vary depending on the amount of light that strikes the sensor. Assuming that the voltage across the light sensor has a range between 0 and 5 volts, the number of voltage values between 0 and 5 volts that can be achieved by the varying degrees of light intensity is infinite. Analog devices are able to convert sound, heat, pressure, light, and other forms of energy into analog signals. These signals are transmitted across conductive wires to provide feedback information within a closed-loop system. A disadvantage of analog signals is noise or unwanted and random variations that sometimes occur in such signals. This noise can inadvertently trigger a false reading and create process errors.

Digital Signals

A **digital signal** is a an electrical signal that has an integral number of discrete levels or values within a given range. Figure 8-6 shows two children pitching coins onto a ramp and a staircase. The ramp represents an analog signal, and the staircase represents a digital signal. The ramp offers an infinite number of points between the ground level and the building's entrance platform. If one were to toss coins onto the ramp, no two coins would land at the same height level. The staircase divides the vertical distance between the ground and the building's entrance platform into discrete levels. Any coin that is pitched onto the staircase must settle at a fixed distance between the ground and the platform.

The simplest digital signal has only two distinct states that are typically characterized as *voltage* or *no voltage*, *on* or *off*, *high* or *low*, or *1* or *0*. The digital sensors that are addressed in this chapter are also characterized in this manner.

If you are using a digital telephone, your voice is converted into a sequence of digital signals and transmitted in the form of a *binary code* (see Figure 8-7). This signal is sent to another device that can receive that signal and convert it into recognizable words or sounds. Digital signals offer distinct advantages over analog signals. They create the highest-definition television pictures and the clearest telephone conversations. An analog signal has a range of information that is harder to duplicate and transmit.

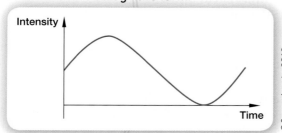

Analog signal:

a continuous electrical signal that varies in intensity over time and is capable of attaining an infinite number of values or levels within a given range.

FIGURE 8-5: Analog wave form.

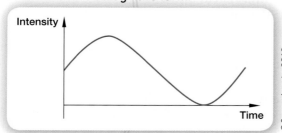

Digital signal:

an electrical signal that has an integral number of discrete levels or values within a given range.

FIGURE 8-6: There are an infinite number of height levels that occur on the ramp between the ground and the building's entrance platform. This is analogous to an analog signal. The staircase divides the same vertical distance into a discrete number of height levels. This is analogous to a digital signal.

Adapted with permission from Master Publishing, Inc., Niles, IL Basic Digital Electronics.

FIGURE 8-7: This is an example of a digital wave form. It has only two different states or positions.

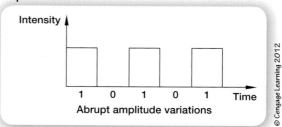

Abrupt amplitude variations

Career Profile

CAREERS IN ENGINEERING

Working with a Net but No Wire

When he was growing up in Brazil, Carlos Cordeiro had no idea he would one day travel the world as a computer engineer. But that's exactly what he's done, connecting with some of the best-known names in international commerce: IBM, Nokia, Philips, Intel. Currently, Cordeiro works on the cutting edge of wireless communications at Intel Labs, the research division of Intel.

"The focus of the research group I work for is the system-on-a-chip, or SOC," which refers to combining all the components of a computer or other electronic system into a single integrated circuit. "I work on the wireless communication side of an SOC," Cordeiro adds, "and the goal is to take wireless to much higher speeds so it will be able to support these applications—high-definition video streaming, instantaneous multigigabit file transfers, wireless USB, etc. Someday, this could all be done without wires."

On the Job

Some of Cordeiro's most interesting work was completed at Philips Research North America. At Philips, Cordeiro worked in the area of cognitive radio, which sounds like an AM/FM device that already knows what station you want to listen to. Actually, cognitive radio refers to the effort to use the electromagnetic spectrum as efficiently as possible.

"Today, because all of the radio spectrum has been statically assigned by the government, space has become scarce," Cordeiro says. "But we figured out a way to harvest those parts of the spectrum that aren't in use at the time—the white space."

Inspirations

As late as high school, Cordeiro had no intention of becoming an engineer. "I was going to be a lawyer," he says, the decision based in part on the respective salaries he expected to make. But then fate took an ironic twist. Cordeiro's brother, who was 6 years older, dropped out of an engineering program during his third year because he decided he wanted to become—a lawyer. But he thought Carlos was made for engineering, so he gave him a tour of the campus.

Carlos Cordeira, Cognitive Radio, Intel Labs.

"I was fascinated, both by the environment and the people," Cordeiro says. "I'd always been good at science and math, but I'd never really put it together. Luckily, my brother saw what I couldn't see."

Education

Cordeiro got his bachelor's and master's degrees in computer science in Brazil, but that proved a difficult place to launch a career. "It's tough," he says. "There isn't a lot of support." But when IBM lured him to California with a job offer, Cordeiro felt his prospects had greatly improved. One day, he received an e-mail from an engineering student. "It said, 'I'm writing a paper and I'm doing this work on this code of yours, and can you send me a source code?' And I wrote back, 'Sure, here it is.' That guy became one of my good friends, and his graduate advisor later became my own graduate advisor," says Cordeiro. He earned his PhD at the University of Cincinnati.

Advice for Students

Having followed his own star as far away as Finland, where Nokia is based, Cordeiro advises today's students to do the same. "Listen to your heart," he says. "Pay attention to anything that tickles you. And talk to people, especially people in the specific field you're interested in. Get a feel for what that field actually is."

Analog to Digital Conversion

Because sampling devices, such as microphones, capture the human voice as an analog signal, digital devices like the modern telephone need to first convert the analog signal into a digital signal before transmitting it. This is done using an analog-to-digital converter (ADC) as shown in Figure 8-8. This device converts continuous signals (analog) to discrete digital signals. A digital-to-analog converter (DAC) is then used to convert a digital signal back to an analog signal.

FIGURE 8-8: This circuit board contains an ADC chip that converts analog signals into digital signals.

Courtesy of Fujitsu Semiconductor.

Analog Input Devices

An analog input device generates an electrical signal that varies across a given voltage range as a result of a change in resistance within its conductive element. Thermistors and potentiometers are two examples of analog devices that are commonly used as sensors in closed-loop control systems. In short, both devices act as variable resistors.

Resistance:
the opposition to the flow of electricity.

THERMISTOR A thermistor [see Figure 8-9(a)] is an analog input device that changes in resistance as it experiences a change in temperature. The change in

FIURE 8-9: (a) As the temperature around this thermistor changes, its resistance to the flow of electricity changes. (b) A thermistor contained within a thermostat will provide an analog signal that changes based on the temperature of the interior space where the thermostat is located.

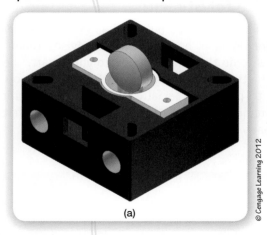

(a)

© Cengage Learning 2012

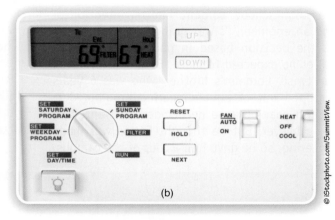

(b)

© iStockphoto.com/SummitView.

resistance results in a change in the analog voltage signal to the controller. Thermistors are used in thermostats, which are responsible for controlling the temperature of a home. A reading taken from a thermistor is not the actual temperature; instead, the resistance range can be correlated to a range of temperatures [see Figures 8-9(b) and 8-10].

POTENTIOMETERS A **potentiometer** is an analog input device that also allows the amount of resistance in a circuit to be varied manually. The most common household application of this device is a rotary or sliding light switch that is used to dim a light above a dining room table.

The 300 ohm potentiometer shown in Figure 8-11 is designed to rotate 270 degrees. As it rotates, its resistance changes. Each 90 degree rotation will increase the resistance by 100 ohms. Therefore, the potentiometer's resistance value will be 200 ohms when the control dial has rotated 180 degrees. When the resistance value for each degree is known, the potentiometer can be used as a position sensor for motion control applications.

Digital Input Devices

The digital input devices identified in this chapter provide information to a controller in the form of an *on* or *off* signal. Limit switches are digital input devices that are activated by physical contact. Other digital input devices such as reed switches and phototransistors are triggered by methods that do not include physical contact.

PUSH-BUTTON SWITCH The limit switch shown in Figure 8-12 can be wired in one of two configurations: **normally-open (NO)** or **normally-closed (NC)**. If you look closely at the figure, you can see a **schematic** diagram that shows how the switch is configured internally. Connecting wires to ports 1 and 3 will result in a

FIGURE 8-10: This heating and cooling system uses the analog input signal from a thermistor to monitor the temperature of a living environment. If the air temperature exceeds a set limit within the thermostat, the air conditioner will be turned on. If the air temperature is below a set limit, the furnace will be activated.

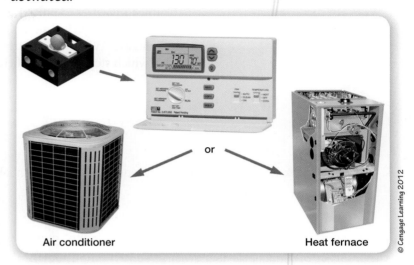

Air conditioner   or   Heat fernace

© Cengage Learning 2012

Schematic:
a diagram that uses symbols to represent components of a system.

FIGURE 8-11: This potentiometer changes resistance as its input dial is turned.

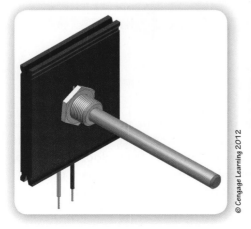

© Cengage Learning 2012

FIGURE 8-12: This limit switch can be wired normally-open, terminals 1 and 3; or normally-closed, terminals 1 and 2.

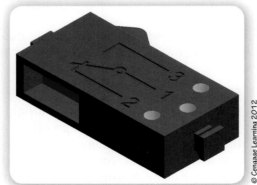

© Cengage Learning 2012

normally-open configuration. In this state, no electrical signal will pass through the switch when the switch is not pressed. A doorbell uses a similar type of switch and wiring configuration. When the switch is pressed, contact is made between terminals 1 and 2, which allows electricity to flow through the switch. The signal continues to flow through the switch as long as the switch is held down. When the switch is let go, the circuit is broken and the signal stops.

Connecting wires to ports 1 and 2 will result in a normally-closed configuration. In this state, an electrical signal is free to pass through the switch when the switch is not pressed. When the switch is pressed, contact is broken between terminals 1 and 2, which stops the flow of electricity through the switch.

Your Turn

Can you think of an application that would rely on a manual switch that is wired in the normally-closed state?

FIGURE 8-13: A reed switch is activated by the presence of a magnetic field. When a reed switch encounters a magnetic field the switch will allow an electrical signal to pass through.

© Cengage Learning 2012

REED SWITCH A reed switch is a type of non-contact digital input device that is often used in alarm systems. The basic reed switch consists of two identical flattened ferromagnetic reeds sealed in a dry, inert-gas atmosphere within a glass capsule (see Figure 8-13). The free ends of the reeds overlap yet do not touch (see Figure 8-14). When a magnetic field is encountered, the reeds are attracted to each other. When the reeds make contact with each other, electricity is allowed to flow through the switch.

Reed switches are used on windows that are linked to home alarm systems. A permanent magnet is placed next to a reed switch, but on separate parts of the window. When the window is opened, the switch is moved away from the magnet, breaking the electric circuit. This change of state sends a message to the alarm system.

FIGURE 8-14: The inner workings of a reed switch.

Courtesy of Reed Relays and Electronics India Limited.

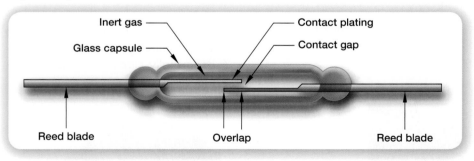

PHOTOTRANSISTOR A phototransistor is another type of non-contact digital input device that functions as a light-sensitive switch (see Figure 8-15). If you've ever heard a buzzer go off as you walk into the front door of a store, then you most likely triggered a phototransistor. As a transistor is a polarity-sensitive electronic component, polarity must be considered when wiring a phototransistor. If wired incorrectly, the phototransistor will not function.

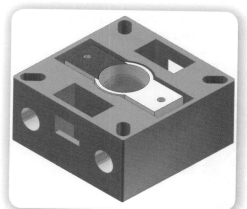

FIGURE 8-15: A phototransistor is a light-sensitive input device. The presence of light makes the phototransistor change states. It will read a 0 or a 1, depending on how much light is being received by the sensor.

© Cengage Learning 2012

FIGURE 8-16: Hydraulic actuators are used in airplane landing gear systems.

© iStockphoto.com/Ratstuben.

OUTPUT DEVICES

An output device is used to perform work or convert energy from one form to another in order to accomplish a specific task. An output device may generate motion, produce light, make sound, or create a magnetic field. Examples of output devices include actuators, lamps, buzzers, and electromagnets.

Actuators

An **actuator** is any output device that generates mechanical motion. Actuators are used in manufacturing applications such as clamping devices, tool spindles, and parts ejectors. As you learned in Chapter 7, hydraulic actuators are used in applications where heavy loads are encountered and where precise movements are required. This makes them ideal for use in airplane landing gear systems (see Figure 8-16).

Figure 8-17 shows an electric motor actuator. One application of the electric motor is to operate a valve. Most of the time the motor is designed to put the valve in either the fully opened position or the fully

FIGURE 8-17: The electric motor functions as an actuator to open and close this industrial shutoff valve.

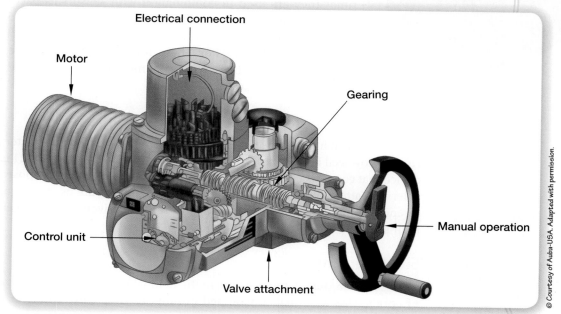

Electrical connection

Motor

Gearing

Control unit

Manual operation

Valve attachment

© Courtesy of Auba-USA. Adapted with permission.

FIGURE 8-18: (a) This car door uses an electric motor actuator. When an electric signal is received by the motor, it powers a mechanism to open or close the car window. (b) Automotive car door window motor.

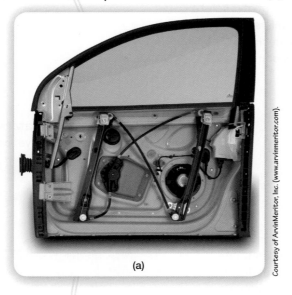

(a)

Courtesy of ArvinMeritor, Inc. (www.arvinmeritor.com).

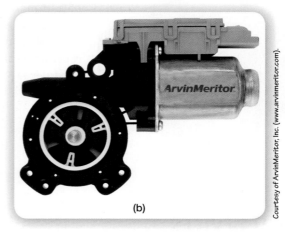

(b)

Courtesy of ArvinMeritor, Inc. (www.arvinmeritor.com).

closed position, but the position of the valve can also be adjusted to any intermediate position.

The car door mechanism shown in Figure 8-18 uses an electric motor to move the glass window up and down.

FIGURE 8-19: Lamp.

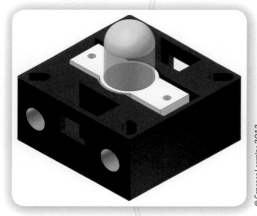

© Cengage Learning 2012

Lamps

In industry, lamps or lights serve a variety of purposes in automated systems. First and foremost, these devices are used for communication: Lights on a control board can indicate when an inspector or operator needs to perform certain actions, or lights such as the lamp in Figure 8-19 can indicate when a process is running or is about to begin or end.

Although lights are traditionally thought of as output devices, sometimes they are used in combination with sensors as part of an input system. For example, Figure 8-20 shows a series of red lights that act as a safety barrier on an industrial workcell. Each light works in conjunction with a phototransistor. These devices will halt machine operations when a person or object accidently moves into an area of the workcell that is deemed hazardous.

Buzzers

Safety is critical in any process. If a system is not safe, it does not matter how fast it runs or how well it produces parts. A buzzer like the one in Figure 8-21 is an output device that converts electrical energy into sound energy for the purpose of drawing a person's attention to a situation. For example, buzzers are used to signal the start or end of a process. They are also used to remind drivers or passengers to fasten their seatbelts.

Electromagnets

Electromagnet:

a device consisting of a coil of conductive wire that is wrapped around an iron core that generates a magnetic field when current passes through the coil.

An **electromagnet** generates a magnetic field when current passes through a coil of conductive wire that is wrapped around an iron core. Unlike a permanent

FIGURE 8-20: This safety curtain is an example of how an input device can be used to increase safety in the work place. When the light curtain is broken, a signal is sent to the processor and the process stops.

FIGURE 8-21: Fischertechnik buzzer.

Courtesy of Banner Engineering Corp.

© Cengage Learning 2012

magnet, an electromagnet can be turned off and on. Electromagnets are used in automatic door locks and on devices that lift ferrous metal objects such as automobile junkyard cranes. They are also used as switching devices in relays. A *relay* is an electromagnetic switch (see Figure 8-22). When the coil inside a relay is energized, it creates a magnetic field that causes the electrical contacts within the switch to close.

Relays are commonly used in electrical systems where a low voltage circuit must control a high voltage circuit's operation. Figure 8-23 shows how a solid-state relay allows a 12-volt DC circuit to control a 120-volt AC lamp circuit. Relays can be found in everything from automobiles to the heating and cooling systems in your school.

FIGURE 8-22: Common electromagnetic relay, used as a switching device between low and high voltage systems.

FIGURE 8-23: Solid-state relay switch diagram.

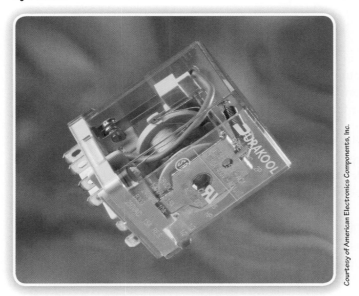

Courtesy of American Electronics Components, Inc.

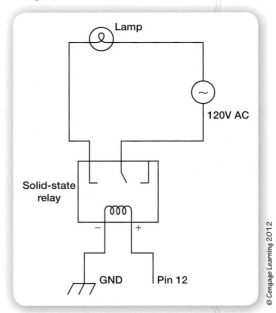

© Cengage Learning 2012

PROCESSORS AND CONTROLLERS

Together, processors and controllers are responsible for telling a control system what to do. Consider them the brains of a system. A processor is a part of a computer that manipulates data and performs calculations. The data is provided by input devices. The *controller* is the component of a control system that provides directions to output devices in order to control the operation of the system. There are a number of different types of processors and controllers to choose from, depending on what the system is designed to do.

Microprocessors

A **microprocessor** like the one shown in Figure 8-24 is usually a single-chip computer element that contains a control unit, central processing circuitry, and all the necessary logic functions to serve as a central processing unit (CPU) of a microcomputer or as a dedicated automatic control system. Microprocessors are capable of performing mathematical and logical operations. Intel introduced the first general purpose microprocessor in 1974. The Intel 8080 was an 8-bit processor capable of executing 290,000 instructions every second with 64 kb of addressable memory. Its development marked the birth of the microcomputer.

A microprocessor is responsible for three tasks:

1. performing math functions,
2. shifting data from one memory location to another, and
3. making decisions and skipping to a new set of commands based on some predefined condition.

The microprocessor's ability to process data is measured in bits. The more bits, the more information or data it is capable of processing. Let's make a comparison between lanes on a highway and bits in a processor. If we want to send a large number of cars down a two-lane highway, it will take a long time because there are only two lanes to handle all of the traffic. However, if the number of lanes is increased to four, and we have the same number of cars traveling, they will be able to use all four lanes and travel much faster. Similarly, increasing the number of bits in a microprocessor allows it to handle much more information in a more efficient way.

The microprocessor's speed is measured in hertz or cycles per second. If a microprocessor is capable of doing a billion operations in one second, then the speed is measured in gigahertz (GHz). Thus, a processor with a speed of 2.6 GHz could process 2.6 billion commands per second.

Microcontrollers

To be functional, a microprocessor must be connected to a series of integrated circuits that provide inputs and outputs. That is where the **microcontroller** comes in. A microcontroller consists of a microprocessor, memory unit, input and output ports, a crystal oscillator for timing, ADCs, and DACs all on one circuit board (see Figure 8-25).

The microcontroller is intended for small dedicated applications such as automobile engine control systems, remote controls, power tools, and toys. These types of applications fall under the category of embedded controllers because the controller is physically embedded in the device.

BASIC STAMPS The **BASIC Stamp** shown in Figure 8-26 is a microcontroller that is programmed using the Parallax PBASIC programming language. The program is stored in an **electrically erasable programmable read-only memory (EEPROM)** or rewritable memory chip that does not need power to keep its

---

Microprocessor:

a single-chip computer element that contains a control unit, central processing circuitry, and all the necessary logic functions to serve as a central processing unit (CPU) of a microcomputer or as a dedicated automatic control system.

FIGURE 8-24: Microprocessors such as this one performed mathematical and logical operations in the first electronic calculators.

© iStockphoto.com/LongHa 2006.

Microcontroller:

an electronic device that consists of a microprocessor, memory unit, input and output ports, a crystal oscillator for timing, ADCs, and DACs all on one circuit board.

FIGURE 8-25: A microcontroller consists of a microprocessor, memory, input and output ports, a crystal oscillator, and an analog-to-digital converter.

Courtesy of Ioan Sameli, wikipedia.org.

FIGURE 8-26: BASIC Stamp.

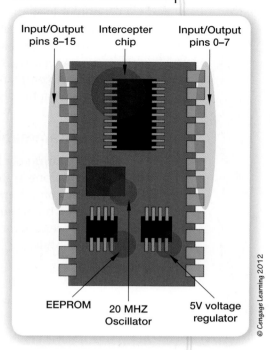

Input/Output pins 8–15   Intercepter chip   Input/Output pins 0–7

EEPROM   20 MHZ Oscillator   5V voltage regulator

© Cengage Learning 2012

content. The PBASIC programming language allows you to control input and output devices, including turning auxilliary devices on and off. The BASIC Stamp has the ability to interface with other integrated circuits, communicate with input and output devices, and operate networks. What makes the BASIC Stamp different is that it has an on-board basic program interpreter, allowing the program to be written in the PBASIC language on a PC and then downloaded directly to the stamp's EEPROM.

Computer-Based Controllers

The computer-based controller uses one of the computer's expansion slots to house an interface card like the one shown in Figure 8-27. These expansion slots are

FIGURE 8-27: Computer expansion card.

Courtesy of Chassis Plans.

directly connected to the motherboard. The interface card has several input and output locations that are used to bridge the gap between the computer and other components of a control system.

Computer-based controllers became feasible with the availability of low-cost computers. Their software systems are graphics-based, which gives the operator the ability to program using an icon-based system and monitor the process graphically. These graphic displays are known as *human–machine interfaces* (HMIs).

Programmable Logic Controllers

Programmable logic controllers (PLCs) like the one shown in Figure 8-28 are stand-alone systems that can be programmed to accomplish a variety of tasks. These controllers contain their own CPU and power supply, and are programmed via microcomputer using a specialized computer language. Once programmed, they are disconnected from the microcomputer and free to operate on a stand-alone basis.

FIGURE 8-28: *Programmable logic controller.*

Courtesy of AutomationDirect.

The controller itself contains several input and output ports that can be customized and monitored by the logic controller circuitry. The controller will constantly check the status of numerous sensor inputs that will, in turn, direct output actuators such as motors, solenoids, lights, displays, and valves.

PLCs have advantages over other types of controllers. They combine aspects of both sequential and combinational logic systems. One PLC can run many machines simultaneously. Correcting errors is as easy as making a program modification. Also, PLCs are small in size and low in cost.

Solenoid:

an electromechanical device that uses the principles of electromagnetism to control the motion of an actuator.

PROGRAMMING

Whether it's a BASIC Stamp, a PLC, or a microcontroller, all controllers need to be programmed. While programming has come a long way in the last few years, some of the older programming languages are still in use. Two common approaches to programming controllers are graphical flowcharting and ladder logic. Text-based programming methods, such as PBASIC, are still used, but such methods are being replaced with more user-friendly icon-based software programming systems.

Graphical Flowcharting

Icon-based softwares use **graphical flowcharting** symbols in place of text-based code. A *flowchart* is a diagram that graphically identifies the sequence of operations that take place within a process. Symbols that are commonly used in flowcharts are shown in Figure 8-29.

FIGURE 8-29: Basic flowchart symbols.

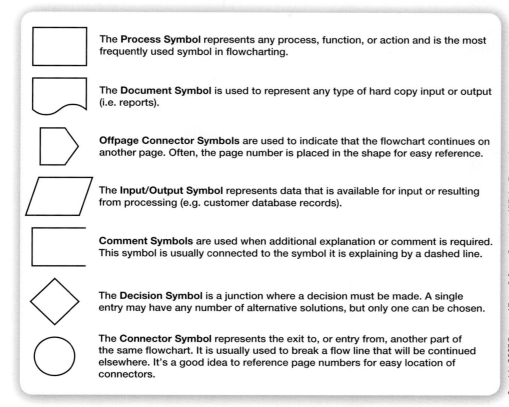

The **Process Symbol** represents any process, function, or action and is the most frequently used symbol in flowcharting.

The **Document Symbol** is used to represent any type of hard copy input or output (i.e. reports).

Offpage Connector Symbols are used to indicate that the flowchart continues on another page. Often, the page number is placed in the shape for easy reference.

The **Input/Output Symbol** represents data that is available for input or resulting from processing (e.g. customer database records).

Comment Symbols are used when additional explanation or comment is required. This symbol is usually connected to the symbol it is explaining by a dashed line.

The **Decision Symbol** is a junction where a decision must be made. A single entry may have any number of alternative solutions, but only one can be chosen.

The **Connector Symbol** represents the exit to, or entry from, another part of the same flowchart. It is usually used to break a flow line that will be continued elsewhere. It's a good idea to reference page numbers for easy location of connectors.

Let's look at the flowchart shown in Figure 8-30. This flowchart was created using a software program called ROBO Pro and is designed to control a motor that drives a linear shuttle back and forth nine times along a short track. The shuttle is outfitted with a permanent magnet. Reed switches are located on either end of the track and function as travel limits. The process flows from the top of the program to the bottom and contains a series of sub-loops. Behind each flowchart symbol is a set of instructions that are written in a language that the computer recognizes.

The program starts by setting a variable called "Counter" to zero until a limit switch (designated I1) is pressed. If the switch is not being pressed, the program will loop continuously until it is pressed. After the switch has been pressed, a motor (designated M1) turns on in a clockwise direction. The motor drives the shuttle along the track toward a reed switch (designated I2). When the magnetic field from the shuttle's permanent magnet comes in contact with the reed switch, the switch will close. The motor then reverses direction by spinning counterclockwise, which moves the shuttle in the opposite direction toward the second reed switch (designated I3). Once this reed switch makes contact with the shuttle's magnetic field, the motor is commanded to stop. The Counter variable is incremented by a value of one and then compared to a mathematical expression which checks to see if the

FIGURE 8-30: This is an example of an icon-based flowchart control program.

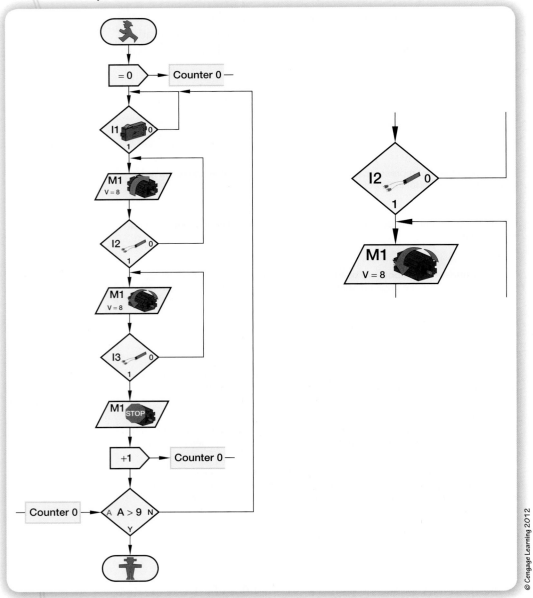

Counter variable is greater than 9. If it is greater than 9, the program stops; if it is not, then the program will loop back and repeat the process until the Counter variable is greater than 9.

FIGURE 8-31: An example of a ladder logic program, showing the rungs of the ladder.

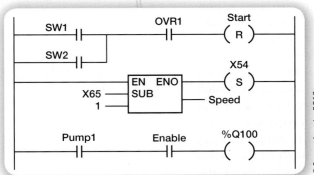

Ladder Logic

The **ladder logic diagram** shown in Figure 8-31 is the most common language used to program a PLC. The programming style is similar to the way in which relay diagrams were drawn by electricians at the time of its introduction.

A ladder logic diagram consists of two vertical lines that represent power rails. One rail is negative and the other is positive. The control logic for each output or coil is placed on a horizontal line called a *rung*, which bridges the negative and positive rails. When a series of rungs are placed one below the next, a ladder is formed, thus giving the programming method its name: ladder logic.

The following symbols are used in ladder logic diagrams:

( ) a regular coil (output); active when the circuit pathway along the rung is closed

(\) a "not" coil; active when the circuit pathway along the rung is open

[ ] a regular contact (input); closed when its corresponding internal bit coil or input device is active

[\] a "not" contact; open when its corresponding internal bit coil or input device is active.

Figure 8-32 shows one rung of a ladder. When the contact labeled input 1 is turned on, it completes the circuit for that rung and energizes output coil 1.

FIGURE 8-32: *One rung of a ladder logic program.*
© Cengage Learning 2012

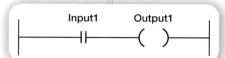

Your Turn

What sequence of operations are involved in the opening and closing of a garage door that is activated by both a keypad and a remote control device? The ladder logic diagram shown in Figure 8-33 offers one alternative. If switch 1 is closed, then the garage door is in the up position; if switch 2 is closed, then the garage door is in the down position. Generate a written explanation of how this ladder logic program operates from top to bottom.

FIGURE 8-33: *Example garage door ladder logic control program.*

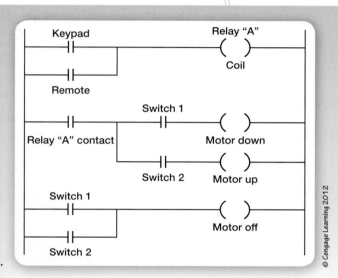

EXAMPLES OF CONTROL SYSTEMS

Most people are unaware of the dozens of automated control systems that they interact with every day. For example, when you go to a large office building, you probably walk through an automated sliding door (see Figure 8-34). Once inside, you may take an elevator to reach the upper floors. That elevator functions as part of an automated control system. Other examples of automated control systems are traffic lights and the heating and cooling systems in our homes.

FIGURE 8-34: *Automated doors.*

Heating and Cooling Systems

We "program" the heating and cooling systems in our homes by setting the thermostat to a desired temperature, say, 70° Fahrenheit. The system will have to first monitor the air temperature in the room; this will most likely be an input from a thermistor. The system's controller will compare the room temperature to the desired temperature and then take action. If the cooling system is on and the temperature is set to 70° Fahrenheit, cool air will be generated by the air conditioner if the room temperature rises above 70° Fahrenheit. If the room temperature is at or below 70° Fahrenheit, nothing will happen. The same process will take place in the heating mode. The furnace will be turned on until the desired temperature is reached, at which point the heat will be turned off.

Process Control System

In the manufacturing environment, a **process control** system is a specific type of closed-loop system that monitors an industrial process so that a uniform product will be produced. This is done by monitoring inputs from the system and then making the appropriate changes in that system to keep things within the **control limits**. The goal is to keep the output consistent and within the manufacturer's specifications.

An example of process control can be seen in a bottling plant. The bottler wants to ensure that each and every bottle it produces is flawless, and this is accomplished through the use of a process control system. The bottling plant has monitoring systems to make sure every bottle is washed and rinsed to remove dust and dirt. The bottles are then soaked in a high-temperature liquid to sterilize them. Next, the bottles go to a hydrowash station before being sent to an electronic inspection station (see Figure 8-35). After inspection, the bottle is filled with the beverage, pressurized, and capped. Finally, the bottles are labeled and sent on for a final inspection. The entire operation is monitored through the use of a process control system.

Control limits:
a detailed and predetermined set of parameters that are acceptable during the manufacturing of a product.

FIGURE 8-35: *Root beer bottling operation.*

© iStockphoto.com/Mitch Aunger.

Your Turn

Use the Internet to research bottling plant procedures and answer the following questions:

1. How are the bottles arranged to be cleaned? Could this be done with a feedback system? Explain your answer.
2. What type of feedback system might be used during a cleaning inspection?
3. How do the labels get placed on the bottles?
4. How is the bottle filled to the same level every time?

Motion-Control System

Motion-control systems, both open and closed, are designed to coordinate the movements of actuators and other components. Such systems may include motors, fluid power actuators, and, in the case of closed-loop systems, feedback sensors. Examples of motion-control system applications include industrial robots, automated manufacturing equipment, and even modern vending machines.

SERVOMECHANISM A servomechanism is a feedback system that consists of a sensing element, amplifier, and a servomotor. Servomechanisms are used in the operation of radio-controlled airplanes. The pilot holds the controller, which sends input signals to the plane wirelessly. Those input signals are addressed to individual servomotors (see Figure 8-36) which in turn make adjustments to the plane's control surfaces. Servomechanisms are also used to adjust television antennas, radar antennas, and the movable axes in numerical control machines.

NUMERICAL CONTROL Numerical control (NC) refers to automation of machine tools, such as milling machines and lathes, that are controlled by encoded commands. Such machines are designed to run without human intervention. Once a machine has been programmed, it will run continuously to manufacture the same part over and over again. Each movable axis on an NC milling machine (see Figure 8-37) is powered by its own individual motor, which is connected to a motion-control system.

Servomechanisms:
a feedback system that consists of a sensing element, amplifier, and a servomotor.

FIGURE 8-36: This is a servomechanism. It is a small electric motor whose position is controlled by an input signal.

Courtesy of Sachin Surendran.

FIGURE 8-37: This is an example of a three-axis CNC milling machine.

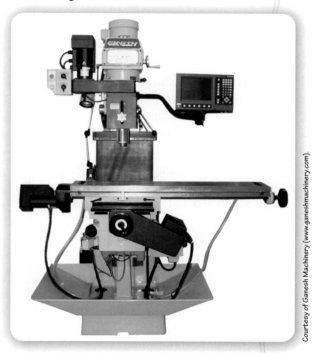

Courtesy of Ganesh Machinery (www.ganeshmachinery.com).

Your Turn

Make a list of five different parts that are made using CNC machines. Explain what type of machine is used to make each part and what other processes, if any, are used.

SUMMARY

- A system is a set of components working together to perform a given task under the direction of a processor or computer.

- The various operations of any system follow a sequential process of input, process, and output. A closed-loop system, unlike an open-loop system, takes information from the output and feeds it back to the input through the use of sensors for the purpose of adjusting the process and keeping the system operating as planned.

- Analog signals are continuous electrical signals that vary in intensity over time and are capable of attaining an infinite number of levels or values within a given range. Digital signals are electrical signals that have an integral number of discrete levels or values within a given range.

- The voltage across an analog input device such as a thermistor or potentiometer will change as a result of a change in resistance in that device's conductive element.

- Switches are digital input devices that may be triggered manually, by exposure to a magnetic field, or by exposure to light.

- Output devices are used to perform work or to change energy from one form to another for the purpose of accomplishing a specific task. Examples of output devices include actuators, lamps, buzzers, and electromagnets.

- A processor is a part of a computer that manipulates data and performs calculations; the data comes to it from input devices.

- The controller is the component of a control system that provides directions to output devices to control the operations of a system.

- A microcontroller is an electronic device that consists of a microprocessor, memory unit, input and output ports, a crystal oscillator for timing, analog-to-digital converters (ADCs) and digital-to-analog converters (DACs) all on a circuit board.

- Computer-based controllers can be customized using graphics-based software programs. These software programs also allow the operator to graphically monitor the process the controller is controlling.

- Programmable logic controllers are small in size and low in cost. They can operate in stand-alone situations to control many machines simultaneously.

- Although text-based programming languages are still used, many control systems use graphical programming methods such as icon-based flowcharts or ladder logic.

- People interact with control systems every day. These interactions occur, for example, when you lower your car's automatic window, ride in an elevator, control the temperature in your home, and adjust a satellite dish to watch television.

- A process control system is a specific type of closed-loop system that monitors an industrial process so that a consistent product will be produced.

- Examples of applications of motion-control systems include servomechanisms and numerical control devices.

BRING IT HOME

1. Identify three examples of control system that you interact with in your everyday world.
2. Which system uses feedback—open-loop or closed-loop?
3. How do analog and digital signals differ?
4. Using a dining room light example, explain why a potentiometer is an analog device.
5. Using a doorbell circuit as an example, explain why wiring a doorbell button switch in the normally-open state is preferred.
6. How does a reed switch work?
7. What function can lights and buzzers serve in control systems?
8. How does a relay switch work as an output device?
9. What does a processor do?
10. Identify all the components on a microcontroller.
11. What are the advantages of programmable logic controllers (PLCs)?
12. Which microcontroller uses the BASIC computer language?
13. Identify two types of graphic programming languages used in control systems.
14. When is a process control system used?
15. Explain how servomechanisms are used to fly remote-control airplanes.

EXTRA MILE

Research another example of a control system and identify the components and how they work.

CHAPTER 9
Materials

GPS DELUXE

Menu | START LOCATION | DISTANCE | END LOCATION

Before You Begin

Think about these questions as you study the concepts in this chapter:

1 What are common domestic and engineering applications of both ferrous and nonferrous metals?

2 How do softwoods differ from hardwoods, and what are the common applications of both materials?

3 What are the different categories of ceramic materials, and what kinds of products are made from the materials in these categories?

4 How do thermoplastics differ from thermosetting plastics, and what kinds of applications are these plastics used in?

5 What is the difference between a matrix and reinforcement within a composite material, and what common applications utilize composites?

Materials are the tangible substances that make up physical objects. The subject of *materials* is so important to our modern technological society that entire disciplines and fields of study within both science and engineering have evolved over the centuries around this topic. Materials engineers and chemists spend their careers exploring the composition and properties of materials. They use the knowledge they gain through these explorations to fashion new materials that exhibit designer properties. As a result, a design engineer has a wide variety of materials to choose from.

This chapter focuses on five types of materials that are commonly used in engineering applications: metals, wood, ceramics, polymers, and composites. Each type of material has special qualities and properties, proven applications, limitations, and advantages and disadvantages over the other material types. As such, a design engineer must have a broad understanding of the various materials so that she can make intelligent decisions regarding the physical makeup of an engineering solution.

Some materials had to be mastered before others could be developed. Such is the case with ceramics, where many of the metal alloys that are commonplace today could not have been produced if ceramic crucibles and brick-lined furnaces had not been developed. The development of composites also follows this pattern. Because material science had its roots in metallurgy, we will begin our exploration of materials with metals.

> **Materials:**
> the tangible substances that make up physical objects.

METALS

Metals have been so important to the development of human civilization and technology that two of the prehistoric periods of human history, the Bronze Age and the Iron Age, are characterized by the type of metal that was most advanced during those time periods.

Metals form the majority of the known elements in the periodic table, and they are arguably the most important materials to the development of modern civilization. To introduce, even briefly, all of the various metals and alloys that are employed in engineering endeavors would take up an entire book. This text will focus on a handful of the metals that have strongly influenced the development of human civilization: those that are readily available through commercial sources for application in engineering design and those that are used extensively in high-profile engineering applications.

There is a difference in the use of the word *metal* between the fields of science and engineering. In science, metals are pure elements as identified in the periodic table. For example, a scientist would not characterize steel, bronze, or brass as metals but would acknowledge that they are a mixture of two or more metal elements. To an engineer, *metal* is a generic term that encompasses all metallic substances, both pure and alloyed. In this text, the term metal will be used (in the engineering sense) to describe a solid material that is typically hard at room temperature (usually between 68° and 77°F [20° to 25°C]), is shiny, and possesses good thermal and electrical conductivity.

> **Metal:**
> a solid material that is typically hard at room temperature, is shiny, and possesses good thermal and electrical conductivity.

Ferrous Metals

Metals are divided into two categories: ferrous and nonferrous metals. *Ferrous* metals are those that contain iron (Fe). People often use the term *ferrous* to describe a metal that is affected by a magnetic field. However, some nonferrous metals also display this effect, such as cobalt (Co), nickel (Ni), gadolinium (Gd), and dysprosium (Dy). A major drawback of most ferrous metals is corrosion. Iron bonds easily with oxygen (O), which is readily available in air and water. This causes ferrous materials such as steel to "rust" when they are exposed to outside environments. To combat this, elements such as nickel (Ni) and chromium (Cr) are alloyed with steel to produce a corrosion-resistant metal called *stainless steel*. If you have a metal kitchen sink, it is most likely made of stainless steel.

IRON Ancient metalworkers in Egypt and Sumeria were the first to create artifacts from iron around 4000 B.C.E. However, the iron they used came from meteorites (see Figure 9-1). Because iron meteorites account for approximately 5% of the meteors that strike Earth's surface, the occurrences of ancient artifacts that were made from meteoric iron are very rare.

FIGURE 9-1: The oldest examples of artifacts made from iron date back to around 4000 B.C.E. (a) Meteor impact sites provided the raw materials for these artifacts in the form of (b) iron meteorites.

(a) Image copyright Walter G. Arce, 2010; used under license from Shutterstock.com. (b) Image copyright Kenneth V. Pilon, 2010; used under license from Shutterstock.com.

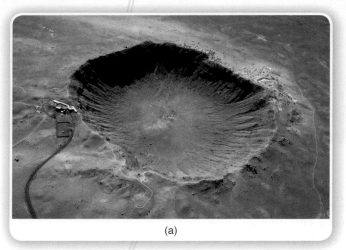

(a)

(b)

The creation of iron artifacts made from iron ores found within Earth's crust would have to wait until humans developed furnaces that could generate the necessary temperatures to smelt iron ore (approximately 2,000°F). Such temperatures are much higher than those required to smelt copper ore (approximately 1,500°F). Smelting is the process of extracting a metal element from an ore by the application of high, sustained heat in the presence of a fluxing material such as limestone. The earliest evidence of iron ore smelting dates back to 1200 B.C.E. in areas of India, Turkey, and Eurasia. This date in human history marks the beginning of the Iron Age.

Smelted iron was repeatedly heated in a charcoal fire and hammered to remove the impurities from the material. The hammering process was also used to fashion the material into useful objects like weapons and building tools. When the craft of metalworking evolved into the science of metallurgy, it was realized that carbon from the charcoal would work its way into the iron, resulting in a very tough material that holds a sharp edge much better than any other metal. This discovery set the groundwork for the development of steel.

Iron is a very important element in the fields of mechanical and civil engineering because most machines, bridges, and buildings are comprised of objects that contain iron. Chemical engineers also use iron in applications such as dies, pigments, and fertilizers. According to the United States Geological Survey (USGS), more than half of the world's supply of iron ore comes from China, Brazil, and Australia (USGS Mineral Commodity Summaries, 2009).

Iron exists as iron oxide in various minerals such as magnetite (Fe_3O_4) and hematite (Fe_2O_3). Collectively, these minerals are referred to as iron ore. Modern methods of iron production involve the extraction of iron ore from Earth's crust through mining operations. The iron ore is then pulverized and mixed with coke (see Figure 9-2). Coke is low-sulfur bituminous coal that has been baked to remove water, coal gas, and coal tar, leaving mostly carbon.

FIGURE 9-2: (a) Iron production begins with mining operations that extract (b) iron ore from Earth. (c) The iron ore is pulverized and mixed with coke. The iron ore and coke are smelted together in a high-temperature furnace that separates the iron from other elements present in the ore.

(a)   (b)   (c)

The iron ore and coke are then smelted in a blast furnace, which causes the oxygen atoms to separate from the iron atoms. The result is molten iron and unwanted impurities called slag. Slag is comprised of undesirable elements that are also contained in the iron ore rock, such as silicon (Si), phosphorus (P), aluminum (Al), and sulfur (S). The slag floats on top of the molten iron, making it easy to separate and remove. However, small amounts of these unwanted elements still remain in the molten iron. The molten iron is then poured into ladles for further manufacturing into other metals or cast into small ingots called *pigs* (see Figure 9-3).

FIGURE 9-3: (a) The unwanted elements within iron form a waste material called *slag* that floats on top of molten iron. (b) The molten iron is either moved on to another furnace for alloying with other elements or cast into small ingots called *pigs*.

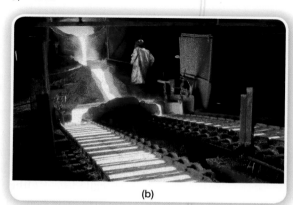

(a)   (b)

Pig iron is used in the production of cast iron, which contains 2% to 4.5% carbon by weight as a result of the coke that is used in the smelting process. Cast iron is very heavy and holds up well against compression forces. It has the ability to produce intricate shapes and is still used in the design of large machine bases and ornate outdoor furniture. Cookware is also a common application of cast iron (see Figure 9-4).

FIGURE 9-4: This skillet is made from cast iron.

Casting iron requires a temperature of approximately 2,800°F. The Chinese were the first to cast iron around 300 B.C.E., having developed furnaces that could produce the necessary temperatures. This was a tremendous technical achievement, especially if one considers the fact that Westerners have only been able to cast iron for the past 700 years.

STEEL Pig iron contains too much carbon and is too brittle to be of much use in designs that must endure high stresses such as the skeletal frameworks of tall buildings. To make pig iron into more useful materials, the excess carbon content is burned off through remelting in either a high-temperature, basic oxygen furnace (see Figure 9-5) or an electric arc furnace. Other materials are added to the molten iron to remove its impurities, resulting in low-carbon steel *alloy*.

FIGURE 9-5: A basic oxygen furnace is used to reduce the carbon content in steel.

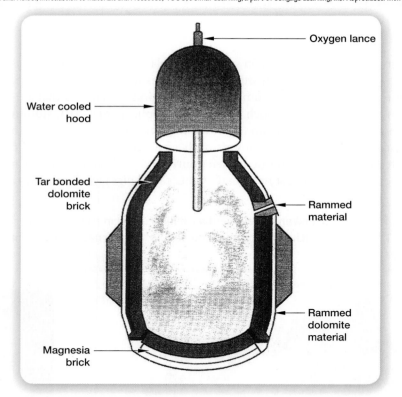

An alloy is a mixture of two or more elements, with one of those elements being a metal. If iron is alloyed with less than 2% carbon by weight, the result is a material called *steel*. Other elements are often added to steel in small amounts to enhance its properties. Steel is a commonly recognized ferrous metal that is applied in building bridges (see Figure 9-6), automobile components, domestic appliances, and countless other engineering and consumer products.

Steels are divided into two categories: plain carbon steels and alloy steels. Three types of plain carbon steels range in carbon content from 0.05% to 1.5% by weight. Mild (low-carbon) steel contains 0.05% to 0.3% carbon by weight. *Cold-rolled steel* is a common type of mild steel that is cheap and often used in the fabrication of machined components that do not require great strength. Example applications of mild steel include chains, rivets, nails, pipe, and automobile car-body panels.

Medium-carbon steel is stronger than low-carbon steel, but it is also more difficult to machine, bend, and weld. Medium-carbon steel contains 0.3% to 0.6% carbon by weight. Example applications include lead screws, gears, axles, and other components that are often heat-treated to increase hardness and reduce wear.

High-carbon steel contains 0.6% to 1.5% carbon by weight. It is harder and more brittle than low- and medium-carbon steel and is known for its ability to hold a sharp edge. High-carbon steel is frequently used in tools such as knives, drills, taps, dies, files, saw blades, and other objects that perform cutting operations.

FIGURE 9-6: Bridges are made from steel alloys.

© iStockphoto.com/ChrisMR

Alloy steels are manufactured for special-purpose applications and are made by combining plain carbon steel with elements such as chromium (Cr), cobalt (Co), manganese (Mn), molybdenum (Mo), nickel (Ni), tungsten (W), and vanadium (V). This process increases certain properties such as tensile strength, hardness, machinability, wear resistance, and corrosion resistance. Alloy steels are also designed to maintain their hardness and strength at high temperatures. One of the most common alloy steels is stainless steel. Figure 9-7 identifies several common steel alloys that are used in engineering applications.

FIGURE 9-7: Common steel alloys and their applications.

Series	Major Alloying Element(s)	Advantages and Disadvantages	Common Applications
A2	Chromium, manganese, molybdenum	Very hard and resistant to abrasion Can hold a cutting edge Resistant to deformation at high temperatures	Metal-forming dies, gauges, punches, injection molds
1018	Manganese, phosphorus, sulfur	Good overall qualities of strength, ductility, and ease of machining	Shafts, sprocket assemblies, and general-purpose machine parts
A36	Manganese, copper, phosphorus, sulfur	Lower cost than 1018 Slightly lower yield strength than 1018 Harder to machine but easily welded	Structural applications that require shapes such as I- and H-beams
4140	Manganese, silicon, chromium, molybdenum	High strength and impact toughness	Piston rods, shafts, metal dies, gears, hand tools, general-purpose machine parts
303 Stainless	Chromium, nickel	Machinability is better than the other common stainless steels Nonmagnetic Not considered weldable	Fasteners, aircraft fittings, shafts
304 Stainless	Chromium, nickel	Most commonly used stainless steel Weldable Not for use in saltwater environments	Cutlery, general-purpose machine components
316 Stainless	Chromium, nickel	Similar in machinability and weldability to 304 stainless Superior corrosion resistance Higher yield strength than 303 or 304 stainless	Marine applications
17-4PH Stainless	Chromium, nickel	Corrosion resistance and very high strength Heat-treatable by many methods Slightly magnetic	Aircraft components

Nonferrous Metals

GOLD Nonferrous metals are simply those metals that do not contain iron. Of all the nonferrous metals, gold (Au) is possibly the most well known because of its status as a precious metal. A precious metal is one that is relatively scarce, highly corrosion resistant, and valued for its color, luster, and malleability. Gold is a heavy, shiny yellow metal that occurs naturally throughout the world in an uncompounded state, though in trace amounts. Gold is one of the more chemically inert metals and does not tarnish as a result of exposure to oxygen. This makes it a highly valued material for the creation of coins, jewelry, and other ornamental objects.

Gold is one of the best conductors of electricity, which is why it can be found as a coating on the electrical contacts of circuit boards and other electronic components (see Figure 9-8). This material is also highly reflective of infrared radiation, which makes it very useful as a protective coating on spacecraft components. In fact, astronauts' helmet visors include a protective gold shield that blocks out harmful levels of radiation from the sun. Gold is also the most malleable of all metals. A malleable metal can be shaped or formed without breaking by means of hammering or some other application of pressure. For example, gold is pounded into extremely thin sheets called *gold leaf* and used as a decorative coating on print media, statues, and architectural features. A 1-gram gold nugget can be pounded into a sheet of uniform thickness (0.0000033 in) that covers an area approximately 6 square feet.

SILVER In the periodic table of elements, silver (Ag) is located in the same vertical group as copper and gold. This helps to explain why all three materials are the best conductors of electricity. Like gold, silver is considered a precious metal and has been used for thousands of years in the design of jewelry and currency. Though it does occur in a native state, the vast majority of the world's silver is a by-product of the refining of copper, lead, and zinc ores.

A compound called *silver halide* has been used for more than a century in photographic film and paper. Silver halide is a salt that forms when silver bonds to a halide material such as bromine (Br), chlorine (Cl), or iodine (I). When exposed to light, the silver atoms separate from their compound state and form clusters. Together, the clusters of metallic silver make up a latent (invisible) image that becomes visible to the naked eye once the film or paper is placed through a chemical development process (see Figure 9-9).

FIGURE 9-9: (a) Silver forms the images that you see in (b) black-and-white photography film and paper.

(a)

(b)

Case Study »→

A new use for silver has been developed that may affect you on a very personal level in the not-too-distant future. A textile material called X-Static is made from pure silver that is permanently bonded to the surface of synthetic fibers such as nylon. The fibers are then knit or woven into a textile fabric (see Figure 9-10).

Fabric made with X-Static fibers is used in towels, blankets, various undergarments, and other clothing articles because it exploits two well-known properties of silver: thermal conductivity (the highest among metals) and its ability to eliminate a wide spectrum of odor-causing bacteria. Athletes and military personnel are now wearing socks, underwear, and other clothing items that are made from X-Static fibers. Think of the amount of time, money, and water that could be saved if your clothes could be worn over and over again before they needed washing.

Figure 9-10: *Clothing containing X-Static fibers exploit silver's antimicrobial properties by neutralizing odor-causing bacteria. Courtesy of Brooks Sports, Inc.*
© Cengage Learning 2012

Silver has the highest optical reflectivity of any metal and is used as a coating on glass and other media to generate mirror surfaces. The ancient Greeks and Romans discovered that silver has an unusual ability to eliminate germs and fungi. They used it, as we still do today, to fight infection when dressing wounds. This quality is also the reason why silver is used in water-purification filters.

COPPER Next to gold, copper (Cu) was one of the first nonferrous metals to be extracted from the earth and hammered into different ornamental and utilitarian objects. Early metalworkers discovered that the hammering process work-hardens copper, allowing it to hold a sharp edge. Like gold and silver, *native* copper is one of the few metals that occur as an uncompounded (chemically pure) element, although it is rare (see Figure 9-11). The oldest known native copper artifacts were found in what is now modern-day Iraq and date back to around 9000 B.C.E.

As with most metals, copper is most plentiful in mineral rocks such as malachite $Cu_2CO_3(OH)_2$. It is believed that ancient peoples used their knowledge of firing ceramics to separate copper from the unwanted materials found in copper ore. Archeological evidence shows that the smelting of copper ores took place as early as 5,000 years ago in the region known today as Thailand.

Today, copper is used extensively in devices that conduct and transmit heat and electricity (see Figure 9-12). Copper is a tough, malleable, and very ductile metal. Ductility is a property that allows copper to be drawn out into long, thin wires without breaking. Freshwater pipes are still made from copper and used extensively in commercial and residential architecture. Copper is the second best conductor of electricity and is used extensively throughout the electronics industry. Countless electronic circuits in industrial and commercial devices rely on copper traces to carry electrical signals back and forth between their electronic components. Copper also serves as the base metal for alloys such as bronze and brass. According

FIGURE 9-11: (a) Native copper is rare and one of the few metal elements that occur in an uncompounded form. For the past 7,000 years, copper has been extracted from minerals such as (b) malachite, which is much more abundant than native copper.

(a) © iStockphoto.com/Terryfic3D. (b) Image copyright Jiri Vaclavek, 2010; used under license from Shutterstock.com.

(a)

(b)

FIGURE 9-12: (a) Copper is used extensively in electrical wiring because of its ductility and ability to conduct electricity. (b) Cooking pans are often plated with copper because it is an excellent conductor of heat.

(a) © iStockphoto.com/Fertnig. (b) Image copyright Joe Gough, 2010; used under license from Shutterstock.com.

(a)

(b)

to the USGS, Chile and the United States are the largest producers of copper (USGS Fact Sheet, 2009–3031).

BRONZE Around 4500 B.C.E., a new type of nonferrous metal alloy was invented that used copper as its base metal. Called *bronze*, this copper alloy consists of approximately 10% tin (Sn). Cassiterite (SnO_2) is a mineral rock that serves as the main source of the metal element tin (see Figure 9-13). Tin ore occurs in the same region of Southeast Asia where the first copper ores were smelted, which explains why the earliest bronze artifacts have also been found in that region.

Bronze ushered in a new era in human civilization that lasted from approximately 3300 to 1200 B.C.E. Once the properties of bronze were understood, bronze tools quickly replaced those made from stone and copper. During this period of history, bronze armor replaced traditional leather armor, and cast bronze weapons replaced their stone counterparts. Bronze forms a surface layer of oxidation that protects the metal underneath from further corrosion. It also has the ability, when

FIGURE 9-13: (a) Cassiterite is a mineral rock that serves as a primary source of (b) tin. When tin is alloyed with copper, the result is (c) a very hard, durable metal called *bronze*.

(a)

Courtesy of photographer Stan Celestian.

(b)

Juergen Bauer, smart-elements.com.

(c)

Courtesy of Bunting Bearings.

FIGURE 9-14: The Thinker, created in 1902 by the French sculptor Auguste Rodin, is one of the most widely recognized bronze sculptures in the world. Though the original is located at the Rodin Museum in Paris, numerous copies of the statue can be found in Asia, Europe, North and South America.

© iStockphoto.com/joecicak.

cast, to take the shape of very fine details. These qualities have made bronze a material of choice for casting sculptures and statues that must weather the elements (see Figure 9-14).

Boat-engine propeller blades are made from bronze because of its ability to withstand the corrosive effects of seawater (see Figure 9-15). Other engineering applications of bronze include gears, bearings, and bushings, which take advantage of the material's low metal-on-metal friction characteristics. Also, bronze alloys have excellent resonant qualities, making them materials of choice in bells and cymbals. Today, the term *bronze* is associated with any copper alloy that does not use zinc (Zn) or nickel (Ni) as its principal alloying element. Aluminum bronze is one such example.

FIGURE 9-15: Bronze is a corrosion-resistant copper alloy, which makes it an excellent material for use in marine applications such as boat propellers.

BRASS Brass is a gold-colored, copper-based alloy that is very easy to machine, can be polished to a high sheen, and is often used in decorative hardware such as doorknobs, locks, and hinges. Pneumatic and hydraulic fittings are also machined from brass alloy. The principal alloying element in brass (as much as 40% by weight) is a bluish-gray metal called *zinc* (Zn). The mineral sphalerite (ZnFe)S is a common zinc ore (see Figure 9-16).

FIGURE 9-16: (a) The dark crystals of the mineral sphalerite contain zinc sulfide and iron. (b) Pure zinc is extracted from zinc ore. (c) Zinc is alloyed with copper to form brass.

(a)

(b)

(c)

Brass is more difficult to produce than other copper-based alloys because of the difference in melting points between copper and zinc (approximately 2,000°F and 800°F, respectively). For this reason, brass does not play as prominent a role in human history when compared to copper or bronze. Though recent archeological evidence shows that brass was being made as early as 1500 B.C.E., the majority of historical brass artifacts date back to the Roman Empire, where brass coins were introduced into the Roman economy during the 1st century B.C.E.

Brass is similar in color to gold but will tarnish over time. Because it is much cheaper than gold, brass is often used to increase the aesthetic appeal of otherwise utilitarian hardware. Like bronze, brass has excellent resonant qualities and is used extensively in the design of musical instruments (see Figure 9-17). Brass also has low metal-to-metal friction characteristics, which is why most people can find brass keys on their key rings.

Some brass alloys are ductile and can be drawn into wire. In the early days of the Industrial Revolution, brass was used extensively to make the much-needed pins for the textile industry. Today, fine brass wire (0.010-in diameter or less) is used in wire electric-discharge machining (wire EDM). The qualities of low friction and ductility also make brass an excellent material for munitions casings.

ALUMINUM Aluminum (Al) is one of the most commonly recognized and most abundant metal elements within Earth's crust. Bauxite is the primary raw material from which aluminum is extracted. During the mid-1800s, aluminum (in its pure form) was worth more than gold because it was very difficult to extract from its various ores. In 1884, when aluminum was selected for the capstone of the Washington monument, the metal had a value that was equivalent to silver.

Pure aluminum is very weak and is therefore not used in traditional engineering applications. It is, however, used in pigments and dies and serves as an alloying element for other materials such as steel and semiconductors. Elements such as copper (Cu), zinc (Zn), magnesium (Mg), manganese (Mn), lithium (Li), and silicon (Si) are added to aluminum to improve its physical properties. Aluminum alloy has been used in the design of aircraft components for the better part of the past 100 years because of its high strength-to-weight ratio.

Aluminum is also applied in situations that call for good electrical and thermal conductivity, corrosion resistance, and where highly reflective surfaces are required. The use of aluminized Mylar film on solar reflectors is one example. The engine block in your family car is most likely made from an aluminum alloy, as are most of the pots and pans found in the average kitchen. Perhaps you've used aluminum foil to cover a baked item (see Figure 9-18).

FIGURE 9-17: Brass has resonant qualities that make it an excellent material for musical instruments.

FIGURE 9-18: Examples of common aluminum alloy applications include (a) internal-combustion engine blocks and (b) thermal wrapping for food.

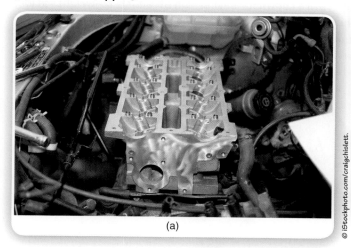

(a)

(b)

There are hundreds of different alloyed aluminums. Figure 9-19 identifies some of the more popular alloys, along with their common applications.

TITANIUM Titanium (Ti) is a metal element that has been known for more than 200 years but has only been used in engineering applications for little more than a half-century. This is because of the complex and cost-prohibitive chemical methods that are required to extract the metal from its various ores on an industrial scale. Consequently, titanium alloys are some of the most expensive metals available. For

FIGURE 9-19: *Common aluminum alloys and their applications.*

Series	Major Alloying Element(s)	Common Applications
2024	Copper (Cu) Magnesium (Mg)	Aircraft wings and fuselage parts that experience tension
5086	Magnesium (Mg)	Transportation equipment, pressure vessels, cryogenics, towers, drilling rigs, gas and oil piping, ordnance, armor plate, and boat and yacht hulls
6061	Silicon (Si) Magnesium (Mg)	Aircraft fittings, camera lens mounts, couplings, marines fittings and hardware, electrical fittings and connectors, decorative and miscellaneous hardware, hinge pins, magneto parts, brake pistons, hydraulic pistons, appliance fittings, valves and valve parts, and bike frames
6063	Silicon (Si) Magnesium (Mg)	Pipe, railings, furniture, architectural extrusions, irrigation pipes, and transportation
7075	Zinc (Zn)	Aircraft fuselages and wings, rock-climbing gear, and bicycle frames

example, a piece of titanium alloy can cost 40 times more than a piece of mild steel or standard alloy aluminum of equal size.

Pure titanium is extracted from minerals such as rutile (TiO_2) and ilmenite ($FeTiO_3$) and alloyed with elements such as aluminum (Al), vanadium (V), and palladium (Pd). Despite their high costs, titanium alloys are growing in application because of three exceptional qualities: high resistance to corrosion, high strength-to-weight ratio, and elevated temperature performance. Next to platinum (Pt), titanium is one of the most chemically inert metals known. It is this property that keeps the human immune system from rejecting medical implants that are made from titanium alloys such as hip joints, knee joints, and dental implants (see Figure 9-20). Chemical manufacturers are replacing stainless steel valves and other components with titanium in an effort to lower maintenance costs, reduce contamination, and generate higher yields.

FIGURE 9-20: **Titanium alloys are known for having high strength-to-weight ratios, and for being chemically inert. Together, these qualities make titanium alloys excellent materials for human joint replacement.**

The aerospace industry is the largest consumer of titanium because it has a strength that is comparable to steel and a weight that is similar to aluminum. This is why titanium parts are frequently found in the design of both military and commercial aircraft (see Figure 9-21). Cast titanium parts have also found their way into the designs of sports equipment such as golf club driver heads.

WOOD

The use of wood in the creation of designs stretches back to the dawn of human-kind when digging sticks were used to forage for food. Even today, primates fashion simple tools from wood for the purpose of extracting termites (which are high in protein) from their nests. Woods have many common applications, including:

- blinds,
- cabinetry,
- caskets,
- cooperage (i.e., wine barrels),
- cutting boards,
- decorative artwork,
- dinnerware items,
- doors,
- electrical utility poles,
- exterior siding,
- fences,
- flooring,
- fuel for fireplace furnaces,
- furniture,
- interior paneling,
- musical instruments,
- pallets,
- paper,
- preservation of meats through smoking,
- railway ties,
- residential framing,
- shingles,
- temporary support for bridge repair, and
- toys.

Woods are divided into two categories: softwoods and hardwoods. Both categories share characteristics and common terminology relating to the makeup of wood. You can think of wood as a mass of tubelike cells that are made from an organic polymer called *cellulose*. These tubes are bonded to each other by another material called *lignin* (see Figure 9-22). The tree trunk provides the main source of wood, wherein the majority of the wood cells are oriented parallel to the center axis of the trunk.

New cells continuously grow during the spring and summer months in a region under the *bark* called the *cambium layer*. It is during this time that a tree's leaves

FIGURE 9-22: Wood is comprised of tightly packed, hollow tubelike cells that are made mostly of cellulose and bonded together by a material called *lignin.*

Courtesy of Dr. Roger Heady. Centre for Advanced Microscopy, The Australian National University.

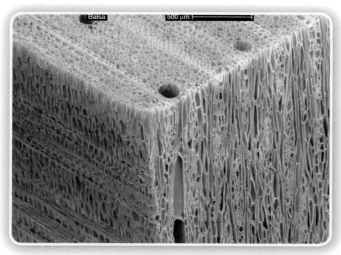

FIGURE 9-23: Cross-section of a tree trunk showing the annular rings and the difference in color between sapwood and heartwood.

Adapted from MacDonald, Woodworking, 2009, Delmar Learning, a part of Cengage Learning, Inc. Reproduced with permission.

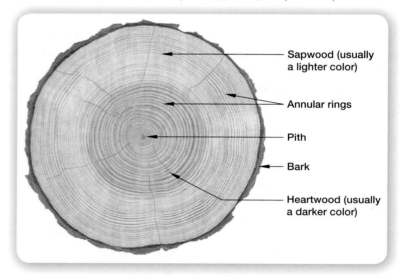

OFF-ROAD EXPLORATION

To learn more about woods, their properties, applications, and design considerations, visit *www.woodweb.com*.

actively produce food in the form of sap. Although mostly water, sap also contains minerals and nutrients from the soil and sugars that are produced by the leaves. The spring months result in the growth of lighter color, thin-walled cells. The summer months result in the growth of darker colored, thick-walled cells. This forms *annular rings* that tend to make wood products so strikingly aesthetic (see Figure 9-23).

The newer cells are used to circulate the sap within the tree and are collectively referred to as *sapwood.* As the tree matures and its girth expands, the older cells that are closer to the *pith* (center of the tree trunk) cease to function as carriers of sap and become the structural backbone of the tree called the *heartwood.* Heartwood is usually darker in color than sapwood as a result of the cells becoming clogged with organic material. Heartwood is harder and more structurally stable than sapwood and will experience less variation in dimensional change from fluctuations in temperature and humidity. For this reason, heartwood is preferred over sapwood in many design applications. Trees are allowed to grow and mature over a series of decades before they are harvested to ensure that significant amounts of heartwood will be present.

Softwoods

Softwoods come from **coniferous trees**, which bear exposed seeds in the form of cones. Most softwood trees are evergreen and have needle-shaped leaves (see Figure 9-24). Commercial softwoods are extracted from forests primarily in the United States, Canada, and Russia. The term *softwood* is deceiving because not all softwoods are soft or weak with respect to their physical properties. In fact, some hardwoods are orders of magnitude softer than most softwood. Balsa, which is a type of hardwood that is used in boatbuilding and model making, is an example of this.

FIGURE 9-24: **(a) Most softwood trees bear their seeds in the form of cones and have needle-shaped leaves. (b) Lumber from such trees are relatively uniform in color, have pronounced grain patterns, and often contain knots.**

(a)

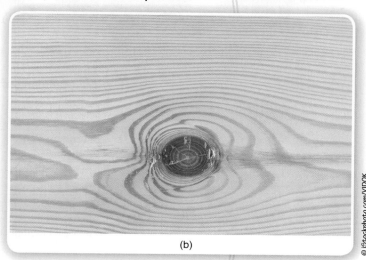

(b)

© iStockphoto.com/Lezh.

© iStockphoto.com/VIDOK.

Generally speaking, softwood trees have straight trunks and take less time to mature than hardwoods. Softwoods are also easier to machine than hardwoods. Together, these qualities explain why softwood is predominantly used for construction lumber. Softwoods tend to generate a greater number of limbs branching off from the trunk than do hardwoods, which is why softwood lumber almost always contains knots. Knots cause weak points in the lumber. If too many knots exist, the lumber is considered undesirable for structural applications and may be harvested for pulpwood. Pulpwood fibers are used to make paper and cardboard (see Figure 9-25).

FIGURE 9-25: **The largest applications of softwood are as (a) construction lumber and (b) pulpwood fiber for paper and cardboard.**

(a)

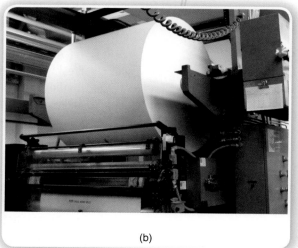

(b)

© iStockphoto.com/jhorrocks.

© iStockphoto.com/PapaBear.

FIGURE 9-26: Silhouette images of (a) pine, (b) spruce, (c) fir, (d) hemlock, and (e) cedar trees.

© Cengage Learning 2012

(a)   (b)   (c)   (d)   (e)

FIGURE 9-27: Common softwoods and their uses.

Softwood	Common Uses
Pine	Furniture, flooring, window frames, doors, interior paneling, general construction, turpentine
Spruce	Paper pulp, boxes, plywood, crates, millwork, general construction, musical instruments
Fir	Paper pulp, plywood, general construction
Hemlock	Paper pulp, tanning leather, plywood, general construction
Cedar	Exterior siding, roof shingles, moth repellent, storage furniture, boatbuilding, musical instruments

Common species of American softwood trees include pine, spruce, fir, hemlock, and cedar; these are distinguishable by their silhouettes (see Figure 9-26). It is difficult to distinguish between the species once they have been harvested and turned into lumber because they have similar colors and grain patterns. The exception to this amongst these species is cedar, which has a distinct color pattern (and aroma). Figure 9-27 identifies the common uses for these softwoods.

Point of Interest

Dimensional lumber is made from various softwoods and comes in standard sizes that can be deceiving. For example, a **nominal size** 2" × 4" stud is actually 1½" × 3½". Before designing objects that use dimensional lumber, it's important to understand where this discrepancy comes from.

After a tree trunk is harvested, it is sent on to a mill to be sawn into lumber. Freshly cut lumber, called *green* lumber, is cut to nominal dimensions. Green lumber is very high in moisture content, which reduces the wood's strength, makes it prone to attack and decay from mold, and makes it difficult to machine. This is why dimensional lumber is placed through a drying process.

As the wood cells dry, they become harder and stronger. Once the lumber has been dried to a moisture content of 19% or less, it is machined flat, straight, and square to its finished actual size. A board's actual length remains the same as its nominal length. Therefore, a nominal size 2" × 4" × 8' pine board will measure 1½" × 3½" × 8'. Figure 9-28 shows common nominal versus actual sizes of dimensional softwood lumber.

Hardwoods

Hardwoods come from **deciduous trees**, which are higher on the evolutionary ladder than softwoods. Hardwood trees have broad leaves that are lost at the end of a growing season and bear their fruit in the form of berries or nuts rather than as cones. Hardwood trees have more complex cell structures than softwood trees and have greater numbers of species. Though there are exceptions, hardwoods are typically harder than most softwoods, which makes them more difficult to machine. Hardwood trees have broad leaves, and their branch structures take on a characteristically dendritic form that contrasts sharply with the typical branch structures of softwood trees.

As the foliage tends to block the branch structure of the trees, it is often easier to identify a hardwood by the shape and color of its leaves or the color and texture of its bark. The color of the wood and the grain pattern that is formed by the annular rings also provide distinguishing features (see Figure 9-29).

Hardwoods have a much warmer look and feel than other rigid materials. They are cheaper and easier to machine and shape than metals and plastics, and their **aesthetic** qualities make them ideal materials for fine furniture and interior flooring (see Figure 9-30). Many types of musical instruments take advantage of the resonant properties of hardwoods.

FIGURE 9-28: Nominal versus actual sizes of dimensional softwood lumber.

Nominal Size (in)	Actual Size (in)
1 × 4	¾ × 3½
1 × 6	¾ × 5½
2 × 2	1½ × 1½
2 × 4	1½ × 3½
2 × 6	1½ × 5½
2 × 8	1½ × 7¼
2 × 10	1½ × 9¼
4 × 4	3½ × 3½

FIGURE 9-29: Different species of hardwoods are identified by the shapes and colors of their leaves as well as by the colors of the woods and the textures of the wood grains. Examples include the (a and d) black walnut, (b and e) cherry, and (c and f) white oak.

(a)

(b)

(c)

(d)

(e)

(f)

FIGURE 9-30: The wide range of colors, aesthetic qualities, and machinability of hardwoods make them excellent materials for fine furniture. This Mission style furniture piece is made from white oak.

Stickley three-drawer nightstand image Courtesy of L. and J. G. Stickley, Inc. (www.stickley.com).

Common species of American hardwood trees include hard maple, cherry, white oak, white ash, black walnut, and yellow poplar. Figure 9-31 identifies the common uses for these hardwoods.

FIGURE 9-31: Common American hardwoods and their uses.

Softwood	Common Uses
Hard maple	Cabinet panels, butcher blocks, fine furniture, flooring, musical instruments
Cherry	Fine furniture, cabinet panels, musical instruments
White oak	Flooring, fine furniture, wine barrels, cabinet panels
White ash	Baseball bats, tool handles, fine furniture, bentwood laminated furniture
Black walnut	Fine furniture, musical instruments, bowls
Yellow poplar	Toys, plywood, boats, framework for furniture and cabinetry

There are fixed amounts of each stable chemical element on Earth. These elements are cycled through different mediums that act as reservoirs. For example, the carbon dioxide in air is absorbed by the leaves of a pineapple fruit in Hawaii, which uses the carbon atoms to form its fruit pulp. When you consume the pineapple, your body converts the carbon in the pineapple into organic polymer molecules such as new skin or hair. This cycling of matter between the various reservoirs is referred to as the *geochemical cycle*.

CERAMICS

Though they are the oldest of all of the engineered materials, *ceramics* are one of the least understood of all the classes of materials. Ceramics are a diverse class of materials based on solid compounds of both metallic and nonmetallic elements (see Figure 9-32) that are prepared by the action of heat and subsequent cooling. Common metallic elements found in ceramic materials include aluminum (Al), silicon (Si), magnesium (Mg), beryllium (Be), titanium (Ti), and boron (B). Common nonmetallic elements include oxygen (O), carbon (C), and nitrogen (N).

Generally speaking, ceramics are weak in tension and extremely rigid in compression. This means they can withstand large compression loads without exhibiting noticeable deformation. Most people use ceramic objects every day because of their other exceptional quality; the ability to act as a thermal insulator. This is why we drink coffee and hot cocoa from ceramic mugs. Ceramics have some of the highest melting points of all engineered materials. Another useful quality is that ceramics are chemically inert, which is why they are used as components in human hip replacements.

Traditional ceramic objects have been and continue to be made from clay, which is composed of mostly silicon (Si) and aluminum (Al). Clay is extracted from the earth at relatively shallow depths, combined with water, and molded into a desired shape (see Figure 9-33). Once the molded part has dried, any residual water is removed through firing in a device called a *kiln*. Firing partially melts and binds the clay materials. The result is a lightweight and very hard material that is classified as ceramic.

Ceramics:

a diverse class of materials based on nonmetallic, inorganic solid compounds that are prepared by the action of heating and subsequent cooling.

FIGURE 9-32: The highlighted elements in the periodic table represent the most common elements in ceramic materials.

Periodic Table of Elements

hydrogen 1 **H** 1.0079																	helium 2 **He** 4.0026
lithium 3 **Li** 6.941	beryllium 4 **Be** 9.0122											boron 5 **B** 10.811	carbon 6 **C** 12.011	nitrogen 7 **N** 14.007	oxygen 8 **O** 15.999	fluorine 9 **F** 18.998	neon 10 **Ne** 20.180
sodium 11 **Na** 22.990	magnesium 12 **Mg** 24.305											aluminium 13 **Al** 26.982	silicon 14 **Si** 28.086	phosphorus 15 **P** 30.974	sulfur 16 **S** 32.065	chlorine 17 **Cl** 35.453	argon 18 **Ar** 39.948
potassium 19 **K** 39.098	calcium 20 **Ca** 40.078	scandium 21 **Sc** 44.956	titanium 22 **Ti** 47.867	vanadium 23 **V** 50.942	chromium 24 **Cr** 51.996	manganese 25 **Mn** 54.938	iron 26 **Fe** 55.845	cobalt 27 **Co** 58.933	nickel 28 **Ni** 58.693	copper 29 **Cu** 63.546	zinc 30 **Zn** 65.39	gallium 31 **Ga** 69.723	germanium 32 **Ge** 72.61	arsenic 33 **As** 74.922	selenium 34 **Se** 78.96	bromine 35 **Br** 79.904	krypton 36 **Kr** 83.80
rubidium 37 **Rb** 85.468	strontium 38 **Sr** 87.62	yttrium 39 **Y** 88.906	zirconium 40 **Zr** 91.224	niobium 41 **Nb** 92.906	molybdenum 42 **Mo** 95.94	technetium 43 **Tc** [98]	ruthenium 44 **Ru** 101.07	rhodium 45 **Rh** 102.91	palladium 46 **Pd** 106.42	silver 47 **Ag** 107.87	cadmium 48 **Cd** 112.41	indium 49 **In** 114.82	tin 50 **Sn** 118.71	antimony 51 **Sb** 121.76	tellurium 52 **Te** 127.60	iodine 53 **I** 126.90	xenon 54 **Xe** 131.29
caesium 55 **Cs** 132.91	barium 56 **Ba** 137.33	57–70 *	lutetium 71 **Lu** 174.97	hafnium 72 **Hf** 178.49	tantalum 73 **Ta** 180.95	tungsten 74 **W** 183.84	rhenium 75 **Re** 186.21	osmium 76 **Os** 190.23	iridium 77 **Ir** 192.22	platinum 78 **Pt** 195.08	gold 79 **Au** 196.97	mercury 80 **Hg** 200.59	thallium 81 **Ti** 204.38	lead 82 **Pb** 207.2	bismuth 83 **Bi** 208.98	polonium 84 **Po** [209]	astatine 85 **At** [210]	radon 86 **Rn** [222]

© Cengage Learning 2012

FIGURE 9-33: When clay is combined with water, it can be easily molded into different shapes, such as this bowl being formed on a potter's wheel.

© iStockphoto.com/CreativeFire.

Fun Facts

Ceramics were the first human-made material. The earliest archeological evidence of fired clay objects date back to 24000 B.C.E., and were found in an area that is now part of the Czech Republic. Such clay materials, which are fired at relatively low temperatures (1,800 to 2,100°F), are called *earthenware*. Pottery, tableware, and decorative works are still made all over the world using the same materials and techniques that our ancient ancestors used.

© iStockphoto.com/sd619.

Clay is only one type of ceramic material. Other ceramic materials, both natural and synthetic, are employed in engineering applications. In fact, there are several categories into which ceramic materials fall, including whiteware, structural clay, refractories, glass, abrasives, and cement and concrete.

Whiteware

The breakfast or dinner that you eat at home is most likely served on a ceramic plate that falls under the category of whiteware. Porcelain is a revered example of a ceramic that falls under this category. It was first created in China during the second half of the first millennia A.D. Two common engineering applications of whiteware are ceramic bathroom tiles and the ceramic component of an engine sparkplug. The term *earthenware* is also used to describe ceramic objects that fall under the category of whiteware, though their color need not be white (see Figure 9-34).

Structural Clay

The ceramic materials contained in bricks must withstand large compression loads. Bricks are the most common and oldest example of structural clay (see Figure 9-35). Clay bricks are believed to have been developed around 3000 B.C.E. by the Sumerians in Mesopotamia. Other examples of structural clay objects include floor tiles, chimney flues, and drain tile.

FIGURE 9-35: Bricks are common examples of structural clay.

© iStockphoto.com/Devonyu.

Refractories

The word *refractory* literally means to resist heat. **Refractories** are nonmetallic ceramic materials that are applied in situations where extreme temperatures (above 1000°F) are encountered. Such materials are used in crucibles, incinerators, kilns, nuclear reactors, and engines. Magnesite ($MgCO_3$) and dolomite [$CaMg(CO_3)_2$] are refractory minerals that are used in the linings of vessels in which steels and other metals are melted. Perhaps the most widely publicized application of refractory ceramics is a space shuttle's thermal protection system: 34,000 individually shaped tiles that make up the shuttle's outer skin. These tiles prevent a shuttle from burning up because of the extreme temperatures that are generated when entering Earth's atmosphere (see Figure 9-36).

As with most ceramic materials, refractories have one major weakness: They are brittle. This disadvantage was a cause of the space shuttle *Columbia* disaster that occurred in February 2003. The shuttle suffered damage to its ceramic heat shield

FIGURE 9-36: The heat-resistant qualities of refractory ceramic materials make them perfect for applications such as the outer skin of a space shuttle.

during liftoff when a piece of foam insulation struck the leading edge of the left wing. The damaged area of the wing caused catastrophic failure upon reentry into Earth's atmosphere two weeks later. The importance of refractory ceramics cannot be overstated because it was the development of this material around the 16th century that led to the Industrial Revolution.

Glass

Glass is a ceramic material that is made from mostly silicon dioxide (SiO_2), which is commonly known as sand. It was first discovered in Egypt around 8000 B.C.E. as an unexpected glaze that would appear on pottery that was fired in an overheated kiln. The glass we are familiar with today, however, was not generated until about 1500 B.C.E. Glass is a chemically inert material that has been applied so extensively to the design of beverage containers (both commercial and domestic) that the word *glass* serves as a formal title for virtually any object that serves such a purpose. Common engineering applications of glass include windows in buildings and automobiles, and lenses in eyewear, microscopes, and telescopes. Other not-so-well known engineering applications of glass include fiber-optic communication lines, insulation in homes, and reinforcement fiber in composite materials (see Figure 9-37).

Abrasives

If you have ever used a piece of sandpaper or sharpened a chisel or a knife with a grinding stone or wheel, then you have used a ceramic material that falls under the category of abrasives. Such objects are composed of materials like aluminum oxide (Al_2O_3) and silicon carbide (SiC). Diamonds, both natural and artificial, are considered ceramic materials and are used as high-quality abrasives in drilling and grinding tools (see Figure 9-38). Another engineering application of abrasives includes ceramic cutting-tool inserts made from silicon nitride (Si_3N_4) that are often used in place of traditional high-speed steel tools in high–production volume machining applications.

Cement and Concrete

Cement, which is also called *mortar*, is a dry, powdery substance that serves as a binder when mixed with water. It is most commonly recognized as the "glue"

FIGURE 9-37: Some of the not so commonly recognized applications of glass materials include (a) fiber-optic communication lines and (b) building insulation.

FIGURE 9-38: Ceramic materials that fall under the category of abrasives are used to make grinding tools.

(a)

© iStockphoto.com/Henrik5000.

(b)

Image copyright Christina Richards, 2010. Used under license from Shutterstock.com.

© iStockphoto.com/Bran.

that holds brick or block walls together. Cement is also an integral component of concrete. The most popular type of cement used today is Portland cement, which is comprised of lime (from limestone or chalk), silica (SiO_2), alumina (Al_2O_3), and iron (Fe). The ancient Romans were pioneers in the development and use of cement and concrete, which is why many of their structures are still in use today. In fact, the Romans were the first to develop cement that cures underwater.

Concrete is a mixture of Portland cement, sand, stone aggregate, and water that can be molded into virtually any shape. Unlike clay, neither cement nor concrete are fired; rather, they require time to cure in an oxygen-rich environment. In fact, a concrete foundation wall must cure for 28 days before a building can be constructed on top of it. By weight, concrete is the most extensively used of all human-made materials, and it is commonly applied to bridge, dam, and tunnel construction where enormous compressive loads must be carried (see Figure 9-39).

FIGURE 9-39: (a) Cement is a binder material that is used to hold objects such as brick walls together. It is also a component of (b) concrete, which can be poured into virtually any shape.

(a) © iStockphoto.com/BirdofPrey. (b) Image copyright Christina Richards, 2010. Used under license from Shutterstock.com.

(a)

(b)

POLYMERS

The words *polymer* and *plastic* are often used interchangeably, but they are not synonymous. The word **polymer** is an umbrella term that encompasses all high-molecular-weight organic compounds either naturally occurring or **synthetic** that are comprised of at least 100 repeating chains of simple molecules called *monomers* (or simply *mers*). The word *polymer* means *many mers*. Polymers are divided into two groups: plastics and elastomers (see Figure 9-40).

Elastomers

All plastics are polymers, but not all polymers are plastic. For example, a rubber band is made from a polymer material that falls under the category of elastomers. **Elastomers** are natural or synthetic polymers that have the ability, at room temperature, to stretch to at least twice their length and return back to their original size after unloading. Some elastomers can be reshaped on reheating, whereas others become set once they have been molded into shape. Natural elastomers, such as gutta-percha, are moldable materials that are made from tree resin called *latex*. Gutta-percha was used as the electrical insulation for the first undersea telegraph wires in the mid-1800s. It was also used from the early 1840s to the early 1930s to mold dozens of utilitarian objects, including golf ball cores (see Figure 9-41).

> **Polymer:**
> a high-molecular-weight organic compound, either naturally occurring or synthetic, that is comprised of at least 100 repeating chains of molecules called *monomers*.

FIGURE 9-40: Polymers are divided into two categories: plastics and elastomers.

© Cengage Learning 2012

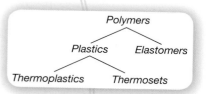

FIGURE 9-41: (a) Latex is a generic term for the sap of several species of rubber trees. It is processed into a natural elastomer material called *rubber*. (b) Among other items, certain types of disposable safety gloves are made from natural elastomer.

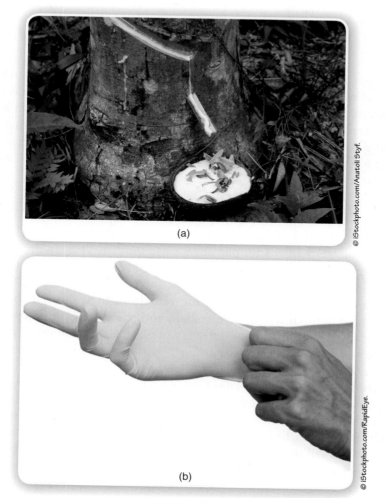

(a)

© iStockphoto.com/Anatoli Styf.

(b)

© iStockphoto.com/RapidEye.

FIGURE 9-42: (a) A flow control valve uses (b) silicone rubber O-rings as seals to prevent liquid or gas from escaping where the different valve components join together.

(a)

Courtesy of East Jordan Iron Works.

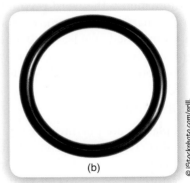

(b)

© iStockphoto.com/prill.

FIGURE 9-43: Long ago, lantern windows were fashioned out of animal horn, a natural plastic.

Courtesy of The Worshipful Company of Horners.

Plastics:

a group of polymer materials that, while solid in the finished state, were liquid at some stage in the manufacturing process and may be formed into various shapes by the application of heat and pressure.

Elastomer materials are used in flexible adhesives, components such as tires that must absorb shock, protective gloves, waterproof jacketing for electrical cables, structural expansion and contraction joints, power transmission belts, and adjustable tension straps. They are also used extensively in objects that serve as seals and gaskets (see Figure 9-42). Natural rubber, silicone rubber, and neoprene are a few examples of common materials that fall under the category of elastomers.

Plastics

Many people consider plastics to be a new material. Although it is true that most synthetic (human-made) plastics we use today have been around for less than 100 years, natural plastics have been used for decorative and utilitarian purposes for hundreds and, in some cases, thousands of years. Natural plastic materials are moldable when heated and remain in their desired shapes when cooled. Amber, animal horn, tortoiseshell, and shellac are all naturally occurring materials that fall under the category of plastics. Such materials were used to make spoons, drinking vessels, combs, buttons, and even lantern windows (see Figure 9-43).

The word *plastic* is both an adjective and a noun. As an adjective, *plastic* describes a material's ability to be molded into different forms. This can be misleading because concrete and plaster are examples of rigid materials that exhibit plastic behavior when they are first mixed and poured into a form or mold. However, these materials do not fall under the category of plastics. As a noun, **plastics** represent a group of polymer materials that, while solid in the finished state, were liquid at some stage in the manufacturing process and may be formed into various shapes by the application of heat and pressure.

FIGURE 9-44: **The plastic wrap that you use to cover and protect food is made from polyethylene plastic.**

Virtually all of the commercial plastic products that we purchase at the store are examples of synthetic polymers. This means that they were formed as a result of human-controlled chemical processes. Synthetic plastics are distant cousins of crude oil and natural gas. Crude oil, or petroleum, and natural gas are made up of a mixture of various **hydrocarbons**, which are molecules that contain only carbon and hydrogen atoms. Some of these hydrocarbons serve as the primary raw material from which the chemicals that are used to make synthetic plastics are ultimately derived. The simplest of all synthetic plastics is polyethylene (PE). If you have ever used plastic wrap to cover leftovers from dinner, then you have used polyethylene (see Figure 9-44).

Polyethylene is comprised of long strings of covalently bonded carbon atoms, each of which is bonded to two hydrogen atoms. The material starts out as ethylene gas (C_2H_4). A single ethylene gas molecule, which serves as a monomer, is comprised of two carbon atoms and four hydrogen atoms. When approximately 1,000 ethylene monomers join together through a process called **polymerization**, they form a solid **macromolecule** of polyethylene (see Figure 9-45).

FIGURE 9-45: **Through the process of polymerization, 1,000 or more molecules of (a) ethylene gas are persuaded to join together to form (b) a solid polyethylene macromolecule.**

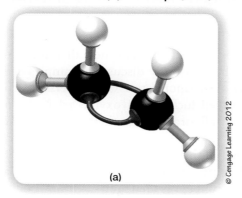
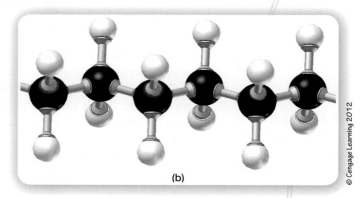

(a)

(b)

To change the properties of plastics, chemical engineers vary the chemical structures of the constituent monomers. One such modification will produce polypropylene (PP). Polypropylene fibers are used to make socks that hikers find especially useful; the material acts like a wick to draw moisture away from the skin. Another common plastic is polystyrene (PS). Polystyrene is a rigid and hard thermoplastic that is well suited for use in disposable cutlery. Hydrogen (H) and carbon (C) are not the only elements that bond to the carbon backbone of a polymer macromolecule. Other elements such as oxygen (O), nitrogen (N), chlorine (Cl), sulfur (S), and fluorine (F) are found within the mers of many plastics. Chlorine is used in the production of polyvinyl chloride (PVC), which is a very corrosion-resistant plastic that is used extensively in plumbing applications (see Figure 9-46).

When several different monomers combine to form repeating chains, plastics such as acrylonitrile-butadiene-styrene (ABS) are formed. ABS is a rigid and easily molded thermoplastic that is used in many engineering and consumer

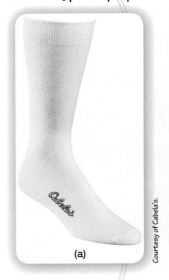

(a)

Courtesy of Cabela's.

(b)

© iStockphoto.com/EricHood.

(c)

© iStockphoto.com/lisafx.

FIGURE 9-47: Lego blocks are made from acrylonitrile-butadiene-styrene (ABS) plastic, which is comprised of repeating chains of three different mers.

© iStockphoto.com/DjordjeZ.

product applications, including children's Lego blocks (see Figure 9-47). Some plastics can be reshaped over and over again, whereas others remain in their initial form and cannot be remolded. It is this quality that defines the two categories into which all plastics are divided: thermoplastics and thermosetting plastics.

THERMOPLASTICS If a plastic can be reheated and reformed, then it belongs to a category known as *thermoplastics*. The previously discussed examples of PE, PS, PVC, and ABS belong to the family of thermoplastics. Other thermoplastics that you have probably used or encountered include polyoxymethylene (also called *acetal*), nylon, polyester, and polytetrafluoroethylene (better known under the brand name Teflon).

A **thermoplastic** is a plastic material that is comprised of disconnected polymer chains that are capable of slipping past each other when heated and return to a rigid state on cooling. The concept is analogous to a bowl of cold but cooked spaghetti. Each strand of spaghetti represents a polymer chain that is finite in length and not connected to any other strand. When reheated, the mass of spaghetti becomes pliable and can be reshaped into different forms. The ability to be reheated and reshaped makes thermoplastics well suited for recycling. Significant efforts have been made by the plastics industry to make the process of recycling easier for the consumer by incorporating standardized recycling symbols into plastic products. Figure 9-48 identifies common recyclable thermoplastics, their advantages and disadvantages, applications, and associated recycling symbols.

Most commercial thermoplastic objects, such as the frame around a computer monitor or the plastic casing of a cell phone, are made from tiny plastic pellets that are reheated and molded using various thermoforming processes. These processes will be discussed in Chapter 11, Manufacturing Processes and Product Life Cycle. Like metals, thermoplastics are also sold in sheet form and common extruded structural shapes such as rods, bars, and tubes from which finished components can be machined, welded, or joined together via mechanical fasteners. Common extruded shapes include circles, squares, rectangles, and hexagons (see Figure 9-49).

FIGURE 9-48: Advantages, disadvantages, and applications of the recyclable thermoplastics.

Thermoplastic	Advantages and Disadvantages	Example Applications	Recycling Code Symbol
Polyethylene terephthalate (PETE)	Tough and rigid Low cost Processable by all thermoplastic methods Low thermal resistance Poor solvent resistance	Synthetic fibers for carpeting; food and carbonated beverage containers	♳ 1 PETE
High-density polyethylene (HDPE)	Tough, impact resistant, chemically resistant, wear resistant Appropriate for contact with food Poor thermal stability Flammable	Milk jugs, detergent bottles, grocery bags, garbage containers, water pipes	♴ 2 HDPE
Polyvinyl chloride (PVC)	Can be made to exhibit both rigidity and flexibility Processable by all thermoplastic methods Nonflammable Inexpensive Good resistance to natural weather elements Subject to degradation by several solvents	Rigid: siding on houses, identification swipe cards, protective cover for wood frame windows, pipe and plumbing fixtures Flexible: clothing, upholstery, flexible tubing, roofing membranes, tool handle coatings, and electrical cable insulation	♵ 3 V
Low-density polyethylene (LDPE)	Processable by all thermoplastic methods Weldable, machinable, flexible, and tough Subject to degradation by solvents containing chlorine	Flexible and squeezable containers, plastic bags, food storage wrap, flexible tubing, six-pack beverage rings, protective coating on paper and toys	♶ 4 LDPE
Polypropylene (PP)	Processable by all thermoplastic methods Flexible Withstands extreme temperatures Can be made into fiber Low friction Appropriate for contact with food Natural moisture wicking properties Subject to degradation by ultraviolet radiation Flammable Difficult to bond	Outdoor and automotive carpet fibers, rope that floats in water, moisture wicking clothing fibers (e.g., Under Armour), containers with integral hinges, electrical wire insulation, food containers, disposable bottles	♷ 5 PP
Polystyrene (PS)	Inexpensive Hard, rigid, transparent, and moisture resistant Flammable Subject to degradation by several solvents Brittle Poor thermal stability	Plastic model assembly kits, thermal foam insulation, plastic cutlery, blister packaging, plastic lenses, brush bristles	♸ 6 PS

FIGURE 9-49: Thermoplastic materials are sold commercially as (a) pellets for molding and as (b) extruded structural shapes for machining.

(a) © iStockphoto.com/Luso; (b) image copyright Jim Barber, 2010; used under license from Shutterstock.com.

(a)

(b)

Your Turn

Though recycling of consumer thermoplastics has been going on for some time now, many people still do not recognize all the categories of plastic products that can be recycled. This activity will allow you to educate others about the different types of plastic products that should be recycled. You will need seven pieces of 30" × 40" foam-core board (seven pieces of recycled cardboard of the same size may be used as a substitute), seven plastic grocery bags, one large garbage bag, a large black marker, access to a hot glue gun, several sticks of hot melt glue, a digital camera, a computer, and a printer.

Step 1: Collect disposable plastic items from home that you would normally throw away or recycle over a period of 2 weeks and place them in a separate garbage bag. Make sure each item is thoroughly washed before placing it in the bag to avoid attracting insects or generating foul odors.

Step 2: Use the black marker to draw a different recycling code on each of the six bags. These bags will be used to store the recyclable items according to their material makeup. The seventh bag will be used for items that are made from an unidentified or unknown plastic material. At the end of the 2-week collection period, separate each item into one of six categories according to the recycling codes shown in Figure 9-48 and transfer them to their designated bags. The recycling symbols are often molded directly into the plastic items, but not all plastic items have identifiable codes. Place any item that is made from an unidentified or unknown plastic material into the seventh bag.

Step 3: During the 2-week collection period, use the Internet to research and identify any unidentified or unknown plastic items and then place them into the correct storage bags. Take digital photographs of any items that still cannot be identified through online research. Locate a plastics or materials engineer and e-mail the pictures to him or her with a polite request for assistance in identification. If you have difficulty finding an engineer, try the home page for the Society of Plastics Engineers (www.4spe.org/).

Step 4: Dedicate each poster board to one recyclable thermoplastic by drawing the recycling symbol at the top of the board. The symbol should fit within an area no smaller than 4 square inches. Add a title to each board by spelling out the name of the plastic followed by its abbreviation as shown in Figure 9-48. The seventh poster should be titled "Unknown Plastic."

Step 5: Use the hot glue gun and sticks of hot melt glue to affix the recyclable plastic items to their respective poster boards. Any item that is still unidentified should be affixed to the seventh poster.

Step 6: Take a digital photograph of each poster, print out the images, and formulate a consumer reference guide that will help educate people about the different types of consumer items that can be recycled. Share this guide with family members and friends and encourage them to recycle.

THERMOSETTING PLASTICS As mentioned earlier, **thermosetting plastics** (also called *thermosets*) are plastics that become "set" in their solid state once they are molded. Consequently, they cannot be remolded, which severely limits their ability to be recycled. Though this may be viewed as a disadvantage, thermosets do have an edge over thermoplastics in that they can withstand higher temperatures, are harder and stronger, and are far less susceptible to chemical decomposition from acids and solvents. Some of the more popular thermosetting plastics include epoxy, polyester resin, and polyurethane.

A thermoset starts out as two different substances called a *resin* and a *hardener*, which are combined together in the proper proportions to generate a chemical reaction. A resin is a gumlike solid or semisolid viscous substance that contains polymer macromolecules that cross-link when combined with a hardener. The hardener is a catalyst that causes the polymer macromolecules in the resin to become *cross-linked* to each other (see Figure 9-50). This process is called curing. When the curing process is finished, the material will remain in a fixed solid state.

The cross-linking of polymers is what chemically separates a thermoset from a thermoplastic. Returning to the analogy of a bowl of spaghetti, imagine that each strand of spaghetti is tied together to form one giant piece of spaghetti. In theory, a thermoset plastic is one giant polymer molecule. As such, the molecular weight of the molecule is so high that the temperature at which the material would melt is beyond its own thermal degradation temperature. This means that a thermoset material will burn and char before it reaches its melting point.

Of all the thermosetting plastics, epoxy is one of the most widely used by the average consumer because of its commercial availability as an adhesive (see Figure 9-51) and as a protective coating. Many thermosets require temperature control because higher temperatures are required to increase polymerization and cross-linking. However, the epoxies found in most hardware stores are designed to cure at room temperature.

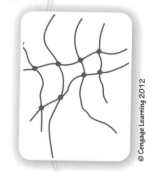

FIGURE 9-50: The polymer macromolecules of a thermosetting plastic become tied together in a cross-linked pattern, which prevents them from being remolded.

© Cengage Learning 2012

FIGURE 9-51: Commercially available epoxy adhesive is often sold in a dual-vial dispenser that allows the user to mix the resin and hardener in the correct proportions. Different epoxies are available for bonding a wide variety of materials.

© iStockphoto.com/jas0420.

Though they are relatively expensive, epoxies have excellent adhesive qualities that make them useful as glues for both similar and dissimilar materials in every material category. For example, epoxies are used to bond the copper pads and traces to the plastic substrate material on printed circuit boards. Epoxy material is often sold in two-part syringes that allow for equal parts of resin and hardener to be mixed. When cured, epoxy forms a nonmeltable and insoluble material. Both quick- and slow-curing epoxies are available, with the difference being that slow-curing epoxies tend to be stronger. Epoxy is applied as a protective coating on concrete surfaces, as well as steel framework on buildings and bridges. This prevents oxidation and corrosion, increases wear resistance, and makes the surfaces easier to paint and wash. Epoxies are also used extensively as the matrix material in composites, which will be discussed further on in this chapter.

Epoxies exhibit good chemical resistance and have high strength-to-weight ratios. They also have good resistance to fatigue, which is material weakness that results from repeated loading and unloading. However, they are susceptible to degradation

Safety Note

You should wear rubber gloves when working with epoxy. If accidental contact occurs with the skin or a work surface, vinegar is safe to use as a solvent on nonhardened epoxy chemicals.

from exposure to ultraviolet (UV) radiation and exhibit lower thermal degradation limits than other thermosets. Epoxies are generally more expensive than other thermosets. Also, the chemicals that make up the resins and hardeners have a shelf life and must be used within a specified time period after they are purchased.

Unsaturated polyester is a common thermosetting resin material that should not be confused with thermoplastic polyester fibers such as those commonly used in wrinkle-resistant clothing. Unsaturated resin hardens when mixed with a catalyst such as methyl ethyl ketone peroxide (MEKP), and are used for castings, encapsulations, auto-body repair work, wood filling, and as adhesives.

The most popular application of polyester resin is in combination with either polyurethane foam or reinforcements such as fiber glass, Kevlar, and carbon fiber. The result is a *composite material* that is used to make boat hulls, automobile and aircraft body panels, golf carts, tub and shower units, spas, helicopter and wind-turbine rotor blades, furniture, and even landscapes for stage and film productions (see Figure 9-52). The exterior applications take advantage of the material's good resistance to water and weather conditions.

FIGURE 9-52: The largest application of polyester resin is as a matrix material for fiber-reinforced composites. Such materials are used in the construction of (a) boat hulls and (b) wind-turbine rotor blades.

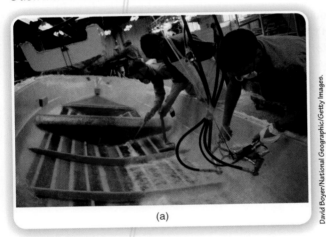

David Boyer/National Geographic/Getty Images.

(a)

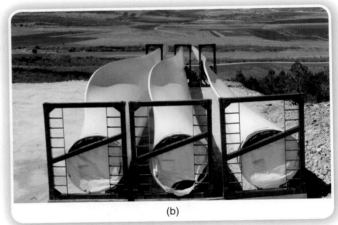

© iStockphoto.com/asterix0597.

(b)

Polyester resins are often mixed with pigments and will maintain their bright colors over time. They are cheaper than epoxies, have high strength, provide good resistance to chemicals, stand up against wear, and are more UV resistant than other plastics. Mold-release agents must be used to keep polyester resin castings from sticking to their molds. It should be noted, however, that polyester resins are not as durable and tend to be more brittle than other thermosets.

Polyurethane (PU) is one of the most versatile and widely used thermosetting plastics. It is produced in many forms such as a protective coatings, elastomers, rigid plastics, adhesives, foams, and fibers. A polyurethane coating serves as a tough, hard, and abrasion-resistant seal that preserves and protects wood from moisture and damage. If you have a hardwood floor or a piece of wood furniture in your home, then it was most likely coated with a layer of polyurethane.

Polyurethane is also used as a water-resistant adhesive for gluing wood items and other items of dissimilar material. Sensitive electrical components on printed circuit boards are sometimes coated with hard polyurethane plastic to prevent short circuits that are caused by dust or moisture. This material is also used to make wheels for grocery carts (see Figure 9-53) and tires for machines such as forklifts.

Foamed polyurethane is available in different densities. The most flexible foams are used in furniture cushions, mattresses, and automobile seats. Semirigid foams

are used for items such as armrest padding on office chairs. High-density polyurethane foam is machinable like wood and has excellent dimensional stability. Designers and architects use this material to create mock-ups of their ideas. It is also used to create replicas of wood carvings and architectural moldings. Low-density rigid foam is employed as thermal insulation in refrigerators, freezers, and even buildings (see Figure 9-54).

FIGURE 9-54: Spray foam insulation is made from polyurethane.

OFF-ROAD EXPLORATION

To learn more about plastics, read *Industrial Plastics: Theory and Applications* by Erik Lokensgard.

Polyurethane foams will discolor when exposed to light and will deteriorate with prolonged exposure to UV radiation. Commercial foaming agents are often used to seal gaps and cracks in residential structures and to act as spray insulation. Such materials are both flammable and toxic and should be handled carefully according to the manufacturer's instructions.

Fun Facts

Question: What do skateboard wheels and spandex have in common?
Answer: They are both made from polyurethane plastic.

Composite material:

a solid that is comprised of two or more distinctly different materials that are intentionally combined in layers or mixtures to exploit and enhance certain material properties.

Matrix:

a binder of continuous phase that surrounds a reinforcing material and holds it in place.

COMPOSITE MATERIALS

The subject of composites has been saved for last because they represent blends of all the other categories of materials. A **composite material** is a solid that is comprised of two or more distinctly different materials that are intentionally combined in layers or mixtures to exploit and enhance certain material properties. Plywood, for example, is a composite material that is formed by thin sheets of wood that are glued together using high-strength thermosetting plastic adhesives. The result is a material that has greater strength-to-weight ratio than a board of equal geometry made from a single species of wood.

Many modern boat hulls, which have been made using composite materials since the mid-1940s, consist of a fiber cloth material that is encased within a thermosetting plastic. Cloth material such as fiberglass or Kevlar serves as the **reinforcement**, and thermosetting plastics such as epoxy or polyester resin serve as the **matrix** material. The material properties are a result of the interaction between the matrix and the reinforcement.

The matrix material is selected based on its ability to withstand the environmental conditions, be they thermal, chemical, electrical, or combinations thereof. The matrix material gives the design its shape. It is the first to encounter an applied load, which it must transfer to the reinforcement. The reinforcement, which often takes the form of fibers, whiskers, or particles, is usually stronger than the matrix material. Its job is to carry and disperse the load. As a result, composite materials are able to withstand extreme conditions that would cause failure in traditional materials.

Modern composites are the newest of all the material categories, though history does show that the concept of composite materials is not a new idea. The Mongols who lived approximately 1,000 years ago were experts at making composite bows out of natural materials. Their creations rival the performance of any bow that has been made since that time. These bows were constructed of layers of wood, animal tendon, and horn that were bonded together using natural fish or animal glues. The assemblies were then steamed, bent into shape, wrapped in silk thread, and allowed to dry very slowly. The bows allowed well-trained Mongolian archers to hit targets more than 1,000 feet away (see Figure 9-55).

Modern composites are replacing traditional materials within the aerospace, automotive, and high-end sporting goods industries because they often exhibit qualities that cannot be matched by traditional materials. For example, the carbon-fiber

composite skin on the B-2 bomber is lighter and stronger than traditional aircraft aluminum. This material also does something that aluminum cannot: It absorbs radar energy. In combination with the plane's geometry, the composite material gives the aircraft a radar cross-section that is equal to that of a small bird (see Figure 9-56).

Modern composites are generally lightweight and moldable into complex three-dimensional forms, and they exhibit high degrees of strength and stiffness. The aerospace industry serves as the greatest consumer of composite materials because of extreme design specifications that involve strength, weight, flexure, temperature, and chemical resistance that make the use of traditional materials either impractical or impossible. Composites are experiencing increased use in the automotive industry as consumers demand more fuel-efficient cars that require less maintenance.

FIGURE 9-56: *The carbon-fiber composite skin in combination with the geometry of the B-2 bomber gives it an equivalent radar cross-section equal to that of a small bird.*

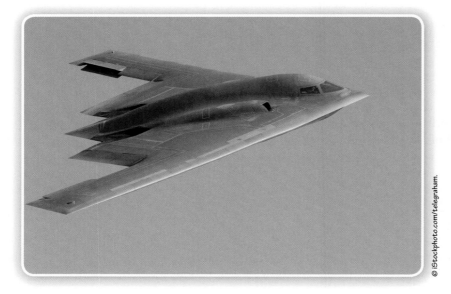

Composite materials are not without their disadvantages. Generally speaking, they incur higher material costs and are harder to manufacture than traditional materials. They tend to utilize materials that are not recyclable and are therefore environmentally problematic. Detecting defects in components that are made from composite materials is harder because the materials are not homogenous. Also, given the wide variety of available composite materials and the myriad ways in which they are formulated, there is substantially less data available to the design engineer in comparison to other traditional materials.

Modern composite materials are divided into several categories. These include polymer-matrix composites (PMCs), metal-matrix composites (MMCs), ceramic-matrix composites (CMCs), and carbon–carbon composites (CCCs). Each group has its advantages over the traditional materials on which they are based. Figure 9-57 identifies these advantages and gives example applications for each category.

FIGURE 9-57: *Categories, advantages, and applications of modern composite materials.*

Composite Material	Advantages over Traditional Materials	Applications
Polymer-matrix composites (PMCs)	Higher tensile strength Greater flexural stiffness Can utilize either thermoplastic or thermosetting plastics, depending on required properties	Aircraft and automobile body panels, boat hulls, golf club shafts, fishing rods, tub and shower units, spas, utility poles, helicopter and wind-turbine rotor blades
Metal-matrix composites (MMCs)	Lower thermal expansion Higher strength and stiffness Less vibration	Pistons, piston rings, connecting rods, cutting tools, bearings, metal components in brake systems, tennis racquets, golf club heads and shafts
Ceramic-matrix composites (CMCs)	Higher-temperature applications Good wear resistance Reduced brittleness and increased toughness Higher abrasion resistance Greater dimensional stability Longer wear life	Gas-turbine blades, aircraft surfaces exposed to hot exhaust gases, heat shields, nose cones, rocket nozzles, sensor covers, metal-cutting tools, sandwich layer in military tank and personnel armor
Carbon–Carbon Composites (CCCs)	Very high thermal resistance; dimensional stability at high temperatures; able to withstand thermal shock.	Rocket nose cones; exit throats on rocket engines; flame barriers, hot gas ducting, and exhaust system components in high temperature engines; brake linings

SUMMARY

In this chapter you learned the following.

- Common engineering materials include metals, woods, ceramics, polymers, and composites.

- Engineers think of metals as solid materials that are typically hard at room temperature, are shiny, and possess good thermal and electrical properties. Metals are divided into two categories called *ferrous* and *nonferrous* metals based on whether they contain iron. Steel is a common ferrous metal whose alloys contain various trace amounts of carbon. Common nonferrous metals that have engineering applications include gold, silver, copper, bronze, brass, aluminum, and titanium. These metals are exploited for their various properties, which include electricity and heat conductivity, low metal-on-metal friction, resonant qualities, high strength-to-weight ratios, high melting temperatures, and corrosion resistance.

- Wood is divided into two categories based on the types of trees from which it is harvested: softwoods and hardwoods. Softwoods come from evergreen trees that bear their seeds in the form of cones and have needle-shaped leaves. Hardwoods have broad leaves that they lose at the end of a growing season. Hardwoods are typically stronger and have a wider range of colors and grain patterns.

- Ceramics are a diverse class of materials based on solid compounds of both metallic and nonmetallic elements that are prepared by the action of heat and subsequent cooling. Ceramic materials are very hard and brittle, have high melting points, exhibit excellent dimensional stability, and are unmatched when it comes to providing insulation from extreme temperatures. These materials are divided into the following categories: whiteware, structural clay, refractories, glass, abrasives, and cement and concrete.

- Polymers are high-molecular-weight organic compounds that are comprised of repeating chains of molecules called *monomers*. Natural polymers are made from tree sap and other organic materials. Synthetic polymers are derived from petrochemicals. Polymers are divided into two general categories: plastics and elastomers. Elastomer materials have the ability to stretch at room temperature to twice their size and then return to their original dimensions on unloading. Plastics are solid polymer materials that were once liquid during their manufacture; they are formed by the application of heat and pressure. Plastics are divided into two categories: thermoplastics and thermosets. Thermoplastics are able to be remelted and reshaped after their initial polymerization. Thermosets are created when a resin and a hardener are mixed together in the proper proportions. A chemical reaction occurs that makes the thermoset incapable of being remelted and reshaped.

- Composites are solid materials that are comprised of two or more distinctly different materials that are intentionally combined in layers or mixtures to exploit and enhance certain material properties. Modern composites are made of reinforcing materials such as fiberglass, Kevlar, and carbon fiber that are embedded within a matrix material. The matrix material gives a design its shape, withstands the exposure of the working environment, and is able to transfer an applied load to the reinforcement.

1. What are the common mineral ores from which iron, copper, tin, zinc, aluminum, and titanium are extracted?
2. What is the difference between a ferrous and a nonferrous metal?
3. Explain how iron is smelted.
4. Identify an example application of cast iron.
5. Explain how steel is made.
6. What is the difference between a metal element and an alloy?
7. What are the three types of carbon steel, and what products are made from each type?
8. Give an example of an alloy steel, and where it is used.
9. What type of steel alloy would be used for structural applications?
10. Identify two engineering applications of gold.
11. Identify two engineering applications of silver.
12. What qualities does copper possess that makes it a good material for engineering applications?
13. Identify two engineering applications of copper.
14. What is the difference between bronze and brass?
15. Describe two qualities that make bronze an excellent material for engineering applications.
16. Identify one engineering application of brass.
17. What kinds of elements are alloyed with aluminum?
18. Identify two engineering applications of aluminum alloys.
19. Identify three engineering applications of titanium alloys.
20. What is the difference between hardwoods and softwoods?
21. Identify three tree species of softwoods and three species of hardwoods.
22. How do trees grow in size?
23. What is the difference between heartwood and sapwood?
24. Give two engineering applications of softwoods.
25. What are hardwoods used for?
26. What is a ceramic material?
27. What are the common metallic and nonmetallic elements that make up ceramic materials?
28. Identify the different categories to which ceramic materials belong, and give an example application of each.
29. What is a polymer?
30. How does an elastomer differ from a plastic?
31. What are elastomers used for?
32. How is polyethylene plastic formed?
33. What is the difference between a thermoplastic and a thermoset?
34. Identify three examples of thermoplastics and an application of each.
35. Identify three examples of thermosets and an application of each.
36. What are composites, and why are they used?
37. What is the difference between a matrix material and a reinforcement?
38. Identify a historical example of a composite material.
39. Identify the four categories of modern composites, and give an example engineering application for each.

CASEIN: MAKING NATURAL PLASTIC GLUE FROM COMMON MATERIALS

Have you ever thought about how plastics are made? The processes involved in making modern plastics are complex and often involve dangerous chemicals. However, in this experiment you will make a form of natural plastic using materials and processes that are common to the average kitchen. The experiment involves the polymerization of a natural plastic called *casein* from skim milk using heat and vinegar. The casein will be dried, pulverized into powder, and eventually mixed with baking soda and water to form glue that can be used with class projects that involve wood or paper.

For this experiment, you will need a saucepan, a large spoon, a baking sheet, an open glass or plastic container, a popsicle stick or small piece of scrap wood, paper towels, a mortar and pestle, a strainer, 2 cups of fresh skim milk, 4 tablespoons of white vinegar, 1/2 teaspoon of baking powder, a cooking thermometer, a sandwich-size sealable storage bag, and access to a kitchen stove and sink.

Step 1: Pour the skim milk into the sauce pan. Place the sauce pan on the stove and set the burner or heating element to medium heat. Heat the milk for approximately 5 to 7 minutes, stirring occasionally. If the milk reaches a temperature that is higher than 140°F, the chemical reaction will not work. Use the cooking thermometer to monitor the temperature. Do not allow the milk to boil.

Step 2: Once the milk has been heated, add the vinegar and stir the contents continuously. Within moments, you should see the milk separate into large clumps of material within a clear liquid. The clumpy material is formed by a milk protein called *casein* that precipitates out of the milk because of a chemical reaction with acetic acid in the vinegar.

Step 3: When the casein has completely separated from the liquid, pour the contents of the sauce pan into the strainer over the kitchen sink. After allowing the casein to cool some, use your hands to squeeze and extract as much liquid from it as possible while it is still in the strainer. Rinse the saucepan and fill it with cold tap water. Place the casein back in the saucepan. Stir and knead the casein in the water bath for approximately 1 minute. Repeat the straining and rinsing process about five times. This is needed to remove as much of the vinegar as possible. Place the casein onto the paper towels and press out as much of the water as you can.

Step 4: Set the oven to 170°F. Use your hands to break off very small pieces of the casein and spread them across the baking sheet. Once the oven has reached the appropriate temperature, place the baking sheet in the oven and allow the casein to dry for approximately 2 hours.

Step 5: Remove the baking sheet from the oven. The pieces of casein should have dried to smaller yellow crystals. Place a small number of the casein crystals in the mortar and use the pestle to grind them into a powder. Place the powdered casein into the sealable plastic bag. Repeat this process until all of the casein crystals have been pulverized into powder. You should end up with approximately 4 teaspoons of powder. Mix the casein powder with ½ teaspoon of baking soda (sodium bicarbonate), and store the powder in the sealable plastic bag until you are ready to use it as glue for a class project involving wood or paper.

Step 6: When you are ready to form the casein glue, pour the powder mixture into an open glass or plastic container. Add to the mixture an equal volume of hot tap water and stir it continuously for several minutes using a popsicle stick or small piece of scrap wood. Add small amounts of hot water to thin the mixture to the desired consistency. Apply the glue to the wood or paper surfaces and leave the glued parts undisturbed to set up for approximately 24 hours.

CHAPTER 10
Material Properties

GPS DELUXE

| START LOCATION | DISTANCE | END LOCATION |

Menu

Before You Begin

Think about these questions as you study the concepts in this chapter:

1 What are the four types of externally applied forces, and how are they generated?

2 What is the difference between stress and strain, and how do they relate to a material's modulus of elasticity?

3 Why would an engineer who designs structural components be most interested in the elasti region of a material's engineering stress–strain diagram?

4 Where are the proportional limit, yield stress, ultimate stress, and fracture stress located on an engineering stress–strain diagram?

5 Why do engineers apply factors of safety to their designs, and how do they use them to determine allowable stresses?

6 What is shear stress, and how is it generated?

7 What is a mechanical property, and what kinds of mechanical properties do materials exhibit?

8 What kinds of destructive testing methods are materials put through, and what do these tests reveal?

Materials exhibit different types of properties that are of particular interest to the engineer. The primary focus of this chapter is the *physical properties* of materials. A physical property is any aspect of a material that can be measured or perceived without changing its chemical identity.

As part of their engineering and engineering technology programs, college students are educated in several areas of applied physics, one of which is *deformable-body mechanics*. All solid bodies experience some degree of deformation when subject to externally applied forces such as tension or compression. Deformation is a change in dimension that is caused by the application of force, pressure, heat, or any other physical phenomenon. Deformations are also accompanied by stress within the object, which increases as the load increases. Engineering stress–strain diagrams show the relationship between stress and the amount of unit deformation that a material will exhibit under load. Engineers use these graphs to make important decisions about which materials to use, and to determine how much loading the materials can withstand without suffering permanent damage.

Engineering stress–strain diagrams are created when materials are subject to various types of destructive testing methods. The destructive testing methods focused on in this chapter include tensile testing, compression testing, direct-shear tests, torsion testing, flexure testing, and hardness and impact-strength tests.

> **Physical property:**
> any aspect of a material that can be measured or perceived without changing its chemical identity.

> **Deformation:**
> a change in dimension(s) that is caused by the application of force, pressure, heat, or any other phenomenon.

DEFORMABLE BODY MECHANICS

Engineers design structures and mechanical components to perform specific functions. For example, the function of a building is to provide shelter from the weather and spaces in which industrial, commercial, or residential activities can safely take place. However, a building cannot serve its function if it cannot support its own weight and the weight of its occupants, or stand up to natural forces such as wind and snow loads. Bridges serve as another example of this structural–functional relationship. Like a building, the first function of a bridge is to support its own weight. If it cannot do this, then it has no chance of serving as a connection between two locations.

How do engineers know how big a building's support columns should be, or why its support beams should be made from steel instead of glass? How do engineers know what weight limit to place on an elevator, or how to determine a maximum payload for an automobile? The answer to these questions can be found within a branch of applied physics called deformable-body mechanics (also called *mechanics of materials*), which is concerned with the deformations of solid bodies as a result of externally applied forces.

Types of Externally Applied Forces

As mentioned in Chapter 3, a *force* can be thought of as a push or a pull. Gravity is a force that we experience all the time because it constantly pulls us toward the center of Earth. In fact, your weight is a function of gravity. If you were to stand on the moon, which exerts a much smaller gravitational force than Earth because of its smaller mass, you would weigh 1/6th the amount that you weigh

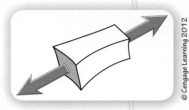

here on Earth. When an external force is applied to an object, it is referred to as a **load**. Engineered objects are designed to withstand specific types of loads such as tension, compression, shear, and torsion. **Tension** results when two forces pull on an object in opposite directions, causing that object to stretch (see Figure 10-1).

When you walk into an elevator, your weight in combination with the weight of the elevator places a *tensile load* on the elevator pulley cables (see Figure 10-2). This causes the cables to stretch, even though the amount that they stretch may not be visible to the naked eye.

FIGURE 10-2: **(a) The weight of the occupants and the elevator structure in combination with the resistance from the counterweight serve to place (b) a tensile load on each cable.**

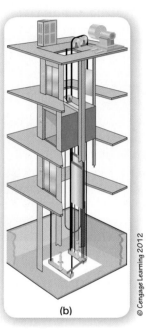

FIGURE 10-3: *Compression causes an object to reduce in length and increase in girth.*

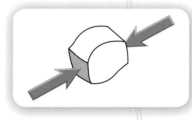

Compression results when two forces push on an object in opposite directions, causing that object to compact (see Figure 10-3). A masonry or concrete foundation is in a constant state of compression because of the weight of the house or building that sits on top of it and the reaction force of the ground pushing upward. When you sit on a chair, your weight serves as the *compressive load,* which is transferred to the floor through the legs of the chair. Image how much compressive force must be generated on the landing gears of an airplane every time it touches down on the tarmac.

Direct shear occurs when two offset forces act on an object in opposite directions, trying to split the object in two along a plane that is parallel to the direction of those forces. You apply a *shear load* to a piece of paper when you cut it to a custom size using a paper trimmer (see Figure 10-4).

FIGURE 10-4: *Direct shear occurs when an object is subject to two offset and opposing forces.*

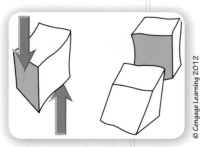

Some mechanical systems are designed to fail in shear in order to protect a machine from damage. If a motor-driven shaft is used to power a mechanical element that could jam, then there is a risk of damage to the motor (which is more expensive to replace). If the two shafts are connected to each other by way of a shear pin and the driven shaft becomes jammed, then the resulting shear force will cause the shear pin to fracture. This will allow the motor shaft to continue spinning without experiencing damage. The auger shafts on a snow blower are designed with this safety feature (see Figure 10-5).

(a)

Troy-Bilt, LLC.

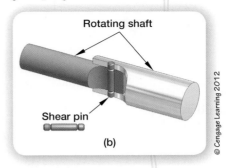

Rotating shaft

Shear pin

(b)

© Cengage Learning 2012

You may recall that the concept of *torque* was addressed in Chapter 4. **Torsion** is a rotational effect that occurs when two forces, separated by some distance, act on an object in opposite directions, causing the object to twist along its longitudinal axis (see Figure 10-6). Torsion is a more complex form of shear and is sometimes referred to as *torsional shearing*. The shafts on gear and chain drive mechanisms experience *torsional loads,* as do automobile drive shafts. The common twist drill is made from materials that can withstand large torsional loads.

When you tighten a screw or bolt, you are applying a torsional load to that fastener. If the load is too great, then the fastener or the object to which it is applied may be damaged. If the load is too small, then vibration during operation may cause the fastener to loosen over time. In critical fastener applications such as an automobile engine assembly, an engineer will specify the optimum torque that should be applied to each fastener. In such cases, a technician will use a torque wrench to ensure that the proper amount of preload is placed on a fastener. Figure 10-7 shows a beam-type torque wrench, which is a type of socket wrench. The beam is made of an elastic metal that will bend as torque is applied to the fastener. A smaller pointer bar remains stationary and points to the resulting torque value on a scale. Some

FIGURE 10-6: An object experiences torsion when two forces act in opposite directions to generate a torque or twisting action.

© Cengage Learning 2012

FIGURE 10-7: A beam torque wrench is a special type of socket wrench that allows the user to measure the amount of torque applied to a fastener.

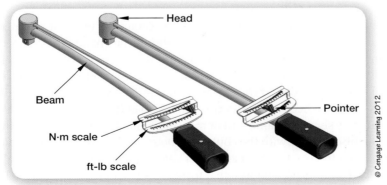

Head

Beam

Pointer

N·m scale

ft-lb scale

© Cengage Learning 2012

torque wrenches are outfitted with two scales: One indicates torque in foot-pounds, and the other indicates torque in newton-meters.

Stress

Stress:

a reaction that occurs within an object that provides resistance to dimensional changes brought on by an applied load.

Stress is a reaction that occurs within an object that provides resistance to dimensional changes brought on by an applied load such as tension, compression, shear, and torsion. Several different types of stresses can form within a structural member, so it is important to identify the type of stress that is being analyzed or calculated. *Uniform stress* will occur in a structural member that has a constant *cross-sectional area* and experiences a **normal force** (i.e., a force that acts along the central axis of a straight structural member) such as tension or compression (see Figure 10-8).

FIGURE 10-8: To avoid excessive amounts of stress in any one support column of a building, a civil engineer will use multiple columns to support the weight of a building's upper floors and roof.

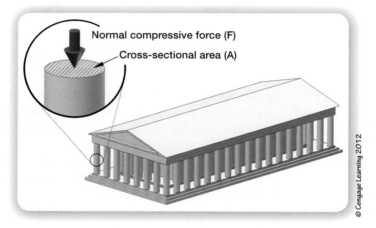

Uniform axial stress (*s*) is defined mathematically as the intensity of a normal force (*F*) per unit cross-sectional area (*A*) within a structural member (see Equation 10-1). The common U.S. customary system unit of measurement for stress is pounds per square inch (lb/in² or psi). The unit ksi, or kips per square inch (1 kip = 1,000 lb), is a larger unit that is often used in structural situations where large loads are present. The common unit of measurement for stress in the SI system of units is the pascal (Pa), although it is such a small unit that megapascals (MPa) and gigapascals (GPa) are often used instead.

$$\text{Uniform axial stress} = \frac{\text{Normal force}}{\text{Cross-sectional area}} \qquad \text{(Equation 10-1)}$$

$$s = \frac{F}{A}$$

We return to the example of the elevator cables. A team of engineers determined what kind of material those cables should be made from, and how large their diameters would need to be to withstand the stresses that build up inside the cables. If the cables are too thin, they will snap because of excessive stress levels that result from too much load. If the cables are too thick, then they will not easily bend around the pulley wheel as the elevator moves up and down.

Though they share the same unit of measurement, it is important that the concept of stress not be confused with the concept of *pressure*. Pressure represents an external load, wherein a force acts across an object's surface like the wind blowing on a sail. Stress is a type of internal pressure that forms as a reaction to an external load.

Problem: A 1.50-inch diameter extension pipe is used to connect a 30-lb fan to a ceiling plate (see Figure 10-9). The wall thickness of the pipe is 0.1 inch.

Solve for the following:

FIGURE 10-9: Ceiling fan extension pipe.

a. What is the cross-sectional area (A) of the pipe?

$$A = \frac{\pi (d^2_{outside} - d^2_{inside})}{4}$$

$$A = \frac{\pi \times \left[(1.50 \text{ in})^2 - (1.30 \text{ in})^2\right]}{4}$$

$$A = \frac{\pi \times 0.56 \text{ in}^2}{4}$$

$$A = 0.44 \text{ in}^2$$

(a)

b. Neglecting the weight of the pipe, how much tensile stress exists within the nonthreaded section of the extension pipe?

$$s = \frac{F}{A}$$

$$s = \frac{30 \text{ lb}}{0.44 \text{ in}^2}$$

$$s = 68.18 \text{ lb/in}^2$$

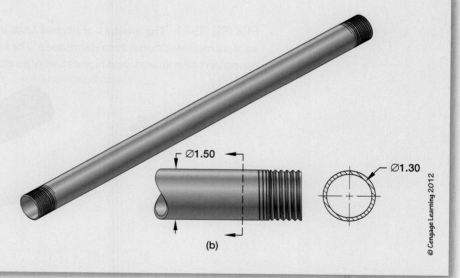

Ø1.50

Ø1.30

(b)

Photo courtesy of Craftmade International, Inc.

© Cengage Learning 2012

When too much stress builds up inside of an object, the end result is either a permanent change in shape or fracture. The average 10-year-old child understands that if you want to hang a swing from a tree limb, then you will need to use a rope that is much larger in diameter than a piece of dental floss. It's easy to see that it takes a relatively small amount of pulling force to tear a piece of dental floss in half. But that same child might be inclined to test out a larger diameter piece of twine only to find out that it wasn't such a good idea (see Figure 10-10). So, when it comes to swings, a larger rope equals a better chance of not having the wind knocked out of you. Why is this so?

Given a constant load, the amount of stress that builds up within a structural member will decrease as the cross-sectional area of that structural object increases. Therefore, spreading a force out over a larger area will result in less force per unit area and therefore less stress (see Figure 10-11).

FIGURE 10-10: What makes (a) twine a poor choice for (b) hanging a swing from a tree branch?

(a)

(b)

FIGURE 10-11: The amount of stress that builds up inside an object will increase as its cross-sectional area decreases. The blue and red colors in the after image represent the lowest and highest stress concentrations, respectively.

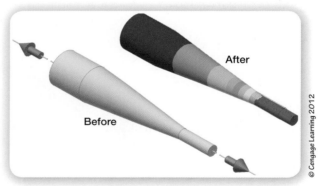

After

Before

© Cengage Learning 2012

Deformation versus Strain

All objects deform when they experience an external load, even though the amount of deformation may be unrecognizable to the naked eye. If a 10-in-long plastic rod (original length, L_o) is subject to a tensile force that stretches it to 11 inches (final length, L_f) then the resulting deformation is a change in length (δ) that is equal to 1 inch (see Figure 10-12). A change in length that results from tension is recorded as a positive value because the object is expanding. A change in length that results from compression is recorded as a negative value because the object is decreasing in length.

FIGURE 10-12: A structural member that experiences a normal tensile force will also experience a change in length.

Before

After

$L_{original}$

L_{final}

δ

© Cengage Learning 2012

Change in length = Final length − Original length (Equation 10-2)

$$\delta = L_f - L_o$$

Determining a material's mechanical properties in tension or compression involves analyzing the relationship between the growing amount of stress that builds up within a material as the applied load increases and the concurrent change in *strain* that the material exhibits. Strain (symbol *e*), also called *unit deformation,* is how much a material deforms for every unit of original size. Materials exhibit different types of strain, depending on the type of load the material is encountering. Tension and compression result in *axial strain,* which is mathematically defined as the change in length divided by the original length (see Equation 10-3). The unit of measurement for axial strain is inch/inch in the U.S. customary system, and mm/m in the metric system.

$$\text{Axial strain} = \frac{\text{Change in length}}{\text{Original length}} \qquad \text{(Equation 10-3)}$$

$$e = \frac{\delta}{L_o}$$

Returning to the previous example of the 10-in-long plastic rod, if the total change in length is 1 inch, then the axial strain will equal 0.1 in/in, or 10% (see Figure 10-13). This means that 0.1 inch of elongation occurred for every inch of original length, which represents a 10% increase in the length of each unit as well as the overall length of the member.

Strain values can be very small, especially for metals. They are often represented in engineering or scientific notation or as a percentage. For example, a typical structural steel will exhibit an increase in strain of approximately 0.000000033 in/in (33×10^{-9}) for each unit of stress. In other words, if you took a 1-in diameter by 10-in-long circular rod made from a typical structural steel and subjected it to a 100-lb normal tensile force, the rod would increase in length by 0.000042 in (42 millionths of an inch). Comparatively speaking, 42 millionths of an inch is to 1 inch as the length of a school bus is to the distance between New York City and Baltimore, Maryland.

FIGURE 10-13: Axial strain is the amount of deformation that occurs for every unit of original length of the structural member.

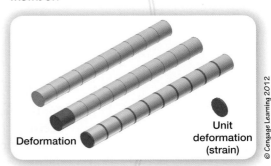

Deformation

Unit deformation (strain)

© Cengage Learning 2012

Engineering Stress–Strain Diagram

Material engineers perform standardized tests on material samples to determine the amount of deformation they will exhibit when subject to specific types of load conditions. The results of such tests often take the form of a **force-displacement diagram**, which is a graph that shows the relationship between force (vertical axis) and deformation (horizontal axis) [see Figure 10-14(a)]. Specific types of material tests are discussed in greater detail further on in this chapter. It is inefficient and cost prohibitive to perform force-displacement tests on every structural component that makes up an engineering design. Therefore, engineers communicate the force-displacement test results of standard-sized material test samples in a form that is applicable to designs of any size. By taking into consideration the original cross-sectional area and the original test length (also called the *gage length*) of the test sample, force and deformation values are converted to stress and strain values, respectively. When plotted on a graph, they form an **engineering stress–strain diagram** [see Figure 10-14(b)]. The word *engineering* is placed in front of the term because a true stress–strain diagram takes into account the constant change in cross-sectional area that occurs throughout a test sample as the value of the applied load increases. The curve on an engineering stress–strain diagram is based on a test sample that has a constant cross-sectional area up to the point of fracture.

Though the shape of an engineering stress–strain curve is identical to its corresponding force-displacement diagram, the values of stress and strain are independent of the test sample's original geometry. Therefore, two test samples that are made from the same material but different sizes (i.e., different cross-sectional areas and original gage lengths) will generate force-displacement diagrams of

FIGURE 10-14: When the cross-sectional area and original test length of a tensile test sample are taken into consideration, then (a) a force-displacement diagram may be converted to (b) an engineering stress–strain diagram. Note that the curves are identical in their shape.

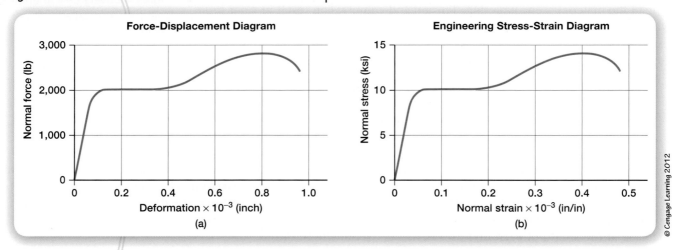

Force-Displacement Diagram

Engineering Stress-Strain Diagram

(a)

(b)

© Cengage Learning 2012

identical shape but of different magnitudes of force and displacement. When converted to engineering stress–strain diagrams, both samples will (theoretically) display identically shaped curves with identical magnitudes of stress and strain. For this reason, designers use engineering stress–strain curves to make informed decisions on the types of materials to use in the design of structural components. Different materials produce different stress–strain curves, many of which are very different in appearance. The following explanations of the important points on an engineering stress–strain diagram relate to a curve that is produced by structural steel under uniform tension.

ELASTIC DEFORMATION As a material is loaded in tension, it will exhibit a buildup of normal stress accompanied by an increase in normal strain. At the early stages of this loading process, a material will experience *elastic deformation*. **Elastic deformation** is the change in size or shape of an object under an applied load from which the object will return to its original dimensions on unloading. The idea is analogous to the behavior that a rubber band exhibits when stretched (see Figure 10-15). Regardless of the time over which the load occurs, if the load is removed from the material while it is experiencing elastic deformation, the material will return to its original dimensions. Virtually all mechanical and structural

FIGURE 10-15: Most materials will exhibit behavior analogous to that of a rubber band when first subject to an applied load. If the load is small enough, the material will return to its original size on unloading.

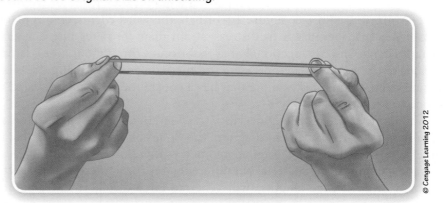

© Cengage Learning 2012

designs that must withstand temporary loads are designed to take advantage of this mechanical property.

For most engineering materials such as metals and plastics, the relationship between stress and strain starts off as a constant ratio, resulting in a straight line from the origin to a point known as the **proportional limit**. The point is so named because proportionality between stress and strain no longer exists beyond it. Elastic deformation takes place between the origin of the graph and another point called the *yield stress*. Consequently, this region of the graph is referred to as the **elastic region** (see Figure 10-16).

The *slope* of the engineering stress–strain curve within the elastic region of the graph is called the **modulus of elasticity** (symbol E). It is also referred to as **Young's modulus** in honor of English scientist Thomas Young (1773–1829), who was the first person to introduce the concept of modulus of elasticity by name. The modulus of elasticity can be found by dividing a change in stress (Δs) by the corresponding change in strain (Δe) (see Equation 10-4). The unit of measurement for the modulus of elasticity is psi or ksi in the U.S. customary system and MPa or GPa in the metric system.

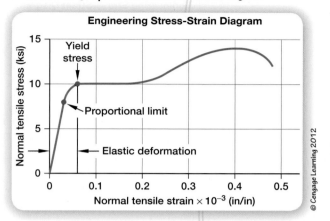

FIGURE 10-16: Many metals and plastics exhibit a straight-line relationship between stress and strain up to a point on a stress–strain curve called the *proportional limit*. Materials will exhibit elastic deformation up to the yield stress. This area of the graph is called the *elastic region*.

© Cengage Learning 2012

$$\text{Modulus of elasticity} = \frac{\text{Change in stress}}{\text{Change in strain}} \qquad \text{(Equation 10-4)}$$

$$E = \frac{\Delta s}{\Delta e}$$

Point of Interest

Robert Hooke (1635–1703) was a famous English scientist who served as the Royal Society's Curator of Experiments for more than 40 years. He conducted many experiments regarding the elastic nature of materials in tension, which included metal, wood, and even bone and sinew. Through his experiments, Hooke discovered the law of elasticity, which he published in 1678. Today, we refer to Hooke's discovery as **Hooke's law** (see Equation 10-5). This law expresses the linear relationship between stress and strain up to the proportional limit for materials in tension.

$$\text{Axial stress} = \text{Modulous of elasticity} \times \text{Axial strain}$$

$$s = Ee \qquad \text{(Equation 10-5)}$$

History has not been kind to Robert Hooke. He was described by biographers for two centuries following his death as being jealous, vane, difficult, and irritable. Unfortunately, no image of the man exists today. It is rumored that Sir Isaac Newton, while serving as president of the Royal Society, failed to preserve the only known portrait of Robert Hooke. In fact, the two men became bitter enemies over intellectual disputes regarding which of them was the first to develop the inverse square law of gravitation and the wave theory of light.

Using the change in stress and strain as measured from the origin of the graph up to the proportional limit is an easy method for calculating the modulus of elasticity (see Figure 10-17).

FIGURE 10-17: The simplest way to calculate the modulus of elasticity is to start at the graph origin and divide the change in stress up to the proportional limit (s_{pl}) by the change in strain up to the same point (e_{pl}).

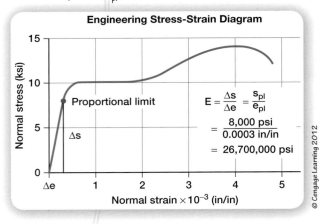

© Cengage Learning 2012

FIGURE 10-18: Approximate modulus of elasticity (E) values for common metal alloys.

Material	Modulus of Elasticity (psi)
110 copper	17,500,000
660 bronze	14,500,000
360 brass	14,100,000
6061 aluminum	10,000,000
7075 aluminum	10,400,000
1018 steel	29,700,000
4140 "chromoly" steel	29,700,000
A36 structural steel	29,000,000
303 stainless steel	28,000,000

It is not unusual to see elastic modulus values in the millions of psi for materials like aluminum, brass, and steel. Figure 10-18 gives the modulus of elasticity for various metals that are commonly used in engineering applications.

The point on an engineering stress–strain curve beyond which a material will begin to experience measurable permanent deformation is called the **yield stress** (s_y). If you have even bent a paperclip out of shape, then you have generated a stress within the paperclip that caused the material to yield. If you tried to return the paperclip to its exact original shape, then you probably realized that it could not be done. When a material is loaded beyond its yield stress, it will experience permanent dimensional change that will become evident once the load is removed (see Figure 10-19).

Few materials have yield points that are easily identified on their engineering stress–strain curves. For most materials, the transition from elastic to plastic behavior is gradual. For example, a typical engineering stress–strain curve for an aluminum alloy has no discernible point where yielding is evident (see Figure 10-20).

FIGURE 10-19: This graph shows that when a material is stressed beyond its yield point, it will experience a permanent change in geometry. As a result, the tensile test specimen is now longer than it was before loading occurred even after the removal of the load.

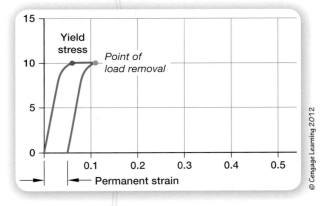

© Cengage Learning 2012

FIGURE 10-20: Usually the proportional limit, ultimate stress, and fracture point on a stress–strain curve are easily identified. A standard 0.2% offset method is used for materials such as aluminum that do not exhibit clearly identifiable yield points on their stress–strain curves.

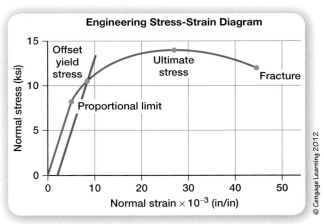

© Cengage Learning 2012

It is therefore necessary to use a standard mathematical process, called the offset method, to determine an arbitrary yield point on a stress–strain curve. The process involves drawing a line parallel to the linear portion of the engineering stress–strain curve, beginning at the 0.002 in/in strain value (or 0.2%) on the X-axis. The point where the offset parallel line crosses the engineering stress–strain curve is known as the offset yield stress (also called the *proof stress*).

The yield point is very important to engineers who design mechanical and structural components because it serves as a threshold limit. Engineers understand that all materials change shape under load. By predicting the greatest load that a structural member will encounter and keeping the material's yield point in mind, an engineer can ensure that his or her design will act like a spring and return to its original shape when the load is removed. Figure 10-21 identifies the yield stresses for several common engineering materials.

FIGURE 10-21: Yield stresses for common engineering materials.

Material	Yield Stress (psi)
Titanium Ti-6Al-4V (grade 5)	128,000
4140 "chromoly" steel	60,500
A36 structural steel	36,000
6061 aluminum	35,000
Acetal	9,000
High-density polyethylene (HDPE)	4,300
Ultra-high-molecular-weight polyethylene (UHMW)	3,000

Factors of Safety and Allowable Stress Imagine you were given an all-day task of moving firewood from a giant outdoor woodpile to a garage, where it must be neatly stacked into rows. Assume that the maximum amount of weight that you can lift without harming your back is 50 lb. This value serves as your actual strength limit. For reasons of personal safety, you might decide to limit yourself to lifting no more than 25 lb of firewood. This value serves as your allowable strength limit. By exerting half of your actual strength limit, you are sure to avoid personal injury and overexertion. Engineers do this same thing when designing bridges, buildings, machines, and anything that incorporates structural components.

If a component is intended to return to its original geometry on unloading, then the yield stress of the material from which the component is made will usually serve as the actual stress limit. The engineer will decide on an allowable stress limit within the elastic region of the material's engineering stress–strain diagram to ensure that the design will avoid structural failure from situations such as accidental overloading. The ratio between a material's actual stress limit, and its allowable stress limit is referred to as a factor of safety (*n*) (see Equation 10-6).

$$n = \frac{\text{Actual stress limit}}{\text{Allowable stress limit}} \qquad \text{(Equation 10-6)}$$

The structural failure of an aircraft component can have far more disastrous results than the structural failure of a component on a piece of farm equipment. As such, factors of safety vary, depending on the application. Some factors of safety are dictated by codes, which are standards that are enforceable by law. In many cases, factors of safety are based on many years of experience and time-tested practice within a field of engineering.

If an appropriate material and factor of safety have been determined, then an allowable stress can be calculated using Equation 10-7. The geometry of the structural component along with the allowable stress are then used to calculate the maximum allowable load.

$$\text{Allowable stress} = \frac{\text{Yield stress}}{\text{Factor of safety}} \qquad \text{(Equation 10-7)}$$

$$s_{\text{allowable}} = \frac{s_y}{n}$$

Every component in a mechanical system acts like a link in a chain, with the weakest link determining the maximum load that the system can withstand. This is why engineers establish maximum weight limits for their designs. The next time you are in an elevator, look for the plaque that identifies the elevator's maximum load capacity. Also, look on the driver's side doorjamb of your family automobile to find the sticker that gives the gross vehicle weight rating (GVWR) value (see Figure 10-22). Both of these maximum weight limits were established by engineers to ensure that the stress values that are experienced by the various mechanical components will not exceed their yield stresses.

FIGURE 10-22: (a) Occupational Health and Safety Administration (OSHA) standard maximum (load) capacity sign and (b) an example of an automobile's GVWR sticker. This sticker identifies the maximum allowable weight of the fully loaded vehicle (including liquids, passengers, cargo, and the tongue weight of any towed vehicle).

© Cengage Learning 2012

Example

Problem: A 0.505-inch diameter structural component in a mechanical system is subject to a tensile load. The material from which the component is made has a yield stress of 10,000 psi. To ensure that this structural member does not fail, a factor of safety of 2 has been established. Use this information to solve for the following:

Step 1 What is the allowable stress for this component?

$$s_{\text{allowable}} = \frac{s_y}{n}$$

$$s_{\text{allowable}} = \frac{10,000 \text{ psi}}{2}$$

$$s_{\text{allowable}} = 5,000 \text{ psi}$$

(continued)

Step 2 Calculate the cross-sectional area of the structural member?

$$A = \frac{\pi d^2}{4}$$

$$A = \frac{\pi \times (0.505 \text{ in})^2}{4}$$

$$A = \frac{\pi \times 0.255 \text{ in}^2}{4}$$

$$A = \frac{0.801 \text{ in}^2}{4}$$

$$A_o = 0.2 \text{ in}^2$$

Step 3 Calculate the maximum tensile load that the component should not exceed during its service life?

$$s_{allowable} = \frac{F_{maximum}}{A} \Rightarrow F_{maximum} = s_{allowable} \times A$$

$$F_{maximum} = 5,000 \text{ lb/in}^2 \times 0.2 \text{ in}^2$$

$$= 1,000 \text{ lb}$$

PLASTIC DEFORMATION **Plastic deformation** occurs beyond the yield stress and continues to the point of **fracture**, or total structural failure. This means that any change in geometry past the yield point will result in a permanent dimensional change on the part of the structural member even if the load is removed. Some designs require that a material be permanently deformed in order to achieve a desired shape. Bending conduit for electrical wires, drawing stainless steel sheet metal to form kitchen sinks, and pressing sheet metal into the form of car-body panels are all examples where materials must be stressed beyond their yield points to achieve a desired geometry. Plastic deformation is characterized by a noticeable amount of unit elongation with relatively little increase in stress (see Figure 10-23). As this occurs, the material becomes harder.

FIGURE 10-23: A material experiences plastic deformation from its yield point up to the point of fracture.

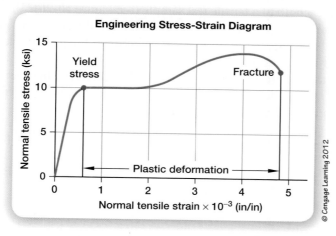

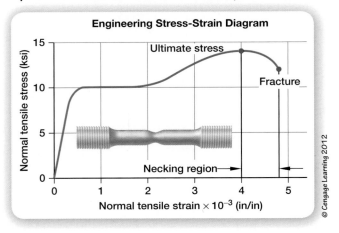

Eventually, the material reaches its **ultimate stress** (s_u), which is the maximum amount of stress that a material can withstand without failing. Beyond this point, the material will exhibit **necking**, which is a visible decrease in cross-sectional area at the location where fracture will eventually occur. Consequently, the region on an engineering stress–strain diagram between the ultimate stress and the point of fracture is known as the **necking region** (see Figure 10-24).

FIGURE 10-25: A tensile test specimen will experience a reduction in size from (a) its original area to (b) its final area at the point of fracture.

Percent Reduction in Area and Percent Elongation Information about the change in diameter and the change in length of a fractured test specimen is used to calculate the *percent reduction in area* and the *percent elongation*. Together these percentages indicate the degree to which a material exhibits brittle or ductile qualities. The **percent reduction in area** is a measure of the amount of necking that occurs in a tensile test sample at the cross-sectional area where fracture takes place (see Figure 10-25).

To calculate the percent reduction in area, divide the difference between the original and final areas of the test specimen by the original area and then multiply the quotient by 100 (see Equation 10-8).

$$\%\text{Area reduction} = \frac{(\text{Original area} - \text{Final area})}{\text{Original Area} \times (100)} \qquad \text{(Equation 10-8)}$$

$$= \frac{(A_o - A_f)}{A_o \times (100)}$$

FIGURE 10-26: The original and final gage lengths of a tensile test sample are used to calculate the material's percent elongation.

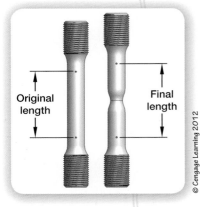

The **percent elongation** is a measure of the amount of stretching that occurs within the gage length of a tensile test sample (see Figure 10-26). The original gage length is marked by two points that are set a standard distance apart. The fractured sample is put back together and the distance between the gage marks is measured so that the percent elongation can be calculated.

To calculate the percent elongation, divide the difference between the final (L_f) and original (L_o) gage lengths of the test specimen by the original length and then multiply the quotient by 100 (see Equation 10-9).

$$\%\text{Elongation} = \frac{(\text{Final gage length} - \text{Original gage length})}{\text{Original gage length}} (100)$$

$$\text{(Equation 10-9)}$$

$$= \frac{(L_f - L_o)}{L_o} (100)$$

Example

Problem: A tensile test sample was made from an unknown aluminum alloy. The sample had an original test length (L_o) of 1 inch and an original diameter (d_o) of 0.125 inch. After the sample was tested in a tensile test machine to the point of fracture, its final diameter (d_f) at the point of fracture measured 0.070 inch. The resulting force-displacement diagram from the tensile test is shown in Figure 10-27. The material exhibited the following mechanical properties:

▶ modulus of elasticity (E) = 10,000,000 psi

▶ 0.2% yield stress (s_y) = 40 ksi

▶ ultimate stress (s_u) = 45 ksi.

FIGURE 10-27: Force-displacement diagram from an aluminum tensile test sample showing (a) the offset yield point, (b) the ultimate force, and (c) the fracture point.

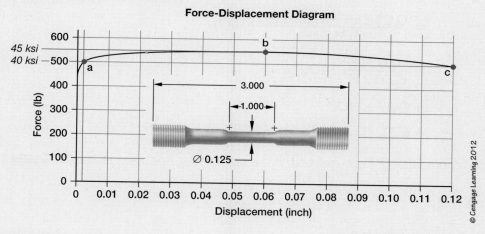

Solve for the following:

a. What was the original cross-sectional area (A_o) of the tensile test sample within the 1-inch test length?

$$A_o = \frac{\pi d_o^2}{4}$$

$$A_o = \frac{\pi \times (0.125\ \text{in})^2}{4}$$

$$A_o = \frac{\pi \times 0.0156\ \text{in}^2}{4}$$

$$A_o = \frac{0.049\ \text{in}^2}{4}$$

$$A_o = 0.012\ \text{in}^2$$

b. What was the final cross-sectional area (A_f) of the tensile test sample?

$$A_f = \frac{\pi d_f^2}{4}$$

(continued)

$$A_f = \frac{\pi \times (0.07 \text{ in})^2}{4}$$

$$A_f = \frac{\pi \times 0.0049 \text{ in}^2}{4}$$

$$A_f = \frac{0.0154 \text{ in}^2}{4}$$

$$A_f = 0.004 \text{ in}^2$$

c. What was the percent reduction in area of the tensile test sample?

$$\%\text{Area reduction} = \left| \frac{A_o - A_f}{A_o} \right| (100)$$

$$= \left| \frac{0.012 \text{ in}^2 - 0.004 \text{ in}^2}{0.012 \text{ in}^2} \right| (100)$$

$$= \left| \frac{0.08 \text{ in}^2}{0.012 \text{ in}^2} \right| (100)$$

$$= 0.67 \times 100$$

$$= 67\%$$

d. How much tensile force (F_y) was exerted on the tensile test specimen at the 0.2% yield stress point?

$$s_y = \frac{F_y}{A_0} \Rightarrow F_y = s_y A_o$$

$$F_y = 40,000 \text{ lb/in}^2 \times 0.012 \text{ in}^2$$

$$F_y = 480 \text{ lb}$$

e. How much tensile force (F_u) was exerted on the tensile test specimen at the ultimate stress point?

$$s_u = \frac{F_u}{A_0} \Rightarrow F_u = s_u A_o$$

$$F_u = 45,000 \text{ lb/in}^2 \times 0.012 \text{ in}^2$$

$$F_u = 540 \text{ lb}$$

f. What was the percent elongation of the tensile test sample at the point of fracture?

$$\%\text{Elongation} = \frac{(L_f - L_o)}{L_o} (100)$$

$$= \frac{(1.12 \text{ in} - 1.00 \text{ in})}{1.00 \text{ in}} (100)$$

$$= \left| \frac{0.12 \text{ in}}{1.00 \text{ in}} \right| (100)$$

$$= 0.12 \times 100$$

$$= 12\%$$

FIGURE 10-28: The two bolts in this structural connection are experiencing shear, which is the result of the force of the horizontal beam acting downward and the vertical beam remaining stationary. If the shear force that is transmitted through the cross-section of each bolt is too great, then the bolts will permanently deform or break.

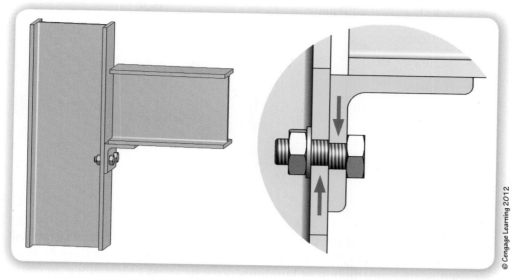

© Cengage Learning 2012

Shear Stress

Welds, bolts, and rivets are used to join structural components within machines, bridges, and the skeletal framework of buildings. These structural connections are subject to **shear force (V)**, which generates a slicing action through the material as a result of **shear stress (τ)**. Such connections must be designed to withstand shear stress without experiencing permanent deformation (see Figure 10-28).

Like normal stress, shear stress is a reaction that occurs within an object in resistance to an external load. It is a function of the amount of shear force per unit of shear area (A_s) (see Equation 10-10). The unit of measurement is the same as that of normal stress. In the case of a *single*-bolted connection, the force is transmitted across one plane of shear. Therefore, the shear area is equal to the cross-sectional area of the bolt where the plane of shear occurs.

$$\text{Shear stress} = \frac{\text{Shear force}}{\text{Shear area}} \qquad \text{(Equation 10-10)}$$

$$\tau = \frac{V}{A_s}$$

The bolted connection shown in Figure 10-29 will generate two planes of shear. Therefore, the shear area is equal to twice the cross-sectional area of one bolt. Metals are approximately half as strong in shear as they are in tension. This is why a large shear area is needed to ensure that a mechanical system does not fail where a bolted connection occurs. To accomplish this, an engineer will either increase the cross-sectional area of the bolt in the case of a single-bolted connection or increase the number of shear planes by adding more bolts. If the shear force remains constant and the shear area increases, then the average shear stress that occurs across each shear plane will decrease.

FIGURE 10-29: This bolted connection generates two planes of shear through the bolt.

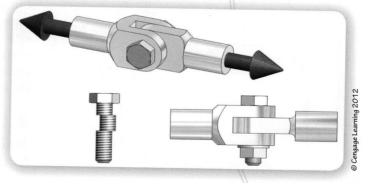

© Cengage Learning 2012

Mechanical properties:

behaviors that a material exhibits when subject to an external force, including ductility, elasticity, hardness, malleability, strength, and toughness.

MECHANICAL PROPERTIES OF MATERIALS

Engineers look for materials that exhibit certain *mechanical properties* when they design solutions to structural problems. Mechanical properties are attributes that all materials exhibit to some degree when subject to external forces. Such properties include, but are not limited to:

▶ brittleness,

▶ ductility,

▶ elasticity,

▶ hardness,

▶ malleability,

▶ strength, and

▶ toughness.

Strength

Strength is a general term that refers to a material's ability to resist applied loads. Materials such as stone and steel exhibit great strength, which is why they are used in the construction of buildings. Various stress points along an engineering stress–strain curve are often referred to with the term *strength*, such as *yield strength* and *ultimate strength*. However, this practice is not technically correct because the concept of strength is directly related to force, not stress. It is more appropriate to refer to points on a force-displacement diagram using the term *strength* and points on an engineering stress–strain diagram with the term *stress*.

Ductility

As mentioned in the previous chapter, ductility is a term that describes a material's ability to experience *plastic deformation* through tension without breaking. The most common example of this is drawing a material out into the shape of a long, thin wire. You learned in Chapter 9 that metals such as copper and brass are very ductile materials that are made into wires for use in electrical applications. Other metals such as low-carbon steel and aluminum are also fashioned into long wires for use in both electrical and structural applications. Elastomer materials, such as rubber, are not considered ductile because the large deformations they experience under load are not permanent.

Brittleness

Materials such as gray cast iron, concrete, and glass do not exhibit plastic behavior. Such materials are characterized as *brittle* because they experience fracture at, or shortly after, their yield point. Therefore, brittleness is a mechanical property that causes a material to fracture at relatively low strain values. Brittle materials like glass, concrete, and cast iron have stress–strain curves that show fracture occurring shortly after the yield point (see Figure 10-30).

Malleability

Malleability is another mechanical property that is similar in concept to ductility. A malleable material is able to

FIGURE 10-30: Brittle materials such as glass experience fracture close to their yield points and do not exhibit plastic flow, strain hardening, or necking. Ductile materials such as aluminum and low-carbon steel experience large amounts of permanent deformation before they fracture.

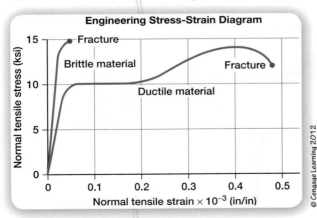

© Cengage Learning 2012

resist permanent change in shape under an applied compressive load without breaking. Therefore, **malleability** is a mechanical property that allows a material to experience *plastic deformation* through compression without breaking. The most common example of this is pounding or rolling a material into the shape of a thin sheet. Gold, copper, and aluminum are naturally malleable materials. Note, however, that not all materials that exhibit high degrees of malleability will be equally ductile. Lead, for example, is a malleable metal that is not ductile.

Elasticity

Materials such as rubber, neoprene, and silicone offer the most extreme examples of a mechanical property called *elasticity*. All materials experience some degree of deformation when subject to an applied load. If the load is small enough, then the material will return to its original shape once the load is removed. Therefore, **elasticity** is a mechanical property that allows a material to return to its original dimensions once its load has been removed. Think of it like this: Every time you drive over a bridge, the steel and concrete materials are deforming to some small degree as a result of tension, compression and shear forces. In other words, the bridge is changing shape. Once you have crossed the bridge, the bridge's components return to their original sizes as a result of the elastic nature of their materials.

A material that exhibits a large yield stress and a small modulus of elasticity will behave like a rubber band. In lieu of an engineering stress–strain diagram, a material's elastic quality is represented by a value known as the *modulus of resilience*. This value represents the area under a material's engineering stress–strain curve between the origin and the yield point (see Figure 10-31).

The **modulus of resilience** (U_R) is defined as a measure of the amount of strain energy per unit volume that a material will absorb without experiencing permanent deformation. The value of the modulus of resilience is usually expressed in psi or MPa, and its formula is given in Equation 10-11.

Modulus of resilience = ½ (Yield stress × Yield strain)

(Equation 10-11)

$$U_R = \tfrac{1}{2}\,(s_y\,e_y)$$

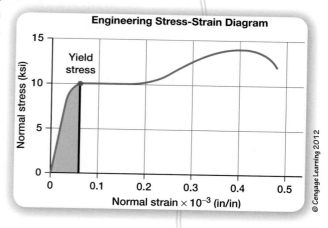

FIGURE 10-31: The area under an engineering stress–strain curve from the origin to the yield point is called the *modulus of resilience*. It gives the engineer an indication of the amount of energy that a material can absorb without experiencing permanent deformation.

© Cengage Learning 2012

Hardness

Hardness is a mechanical property that gives a material the ability to resist surface penetration, scratching, or wear. It is most associated with metals and ceramics, though it is also a property that is exploited in the design of some composites materials. Hard materials, such as glass, also tend to be brittle.

You may have heard that a diamond is a girl's best friend, but did you know that a diamond is the hardest known naturally occurring mineral? Diamonds are so hard that they can scratch almost all other materials, which is why they are used on the ends of drill bits. Diamond-tipped drills are often used for mineral exploration and core sampling (see Figure 10-32).

FIGURE 10-32: Oil drills are coated with diamonds, which allows them to cut through very hard materials.

© Cengage Learning 2012

Toughness

Toughness refers to a material's ability to absorb energy that results in permanent deformation. This material property can be thought of as a combination of both strength and ductility. The head of a hammer must be made from a tough material or it would shatter on impact with another hard material. Though it does not provide the accuracy of an impact test, which we discuss later in this chapter, the total area under a material's engineering stress–strain curve will give the engineer an indication of how a material will perform under impact conditions. This area is known as the **modulus of toughness** (U_T), which is defined as the amount of work that can be done on a material per unit volume without causing fracture (see Figure 10-33).

FIGURE 10-33: The modulus of toughness gives an indication of how tough a material is. The value of the modulus of toughness is equal to the area under a stress–strain curve from the origin to the fracture point. Brittle materials are not as tough as ductile materials.

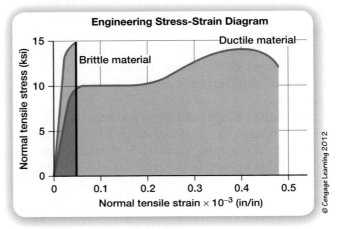

A material that exhibits a large modulus of toughness is more likely to withstand the force of a sudden impact without failure. The value of the modulus of toughness shares the same unit as the modulus of resilience—usually psi or MPa. Because of the infinite variety of shapes of engineering stress–strain curves, engineers use a mathematical approximation of the area under the curve. For *ductile* metals, the formula is given in Equation 10-12.

$$\text{Modulus of toughness} \approx \left[\frac{(\text{Yield stress} + \text{Ultimate stress})}{2}\right](\text{Fracture strain})$$

(Equation 10-12)

$$U_T \approx \left[\frac{(s_y + s_u)}{2}\right](e_f)$$

For *brittle* metals, the shape of the engineering stress–strain curve is assumed to be similar to one-half of a parabolic curve. As such, the formula used to calculate the modulus of toughness for a brittle metal is given in Equation 10-13.

$$\text{Modulus of toughness} \approx \tfrac{2}{3}(\text{Ultimate stress} \times \text{Fracture strain}) \quad \text{(Equation 10-13)}$$

$$U_T \approx \tfrac{2}{3}(s_u\, e_f)$$

Career Profile

CAREERS IN ENGINEERING

A Bridge to Tomorrow

On occasion, Mother Nature likes to wreak havoc on her handiwork, and that's when Jamie Padgett puts her hard hat on. Actually, that's when she puts her thinking cap on. An assistant professor in Rice University's Department of Civil and Environmental Engineering, Padgett went straight into academia out of graduate school, and she hasn't regretted a minute of it. "I just love doing research in my field," she says. And what this Florida native loves to research, more than anything else, is the effect of extreme events such as hurricanes and earthquakes on infrastructure, particularly bridges.

"I look at how bridges have performed in the past, and I try to predict how they'll perform in the future," says Padgett, who uses both computer modeling and laboratory simulations to test new materials and designs. "The goal is to identify and minimize the risk that a structure faces, everything from a hurricane to the usual wear and tear." That's a time-honored goal that reaches back to the roots of civil engineering in the ancient pyramids and beyond. Padgett thinks this is a particularly exciting time to be a civil engineer.

"There are a lot of innovative designs," she says, "and a lot of new materials with unique properties. Traditionally, bridges were made out of steel, concrete, masonry or wood. Today, we're adapting materials from the aerospace industry and biomedical research—so-called smart materials like shape-memory alloys. With sustained funding, we can continue to come up with creative solutions to these problems."

On the Job

As a college professor, Padgett teaches courses and advises students. She also conducts research through the Padgett Research Group, which provides her students with the opportunity to do meaningful, rewarding work toward their degrees. One of Padgett's most memorable experiences as an engineer, however, came on an outside assignment as a member of a special reconnaissance team.

"We were sent by the American Society of Civil Engineers to investigate all of the bridge damage from

Jamie Padgett, Civil and Environmental Engineering, Padgett Research Group

© Cengage Learning 2012

Hurricane Katrina," Padgett says. "We followed the path through Alabama, Mississippi, and Louisiana, visiting sites and speaking with local engineers and transportation officials. For me, it was like being hit in the gut, witnessing the kind of damage that a hurricane is capable of inflicting."

Inspirations

Having grown up in Florida, Padgett was familiar with hurricanes. In fact, Florida's frequent hurricanes helped motivate her to pursue an engineering career. "The idea of protecting people was already there," she says, "because we were living under the constant threat of disaster."

Another factor in Padgett's career choice was her family's general-contracting business. "It was everything from commercial office buildings to industrial warehouses and medical facilities," she says, "and it was always fascinating to see something go from a piece of paper all the way through the construction phase. On weekends, my father would take us to job sites."

Education

Padgett stayed close to home to do her undergraduate work, earning a BS at the University of Florida, where she undertook some of her first research projects. "I was a member of an undergraduate research team that would set up towers and collect wind data as a hurricane approached," she says. "We were hurricane chasers, essentially."

Padgett completed her education with a PhD from the Georgia Institute of Technology, focusing on structural and earthquake engineering.

Advice for Students

Having experienced both sides of the teacher–student divide, Padgett can't emphasize enough the need to study hard. "Do your best," she says. "Learn all you can, especially in science and math. Also pursue opportunities outside the classroom. Get a summer job as an intern, even if you're not asked to do any technical work. Just by hanging around these people, you'll absorb things."

TESTING MATERIALS

If you enjoy cooking, then you probably realize that a good meal is the product of fresh ingredients that are carefully combined according to specific recipes and made with great attention to factors such as time and temperature. Following the same recipe under the same conditions will almost always yield the same quality results. Such is the case with the manufacture of engineered materials. When a civil engineer designs a skyscraper out of structural steel, he or she places a great deal of trust in the steel manufacturer by assuming that each steel member is made to the same specifications and will exhibit the same material properties. Steel beams are expected to carry immense loads and exhibit predictable deformations that are accounted for in a complex architectural design (see Figure 10-34). These are only a few of the reasons why materials are tested and analyzed to ensure that they meet specifications and quality standards before they leave the manufacturer.

Engineers not only design parts using materials that have known properties but also are responsible for designing new materials that exhibit more desirable properties. For example, Japanese engineers developed the still popular aircraft aluminum alloy 7075 during the mid-1930s. The material found its first aeronautical application in the Imperial Japanese Navy's Zero fighter aircraft. In this case, the Japanese engineers took a known material that had already desirable characteristics such as light weight and modified it to have greater tensile strength.

Materials are subjected to various standardized tests for the purpose of gathering data. The data is then analyzed to verify that the material properties meet established standards. As you learned in Chapter 7, a **standard** is a reference that is developed by an authority or through general consent and used as a basis for comparison and verification. In the case of new materials, the tests are performed to identify the materials' properties and to establish standards that engineers can rely on. Once fashioned into a finished product, materials are often tested in order to guarantee that the product was made without flaws and according to specifications. It is also common for objects that have been in service for some time to be routinely tested to make sure that operation will continue unabated

FIGURE 10-34: All of the structural steel members that make up the skeletons of these skyscrapers must exhibit consistent and uniform material properties if they are to withstand the immense loads that they encounter.

or to determine if maintenance is necessary. In the United States, standardized material testing methods and procedures are formulated, defined, published, and updated by the American Society for Testing and Materials (ASTM).

To maintain a competitive edge in the marketplace, companies invest considerable amounts of money in the research and development (R&D) of new materials. R&D efforts are built on previous knowledge and often result in the generation of new knowledge. Engineers use these bodies of knowledge to develop new technological applications or to innovate existing technologies. For example, Thomas Edison's research and development of the incandescent lightbulb during the late 1870s involved the creation of heat-resistant carbon-fiber filaments made from bamboo. This technology was expanded on in the late 1950s when the Union Carbide Corporation developed high-strength carbon fibers from polymer filaments. Today, carbon-fiber–reinforced plastic materials are replacing traditional materials such as aluminum in high-profile engineering applications. For example, the new Boeing 787 Dreamliner passenger airplane is made from 50% composite materials by weight (mostly carbon-fiber–reinforced plastic) to reduce weight, maintain strength, and increase fuel efficiency.

Like a cook, a manufacturer uses specific recipes and procedures when creating materials. Manufacturers conduct various destructive tests to verify that the end results fit within the specifications and standards that are associated with a particular material. **Destructive testing** involves loading a test specimen to the point of structural failure while collecting data that indicates a material's mechanical properties. There are many different kinds of destructive tests, with the most common being those that measure deformation, hardness, impact strength, and fatigue characteristics. Test loads are classified as either static or dynamic. *Static* tests involve applying a load gradually up to the point of failure. This type of testing helps eliminate unwanted effects such as vibration. Tension, compression, shear, and torsion testing usually involves the application of static loads. *Dynamic* tests involve sudden and extreme loading or repeated loading over large numbers of cycles.

Destructive testing: a type of material test that involves loading a test specimen to the point of structural failure while collecting data that indicates the material's mechanical properties.

Tensile Testing

Woods, ceramics, and especially plastics and metals are placed through a common type of destructive stress test, called a *tensile test*, that measures a material's strength in tension. A *tensile test specimen*, sometimes referred to as a "dog bone," is fabricated according to standard geometries that are defined by the ASTM (see Figure 10-35). The specimen takes the form of a **prismatic bar**, which is a straight member of uniform cross-sectional area along its *gage length*. The **gage length** is often defined by two small marks that are placed a specific distance apart according to the ASTM standards. The distance between these

FIGURE 10-35: *Standard geometries for "dog bone" tensile test specimens with gage marks shown.*

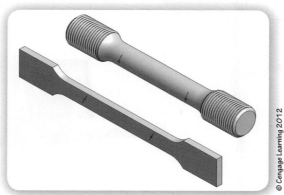

© Cengage Learning 2012

© Cengage Learning 2012

FIGURE 10-36: Flat tensile test samples are held on either end by a tapered collet system that grips the sample tighter as the tensile force increases.

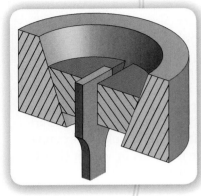

marks serves as the original length of the test sample. The gage marks are important for determining the amount of elongation that occurs up to the point of structural failure.

The ends of a tensile test specimen are enlarged so that they may be gripped by the tensile test machine. Their larger size also prevents fracture from occurring at those locations because higher stress concentrations will occur along the gage length. Failure tends to occur where an abrupt change in cross-sectional area occurs, so fillets are used to transition from the enlarged ends to the test area. Cylindrical test specimens may have threaded ends that allow a sample to be held through mechanical fastening. A test specimen that does not have threads is held in place using a special tapered collet that grips the sample tighter as the tensile force increases (see Figure 10-36).

The test sample is fitted with an *extensometer* and placed into a universal testing machine. This is a multipurpose machine that is capable of performing several different types of destructive stress tests. The universal testing machine subjects the specimen to an axial tensile load (see Figure 10-37). An extensometer may be a mechanical gauge but usually takes the form of an electromechanical sensor that continuously measures the amount of axial deformation that occurs during the test and feeds this information to a computer. Universal testing machines are also used to test samples in compression and flexure, which will be discussed later in this chapter. As the tensile test specimen is pulled apart, the computer continually samples the machine's load force and the test sample's elongation values, converts them to stress and strain values, and graphs the information in the form of a stress–strain curve. As discussed earlier, the stress–strain curve is then analyzed to determine the material's modulus of elasticity, yield stress, ultimate stress, fracture stress, and other mechanical properties.

Dynamic tensile tests are performed to determine how many times a material can be loaded before it will fail due to fatigue. Fatigue is the deterioration of a material as a result of repeated loading and unloading. Materials that experience dynamic loading will develop microscopic fractures at stress concentration points over time. These microfractures grow until the material exhibits total failure.

FIGURE 10-37: Tensile test specimen outfitted with (a) an extensometer and loaded into (b) a tensile test machine.

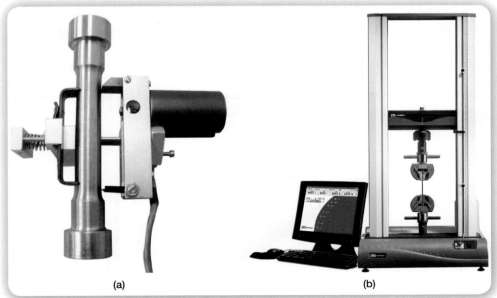

(a) (b)

Photos courtesy of Instron.

Point of Interest

On April 28, 1988, a Boeing 737 en route to Honolulu, Hawaii, experienced an engineering disaster when the top half of the fuselage over the 1st-class section tore away as the aircraft was flying 24,000 feet above the ocean (see Figure 10-38). The resulting explosive decompression resulted in the death of one flight attendant. The reason for the damage to Aloha Airlines Flight 243 was attributed to metal fatigue.

Because there is very low air pressure and dangerously low oxygen levels outside an airplane above 10,000 ft, an aircraft's interior is gradually pressurized to provide a safe and comfortable environment for the passengers and crew. This causes the plane's metal skin to stretch like the walls of a balloon as it fills with air. As the plane descends for a landing, the cabin is gradually depressurized. In the case of Aloha Airlines Flight 243, the plane had experienced more than 89,000 compression and decompression cycles during its 19-year service. Investigators determined that corrosive saltwater vapor from the ocean in combination with the extreme amount of dynamic loading on the aircraft's aluminum skin caused microfractures to form in the joints where the sheets of aluminum overlapped. Eventually, these microfractures expanded and joined, which led to total material fracture.

Such disasters are used as case studies to educate engineers and to develop standards and codes for the prevention of similar events in the future. Aircraft manufacturers now place limits on the number of compression and decompression cycles that their aircraft can safely experience, and airlines perform routine inspections that involve nondestructive testing methods designed to detect microfractures that cannot be seen with the naked eye.

FIGURE 10-38: Aloha Airlines Flight 243 just after landing, showing a large section of the fuselage that was ripped away because of metal fatigue.

Bruce Asato/Time and Life Pictures/Getty Images.

Tensile test specimens that are made from ductile metals such as aluminum and copper will experience significant reduction in cross-sectional area and exhibit a so-called cup-and-cone fracture pattern (see Figure 10-39). Brittle materials show little if any reduction in the cross-sectional area where fracture occurs, and exhibit a more flat and uniform fracture pattern.

OFF-ROAD EXPLORATION

Perform an Internet search using the key phrase "steel tensile test" to find video clips that show an actual steel test sample loaded to failure.

FIGURE 10-39: This tensile test sample was made from a ductile metal, and exhibited a classic (a) cup and (b) cone fracture pattern.

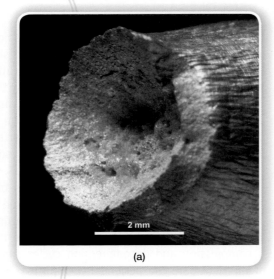

(a)

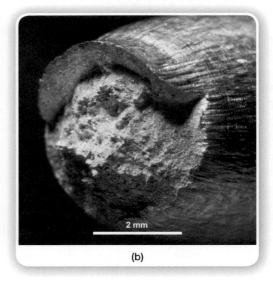

(b)

Compression Testing

Some materials can withstand much larger compression stresses than tensile stresses. One example is concrete, which is used extensively in the foundations of buildings and bridges. Compression tests are also conducted on a universal testing machine. Compression samples are usually made in the form of cylinders that have a 2:1 length-to-diameter ratio. For example, a standard concrete test sample is 6 inches in diameter, 12 inches long, and has cured for 28 days.

The compression sample may be outfitted with one or more extensometers to measure change in length or radial deformation (see Figure 10-40). As the material

FIGURE 10-40: (a) A compression test specimen, such as this concrete cylinder, is outfitted with one or more extensometers and placed in a universal test machine. (b) When failure occurs, the results can be explosive.

(a)

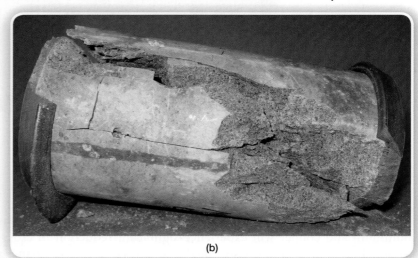

(b)

is loaded in compression, the change in force and deformation is converted and graphed on a stress–strain diagram. The shape of the curve is very similar to that of a tensile test, wherein a distinct linear region is evident up to a proportional limit. From brittle materials, like concrete, fracture occurs around the yield point. If you ever have the chance to watch a real concrete sample being tested to failure in compression, make sure you keep your distance. The material has a tendency to explode!

Though concrete is by far the most tested material in compression, this form of destructive testing is also performed on other types of materials. Cylindrical metal compression samples are generally made 3/8 inch diameter by 1 1/8 inch long. Ductile metals exhibit similar behavior in compression as they do in tension up to their yield points, wherein the relationship between stress and strain is linear. Beyond this point, the stress–strain curve takes a different shape. Friction restricts the amount of increase in girth that occurs where the top and bottom surfaces of the test specimen make contact with the testing machine. However, the middle section of the test sample will bulge outward, resulting in a barrel shape (see Figure 10-41). Once the material has flattened out significantly, it will resist further change in axial strain, which causes the stress strain curve to bend upward until failure occurs.

OFF-ROAD EXPLORATION

Perform an Internet search using the key phrase "compression testing of concrete" to find video clips that show an actual concrete test sample loaded to failure.

FIGURE 10-41: (a) A compression test specimen made from a ductile metal will become barrel shaped once it is loaded beyond the yield point. (b) The resulting stress–strain curve is noticeably different from a tension curve beyond the yield point.

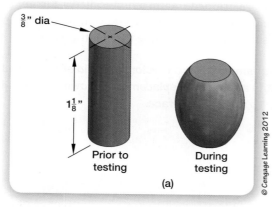

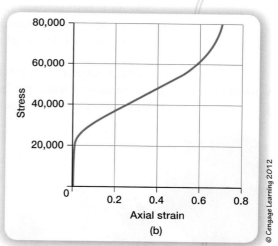

Shear Tests

Fasteners and other cylindrical components that are subject to direct shear forces are often tested in double-shear using specially designed fixtures on a universal testing machine. Test specimens are usually 3/16 inch diameter or larger. The accuracy of the test results from this type of shear test rely on the hardness of the material from which the test fixture is made, the fit between the test specimen and the holes in the fixture plates, and the sharpness of the edges of the holes where the shear planes occur. Sheet and thin plate materials are cut to form a slotted sheet specimen, and tested in single shear using the same setup as a flat tensile test specimen (see Figure 10-42).

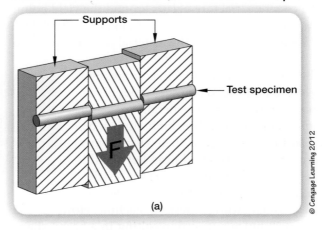

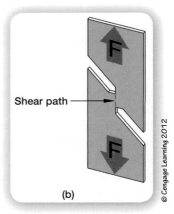

(a)

(b)

© Cengage Learning 2012

Torsion Testing

As mentioned earlier in this chapter, engineered objects such as drive shafts, fasteners, and twist drills must withstand torsional loads that generate shear stresses. The materials from which such objects are made are subjected to a type of destructive test called a *torsion test*. Torsion tests provide the engineer with information about a material's modulus of elasticity in shear, yield stress in shear, ultimate shear stress, fracture stress, and other important information. A torsion test sample may take the form of a solid or hollow cylinder of uniform diameter (d) and cross-sectional area. Both ends of the sample are gripped in a machine called a *torsion tester*. The sample is then placed in pure torsion by having one end of the sample remain stationary while the other end is subjected to a torque (see Figure 10-43).

FIGURE 10-43: (a) A torsion tester subjects a cylindrical test sample to pure torsion. One end of the sample is held in place while the other end experiences a torque. (b) The result is an angular displacement (shown in red) and an increasing shear stress from the center axis of the test sample to its outer surface.

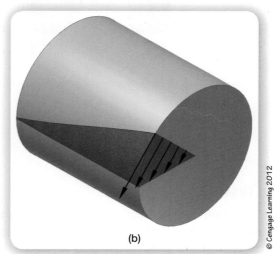

(a)

(b)

Photo courtesy of Instron.

© Cengage Learning 2012

The shear stress that is generated within the sample increases from zero along the central axis to a maximum value at the sample's surface. As the sample is tested to fracture, the torque (T, measured in inch-pounds) and angle values are recorded and plotted on a graph (see Figure 10-44). Using information about the test sample's original geometry, these values are then converted and plotted on an engineering shear stress–strain diagram.

The formula in Equation 10-14 is used to calculate the maximum shear stress on the outer surface of a solid cylindrical test sample at any point along the torque-angle curve.

$$\tau_{\text{max shear stress at the outer surface}} = \frac{16\,T}{\pi\,d^3}$$ (Equation 10-14)

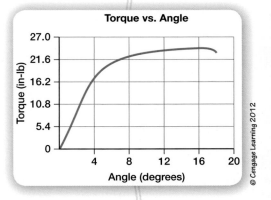

© Cengage Learning 2012

FIGURE 10-44: As the torsion test sample is twisted apart, information regarding torque and angular displacement is gathered and plotted on a graph.

Flexure Testing

Engineers realize that structural components such as the wood floor joists in a residential home or the I-beams within a large building will *deflect* under load. Deflection is a type of deformation that occurs in beams in which one side of the beam is placed in compression and the other side is placed in tension. If you have ever stood on the end of a diving board, then you have experienced the concept of deflection. Flexure tests are performed on materials to evaluate how well they withstand the complex types of stress that are generated during deflection and how they will experience failure. Beam deflection generates both tension and compression on opposite sides of a neutral plane, which is an imaginary plane that runs through the center of a structural member and along which no stress occurs (see Figure 10-45). A universal testing machine is often used to conduct flexure tests. Flexure test specimens are made in various shapes, depending on the application, but the length of the sample should be between 6 and 12 times its thickness or height in order to avoid buckling or direct shear.

FIGURE 10-45: A beam that is subjected to a point load at its midspan will experience deflection. The material above the neutral plane will experience compression, while the material below the neutral plane experiences tension.

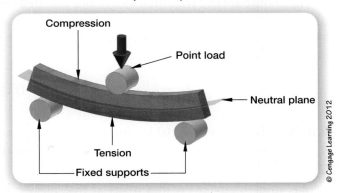

© Cengage Learning 2012

CENTROIDS AND AREA MOMENT OF INERTIA A beam's cross-sectional shape is very important because some shapes provide greater stability than others. The center of a shape is called a centroid. For a simple rectangular beam, the centroid is located one-half of the base distance (b) and one-half of the height distance (h) as measured from a common origin point. A centroidal axis is an imaginary line that passes through the centroid and is appropriately aligned with one of the three principle Cartesian coordinate axes (X, Y, or Z), depending on the orientation of the structural member in three-dimensional space. The centroidal axis is therefore labeled X–X, Y–Y, or Z–Z (see Figure 10-46). As the cross-sectional shape is drawn out into the three-dimensional beam, the centroidal axis is also drawn out to form the beam's neutral plane.

The idea that one shape may be stiffer in bending than another can be a bit confusing. The way that engineers can determine this is through calculating a shape's

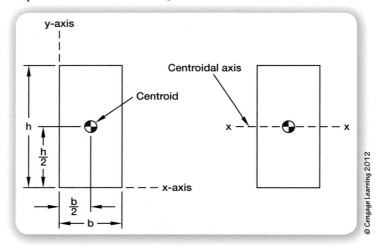

FIGURE 10-46: The location of a structural shape's centroid dictates the location of the neutral plane, which follows along a centroidal axis.

area moment of inertia. **Area moment of inertia** (symbol I_{XX}), also referred to as the *second moment of an area*, is a geometrical property of a structural member's cross-sectional shape, which is measured in inches raised to the power of 4 (in^4). This property reflects how the area of a shape is distributed in relation to its centroidal axis. When compared to the value of other structural shapes, the area moment of inertia will indicate which shape has the highest capacity to resist rotating in the plane of its centroidal axis (such as X–X axis). If the majority of the shape's area is distributed a large distance away from the centroidal axis, then the shape will provide greater stiffness. If the shape's area is concentrated around the centroidal axis, it will provide less resistance to rotation in the plane of that axis. This is why the wood floor joists in a house are laid with their longer side oriented vertically rather than laid flat like a diving board. The formula in Equation 10-15 is used to calculate the area moment of inertia (I_{XX}) for a simple square- or rectangular-shaped beam given its base (b) and height (h) dimensions.

$$I_{XX} = \frac{bh^3}{12}$$ (Equation 10-15)

CALCULATING DEFLECTION Several variables must be known in order to calculate the maximum deflection of a beam that experiences a point load at its midspan. These variables include the following:

▶ the modulus of elasticity (E) of the material that makes up the flexural sample,

▶ the span of the unsupported section of the flexural sample (L) in inches

▶ the magnitude of the point load (F) in pounds, and

▶ the area moment of inertia (I_{XX}) of the sample's cross-sectional shape in inches⁴.

The maximum deflection formula shown in Equation 10-16 can be algebraically manipulated to calculate the modulus of elasticity for a material such as wood, which naturally exhibits large variation because of the different circumstances under which trees grow and the different ways in which lumber is cut from a tree trunk.

$$\Delta_{max} = \frac{FL^3}{48\, EI_{XX}}$$ (Equation 10-16)

Equation 10-17 is used to calculate the flexural stress that is generated in the test specimen, which is measured in lb/in².

$$s_{\text{flexure}} = \frac{3FL}{2bh^2} \qquad \text{(Equation 10-17)}$$

Destructive testing of materials relies heavily on mathematical principles. For example, destructive test samples are made according to standard geometric dimensions so that accurate comparisons can be made between different types of materials. Data related to measurable attributes is collected using standard measurement techniques. Common units are associated with these attributes, and they may be converted from one measurement system to another. Standard methods of communication, such as the engineering stress–strain diagram, are used to present the data that is gathered through destructive testing of materials. Over time, algebraic relationships between material attributes have been formulated that allow engineers to generate models that predict how materials will behave under working conditions.

Hardness Testing

Gears, bearings, and cutting tools used in manufacturing are only a few examples of objects that must be made from very hard materials. To verify that a material has the proper hardness, it may be placed through one of several types of hardness tests. Most of these tests involve indenting the material using a specially shaped tool under a given static load and measuring the resulting indent geometry. Three of the most common hardness tests are the Brinell test, the Rockwell test, and the Vickers test. The purpose of this text is to provide an overview rather than a detailed analysis of the process of hardness testing, so we only focus here on the Brinell test. A Brinell hardness test involves using a special machine to indent a smooth surface on a test specimen (see Figure 10-47).

FIGURE 10-47: (a) A Brinell hardness tester is used to push a hardened steel or tungsten sphere into a test sample. (b) The diameter of the resulting indentation is used to calculate the material's hardness.

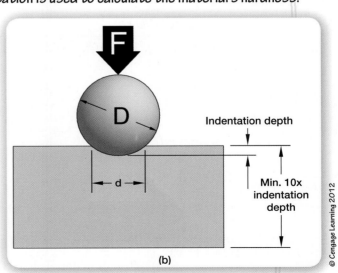

Photo courtesy of AMETEK, Inc.

(a)

(b)

© Cengage Learning 2012

As Figure 10-47 indicates, the most common indent tool used on a Brinell hardness tester is a 10-mm diameter sphere (D) made from either hardened steel or tungsten carbide. There is no standard size for a test sample, though a general rule is to use a thickness that is 10 times the indentation depth. Test samples that are too thin will give inaccurate test results. The machine places a static compressive force (F) on the sphere once it has made contact with the test sample. This force is equal to 3,000 kg (6,614 lb) for hard metals, 1,500 kg (3,307 lb) for softer metals, and 500 kg (1,102 lb) for soft materials. The sphere remains fully loaded for a period of 30 seconds for ferrous materials and 60 seconds for softer metals. When finished, the ball is removed and the diameter of the resulting indent (d) is optically measured. This information is then used to calculate the Brinell hardness number (BHN) using the formula shown in Equation 10-18. Though the proper unit of measurement for a Brinell hardness value is kg/mm^2, it is standard practice to identify the number as a unitless value, followed by initials that identify the type of hardness test that was used (HBW or HBS).

$$BHN = \frac{2F}{\pi D (D - \sqrt{D^2 - d^2})} \qquad \text{(Equation 10-18)}$$

Because there may be variation in the size and material makeup of the indent tool, as well as the magnitude of the compressive load, it is good practice to follow the Brinell hardness number with information that identifies the conditions under which the number was derived. For example, 95 HBW 10/500 identifies a Brinell hardness value of 95 kg/mm^2, which resulted from a 10-mm diameter tungsten (chemical symbol W) carbide indent sphere under 500 kg of force. If a hardened steel sphere is used, then the initials HBS would follow the hardness value.

Example

Problem: A Brinell hardness test was performed on a piece of 6061 alloy aluminum using a 10-mm diameter tungsten carbide indent tool applied with 500 kg of force for 60 seconds. The diameter of the indent that remained was 2.57 mm. What is the Brinell hardness number for this material to the nearest integer?

$$BHN = \frac{2F}{\pi D (D - \sqrt{D^2 - d^2})}$$

$$BHN = \frac{2 \times 500 \, kg}{(\pi \times 10 \, mm) \times 10 \, mm - \sqrt{(10 \, mm)^2 - (2.57 \, mm)^2}}$$

$$BHN = \frac{1,000 \, kg}{31.4 \, mm \, (10 \, mm - \sqrt{93.395 \, mm^2})}$$

$$BHN = \frac{1,000 \, kg}{31.4 \, mm \times 0.336 \, mm}$$

$$BHN = \frac{1,000 \, kg}{10.55 \, mm^2} = 95 \, HBW \, 10/500$$

OFF-ROAD EXPLORATION

To learn more about other methods of testing a material's hardness, perform an Internet search using the key phrase "material hardness tests."

Toughness and Impact Strength Testing

Tough materials have the ability to absorb large amounts of energy before they fracture. This material property is a very important consideration in the design of guardrails, safety shields, body armor, and other equipment that is made to protect people from sudden impacts. It is also a consideration in the design of objects that must be pressed into their

(continued)

26. What is the difference between normal stress and shear stress?
27. Identify two situations that result in direct shear of a mechanical component.
28. What is a mechanical property?
29. Identify seven types of mechanical properties that all materials exhibit to some degree.
30. What is the difference between strength and stress?
31. What is the difference between a ductile and a malleable material?
32. How do brittle materials differ from ductile materials?
33. What constitutes a tough material?
34. What is the difference between a material's modulus of resilience and its modulus of toughness?
35. Give an example of application of hard materials.
36. What is a standard?
37. Who is responsible for the development of material testing standards in the United States?

38. What kind of machine is used to perform destructive tests of materials in tension or compression?
39. What is an extensometer?
40. What is the difference between a static and dynamic tensile test?
41. What is fatigue?
42. What kind of fracture pattern do ductile metals exhibit?
43. How are materials tested in shear?
44. Where does the maximum amount of shear stress occur for an object that experiences a torsional load?
45. What is a neutral plane?
46. What is a centroid?
47. What is area moment of inertia, and what is it used for?
48. What is the process for conducting a Brinell hardness test?
49. Why would an engineer be interested in a material's test results from a Charpy impact test?

EXTRA MILE

Problem Statement: To reduce the overall weight of its products, an ultralight aircraft design company is considering replacing the majority of the steel fasteners that are used on its aircraft with plastic fasteners. To determine whether this is even possible, the engineers need to analyze data related to the mechanical properties of various plastic materials under direct shear.

Design Statement: Design, build, and test a device that will hold a 0.25-inch-diameter plastic rod during a destructive test for the purpose of gathering data on the material's shear strength. The device must be made from aluminum and must interface with a commercially available material testing system such as the Structural Stress Analyzer 1000.

CHAPTER 11
Manufacturing Processes and Product Life Cycle

GPS DELUXE

Menu	START LOCATION	DISTANCE	END LOCATION

Before You Begin

Think about these questions as you study the concepts in this chapter:

1. How are raw materials converted into finished products?

2. How does metal casting work?

3. What are the methods for forming metal?

4. How does metal forging work?

5. How are extruded metal parts manufactured?

6. What manufacturing processes are classified as cold forming?

7. How are plastics processed?

8. What manufacturing processes are ideal for producing holes?

9. What is meant by a product's *life cycle*?

①

Think about the products you use every day. Some of these items might be made of only one material, whereas others might be made using a combination of materials. We use some materials in the same form that they came out of the ground such as sand, stone, or coal. Other materials such as aluminum, plastic, and steel are made from a combination of raw materials that require modification before they can be used. Materials processing is the procedure of converting raw materials into finished products.

Even those materials found and used in their natural form usually need some degree of *primary processing*. For example, stone needs to be cut or crushed, and wood needs to be cut into dimensional lumber and dried before being shipped to the customer. Other materials such as plastics require multiple stages of primary processing. Plastics are manufactured using by-products from the oil industry that are further refined to get them ready for use. The procedure of converting primary materials into finished products is called *secondary processing*. An example of secondary processing is the conversion of plastic pellets into the body of a cell phone.

Combining your knowledge of materials with an understanding of materials processing will allow you to see how engineers choose materials that have outstanding properties and can be processed into useful forms in an economical way. For example, a bulk roll of steel would not be useful to the average person. However, if that rolled steel was run through a punch press and turned into forks, spoons, and knives, it now has become products that the average consumer would purchase.

> **Materials processing:** the procedure of converting raw materials into finished products.

PROCESSING METALS

Metal Casting

②

Casting is one of the oldest manufacturing processes used to inexpensively produce intricate parts. The casting process allows parts to be formed to near final shape that requires only minor finishing or machining. The procedure for casting metals typically involves filling a mold with a molten material, waiting for it to solidify, and then removing the part from the mold. Metals and polymers are most commonly used for casting, but some oxide glasses can also be cast using the same process. A **foundry** is a workshop in which metal castings are produced.

> **Casting:** a procedure for producing metal parts by filling a mold with a molten material, waiting for it to solidify, and then removing the part from the mold.

SAND CASTING *Sand casting* is one of the oldest and most common methods of casting metals into useful forms (see Figure 11-1). This procedure can be used to produce parts as small as a pair of earrings to objects that weigh thousands of pounds. The sand-casting process involves pouring molten metal inside a hollow two-part mold made of sand. The molten metal takes the shape of the cavity inside the mold and rapidly solidifies as it comes in contact with the sand. Because the molten metal cools against the sand, all sand-cast parts have a rough texture when removed from their molds. Additional machining must be done if a smooth finish is required.

The making of a sand-casting mold begins with the creation of a split mold pattern of a desired geometry. One half of the pattern is called the *drag pattern* and

FIGURE 11-1: This axle bracket for logging industry equipment was produced using the sand-casting process.

Courtesy of Kubota Metal Corporation.

FIGURE 11-2: A split mold pattern is used in the sand casting process.
Courtesy of CustomPartNet, Inc.

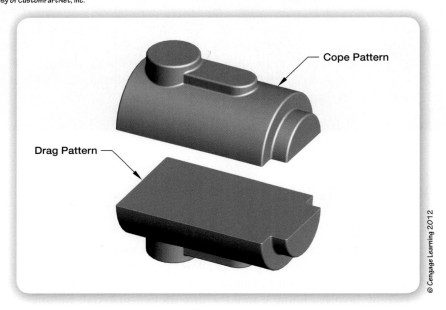

Cope Pattern

Drag Pattern

© Cengage Learning 2012

the other half is called the *cope pattern* (See Figure 11-2). The sand mold is packed within a two-piece container called a *flask*. The bottom half of the flask is called the *drag*, and the top half is called the *cope*. The process of making a sand-casting mold involves the following:

1. The bottom half of the flask (the drag) is placed upside down on a flat surface with the drag pattern positioned flat side down and centered in the flask.

2. A parting compound is sprinkled over the table and the pattern to prevent the sand from sticking. Sand is riddled (sifted) over the pattern and then compacted. The drag is then filled, compacted, and leveled off.

3. The drag is then flipped over, and the top half of the flask (the cope) is placed on top of the drag. The cope pattern is placed over the drag pattern, which was embedded in the sand in step 2. The two halves of the pattern have locator pins that allow the halves to perfectly align.

4. Parting compound is sprinkled over the cope pattern and sand in the drag to prevent the sand from sticking when the two halves are separated. Sand is riddled over the cope pattern and then compacted. The cope is then filled, compacted, and leveled off.

5. Before the halves are separated, two holes are cut into the top of the cope. The *sprue* is the hole where the molten metal will be poured, and the *riser* is the hole that allows the molten metal to come back to the surface.

6. The flask is separated, and the pattern halves are removed. Channels are cut into the halves connecting the sprue and the riser to the mold cavity. If the part has any internal features, such as a large hole, a core may be added.

7. The mold is closed, and the molten metal is poured into the sprue, filling the mold cavity (see Figure 11-3). The metal is allowed to cool and solidify.

8. The only way to get the part out of the mold is to destroy the mold.

FIGURE 11-3: *Sand-casting mold: closed and filled.*

Courtesy of CustomPartNet, Inc.

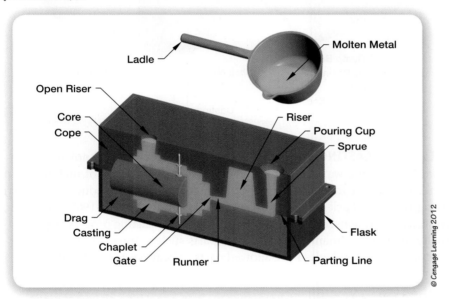

© Cengage Learning 2012

SHELL-MOLDING PROCESS The *shell-molding process*, also known as the *croning process*, was developed during World War II by Johannes Croning, a German inventor. Shell molding is very similar to sand casting, except the formation of the mold is different. The process requires two permanent patterns, one for the cope and one for the drag to be manufactured. The sprue, riser, and channels are also included in the pattern.

1. The pattern is heated to 400°F to 500°F and clamped upside down in a dump box (see Figure 11-4). The dump box is filled with a sand–resin mixture and flipped upside down. When the sand comes in contact with the pattern, the resin melts and bonds the grains of sand together, forming a thin shell of hardened sand over the pattern.

2. The box is flipped upright again, and the pattern and shell are removed together. This process must be repeated again to form the cope and the drag. The patterns with shells still attached are placed in an oven to finish the curing process.

3. The molds are removed from the oven, and the shell is separated from the pattern. The final step before pouring is to bond the two halves of the mold (cope and drag) together.

4. The mold is placed inside a flask and filled with a backing material to prevent the shell mold from cracking. Molten metal is poured into the mold and allowed to cool and solidify.
5. The sand–resin shell is removed to expose the finished casting.

FIGURE 11-4: Shell-molding process.

Courtesy of CustomPartNet, Inc.

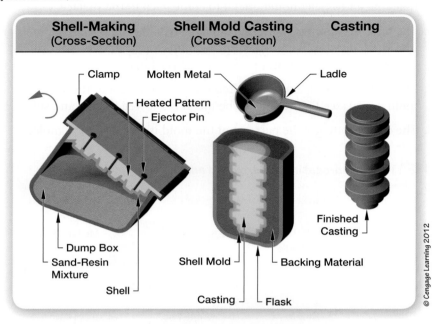

INVESTMENT CASTING Investment casting is commonly known as *lost wax casting* because the wax pattern is melted away in the process of making the mold. Archeologists have found jewelry and ornaments that were produced by investment casting that date back thousands of years. Not until the mid-1900s, however, was investment casting used by industry. The investment-casting process (Figure 11-5) is usually reserved for manufacturing small parts such as those shown in Figure 11-6; however, new technologies have allowed parts as heavy as 600 lb to be cast.

Investment casting:

a manufacturing process commonly known as *lost wax casting* because the wax pattern is melted away in the process of making the mold.

FIGURE 11-5: A foundry worker pours molten steel into an investment casting mold.

FIGURE 11-6: These parts were produced using the investment-casting process.

© iStockphoto.com/Chuck Rausin.

FIGURE 11-7: Investment-casting tree.

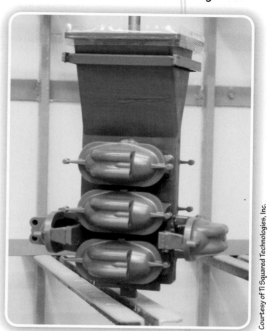

Courtesy of TI Squared Technologies, Inc.

Investment casting uses the following process:

1. An injection-molding machine is used to produce a wax replica of the part to be cast. The wax parts are fastened to a "tree" that allows multiple parts to be cast in one pour (see Figure 11-7). The center of the tree serves as the sprue in which the molten metal is poured.

2. The tree is dipped into a container filled with ceramic slurry to evenly coat all surfaces. The tree is removed from the slurry, and a layer of fine refractory material called *stucco* is sprinkled onto it. This process is repeated until a thick shell builds up around the wax tree.

3. The wax is removed by placing the entire tree upside down into a steam autoclave (a fully enclosed pressurized, heated vessel). The majority of the wax melts away, leaving the ceramic shell behind. It is this step that gives the process the name *lost wax casting*. The tree is placed in a kiln and fired to burn away any remaining wax.

4. The heated tree is removed from the kiln, and molten metal is poured into the top, producing an exact replica of the tree.

5. The shell is removed from the tree, and the individual parts are cut away.

6. The attachment points (where the parts attached to the tree) are ground away, and the entire part is sandblasted to produce a finished part (see Figure 11-8).

LOST-FOAM CASTING *Lost-foam casting* is almost identical to lost wax casting except that the pattern is made of foam instead of wax and the foam stays inside the shell while being cast. The molten metal is poured directly over the foam, causing it to vaporize.

CENTRIFUGAL CASTING Also known as *rotocasting*, **centrifugal casting** is a process used to produce parts that are smooth and clean, such as water pipes, brake drums, and many other symmetrical parts (see Figure 11-9). The process uses centrifugal force to hold the molten metal against the inside of a rotating mold. The molds are typically made of steel, graphite, or cast iron. Centrifugal casting is done on either horizontal or vertical machines.

> **Centrifugal casting:**
> a manufacturing process that uses centrifugal force to hold the molten metal against the inside of a rotating mold.

FIGURE 11-8: Investment casting.

Courtesy of CustomPartNet, Inc.

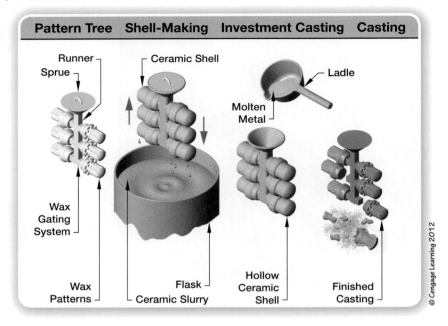

FIGURE 11-9: Water pipes produced by centrifugal casting.

Centrifugal casting uses the following process:

1. As shown in Figure 11-10, the mold is placed inside a centrifugal machine that consists of rollers that support the mold and a motor that spins it.
2. The inside of the mold is prepared by applying a thin ceramic shell. The mold is spun while being heated to harden the ceramic shell.
3. Once the coating is complete, the mold is spun at a high rate of speed (300–3,000 RPM) while the molten metal is poured into one end of the mold. The thickness of the part can be controlled by the amount of molten metal added.
4. When the pour is complete, water is sprayed on the outside of the mold to speed the cooling process.
5. After the part is completely cooled, it can be removed from the mold.
6. The centrifugal force leaves most of the impurities in the molten metal on the inside of the part. Before the part can be shipped to the consumer, the inside must be machined or sandblasted to remove these impurities.

FIGURE 11-10: Centrifugal-casting process.
Courtesy of CustomPartNet, Inc.

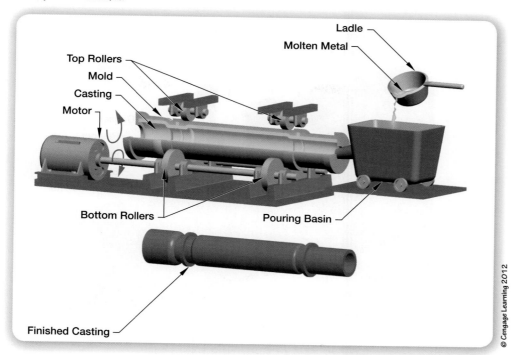

DIE CASTING Die casting is very similar to the injection-molding process used with plastics. Molten metal is injected directly into a metal die (mold), rapidly cooled, and quickly ejected from the die. A typical die-casting machine can produce upward of 800 pieces per hour. Die-cast parts are usually made of aluminum, zinc, brass, tin, lead, magnesium and other nonferrous alloys (see Figure 11-11).

FIGURE 11-11: Die-cast model airplane.

Image copyright Morgan Lane Photography, 2010. Used under license from Shutterstock.com.

Die casting:

a manufacturing process similar to injection molding in which molten metal is injected directly into a metal die, rapidly cooled, and quickly ejected from the die.

Die:

a mold used in manufacturing processes.

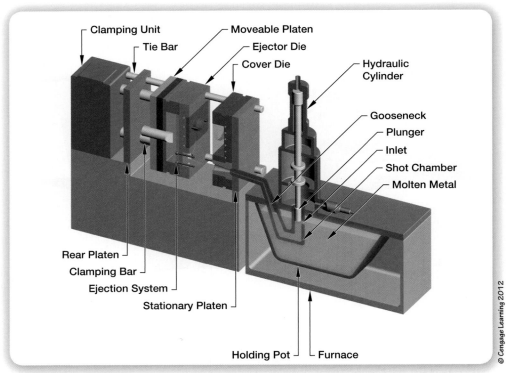

The machines used for die casting are divided into two categories: hot-chamber machines and cold-chamber machines. Hot-chamber machines are used with low-melting-temperature alloys. Cold-chamber machines are used with high-melting-temperature alloys. The process of creating a part is nearly identical on both types of machines:

Die casting uses the following process.

1. A two-part die consisting of a movable ejector die and a stationary cover die are first cleaned and prepped using a lubricant to prevent the parts from sticking. The cleaning and prepping process does not need to be done on every cycle; it is commonly done every two to three cycles. The two dies are then firmly clamped together using a hydraulic press. Because the molten metal is injected into the mold at a very high pressure, the clamping force on the die needs to be sufficient to prevent the metal from escaping (see Figure 11-12).

2. The molten metal in the holding pot is heated by a furnace. The metal flows through the inlet in the gooseneck, where a hydraulic cylinder pushes the plunger down to force the molten metal farther up the gooseneck and into the die (see Figure 11-13). The amount of molten metal that is injected into the die is referred to as the *shot*.

3. The molten metal will solidify almost immediately after it comes into contact with the die. The dies have a network of water-cooling channels running through them to speed the cooling process.

4. Once the parts have solidified, the die halves are opened and the parts are ejected using an ejection system. Once the parts have come out, the die is quickly clamped back together and the process cycles again.

5. Die-cast parts usually have some *flashing* on them. Flashing is excess material that must be trimmed from the casting. Low-volume die-casting operations will trim the flashing from the parts by hand. High-volume die-casting operations use a trimming press that automatically trims the flash from the part. The waste material can be reprocessed and reused in a future die cast.

FIGURE 11-13: Hot-chamber die-casting machine: die closed.

Courtesy of CustomPartNet, Inc.

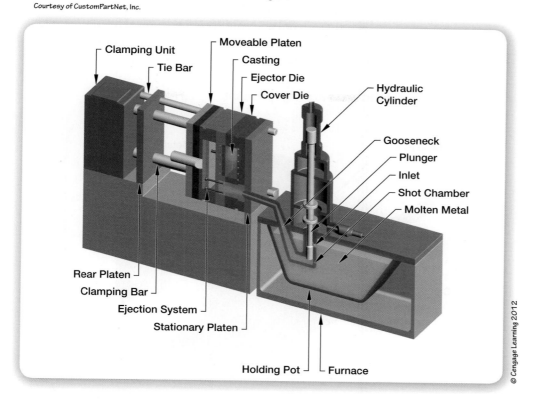

Clamping Unit
Tie Bar
Moveable Platen
Casting
Ejector Die
Cover Die
Hydraulic Cylinder
Gooseneck
Plunger
Inlet
Shot Chamber
Molten Metal
Rear Platen
Clamping Bar
Ejection System
Stationary Platen
Holding Pot
Furnace

© Cengage Learning 2012

FIGURE 11-14: Several of the metal components that make up this motorcycle engine were made using the permanent mold casting (gravity die casting) method.

© iStockphoto.com/Zoe Yau.

PERMANENT MOLD CASTING Permanent mold casting is very similar to both die casting and sand casting. Figure 11-14 shows motorcycle engine parts that were manufactured using this technique. This process uses the permanent die halves similar to those used in a die-casting machine. As shown in Figure 11-15, the molten metal is poured into the top of the mold, similar to the way a sand casting is done. For this reason, permanent mold casting is often referred to as *gravity die casting*.

Permanent mold casting uses the following process:

1. The mold is prepped by preheating it to 300°F to 500°F to allow the metal to flow through the mold easier. A ceramic coating is evenly applied to all surfaces of the mold, making it easier to remove the part as well as extending the life of the mold.
2. If the mold has internal features, a core may be used. Sometimes these cores are made of steel and sometimes of sand.

3. The two halves of the mold are clamped together, and the molten metal is poured through the sprue in the top of the mold.
4. After the part has had time to cool and solidify, it is removed from the mold. Any excess metal from the sprue and runners is cut away to produce a finished casting.

FIGURE 11-15: *Permanent-mold casting operation.*

Courtesy of CustomPartNet, Inc.

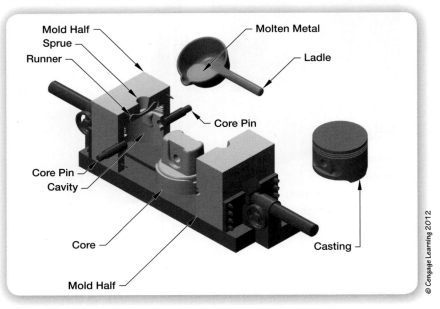

© Cengage Learning 2012

Metal Forming

Forming encompasses many different manufacturing processes in which a material is permanently deformed. Forming metals involves permanently changing the shape of the material by applying pressure. As you learned in the previous chapter, any material that experiences a permanent change in shape is said to have been *plastically deformed*. Ductile and malleable materials that are capable of withstanding a large amount of plastic deformation before fracturing are ideal for forming. Forming techniques are divided into two categories: hot forming and cold forming. *Hot forming* involves heating a metal to its *plastic* state and applying tremendous pressure by using devices that cause the metal to take on a new shape. *Cold forming* is usually done at room temperature or when the metal is slightly warmed.

Forming:

a manufacturing process that involves permanently changing the shape of the material by applying pressure.

Your Turn

Pick five items from your kitchen and determine what manufacturing process or processes were used to create those products.

HOT FORMING METAL Hot forming metal is always done above the metal's **recrystallization temperature** so that the metal's strength decreases and it is more easily formed. It is not uncommon for the part being worked to be reheated several times during the process to maintain the metal's temperature. Iron-based alloys are usually formed at temperatures ranging from 1,700°F to 2,500°F. Nonferrous metals can be hot worked at lower temperatures.

METAL FORGING Like casting, forging techniques have been used for centuries to make jewelry, coins, horseshoes, and other commonly used items. Using only a hammer, an anvil, and a furnace, blacksmiths over time have created a wide variety of products (see Figure 11-16). Today, forging is widely used to make tools, aircraft components, and other parts where strength is a concern. Today's forging machines have come a long way from the days of the blacksmiths. Most machines today are computer controlled and fully automated.

Forging is a form of metal forming that involves shaping a metal part through controlled plastic deformation. If the forging is done when the metal is cool, it is called *cold forging*. If the metal is hot when being forged, then it is called *hot forging*. Most forging operations are done in a multistep process that allows the material to be gradually worked into the final shape. Forging techniques can be further broken down into three categories: draw forging, upset forging, and pierce forging. *Draw forging* tends to flatten parts through compression [see Figure 11-17(a)]. *Upset forging* produces parts by decreasing the length of the stock and increasing its cross-section using compression [see Figure 11-17(b)]. *Pierce forging* creates a hollow part by forcing a peg on the upper die into the part [see Figure 11-17(c)].

Forging:

a form of metal forming that shapes a metal part through controlled plastic deformation.

FIGURE 11-17: Three types of forging: (a) drawing, (b) upset, and (c) pierce.

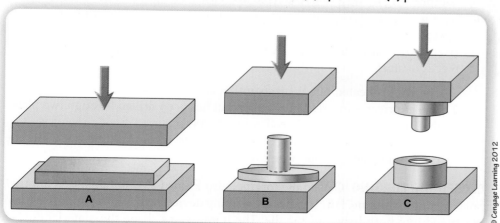

© Cengage Learning 2012

FIGURE 11-18: These three curved rods are all made of the same type of metal but were processed using different methods: (a) machined from a block of metal, (b) cast out of molten metal, and (c) hot forged.

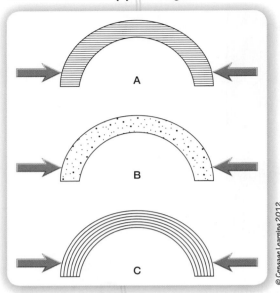

The grain of the metal can be controlled during the forging process. Figure 11-18 shows three curved rods made of the same type of metal but processed using different methods.

▶ The rod in Figure 11-18(a) was cut from a solid block of metal using various machining operations. Note that the grain structure is going in the same direction throughout the rod. When the forces are applied as shown, the rod will fracture along one of the grain lines.

▶ The rod in Figure 11-18(b) was made by sand casting. Cast parts have individual grains, also known as *crystals*, that are formed during the casting process. Controlling the grain direction during casting is almost impossible. When the forces are applied as shown, the casting will usually fracture along imperfections within the casting.

▶ The rod in Figure 11-8(c) was made using the forging process. The grain direction can be controlled during the forging process. The grain direction will form in the same direction as heat flow. Through proper design and innovative forging techniques, the direction of the grain can be controlled. When the forces are applied as shown, the forged part will bend along the grain lines, much like a paper clip is able to be bent without immediately failing.

FIGURE 11-19: This 2,000-ton forge press is computer controlled and can manipulate parts weighing as much as 20 tons.

OPEN DIE FORGING Open die forging (Figure 11-19), also known as *hammer forging,* involves squeezing or impacting the metal between flat or rounded dies until it takes on the desired shape. The open die forging process used today is very similar to the technique blacksmiths used years ago. The part being forged is never completely enclosed at any point during the process. Open die forging is a hot forging operation and is commonly used to produce large and simply shaped parts. Recent advancements in the forging industry have allowed greater flexibility in the types of parts that can be manufactured using this process. Open die forged parts have high strength and uniform characteristics throughout, and they are very durable. Parts can range in weight from a few pounds to several thousand pounds. The open die forging process is usually reserved for low production runs or custom forgings. It is commonly used as a primary processing technique to prepare the material for secondary processing.

UPSET FORGING Upset forging, also known as *machine forging,* is widely used in manufacturing today. As noted earlier, upset forging involves compressing metal along its axis to decrease its length and increase its cross-section. In upset forging, the entire part or just a section of the part can be upset. The process of upset forging can be seen in the installation of a hot rivet shown in Figure 11-20. A rivet is a fastener used to join two or more pieces of steel.

1. The rivet is heated in a furnace until it is glowing red.
2. The hot rivet is inserted into a hole that has been drilled through two plates that are to be connected [see Figure 11-20(a)].
3. The rivet is backed up by an anvil to prevent it from sliding out while the rivet gun upsets the opposite end of the rivet.

CLOSED DIE FORGING Closed die forging, also known as *drop forging,* is another hot-forming technique. Forging dies are shaping devices with impressions (cavities) machined into both halves of the die. The impressions shape the material during

FIGURE 11-20

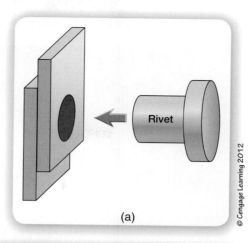

(a)

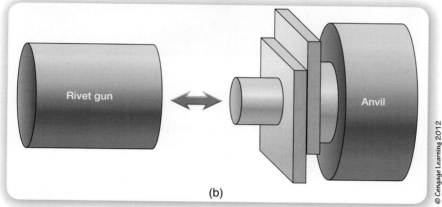

(b)

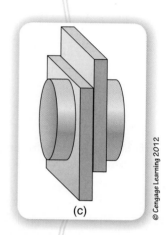

(c)

FIGURE 11-21: *Carbon-steel billet.*

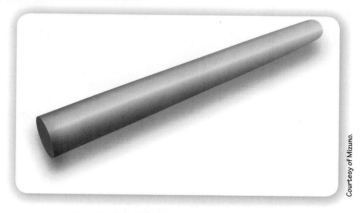

the forging process. The process of closed die forging can be seen in the making of a golf club head.

1. A carbon-steel billet (see Figure 11-21) is prepared by stretching and squeezing it to align the grain.

2. The primary forging is done by a 1,000-ton air hammer (see Figure 11-22). The steel billet is placed in the lower (stationary) die, and then the upper (movable) die is slammed into the billet several times, forcing the metal to take the shape of the die. The steel billet is gradually formed into the rough shape of the golf club head. The metal needs to be reheated many times during this operation. The end product of the primary forging process can be seen in Figure 11-23. The extra material surrounding the head is called *flashing,* and it was caused by the excess metal escaping from the impressions in the die. Most drop forged parts have some flashing after the primary forging. This ensures that the impressions in the dies have been completely filled.

FIGURE 11-22: Hot metal being formed by drop forging.

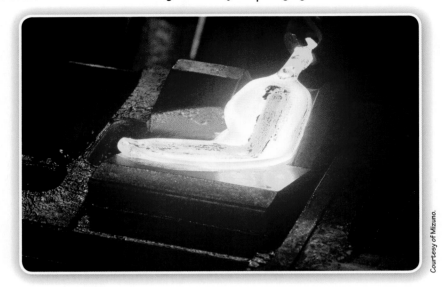

Courtesy of Mizuno.

FIGURE 11-23: Primary forging.

Courtesy of Mizuno.

FIGURE 11-24: First precision forging.

Courtesy of Mizuno.

FIGURE 11-25: Second precision forging.

Courtesy of Mizuno.

3. The flashing is removed, and the head is placed in a 650-ton hydraulic press, where the face of the head is made perfectly flat and the grains are tightened and aligned (see Figure 11-24).

4. The flashing from the first precision forging is removed, and the head is placed in the second and final die to receive the deep U-shaped grooves and to finalize the shape of the head (see Figure 11-25).

HOT ROLL FORGING Hot roll forging, also known as *forge rolling*, involves placing either round or flat stock between roller dies that reduce or change the cross-section and increase the length of the material. As shown in Figure 11-26, the roller dies are powered in opposite directions to pull the stock into the dies. This process can be used for primary or secondary processing. Common examples of products made using hot roll forging include hand shovels, chisels, and tapered axles.

FIGURE 11-26: Metal rolling.

FIGURE 11-27: Aluminum extrusions.

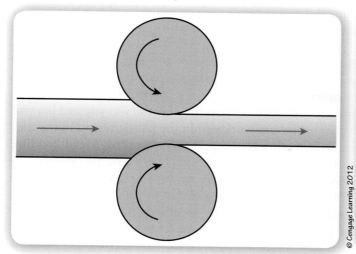

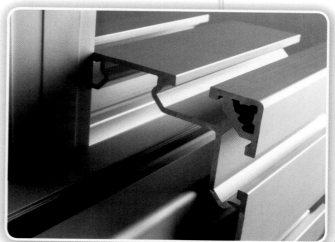

METAL EXTRUSION Metal **extrusion** is a manufacturing process used to create parts with uniform cross-sections. The extrusion process is done by squeezing metal through a die containing the desired shape. The cross-sectional shapes that can be produced with the extrusion process include rectangular, solid round, I-shape, T-shape, and many other common structural shapes (see Figure 11-27).

There are two ways to perform metal extrusion: direct extrusion and indirect extrusion (see Figure 11-28). In *direct extrusion,* the extruded product moves in the same direction as the ram. In *indirect extrusion,* the die moves toward the billet, which remains stationary. Extrusion can be done either hot or cold, depending on the shape and material being processed.

COLD FORMING METALS Cold forming metal is always done when the metal is at room temperature or slightly warmed. The process of working a material that has been warmed is appropriately named *warm forging.* The temperature must be below the

Extrusion:

a manufacturing process used to create parts with uniform cross-sections.

FIGURE 11-28: Direct and indirect extrusion.

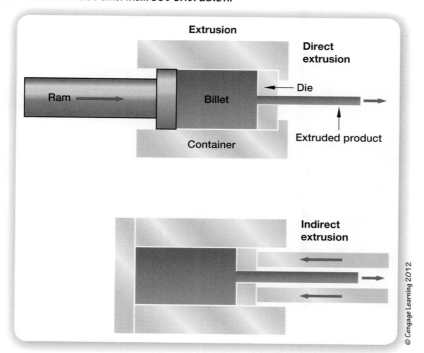

material's recrystallization temperature to prevent the alteration of the material's grain structure. Most cold-forming techniques are geared toward high-volume production. Cold forming can be broken down into four categories: bending, drawing, squeezing, and shearing.

If you were to use a small paper clip to hold 4 or 5 sheets of paper together, chances are good that the paper clip will return to its original shape when removed. If you were to use the same paper clip to hold 50 sheets of paper together, the paper clip is most likely permanently deformed because the material was stressed beyond its yield point. *Metal-bending operations* work within the plastic region of the material to cause permanent deformation to the part being worked. Metal-bending operations result in little or no noticeable change in the surface of the material.

Thinking back to the 50-sheet paper clip example, if you remove the clip from the large stack of papers, it would spring closed a bit but probably not return to its original position. This principle, called *springback,* occurs when the material has undergone partial plastic and partial elastic deformation. The elastic deformation will cause the material to bend back slightly. The amount of springback depends on the material properties, the bending operation, and the temperature of the material.

Most operations to bend flat sheets are done on a machine called a *press brake.* Figure 11-29 shows a typical press brake with partial computer control. Sheet metal or plate stock is placed onto the bed until it hits the back gauge, which is positioned by the computer for the next bending operation. Once the metal is in position, the operator turns the hydraulic ram on sending the punch toward the die. The computer controls how far the punch will go, depending on the type and degree of bend needed. Figure 11-30 shows the press brake in the closed position. Figure 11-31 shows a part that was bent using a press brake.

The technique known as *roll bending* is used to bend bars, sheets, plates, rods, tubes, pipe, and other standard structural shapes. Figure 11-32 shows a computer

FIGURE 11-29: **Press brake open.**

Courtesy of CustomPartNet, Inc.

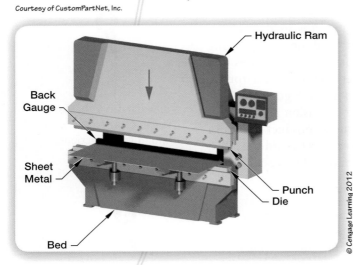

© Cengage Learning 2012

FIGURE 11-30: **Press brake closed.**

Courtesy of CustomPartNet, Inc.

© Cengage Learning 2012

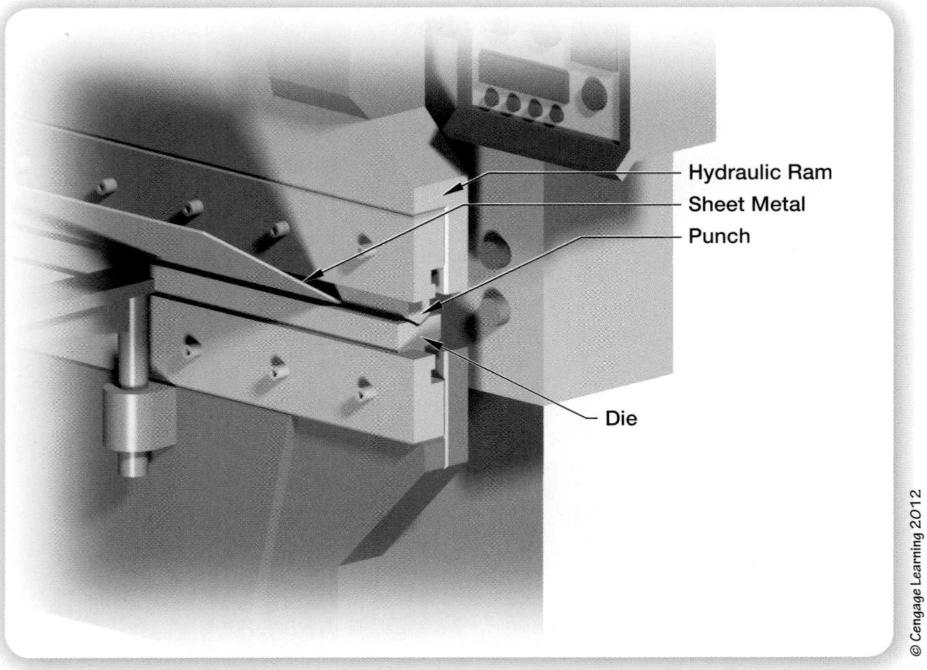

FIGURE 11-31: This sheet metal part was bent using a press brake.

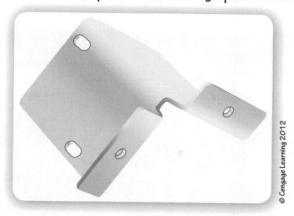

© Cengage Learning 2012

FIGURE 11-32: CNC tubing bender.

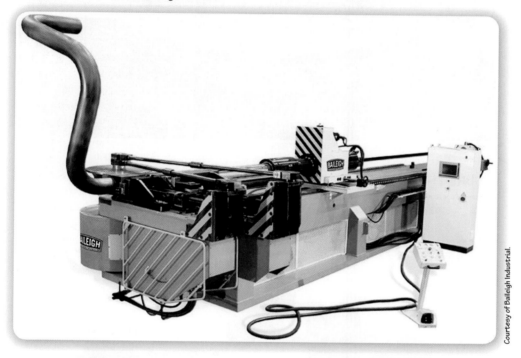

Courtesy of Baileigh Industrial.

numerically controlled (CNC) tube bender, which can be programmed to produce custom bends in long metal pipes. The unique roof of the California Academy of Sciences building in San Francisco is made of bent structural steel members (see Figure 11-33).

Continuous-rolling machines like the one in Figure 11-34 can easily bend long lengths of sheet metal into a desired shape. The process works by feeding sheet metal through a series of roller dies that progressively bend the metal into the final shape. Continuous-rolling machines are used to produce metal roofing, gutters, downspouts, shelving, and many other products.

When metal is forced through or around a die by stretching, compressing, or bending, the process is called **drawing**. Drawing is a cold-forming technique used to produce a wide variety of products. Products formed by drawing have smooth surfaces and accurate dimensions. The major drawing techniques are:

► shallow and deep drawing,

► stretch forming,

► wire drawing, and

► spinning.

Drawing:

a manufacturing process where metal is forced through or around a die by stretching, compressing, or bending.

FIGURE 11-33: Bent structural steel members were used to create a unique roof on the California Academy of Sciences building in San Francisco.

FIGURE 11-34: Roll-forming line.

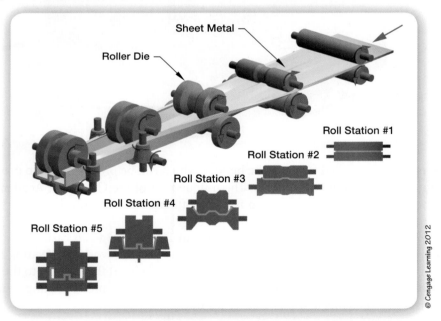

In shallow and deep drawing, thin, flat (sheet) metal is forced into a die using a punch. As with other forming operations, drawing involves working the metal in its plastic range to cause permanent deformation. Shallow drawing is a forming process in which the depth of the draw is less than one-half the diameter of the blank (stock). Deep drawing forms products with a greater depth than the final diameter.

A simple deep-drawing operation is shown in Figure 11-35. A circular piece of flat stock called a *blank* is inserted between the *blank holder* and the *die*. When the blank holder and the die are clamped together, they provide the *holding force* required to

FIGURE 11-35: Deep drawing.

Courtesy of CustomPartNet, Inc.

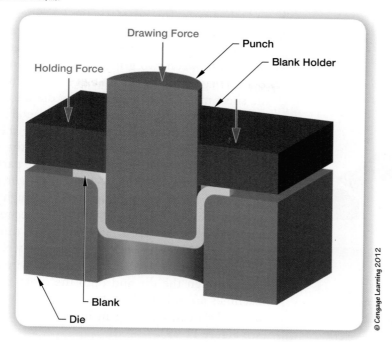

Drawing Force

Punch

Holding Force

Blank Holder

Blank

Die

© Cengage Learning 2012

keep the blank in place during the deep-drawing operation. The punch is forced onto the blank, causing the blank to stretch into the die. As the punch travels downward, the walls of the drawn part become thinner and thinner as the material is stretched more and more. The process of drawing a blank usually occurs in several steps called *draw reductions,* as shown in Figure 11-36. A different die is used for every step, causing the drawn part to stretch to a greater depth each time. When the drawing process is finished, the punch is removed and the part can be removed. The material that was clamped under the blank holder is excess and can be trimmed off.

Aluminum cans like the one shown in Figure 11-37 are made using the deep-drawing process. A giant press cuts circular blanks from a roll of thin aluminum and draws them into shallow cups. The shallow cups are then deep-drawn into a tall cylinder with the familiar ring on the bottom. The narrowed top is formed using a roll-forming process. The lid is assembled onto the can's body after it has been filled with liquid.

FIGURE 11-36: Deep-drawing operation (reductions).

Courtesy of CustomPartNet, Inc.

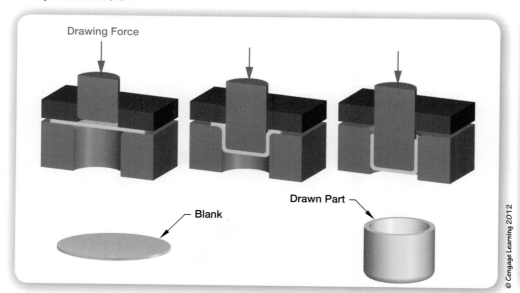

Drawing Force

Blank

Drawn Part

© Cengage Learning 2012

FIGURE 11-37: Aluminum cans are made using the deep-drawing process.

© iStockphoto.com/martinspurny.

FIGURE 11-38: Stretch forming.

Courtesy of CustomPartNet, Inc.

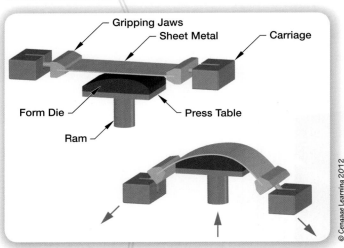

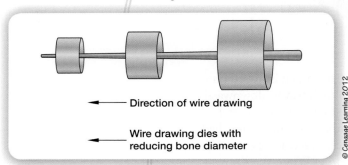

© Cengage Learning 2012

FIGURE 11-39: Wire drawing.

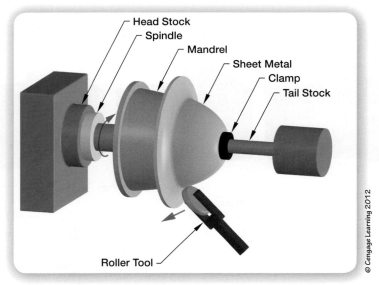

Direction of wire drawing

Wire drawing dies with reducing bone diameter

© Cengage Learning 2012

The process of *stretch forming* was originally developed by the aerospace industry to make aircraft wings and other parts. This process is similar to deep drawing, except the die moves into the blank in stretch forming. Figure 11-38 shows a piece of sheet metal (blank) clamped between two gripping jaws that are attached to a carriage. The die sits on a press table underneath the sheet metal. The ram forces the die into the sheet metal while the gripping jaws keep the metal from slipping out. As the die is forced upward, the sheet metal is plastically deformed into the shape of the die. This process can create large and very accurate parts.

The process of making wire is called *wire drawing*. The process involves pulling stock through a series of progressively smaller dies until the desired size is achieved. The material is fed from a reel, through the die, and then wound back onto another reel. The die is replaced with a smaller one, and the wire is pulled back through the die and onto the other reel. Some machines are able to draw the wire directly from one die to the next, resulting in higher production rates (see Figure 11-39).

Many round or bowl-shaped parts are formed using a process called *metal spinning*. The process begins by inserting a thin metal disc into a lathe that will spin the disc. Figure 11-40 shows a sheet metal disk clamped onto a spinning lathe while the roller tool forms the sheet metal against a part called the *mandrel,* producing the finished part. Examples of spun parts include hubcaps, bowls, and rocket nose cones. There are two types of spinning: conventional spinning and shear spinning (see Figure 11-41). Conventional spinning methods form the sheet metal to the shape of the mandrel without changing the thickness of the metal. Shear spinning uses a roller to form the sheet metal to the shape of the mandrel and stretches (draws) the sheet metal as well. The inner diameter of the parts will remain the same, but the shear spun part will have longer and thinner walls.

FIGURE 11-40: Metal spinning.

Courtesy of CustomPartNet, Inc.

Head Stock
Spindle
Mandrel
Sheet Metal
Clamp
Tail Stock

Roller Tool

© Cengage Learning 2012

FIGURE 11-41: *Conventional versus shear spinning.*

Courtesy of CustomPartNet, Inc.

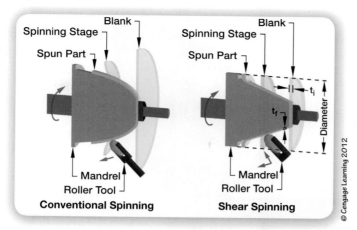

Conventional Spinning

Shear Spinning

© Cengage Learning 2012

FIGURE 11-42: *Thread-rolling techniques.*

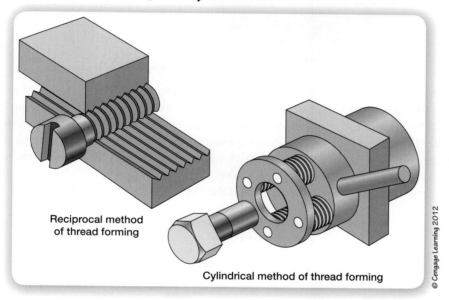

Reciprocal method of thread forming

Cylindrical method of thread forming

© Cengage Learning 2012

Thread rolling is a process used to form threads on screws and bolts. It works by literally pressing threads into the part. There are two major methods of thread rolling: the reciprocating die method and the cylindrical die method (see Figure 11-42). The reciprocating die method forms threads on parts sandwiched between two flat dies, one movable and the other fixed with grooves cut into both dies. The grooves match the desired thread pattern (threads per inch). When the two dies are forced together, they push into the metal, causing some metal to be displaced. The displaced metal is pushed upward to form the crest (top) of the thread. The cylindrical die method uses two or more rotating roller dies with the desired thread pattern cut into them. The part is held in a clamp to prevent it from rotating and then forced into the rotating roller dies. The roller dies press threads into the part, similar to the reciprocating method.

Shearing is a process used to punch shapes into or out of sheet stock or plate stock. Figure 11-43 shows a shearing operation where the upper blade applies a shearing force to the sheet metal, creating shear stress inside the sheet metal. Once the maximum shear stress of the material is exceeded, the material will fail, causing the two halves to separate. The shearing process causes an irregular edge on the sheared part called an *edge burr*. The three types of shearing are shearing, blanking, and punching.

Shearing:
a manufacturing process used to punch shapes out of sheet stock or plate stock.

FIGURE 11-43: Shearing.
Courtesy of CustomPartNet, Inc.

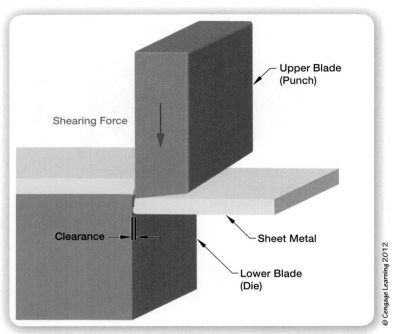

As shown in Figure 11-43, *shearing* involves cutting across a metal sheet to produce smaller sections. This operation is commonly used as a primary processing technique to prepare the metal for the next operation. The blanking process is shown in Figure 11-44. *Blanking* involves shearing a section out of a larger sheet. The blank that is sheared out of the sheet is the desired part, and any leftover material is considered scrap. As shown in Figure 11-45, *punching* involves shearing sections out of a larger sheet. The punched pieces are considered scrap, and the rest of the sheet is the good part.

FIGURE 11-44: Blanking.
Courtesy of CustomPartNet, Inc.

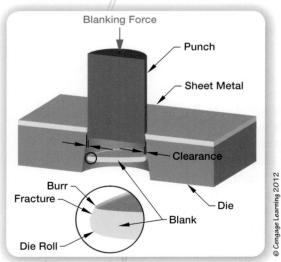

FIGURE 11-45: Punching.
Courtesy of CustomPartNet, Inc.

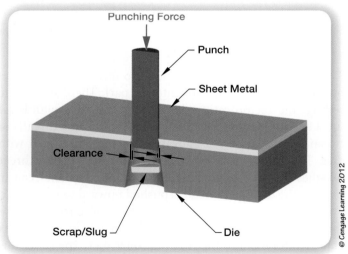

PROCESSING PLASTICS

Learning how plastics and other synthetic materials are processed is just as important as how metals are processed. Look at the car or the bus that took you to school this morning. How many plastic and metal parts are they made of? The answer is probably in the thousands. As you learned in Chapter 9, plastics are a type of polymer derived from a by-product of the oil industry. There are two major classifications of plastics: thermoplastic and thermoset. *Thermoplastic* is a type of plastic that has the ability to soften when heated and harden when cooled. *Thermoset* plastics have undergone a chemical reaction when exposed to heat that caused the polymer to harden. Thermoset plastics do not soften when heated and cannot be reshaped. Like metals, plastics can be formed using a number of different methods including:

- ▶ blow molding,
- ▶ thermoforming,
- ▶ extrusion,
- ▶ injection molding, and
- ▶ metal injection molding.

Blow Molding

Blow molding is a process used to create hollow plastic parts. Figure 11-46 shows a blow-molding operation. Thermoplastic pellets are dumped into a hopper (not shown) and fed into an extruder, where they are melted. The molten plastic is forced into the die head and squeezes out between the blow pin and the die to form a hollow plastic tube. This tube is called the *parison*. Parisons can also be formed using injection-molding techniques. The parison is extruded until it reaches the bottom of the mold cavity. At this point, the mold halves close, instantly sealing the bottom of the parison. High-pressure air is injected into the parison through the blow pin, forcing the molten plastic against the walls of the mold cavity. The part is then allowed to cool while still under pressure. The molds have cooling lines running through them to speed the cooling process. Once the plastic is cooled and hardened, the pressure is released and the mold opens up to release a finished blow-molded part.

Thermoforming

Thermoforming is ideal for shaping flat sheets of plastic into finished parts. The thermoforming process is similar to blow molding because it also uses heat and pressure to obtain a three-dimensional finished part. Thermoforming operations typically use thermoplastic sheets. A thermoforming operation, also known as *vacuum forming*,

FIGURE 11-46: Blow-molding process.

Courtesy of CustomPartNet, Inc.

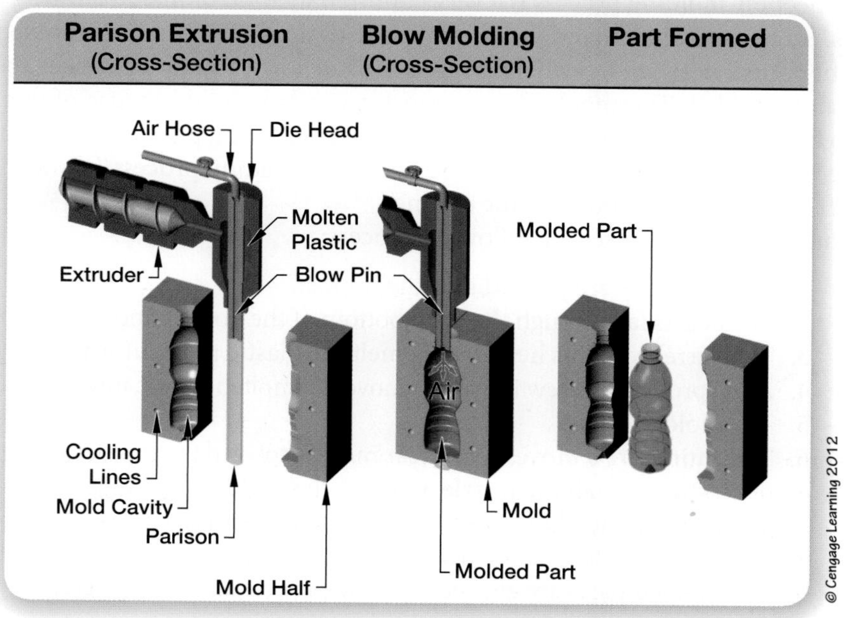

© Cengage Learning 2012

FIGURE 11-47: Thermoforming plastics.

Courtesy of CustomPartNet, Inc.

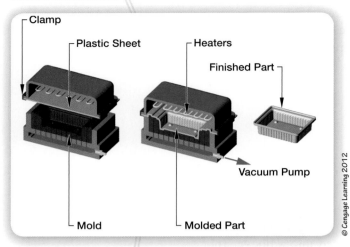

is shown in Figure 11-47. A plastic sheet is securely clamped around the edges of the mold. The plastic is heated until it is easily malleable, and then the vacuum is turned on to draw the plastic toward the bottom of the mold. The heater is turned off, and the part is allowed to cool before it is removed. The excess material needs to be trimmed to remove the material that was in the clamp. The finished part takes the exact shape of the mold.

Pressure forming is a thermoforming technique that is identical to vacuum forming, except compressed air is blown in above the part. The compressed air combined with the vacuum creates additional force that allows thicker materials to be molded and finer detail to be achieved. *Mechanical forming* is a thermoforming technique that is very similar to metal drawing (see Figure 11-48). A preheated plastic sheet is clamped onto the edges of the mold. The core plug is forced down onto the plastic sheet, creating the finished part.

FIGURE 11-48: Mechanical forming of a plastic sheet.

Courtesy of CustomPartNet, Inc.

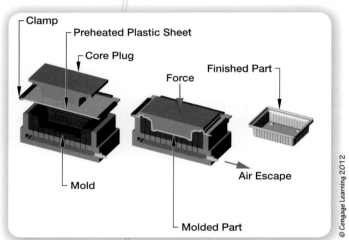

Extrusion

You learned about the extrusion process earlier in this chapter in the metal-forming section. Plastic extrusion is generally done with thermoplastic materials. Figure 11-49 shows common extruded plastic shapes. The process of plastic extrusion involves the following steps:

1. Plastic material is mixed with additives and coloring agents to alter the properties of the final product.
2. The material is heated to achieve a uniform, viscous consistency.
3. The molten plastic is forced through a die to form the desired shape.
4. The plastic extrusion is allowed to cool.

Injection Molding

The injection-molding process is a very common method of manufacturing plastic parts. From toothbrush handles to automobile dashboards, injection-molded parts can be found nearly everywhere. Figure 11-50 shows a laundry basket that was injection molded. Thermoplastics are most commonly used because of their ability to be melted and reshaped. Interestingly, thermosetting plastics are also used in some injection-molding processes. The injection-molding process (shown in Figure 11-51) is nearly identical to die casting.

Injection molding uses the following process:

FIGURE 11-49: Extruded plastic shapes.

© iStockphoto.com/JulianHeaven.

1. Plastic pellets are poured into the hopper.
2. The pellets fall through the open bottom of the hopper and into the barrel.
3. The barrel contains heaters that melt the plastic to a uniform consistency.
4. A reciprocating screw is used to move the molten plastic toward the mold.
5. The mold is closed.
6. The entire screw moves in a linear motion toward the mold, forcing a shot of plastic through the nozzle and into the mold.
7. The plastic is quickly forced into the mold cavities.
8. The plastic is allowed to cool.
9. The molded parts are ejected, and the process starts over again.

FIGURE 11-50: Injection molded laundry basket.

© iStockphoto.com/ozgurdonmaz.

FIGURE 11-51: Injection-molding machine.

Courtesy of CustomPartNet, Inc.

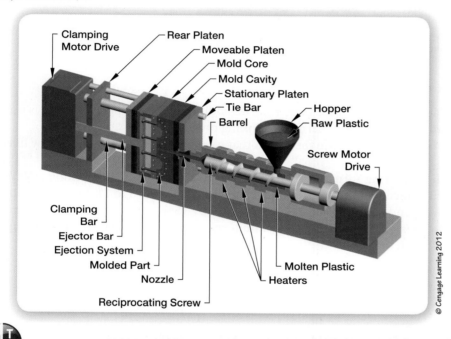

© Cengage Learning 2012

Metal Injection Molding

Metal injection molding (MIM) is a newer technology that allows the injection-molding process to fabricate solid metal parts. Parts produced using MIM are equivalent in strength and finish to sintered, investment cast, turned, and machined parts. Materials include stainless steels and low-carbon ferrous alloys. MIM products are used in the medical, firearm, aerospace, automotive, and dental industries. Metal injection molding follows this process (see also Figure 11-52):

1. Finely ground metal powder and a polymer binder are dumped into a mixer at a ratio of approximately 60% metal to 40% binder.
2. The metal–polymer mix is then granulated to form pellets, called *feed stock,* that is ready for injection molding.
3. The feed stock is placed into the hopper of an injection-molding machine. Traditional injection-molding processes are used to produce the metal–polymer part, which is commonly called a *green part.* The part is cleaned and deflashed before debinding.
4. The polymer can be separated from the metal through a solvent debinding process or a thermal debinding process. The solvent debinding process uses water or other chemicals to dissolve most of the binder.

(continued)

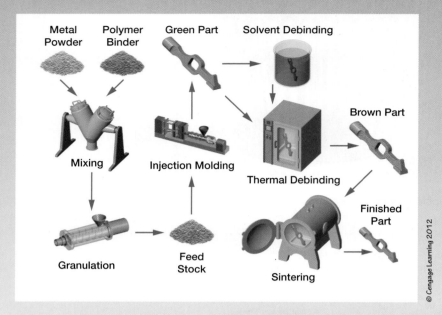

It is usually used in combination with or in place of thermal debinding. A low-temperature oven is used to evaporate the binder from the metal in the thermal debinding process.

5. The resulting *brown* part is 40% porous because the spaces that were filled with polymer are now filled with air.

6. The brown part is placed in a sintering furnace that bakes the part at temperatures reaching 2,500°F. The purpose of this step is to reduce the empty space in the part from 40% down to 1% to 5%, resulting in a very dense part with almost no empty space. The part shrinks by 15% to 25% during the sintering process. Because the shrinkage is predictable and occurs uniformly, the finished part retains the high tolerances formed by the injection mold.

CHIP-PRODUCING MACHINING METHODS

Chip-producing machining methods produce finished parts by cutting away material in the form of chips. Machines in this category include saws, mills, lathes, and grinders. This type of machining is commonly referred to as *conventional machining*. Many modern chip-processing machines are either fully or partially computer controlled.

The performance of any machining process is graded by its removal rate, accuracy, and repeatability. *Removal rate* is how fast material can be cut or ground away. *Accuracy* is the machine's ability to produce single parts with tight tolerances. *Repeatability* is the machine's ability to produce multiple parts with great consistency.

Tools

SAWING *Sawing* is the action of removing a portion of material using a cutting blade that is lined with teeth. The teeth are evenly spaced and work by making a series of narrow cuts in the material. Chips are the waste material that is produced, as a thin path, called a *kerf*, is created by the blade cutting through the material. Three styles of blades are used in sawing machines: straight blades, continuous blades, and circular blades. Examples of each of these can be seen in Figure 11-53.

FIGURE 11-53: *Various blade types.*

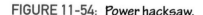

Straight Blade	Continuous Blade	Circular Blade
Straight blades are made of thin strips of steel with teeth along one edge.	Continuous blades are basically straight blades formed into a continuous loop.	Circular blades are made of a thin steel disk with teeth mounted on the outer edge.

Sawing machines can be classified by the type of blade they use. Straight blade machines use a reciprocating (back and forth) movement to cut. The *hacksaw* is a common hand tool used to cut metal in this fashion. A power hacksaw like the one shown in Figure 11-54 works by forcing a reciprocating blade into a piece of stock that is clamped onto the bed of the machine. Because cutting only takes place in one direction, the hacksaw is fairly inefficient but is an inexpensive option for cutting metals.

The bandsaw was originally developed as a woodworking machine. Although still widely used by woodworkers, the band saw has been adopted to cut metals as well. Band saws are categorized by the orientation of their blades. There are vertical band saws and horizontal band saws. Both designs have a continuous blade that is

FIGURE 11-54: *Power hacksaw.*

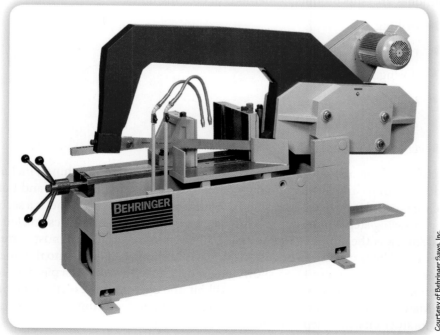

Courtesy of Scotchman Industries.

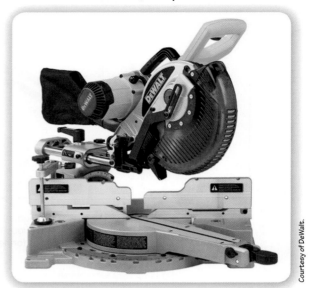

Courtesy of DeWalt.

stretched between two wheels. One wheel is powered, and the other wheel simply guides the blade. Band saws are commonly used to cut wood, metal, plastic, and composites. Figure 11-55 shows a fully automated horizontal band saw used in the metal industry.

Circular saws use a rotating disc with teeth around its circumference to remove stock. Like band saws, the circular saw was originally developed for woodworking but was later adapted to metalworking. Some circular saws move the blade into the stock like the sliding compound miter saw shown in Figure 11-56. Other circular saws like the table saw rely on the stock being moved into the blade.

HOLE PRODUCING Many manufactured parts require holes to be put into them. A hole is a cylindrical feature that is cut into a part by a rotating cutting tool. Holes can be produced using several different manufacturing techniques. The following techniques are chip-producing methods of creating holes:

▶ drilling,

▶ reaming,

▶ boring,

▶ tapping,

▶ counterboring, and

▶ countersinking.

Hole-making operations can be performed on a variety of machines, including traditional mills and lathes. Specialty equipment such as drill presses and boring machines can also be used. Typically, hole-making operations are done as a secondary processing step. For example, a forged part may need several holes created.

Drilling is the simplest way to make a hole in a material. The most common type of drill bit is called a *twist drill*. As the twist drill spins, it is fed into the stock to the desired depth. Figure 11-57 shows a twist drill making a hole in a piece of stock. When a hole is enlarged with a larger drill bit, the operation is called *counterdrilling*. Many different styles of drill bits are used in manufacturing, including spade bits, forstner bits, and step-drill bits.

FIGURE 11-57: Drilling.
Courtesy of CustomPartNet, Inc.

FIGURE 11-58: Reaming.
Courtesy of CustomPartNet, Inc.

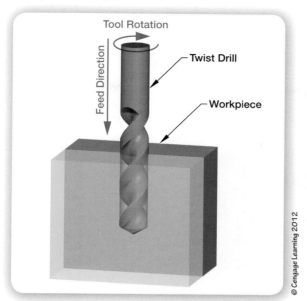

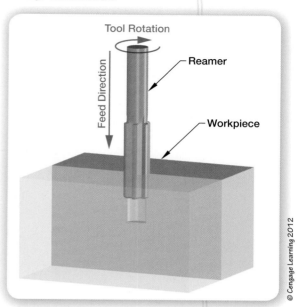

Reaming is done after a drilling operation with a multiple-tooth cutting tool called a *reamer* (Figure 11-58). The reamer is a tool that precisely machines holes to a standard diameter. Reamers are available in many different sizes, styles, and materials depending upon the application.

Boring is a machining process that is used to accurately enlarge an existing hole to a diameter that can't be achieved by standard diameter drills. As the boring tool is fed into the stock, the boring tool cuts the internal surface of the hole using a single-point cutter (Figure 11-59). Cylinder boring is a manufacturing operation commonly used in the automotive industry to accurately cut the cylinders within an engine block.

Tapping is a method of producing internal threads. A tap is a tool with the desired thread pattern cut into it as well as flutes to help remove the chips. Some taps require a predrilled hole, and other self-drilling taps are able to perform the entire operation in one step. The rotating tap is pushed into the hole, causing the teeth of the tap

Boring:
a machining process that accurately enlarges the diameter of a hole.

Tapping:
a machining process used to produce internal threads.

FIGURE 11-59: Boring.
Courtesy of CustomPartNet, Inc.

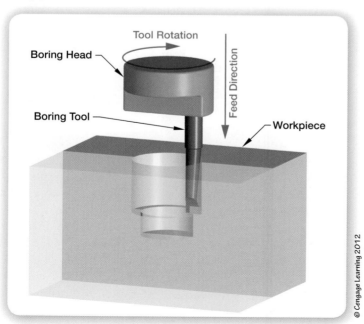

FIGURE 11-60: Tapping.

Courtesy of CustomPartNet, Inc.

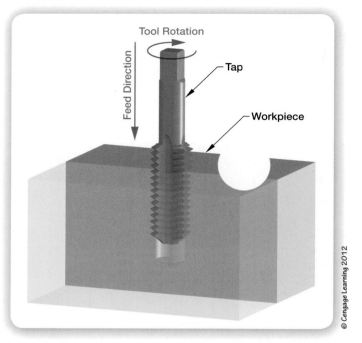

© Cengage Learning 2012

Threading:

a machining process used to produce external threads.

to cut into the internal surface of the hole forming threads as it twists downward. Each tap has a corresponding die that is used to cut external threads. **Threading** is a machining process used to produce external threads. Figure 11-60 shows a tapping operation.

A *counterbore* is frequently used to hide bolt heads, nuts, and other round objects below the material's surface. Counterboring tools are similar to twist drills, except they produce a flat-bottomed hole in the stock. Figure 11-61 shows a counterboring operation. Notice the pilot on the end of the counterbore that guides it into the original hole.

FIGURE 11-61: Counterboring.

Courtesy of CustomPartNet, Inc.

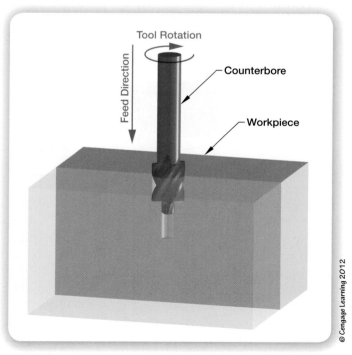

© Cengage Learning 2012

Countersinking (Figure 11-62) is similar to counterboring except the counterbore produces a sloped face in the hole. Counterbores are commonly used to hide flat-head wood and machine screws below the surface of the material. The most common angles for countersinking are 60°, 82°, 90°, 100°, 110°, and 120°.

TURNING *Turning* produces round, symmetrical parts using a single-point cutting tool. A lathe is a machine used to turn wood, metal, and plastic parts. Figure 11-63 shows the parts of a typical metal lathe. Turning is a machining technique that creates cylindrical parts by cutting away unwanted material. Figure 11-64 shows a magnified view of the turning operation. The workpiece is a section of material that is securely fastened to the spindle using a clamping device called a *chuck*. The workpiece is spun while the cutting tool is forced into the workpiece, causing material to be removed in the form of chips or helical spirals. Modern computer-controlled lathes can produce parts with complex 2-dimensional curved surfaces.

FIGURE 11-62: Countersinking.
Courtesy of CustomPartNet, Inc.

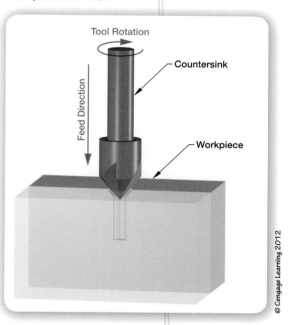

© Cengage Learning 2012

FIGURE 11-63: Metal lathe.
Courtesy of CustomPartNet, Inc.

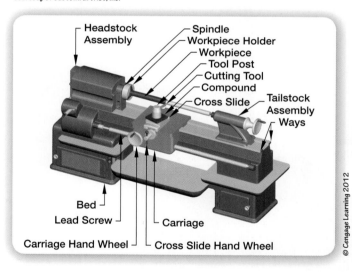

© Cengage Learning 2012

FIGURE 11-64: Turning.
Courtesy of CustomPartNet, Inc.

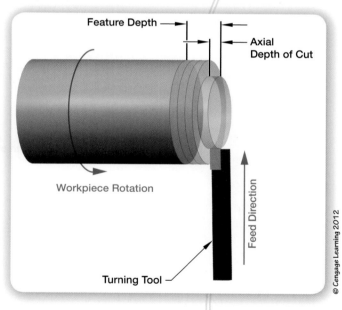

© Cengage Learning 2012

MILLING *Milling* is a very common form of machining in which unwanted material is removed by using a rotating cutting tool. The part being machined is either securely fastened directly to the table or clamped into a vise. The bed (worktable) is able to move in three axes (*X, Y,* and *Z*), allowing the part to be precisely positioned. CNC mills use special motors to precisely move each axis, resulting in a highly accurate finished product. The parts of a simple vertical mill are labeled in Figure 11-65. A magnified view of a milling operation can be seen in Figure 11-66. The part is forced into the spinning end mill (cutting tool), causing a slot to be created. Mills can perform many operations, including surface planing, curved surface milling, drilling, and slotting.

FIGURE 11-65: Vertical mill.

Courtesy of CustomPartNet, Inc.

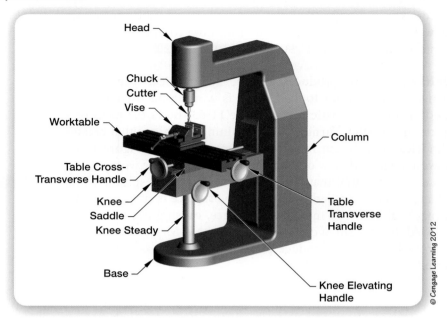

FIGURE 11-66: End milling.

Courtesy of CustomPartNet, Inc.

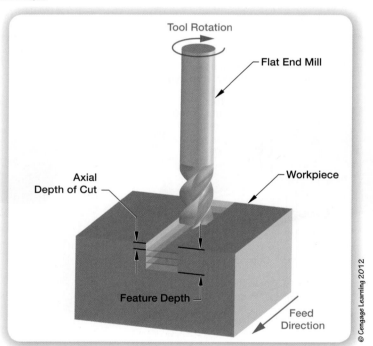

 ## PRODUCT LIFE CYCLE

All products have a *life cycle*, which is the time from when a product is introduced to the market to the time it is taken off the market (see Figure 11-67). Some products have life cycles that last a few years, and others last for centuries.

Design Phase

The design phase utilizes a design process such as the one that was introduced in Chapter 2. In this phase of the life cycle, teams of designers utilize their creative talents to brainstorm ideas for solutions to problems. This often involves sketching, discussing, combining, and refining ideas (see Figure 11-68).

MARKET RESEARCH The marketing phase begins while the product is being developed. In some cases, market research drives the development of new product designs. Before a new product is launched, companies want to find out if consumers are likely to purchase their product. Market research is the study of purchasing habits of various groups of people. A company can determine the consumer demand for its product based on this research. Ideas that emerge from market research are used to refine the product to make it more appealing to the consumer. The likelihood of a product succeeding in the marketplace can be predicted by conducting research to determine the following factors:

FIGURE 11-67: *Product life cycle.*

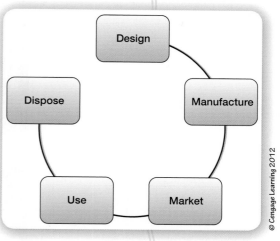

© Cengage Learning 2012

1. The breakeven point is the point at which the profit from products sold equals the total cost of manufacturing the parts. If the breakeven point is too high, it means the market might be too small and manufacturing might not be worthwhile. This stage is done before the product is put into production.

2. Supply and demand is the relationship between the number of products of a certain type on the market and the number of consumers that want to purchase them. If the demand is high and the supply is low, a company can raise its prices and increase profits. However, if the supply is high and the demand is low, the company will struggle to make a sale and be forced to lower its prices. Supply and demand helps companies determine what and how many products to produce.

Market research:
the study of the purchasing habits of various groups of people.

3. Feasibility studies can tell a company how practical it would be to produce a certain product. Questions raised in a feasibility study might include the following:
   - How much storage space will be needed for the finished products?
   - Is the equipment available to manufacture this product?
   - What labor costs will be associated with manufacturing this product?

FIGURE 11-68: *Design sketch.*

4. *Consumer surveys* are a marketing tool that can help determine customer wants, needs, likes, and dislikes. The survey might be done in person, online, by mail, or over the phone. The results of customer surveys can often drive companies to develop new products or improve existing ones.

5. *Trend analysis* is a historical perspective of the purchasing history for a type of product. Companies can use this analysis to predict future sales based on past data.

6. *Sizing up the competition* can help a company decide how much of the product to produce. If another company is going to produce large quantities of a similar product or even the same product, the production numbers need be scaled back because of the rules of supply and demand.

© iStockphoto.com/Nicolas Loran.

7. *Marketing reports* give manufacturers current information on product sales. Sometimes these are referred to as *sales reports*. The report will tell the manufacturer how well the product is selling and if the company is making a profit.

PRODUCT DEVELOPMENT Product development is a time-consuming and costly but necessary operation. The steps of the product development are to:

▶ finalize the product design,

▶ decide on materials,

▶ test for function,

▶ test for safety, and

▶ engineer production.

FIGURE 11-69: Workers install automobile parts on an assembly line.

Courtesy of AP/Wide World Photos.

Manufacturing Phase

During the manufacturing phase, the product will be prepared for production by a manufacturing engineer. Figure 11-69 shows an automobile assembly line designed by a manufacturing engineer.

PREPARING FOR PRODUCTION Proper preparation will prevent wasted time, money, and materials later on. Important decisions must be made during the planning phase such as what the product will look like, and how it will function. Designing the product for manufacturability is critical to controlling manufacturing costs and keeping profits high.

Manufacturing engineers specialize in the planning and coordination of every step of a production process. They are knowledgeable about materials processing techniques, including the tooling, assembly, and sequencing of operations. They may also plan the layout of factory production lines. Manufacturers use *production flowcharts* to visualize how and in what order every component of the product will be made and assembled. Production flowcharts are graphical representations of materials processing, inspection, delays, and storage. Production flowcharts often use graphic symbols to represent each step. Additional factors that must be considered during the production planning phase are:

- ▶ machine and tool requirements,
- ▶ availability of production materials,
- ▶ packaging,
- ▶ factory layout,
- ▶ quality control, and
- ▶ personnel requirements.

Production

At this point, the factory is set up to produce the product, workers have been trained, and the production materials are in stock. The next step is to make a trial production run to observe how the entire process works. During this process, problems can be identified and solved before the actual production begins. This also gives workers a chance to get used to their new jobs and to receive additional training. The timing of the production process can be changed to eliminate delays and to minimize production time.

Once the trial production run has been completed and the necessary adjustments have been made, it's time for production to begin. The production process is constantly monitored by quality-control personnel to ensure the product being manufactured meets both the company standards as well as any applicable standards dictated by the government. When multiple components need to be pieced together, it is called an *assembly*. Subassemblies are assemblies that will become part of a larger assembly. The LCD screens on cell phones are manufactured as subassemblies. During the final assembly of the cell phone, individual components are combined with the subassemblies to produce a finished cell phone. The finished products are then inspected, packaged, and stored in a warehouse before being shipped to a distributor or to consumers.

Marketing Phase

Marketing is the process of preparing and performing the research, pricing, promotion, and distribution of ideas, goods, and services to generate sales. In short, this means that marketing influences the design, manufacturing, and sales of products. Most companies have marketing departments that are responsible for developing a marketing plan. Marketing plans are composed of advertising methods, product distribution, market research, and sales reports. Every marketing plan looks slightly different, but most will have the following elements:

▶ an executive summary that briefly summarizes the company's marketing plan,

▶ a snapshot of the current state of the market (market situation),

▶ an overview of the good and bad aspects of the company and the marketplace (threats and opportunities),

▶ marketing objectives (what the company wants to accomplish over the course of the plan and how it will be done),

▶ a budget that allocates money to each marketing activity, and

▶ a method of monitoring the progress of the marketing plan (tracking effectiveness).

User Phase

The user phase begins when the product is first introduced to customers. It is then promoted during the product growth period (see Figure 11-70). After the product has been on the market for a while, it enters a maturity period where sales begin to level off. The product will eventually experience a period of decline in sales towards the end of its lifecycle.

PRODUCT INTRODUCTION PERIOD When a new product is first introduced to customers, sales may be slow until the customer learns about the benefits of the product. The primary goal of product introduction is to establish a market and to build demand for a product. When products are first introduced, the price is generally much higher to allow the manufacturer to recover some of the design and production costs.

PRODUCT GROWTH PERIOD The product growth period is defined by a rapid increase in sales as customers become aware of the product and its benefits. Building customer loyalty to the product is important during

FIGURE 11-70: **The sales of new products usually start off slow and gradually increase as more customers find out about the product.**

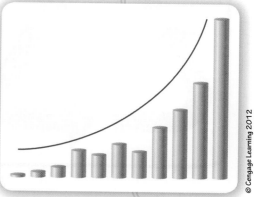

© Cengage Learning 2012

this period. The price of the product may be kept high if the demand is high or it may be reduced if more customers are needed.

PRODUCT MATURITY PERIOD The product maturity period is the most stable and the most profitable. Sales continue to increase, even if only slightly. Because most customers are aware of a product, advertising costs may be reduced. The goal of any company during this period is to extend the life cycle of the product. Commonly, a slight modification or new features will be added to an existing product to make it stand out against the competition. The price of the product may be reduced to align with the competition.

PRODUCT DECLINE PERIOD Eventually, every product will run its course and sales will begin to decline. The reasons for the decline could be that the market has been saturated with similar products, the product became technologically obsolete, or customer preferences have changed. During this period, a company can decide to reduce advertising and other expenses to make up for lower sales. A company may choose to discontinue the product when profits drop too low. The price of the product may be dropped significantly during this period to reduce inventory and encourage customers to make a purchase.

Disposal Phase

The user phase is not the end of the story. No product life cycle lasts forever. In the disposal phase, the product is thrown away or recycled. Today, designers must think about how to avoid the use of hazardous materials or design the product so that any hazardous materials can be removed and disposed of safely. For example, if a product such as a battery is improperly discarded, the toxic chemicals can cause environmental damage. Sustainable design, also called *green* or *eco-design,* occurs when designers account for the ecological impacts of the design solution. Examples of sustainable designs include the use of recycled materials, and the use of wind power to generate electricity for a manufacturing process.

Individuals, communities, and companies are now looking for better ways to manage products at the end of their life. Rather than sending used products to a landfill, infrastructures are being put in place to reuse, recycle or remanufacture them. A common mark found on products made of plastic is the recycling logo (see Figure 11-71). Today, plastics are recycled to make new products such as recycled plastic lumber. Most American states have passed laws to reduce the amount of garbage sent to landfills. Plans typically include the following:

1. *Waste prevention* means finding ways to reduce the size of a product, reduce the amount of material used in the product, or use materials that are more environmentally friendly. Today, new materials are being used to reduce the weight of cars and airplanes and most other consumer products. In packaging, new manufacturing processes use less material to make bottles and cans by reducing wall thickness. Some disposable plastic goods are made to biodegrade.

2. Using *recycling,* most cities have plans to recover materials from the waste stream for use in new products. Of the items recycled, typically 35% is paper, 27% is plastic, 22% is glass, and 6% is metal.

3. With *composting,* many towns and cities collect garden and lawn clippings and fall leaves so they can be composted and reused as fertilizer or mulch.

Sustainable design: occurs when designers account for the ecological impacts of the design solution; also called *green* or *eco-design*.

FIGURE 11-71: *Recycling logo.*

© Cengage Learning 2012

Your Turn

Choose a simple product (could be a toy, a kitchen gadget, a tool, etc.) to reverse engineer.

1. Assume the product has broken and that you want to redesign it to make it stronger. Describe how you would reverse engineer this product.
2. Develop a hypothesis of what could be improved.
3. Carefully disassemble your product.
4. Create a list of every part.
5. Use your knowledge of materials and manufacturing processes to determine what material each part is made of and how it was made.
6. Use a solid-modeling CAD software to redesign any problem parts.

SUMMARY

- *Materials processing* involves converting raw materials into finished products.

- *Casting* is a process that typically involves filling a mold with a molten material, waiting for it to solidify, and then removing the part from the mold.

- *Forming* encompasses many different manufacturing processes in which a material is permanently deformed. Forming metals involves permanently changing the shape of the material by applying pressure.

- *Forging* is a form of metal forming that involves shaping a metal part through controlled plastic deformation.

- Metal *extrusion* is a process through which long, uniform parts can be fabricated. The extrusion process is done by squeezing metal through a die that causes the material to take on a desired cross-sectional shape.

- *Cold forming* metal is always done when the metal is at room temperature or slightly warmed. Cold-forming techniques include bending, drawing, squeezing, and shearing.

- Like metals, plastics can be formed using several different methods, including blow molding, thermoforming, extrusion, injection molding, and metal injection molding.

- Chip-producing machining methods produce finished parts by cutting away material in the form of chips. Machines in this category include saws, mills, lathes, and grinders.

- The time from when a product is introduced to the market to the time it is taken off the market is considered the product's *life cycle*. The stages of the product life cycle are the design phase, manufacturing phase, production phase, marketing phase, user phase, and disposal phase.

1. Describe manufacturing in your own words.
2. Explain the difference between a primary process and a secondary process.
3. List 10 raw materials.
4. What products are made using the sand-casting process?
5. List the steps involved in creating a sand-cast part.
6. List and describe the various types of sand used for casting.
7. Describe the die-casting process in your own words.
8. What types of parts are best suited for the injection-molding process?
9. Explain the manufacturing process called *forming* in your own words.
10. Explain the difference between hot forming and cold forming.
11. Define recrystallization.
12. Give an example of plastic deformation.
13. Describe the following forming actions in your own words: (a) draw forging, (b) upset forging, and (c) pierce forging.
14. What is the difference between an indirect extrusion and a direct extrusion?
15. Describe the manufacturing process called *drawing*.
16. How is wire made?
17. What is the difference between a thermoplastic and a thermoset plastic?
18. Explain how a plastic bottle is made.
19. List the steps involved in the thermoforming process.
20. Describe the plastic extrusion process.
21. List the chip-producing manufacturing techniques.
22. What is a feasibility study?
23. Describe the duties of a manufacturing engineer.
24. Describe what happens during the marketing phase in your own words.
25. What is a sustainable design?

EXTRA MILE

As a society, we are realizing the value of gardening in urban areas. Growing food and nonfood crops in and around cities contributes to healthy communities by improving nutrition, ensuring food security, promoting exercise, maintaining mental health, and improving urban environments. In many areas, finding a flat surface to grow plants on can be nearly impossible.

Your challenge is to design a vertical gardening system that can be mounted to an exterior wall. What materials and manufacturing processes will you use to construct the product? Create a sketch of your system and write a short technical report on your solution.

CHAPTER 12
Statics

Menu

START LOCATION DISTANCE END LOCATION

Before You Begin

Think about these questions as you study the concepts in this chapter:

1 Why is statics considered fundamental to all fields of engineering?

2 What do civil engineers do?

3 What types of loads can structures be subjected to?

4 What is the difference between a scalar and a vector quantity?

5 What does it mean if an object or structure is in static equilibrium?

6 What are free-body diagrams and when are they used?

7 How do you determine the direction and magnitude of a reaction force that occurs at a beam or truss support location?

8 How do you determine the magnitude and type of force that is transmitted through a truss member?

From the great pyramids of Egypt to the grand coliseums of ancient Greece, magnificent engineering achievements of our forefathers can be found around the world. Those ancient structures fulfilled the needs of the people living then, but our modern society demands structures that are bigger, lighter, cheaper, and stronger than ever before (see Figure 12-1).

Figure 12-1: *Main Street truss bridge in Brockport, New York.*

With every new structure comes a new problem, often requiring an engineer to try something that has never been done before. Consequently, with any new design comes the inherent risk of failure. The history of humankind has been riddled with engineering disasters of various magnitudes. Some are merely annoying, as when a car breaks down because a part was poorly designed and fails. Other engineering disasters such as the space shuttle *Columbia* disaster and the I-95 bridge collapse in Minneapolis have cost the lives of many innocent people who entrusted their safety to the hands of engineers. Lessons can be learned from engineering triumphs but, unfortunately, more can often be learned from failure.

MECHANICS

Engineers have always relied on their experience, ingenuity, creativity, and judgment to develop new designs. Modern engineers also use mathematical equations and software analysis in the development and analysis of their designs. This allows the engineer to study how a design should behave under load, and modify it in ways that bring about more desirable results. When an idea finally works on paper and on a computer screen, it is then ready for actual construction.

Automotive engineers are now able to design safer vehicles by first testing their designs in powerful simulation software programs as shown in Figure 12-2. This type

FIGURE 12-2: A computer simulation of a vehicle frontal crash.

FIGURE 12-3: The forces on the columns of the Parthenon can be calculated using equilibrium equations derived in statics.

of software uses the principles of mechanics to simulate an automotive crash before the vehicle has been built.

The science of mechanics involves the study of **forces** and their effects. Two of the common subject areas within the science of mechanics are *statics* and *dynamics*. **Statics** is the study of objects at rest (see Figure 12-3), and **dynamics** is the study of objects in motion (see Figure 12-4).

Force:

a push or pull that acts on an object to initiate or change its motion or to cause deformation.

STATICS

In a nutshell, statics helps us to understand why things stay still. Think of the last time you drove across a bridge. Perhaps you wondered who designed the bridge or how old it is. Many of us never question the strength of a bridge before crossing it.

FIGURE 12-4: The flight of a football can be predicted using equations of motion derived in dynamics.

Bridges, buildings, and other objects that are intended to be stationary are called static structures. When forces or force systems acting on an object are in a state of balance, the object is said to be in a state of **static equilibrium**.

STRUCTURE DEFINED

A **structure** is a stable assembly of connected components that is designed to withstand loads. We tend to think of structures as bridges and buildings, but structures can also be found in nature. The human skeleton is an amazing structure composed of interconnected bones and joints that work together to form the framework of our bodies.

Structural and civil engineers are responsible for planning and designing buildings, bridges, and other structures. They must have a thorough knowledge of physics and mathematics to analyze and predict the behaviors of their final designs.

In this chapter, you will learn how to analyze trusses, which are structures composed of interconnected parts known as **members**. A **truss** is a triangular arrangement of members commonly made of wood or steel that forms a rigid framework. The open design of a truss makes it very strong for its weight and is often used to span long distances. Figure 12-5 shows a series of roof trusses commonly used in residential and commercial construction.

STRUCTURAL MEMBERS AND LOADS

A structure is only as strong as its weakest link. Engineers must be able to calculate the strength of structural members before calculating the strength of the entire structure. The strength of a structural member is defined as the maximum amount of force it can withstand before failing. Forces that are transmitted through a structural member are called **internal member forces**. If the internal member force exceeds the strength of the material, the member will fail.

Static equilibrium:
a condition wherein all of the forces acting on an object are in a state of balance, causing the object to remain stationary.

Structure:
a stable assembly of connected components that is designed to withstand loads.

Truss:
a triangular arrangement of members commonly made of wood or steel that forms a rigid framework.

Internal member force:
a force that is transmitted through a structural member.

FIGURE 12-5: Roof trusses provide a strong yet lightweight framework to support the roof.

Careers

CAREERS IN THE ENGINEERED WORLD

Civil engineers are responsible for the design and construction of roads, bridges, buildings, airports, waterways, and water- and wastewater-treatment facilities. Some civil engineers work on our roads and bridges while others are involved with environmental issues ranging from hazardous and solid-waste management to lake and river environmental restoration. They also help protect people from catastrophic natural disasters such as hurricanes, floods, and earthquakes. Civil engineers commonly specialize in one of the following areas.

▶ *Environmental engineers* may conduct environmental impact assessments or be involved in natural resource management. Poor air and water quality, noise, and radiation are problems these engineers deal with every day. They develop appropriate disposal strategies for residual waste and defend our environment from the potentially damaging effects of human activity. Environmental engineers may design, operate, and manage systems for environmental protection.

▶ *Geotechnical engineers* perform studies for civil engineering, mining, and oil and gas projects. They collect and evaluate geological data and write reports based on what they discover.

▶ *Structural engineers* design and oversee the construction of all types of buildings, bridges, and industrial structures. They analyze and design structures to ensure that they safely perform their purpose, support their own weight, and resist dynamic environmental loads such as hurricanes, earthquakes, blizzards, and floods. Some examples of their work include stadiums, arenas, skyscrapers, offshore oil structures, space platforms, amusement-park rides, bridges, and office buildings. Structural engineers use their knowledge of the properties and behaviors of steel, concrete, aluminum, timber, and plastics to build new structures. Many structural engineers will often spend a great deal of time on the construction site inspecting and verifying that all aspects of the project are going according to plan.

▶ *Hydraulic engineers* design and oversee the construction of structures such as dams, reservoirs, or wells that control the flow and distribution of water. Hydraulic engineers deal with problems concerning the quality and quantity of water. They work to prevent floods; supply water for cities, industry, and irrigation; treat wastewater; protect beaches from erosion; and manage and redirect rivers. They also design, construct, and maintain hydroelectric power facilities, canals, dams, pipelines, pumping stations, locks, and seaport facilities.

▶ *Transportation engineers* design and oversee the construction of highways and mass-transit systems such as subways and commuter trains. They design and oversee the construction of all types of facilities, including highways, railroads, airfields, and ports. An important part of transportation engineering is to upgrade our transportation capability by improving traffic-control and mass-transit systems and by introducing high-speed trains, people movers, and other new transportation methods.

▶ *Urban planning engineers* develop plans for entire communities. They evaluate physical, demographic, economic, environmental, sociological, and political issues that can affect land use. Urban planners create plans for both public and private land with the focus of maintaining or improving the human environment.

▶ *Construction engineers* use their technical and management skills to turn designs into reality while keeping the project on time and within budget. They apply their knowledge of construction methods and equipment, along with principles of financing, planning, and managing, to turn dreams into reality.

The strength of structural members can be found through experimentation, which was discussed in Chapter 10. The experimentation methods use specialized equipment that can pull, push, or twist structural members up to the point of failure. These machines produce useful data that engineers can use to determine the strength of structural members.

The cross-section of a structural member is a very important factor in determining its strength. You can see a cross-section by looking at the end of a member. Figure 12-6 illustrates shapes that are commonly used in structures. The shaded portion of each member represents its cross-sectional shape.

The cross-sectional area is the surface area of the cross-section of a member. Calculating the cross-sectional area is very simple for solid rectangular shapes but gets more difficult when dealing with complex shapes.

FIGURE 12-6: *Common shapes of structural elements.*

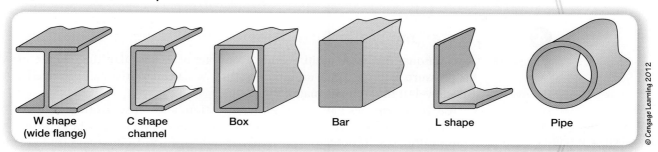

| W shape (wide flange) | C shape channel | Box | Bar | L shape | Pipe |

© Cengage Learning 2012

Types of Structural Loads

Structural loads can be divided into (1) static and (2) dynamic loads. Static loads are usually gravity-type loads. These would include the weight of the structure itself or any other load that is stationary. Dynamic loads are movement-based loads. Jumping on your bed will produce a dynamic load.

Dead loads are static loads that include the weight of the structure itself, flooring, and other stationary items supported by structures. *Live loads* are temporary loads that include vehicles, furniture, and people (Figure 12-7). Dead and live loads are civil engineering terminology for static and dynamic loads. The efficiency of a structure can be measured by comparing the weight of the structure (dead load) to the maximum weight the structure can withstand.

Static loads:
loads that include the weight of the structure itself.

Dynamic load:
objects that are in motion on or within a structure.

FIGURE 12-7: *Dead and live loads.*

Dead load   **Live load**

© Cengage Learning 2012

OVERVIEW OF FORCES

Sir Isaac Newton's first law of motion states that a body at rest will stay at rest and a body in motion will stay in motion in a straight line unless acted upon by an external force. In statics, we are interested in bodies at rest. Newton's second law of motion states that force is equal to the product of mass and acceleration. In his third law of motion, Newton stated that every action has an equal and opposite reaction. In the study of statics, it is important to note that every force is opposed by an equal and opposite force.

Forces have the capability of (1) moving an object that is at rest, (2) changing the velocity of an object that is already in motion, or (3) deforming the object. In the study of statics, we will ignore the deformation of an object. As the name *statics* implies, we will be dealing with bodies at rest. Predicting failure and deformation lies within the field of *mechanics of materials*, which was addressed in Chapter 10.

Vector and Scalar Quantities

A scalar quantity has magnitude only. Examples of scalar quantities include speed, air pressure, volume, time, and temperature. Scalar quantities can be added algebraically, but the sum has magnitude only.

Vector quantities have magnitude, direction, and sense. Examples of vector quantities include force, velocity, and acceleration. A vector can be represented graphically as a straight line with an arrow at one end. The magnitude of the vector quantity is represented by the length of the line. For example, a 1 inch line may serve as the equivalent of a 100 lb force. The characteristics of a force are fully described by its magnitude, direction, sense, and point of application (see Figure 12-8).

Scalar:
a quantity that can be described by magnitude alone.

Vector:
a quantity that is described by its magnitude, direction, and sense.

FIGURE 12-8: *Characteristics of a force.*

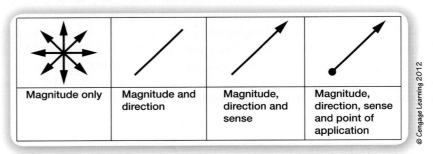

© Cengage Learning 2012

1. The *magnitude* of a force is expressed in units such as pounds or newtons.
2. *Direction* describes the path of the line along which a force acts. This is commonly referred to as the *line of action*.
3. *Sense* is the direction in which the force acts along the line of action. Sense is generally noted by drawing an arrowhead on the line of action.
4. *Point of application* is the location at which the force is applied to an object.

Figure 12-9 shows a baseball player who has just struck a baseball. To completely describe the force applied by the bat to the ball, you must know its magnitude, direction, sense, and point of application.

© iStockphoto.com/RBFried.

Force Systems

A **force system** is two or more forces that are collectively considered. All force systems fall under two broad classifications: coplanar and noncoplanar. If the lines of action of all forces within a system all lie in the same plane, we refer to the system as a **coplanar force system**. If the lines of action of some forces within a system do not lie in the same plane, then we refer to the system as a **noncoplanar force system**.

Coplanar and noncoplanar forces can be further classified into **concurrent force systems** and **nonconcurrent force systems**. Concurrent forces have lines of action that intersect at a common point. Nonconcurrent forces have lines of action that do not share a common point. By combining these terms, we end up with four major classifications of force systems:

> **Coplanar force system:**
> the lines of action of all forces within a system lie in the same plane.

> **Concurrent force system:**
> the lines of action of all forces intersect at a common point.

1. **Coplanar concurrent**: The lines of action of all forces are acting in the same plane and pass through a common point [see Figure 12-10(a)]. Tug-of-war is a good example of a coplanar concurrent forcessystem.
2. **Coplanar nonconcurrent**: The lines of action of all forces are acting in the same plane but do not pass through a common point. A teeter-totter is a good example of a coplanar nonconcurrent force system [see Figure 12-10(b)].
3. **Noncoplanar concurrent**: The lines of action of all forces are not acting in the same plane but do pass through a common point. The frame of a teepee is an example of a noncoplanar concurrent force system.
4. **Noncoplanar nonconcurrent**: The lines of action of all forces are not acting in the same plane and do not pass through a common point. A marionette is an example of a noncoplanar nonconcurrent force system.

FIGURE 12-10: (a) Coplanar concurrent and (b) coplanar nonconcurrent force systems.

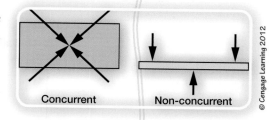

Concurrent   Non-concurrent

© Cengage Learning 2012

Because most structural systems can be reduced to coplanar force systems, this chapter will only cover coplanar force systems.

Vector Addition

Two or more vectors may be added together to produce a *resultant* vector.

If two vectors have the same direction and sense then the magnitude of their resultant vector will be equal to the sum of their magnitudes and will also have the same direction and sense. If the vectors are acting in opposite directions but are opposite in sense the vectors are subtracted to obtain the resultant. In general, because vectors may have any direction, we must use one of the following three methods for adding vectors: (1) parallelogram method, (2) tip-to-tail method, or (3) algebraically adding vector components.

PARALLELOGRAM METHOD In the parallelogram method for vector addition, the vectors are moved to a common origin, and the parallelogram is constructed as follows:

Step 1. Decide on a scale for your diagram and use a protractor and a ruler to draw the forces to scale with the arrows pointing in the proper direction as shown in Figure 12-11.

Step 2. Draw two more lines parallel to the first lines you drew beginning from the tips of the previous vectors to form a parallelogram as shown in Figure 12-12.

Step 3. Draw in the resultant force as shown in Figure 12-13. Determine the magnitude and direction of the resultant force by measuring the length and angle of the line.

FIGURE 12-11: *Vector addition: parallelogram method, step 1.*

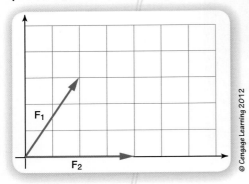

FIGURE 12-12: *Vector addition: parallelogram method, step 2.*

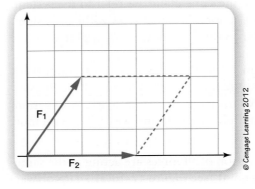

FIGURE 12-13: *Vector addition: parallelogram method, step 3.*

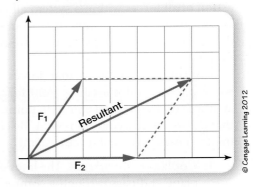

VECTOR ADDITION: TIP-TO-TAIL METHOD Two or more vectors can be added by placing the tail of one vector at the tip of another vector. The resultant force is drawn from the origin to the tip of the last vector.

Step 1. Decide on a scale for your diagram and use a protractor and a ruler to draw forces F_1 and F_2 to scale (see Figure 12-14).

Step 2. Draw force F_1 starting from the tip of force F_2 (see Figure 12-15).

FIGURE 12-14: *Vector addition: tip-to-tail method, step 1.*

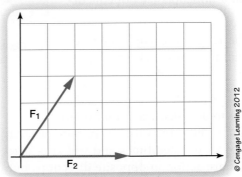

FIGURE 12-15: *Vector addition: tip-to-tail method, step 2.*

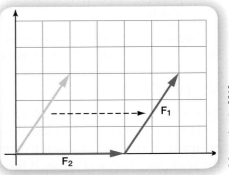

Step 3. Draw in the resultant force and determine its magnitude and direction by measuring the length and angle of the line (see Figure 12-16).

VECTOR ADDITION: TIP-TO-TAIL METHOD—THREE OR MORE VECTORS

Three or more vectors can be added together using the same tip-to-tail method used to add two vectors (see Figure 12-17).

Step 1. Decide on a scale for your diagram.

Step 2. Use a protractor and a ruler to draw each vector to scale. Draw the first force at the proper angle and length starting at the origin. Draw each additional vector from the tip of the previous vector (see F igure 12-18). The order in which you place the vectors is not important as long as you represent each of the forces once.

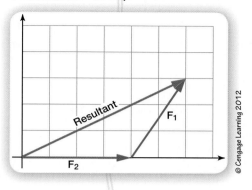

FIGURE 12-16: *Vector addition: tip-to-tail method, step 3.*

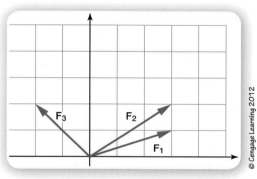

FIGURE 12-17: *Vector addition: tip-to-tail method for three or more vectors, step 1.*

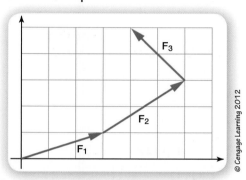

FIGURE 12-18: *Vector addition: tip-to-tail method for three or more vectors, step 2.*

Step 3. Draw the resultant force by connecting the origin to the tip of the last vector drawn. Measure the length and the angle of the resultant force to determine its magnitude and direction (see Figure 12-19).

SOLVING FOR VECTOR COMPONENTS ALGEBRAICALLY

Vectors are easier to deal with when they are expressed in terms of perpendicular components. The Cartesian coordinate system (see Figure 12-20) uses the X, Y, and Z axes to locate points in space. In this chapter, we will only be dealing with the X and Y axes because all of the structures we will be analyzing lie in one plane.

FIGURE 12-20: *Cartesian coordinate system.*

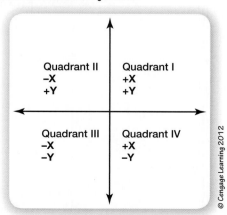

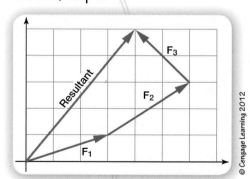

FIGURE 12-19: *Vector addition: tip-to-tail method for three or more vectors, step 3.*

Example 12-1

Use Pythagorean theorem to determine the resultant force for the system of forces shown in Figure 12-21.

$$F = \sqrt{F_1^2 + F_2^2}$$

$$F = \sqrt{(50\ lb)^2 + (20\ lb)^2}$$

$$F = \sqrt{2900\ lb^2}$$

$$F = 54\ lb$$

Now that we know the magnitude of the resultant force, we will use trigonometry to calculate the angle of the resultant force.

$$\theta = \tan^{-1}\frac{F_2}{F_1}$$

$$\theta = \tan^{-1}\frac{20\ lb}{50\ lb}$$

$$\theta = \tan^{-1}(0.4)$$

$$\theta = 21.8°$$

This vector has a magnitude of 54 lb and acts at 21.8°.

FIGURE 12-21: Calculate the resultant force.

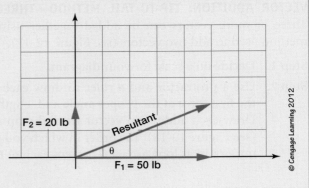

© Cengage Learning 2012

Example 12-2

Find the *X* component of force F_A in Figure 12-22.

$$F_{AX} = (F_A)\cos\theta$$

$$F_{AX} = (50\ lb)\cos 30°$$

$$F_{AX} = (50\ lb)0.866$$

$$F_{AX} = 43.3\ lb$$

Find the *Y* component of force *F* in Figure 12-22.

$$F_{AY} = (F_A)\sin\theta$$

$$F_{AY} = (50\ lb)\sin 30°$$

$$F_{AY} = (50\ lb)0.500$$

$$F_{AY} = 25\ lb$$

FIGURE 12-22: Calculate the X and Y components of force F_A.

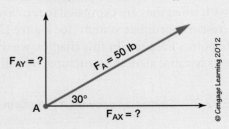

© Cengage Learning 2012

RIGID BODIES

It is hard to imagine a rubber band that can't stretch or a marshmallow that is impossible to compress, but in the study of statics we assume just that. An object that does not experience a change in shape or size when a force is applied to it is referred to as a **rigid body**. This is a theoretical concept because, in reality, all materials undergo some level of deformation when a force is applied to them.

The Principle of Transmissibility

When an external force is applied to a rigid body, the force is transmitted through the body in a straight line called the *line of action*. The line of action is an imaginary line that is collinear with the applied force and passes through the entire object or structure. The *principle of transmissibility* applies only to rigid bodies. In Figure 12-23, the line of action of force F_p extends through the entire object, transmitting the force to the opposite end. The same concept can be used when analyzing a force that is applied to a truss.

The see-saw shown in Figure 12-24 might bring back memories. If you sat across from someone who weighed more than you, chances are you spent a good deal of time in the air. But if you were across from someone who weighed the same as you, you could balance the see-saw so you were both off the ground.

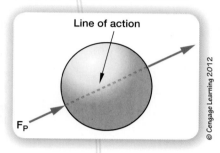

FIGURE 12-23: Line of action.

© Cengage Learning 2012

FIGURE 12-24: A see-saw is a good example of balanced and unbalanced forces.

Image copyright Jane September, 2010. Used under license from Shutterstock.com.

As we learned earlier, statics deals with forces acting on rigid bodies in equilibrium. We can conclude that when the see-saw is perfectly balanced and not moving, it is in equilibrium. If a body is in equilibrium, it means that the force system, no matter how complex, must sum to zero. If you're sitting in a chair reading this book, you and the chair are a force system in equilibrium.

External and Internal Forces

An external force is applied by a different body than the one being analyzed. For example, if you are analyzing the force on the chair you are sitting in, you would be considered the external load. A body or structure that applies a force to another part of the same body or structure is considered an internal force. For example, we can calculate the internal member forces acting within the legs of your chair.

Moments

Moment:

a tendency of a force to twist or rotate an object.

A **moment** (*M*) is a tendency of a force to twist or rotate an object. Next time you close a door, try pushing on the door near the hinges. Now try pushing the door closed using the handle. You probably noticed that it is much easier to close the door pushing at the handle (see Figure 12-25). Now try to close the door by pushing directly on the hinge. Because you are pushing on the point at which the door rotates (the fulcrum of a lever), you can push or pull as hard as you want but the door won't budge.

FIGURE 12-25: A door can be easily opened using the handle because only a small amount of force is applied at a distance away from the door's hinges. It becomes increasingly difficult to open a door the closer the force is to the hinges.

© Cengage Learning 2012

A moment is generated when a force (*F*) acts on an object perpendicular to a moment arm distance (*D*), which originates at the fulcrum. The formula for calculating moment is force times distance ($M = F \cdot D$). In the English system, the units for moment are commonly expressed in foot-pounds (ft · lb). In the metric system, moments are commonly expressed in newton-meters (N · m).

FIGURE 12-26: Sum of the forces can be broken down into *X* and *Y* components.

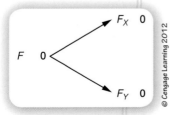

© Cengage Learning 2012

Conditions of Equilibrium

The following two conditions must be satisfied for a body or a structure to exist in a state of static equilibrium:

1. All forces (or components of forces) must sum to zero. In two-dimensions, we must consider both the *X* and *Y* components of these forces. Therefore, we can break down the sum of the forces ($\Sigma F = 0$) into the sum of the forces in the *X*-direction ($\Sigma F_X = 0$) and the sum of the forces in the *Y*-direction ($\Sigma F_Y = 0$) (see Figure 12-26).
2. All of the moments about any given point must sum to zero ($\Sigma M = 0$).

FREE-BODY DIAGRAMS

Drawing free-body diagrams is an essential part of solving equilibrium problems. A free-body diagram (FBD) is a simplified graphic representation of a rigid body, mechanism, or particle that is isolated from its surroundings and on which all applied forces are shown. We will utilize two types of free-body diagrams called *particle* or *point free-body diagrams* and *rigid-body free-body* diagrams when we perform an analysis of the forces acting on and within a truss system.

Particle Free-Body Diagram

Drawing a free-body diagram of a *particle* is very easy because all forces are *concurrent* and intersect at the same point. Figure 12-27 shows a picture that is hung using a piece of wire and a thumbtack. The free-body diagram of particle B is shown to the right. The *reaction force* on point B (F_B) is equal to the total weight of the picture frame. Forces F_{AB} and F_{BC} represent the tension forces in the wires holding the picture frame.

FIGURE 12-27: A free-body diagram of a picture hanging on a wall.

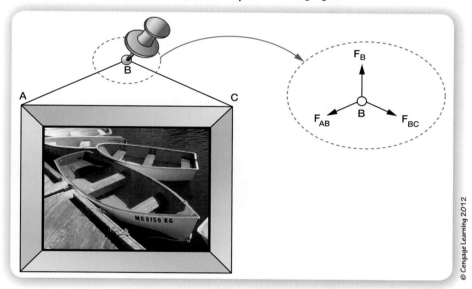

© Cengage Learning 2012

Rigid-Body Free-Body Diagrams

Rigid-body free-body diagrams are usually more complex than particle free-body diagrams because the forces no longer act on a single point. When constructing rigid-body free-body diagrams, it is very important to label all known external forces. Support reactions are external forces that occur where a structure connects to a wall or the ground. The number of reaction forces that occur at these connection points is dependent on the type of support used. The following section will further explain the types of structural connections that are commonly used and their associated reaction forces.

SIMPLE SUPPORTS AND REACTIONS

Figure 12-28 shows a simply supported beam that consists of one (A) *pinned support* and one (B) *roller support*. These supports are commonly referred to as simple supports.

Think of a pinned support as a door hinge. Hinges allow doors to rotate freely, but they do not allow the door to move in any other way. Your knee joints work

FIGURE 12-28: Simple supported beam.
© Cengage Learning 2012

in much the same way as a pinned connection works: They only allow your legs to bend. A pinned support allows a structure to rotate, but it does not allow linear movement. As such, two reaction forces are generated at a pinned support: one in the X-direction and one in the Y-direction. In the study of statics, we assume pinned connections to be frictionless. The free-body diagram of a pinned support includes both horizontal and vertical reactions (see Figure 12-29).

A single pinned support is not enough to make a structure stable. One more support must be placed on the structure to prevent rotation. Roller supports are able to rotate and move along the surface upon which the roller rests (see Figure 12-30). The surface can be horizontal, vertical, or sloped at any angle. Roller supports are always used at one end of a simply supported structure to minimize potentially destructive forces resulting from thermal expansion and contraction. Roller supports can also be rubber bearings that are designed to allow a small amount of lateral movement. The representation of a roller support includes one reaction force that is perpendicular to the surface.

FIGURE 12-29: Pinned support.   FIGURE 12-30: Roller support.

© Cengage Learning 2012

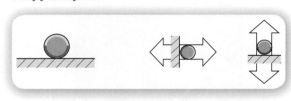

Cantilever:

a structural member that projects beyond a supporting column or wall, leaving it unsupported on one end.

Figure 12-31 shows a **cantilever** beam held in place by a *fixed support*. A cantilever is a structural member with one built-in end and one unsupported end. The fixed support shows the supported object literally built into the supporting surface.

FIGURE 12-31: Cantilever beam.

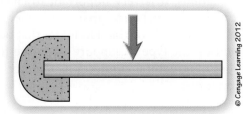

SIMPLE BEAMS

The first step in calculating the reaction forces that act on a beam is to construct a free-body diagram (see Figure 12-32) that includes any applied forces. Once this is done, the equilibrium equations will be used to determine the support reactions.

FIGURE 12-32: (a) Simply supported beam with two supports; (b) overhang with two supports, but one end is not supported; (c) cantilever with one end of the beam fixed.

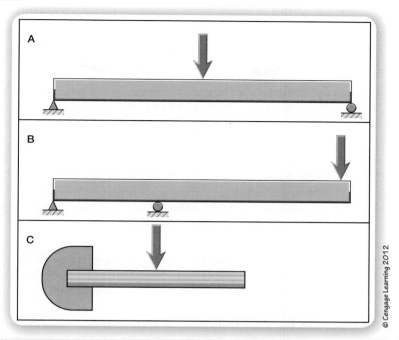

© Cengage Learning 2012

Example 12-3

Draw a free-body diagram of the beam shown in Figure 12-33 and calculate the reaction forces.

FIGURE 12-33: Diagram of beam.

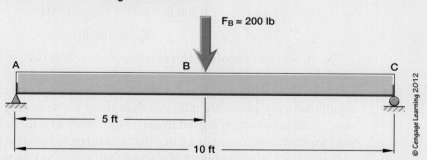

$F_B = 200$ lb

A   B   C

5 ft

10 ft

© Cengage Learning 2012

The beam shown in Figure 12-34 is a simply supported beam. It is supported by a pinned support on the left and a roller support on the right. The force on point B (F_B) is a point or concentrated load. A concentrated load occurs at specific locations on the body being analyzed. The support reactions for this beam can be calculated using the equations of equilibrium.

FIGURE 12-34: Free-body diagram of beam.

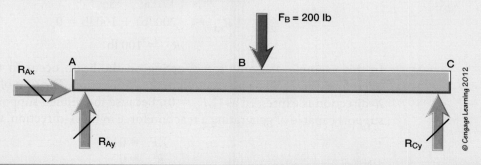

$F_B = 200$ lb

R_{Ax} A   B   C

R_{Ay} R_{Cy}

© Cengage Learning 2012

Step 1. *Draw the free-body diagram.* Draw a free-body diagram of the entire body being analyzed. The support reactions are represented by arrows with a diagonal line through each one. Point A has two reactions because it is a pinned support. Point C only has one reaction because it is a roller support.

Step 2. *Calculate the support reactions.* The magnitude and directions of the reaction forces that occur at the supports are calculated using the equations of equilibrium.

We will begin by using the sum of the moments about a given point ($\Sigma M = 0$). In this case, we will use the pinned support (A) as the point to sum the moments about ($\Sigma M_A = 0$). It is possible to sum the moments about any point, but for consistency we will always sum the moments about the pinned support in this text. Because all forces acting through the point where we are summing the moments are at a distance of zero from it, we can ignore all forces acting through that point.

Before writing the actual equation, place your finger on point A and imagine being able to spin the beam around that point. When you spin the beam clockwise, we will consider that a negative direction. Spinning the beam counterclockwise is considered to be in a positive direction.

All forces that tend to cause a rotation about point A need to go into the moment equation. Remember, a moment results from a force that acts perpendicular to a moment arm distance that originates at the fulcrum. Therefore, equilibrium equations to balance moments must always have a force times a distance. The force on point B acts at a distance of 5 feet away from point A and tends to cause a clockwise rotation about point A, so the moment of F_B is negative. The support reaction at point C acts at a distance of 10 feet away from point A and tends to cause a counterclockwise rotation about point A. Therefore, the moment caused by R_{CY} is positive.

$$\Sigma M_A = 0$$
$$-(F_B \cdot 5 \text{ ft}) + (R_{CY} \cdot 10 \text{ ft}) = 0$$
$$-(200 \text{ lb} \cdot 5 \text{ ft}) + (R_{CY} \cdot 10 \text{ ft}) = 0$$
$$R_{CY} = 100 \text{ lb}$$

Next, we can sum of the forces in the Y-direction ($\Sigma F_Y = 0$) to find the reaction force R_{AY}.

When summing the forces in the X or Y directions, the rules for setting up the equilibrium equations are different than those for summing moments. The Cartesian coordinate system is used to determine whether a force is positive or negative. Any force or component of a force that acts up or to the right is considered positive, and any force or component of a force that acts down or to the left is considered negative. The reaction force at joint A acts upward ($+$) in the Y-direction. The reaction force at joint C acts upward ($+$) in the Y-direction. And, the applied force on point B acts down ($-$). The order in which the forces are arranged is not important as long as all of the forces acting in the Y-direction are accounted for:

$$\Sigma F_Y = 0$$
$$R_{AY} + (-F_B) + R_{CY} = 0$$
$$R_{AY} + (-200 \text{ lb}) + 100 \text{ lb} = 0$$
$$R_{AY} = 100 \text{ lb}$$

In this example, the only forces acting on the beam occur in the vertical axis. Therefore, there will be no reaction in the X-direction. The sum of the forces in the X-direction is expressed as ($\Sigma F_X = 0$). Because the pinned support (A) is the only support capable of generating a reaction force in the X-direction, we write $R_{AX} = 0$:

$$\Sigma F_X = 0$$
$$R_{AX} = 0$$

Example 12-4

Figure 12-35 shows another simply supported beam, but this time the applied force is located off center.

FIGURE 12-35: Diagram of beam.

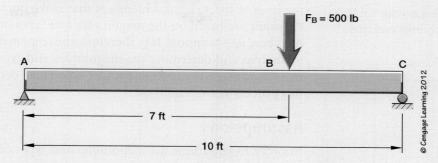

1. Draw a free-body diagram of the beam shown in Figure 12-36 and calculate the reaction forces.

FIGURE 12-36: Free-body diagram.

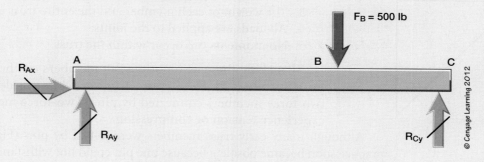

2. Sum of the moments about point A:

$$\Sigma M_A = 0$$
$$-(F_B \cdot 7 \text{ ft}) + (R_{CY} \cdot 10 \text{ ft}) = 0$$
$$-(500 \text{ lb} \cdot 7 \text{ ft}) + (R_{CY} \cdot 10 \text{ ft}) = 0$$
$$R_{CY} = 350 \text{ lb}$$

3. Sum of the forces in the Y-direction:

$$\Sigma F_Y = 0$$
$$R_{AY} + (-F_B) + R_{CY} = 0$$
$$R_{AY} + (-500 \text{ lb}) + 350 \text{ lb} = 0$$
$$R_{AY} = 150 \text{ lb}$$

4. Sum of the forces in the X-direction:

$$\Sigma F_X = 0$$
$$R_{AX} = 0$$

A truss is a structural system that distributes loads to supports through a series of members connected in a triangular pattern. You can see what happens when the triangle shape is not used (Figure 12-37). Engineers are responsible for determining the types of materials that trusses are made from, along with the sizes of the structural members that make up a truss system. These decisions are based on the magnitudes and types of loads that the truss members must support. It is, therefore, the responsibility of the engineer to analyze and determine the magnitudes and types of forces that the truss members will experience under actual load conditions. We refer to this process as truss analysis.

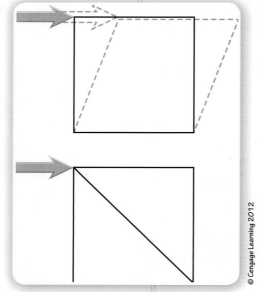

FIGURE 12-37: *Adding a diagonal to this structure greatly improved its stability.*

© Cengage Learning 2012

Assumptions

In order to determine the magnitudes and types of forces that are carried by the members of a truss system, we must make the following assumptions:

1. All members are perfectly straight.
2. All members are pinned and frictionless.
3. The weight of each member and the entire truss are neglected.
4. All loads are applied to the joints.
5. No moments can occur within the truss.

By making these assumptions, all members can be treated as two-force members and the entire truss can be considered a group of two-force members connected by pins. Two-force members can only experience tension or compression.

Although many early truss members were joined by pins (Figure 12-38), this practice soon became obsolete because one pin could not withstand the shear stress that occurs as the truss members transmit their forces through the truss system. The solution to this problem was to use multiple bolts, rivets, or other fasteners to hold each joint together (Figure 12-39).

FIGURE 12-38: *Pinned connection.*

Wes Kinsler, 2206, http://okbridges.wkinsler.com

FIGURE 12-39: *Gusset plate.*

How does the gusset plate connection in Figure 12-39 differ from the pinned connection in Figure 12-38? The gusset plate connection shown in the photo does not allow the members to rotate, so the members can also experience nonaxial forces.

STATIC DETERMINACY

The first step in analyzing a truss is to create a free-body diagram of the entire truss. All forces and support reactions need to be included in this diagram. From this FBD, we can determine the number of equilibrium equations needed to solve for the unknown support reactions.

For example, the truss in Figure 12-40 has one pinned support, one roller support, and two applied forces. The formula to check for static determinacy is $2J = M + R$.

$$2J = M + R$$

J = number of joints
M = number of members
R = number of support reactions

The following calculation shows how to check for static determinacy for the truss in Figure 12-40:

$$2J = M + R$$
$$2 \cdot 6 = 9 + 3$$
$$12 = 12$$

FIGURE 12-40: *Statically determinate truss.*

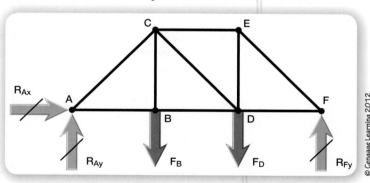

In this case, the numbers are equal and the truss is said to be statically determinate. The next example will guide you through an example of a *statically indeterminate* truss. In this case, the number of unknowns exceeds the number of equations of equilibrium.

Example 12-5

The truss in Figure 12-41 has one additional member added from the previous example. In this case, even though there is one pin and one roller support, the truss is statically indeterminate because there are too many internal members. Our method of truss analysis will not work for a statically indeterminate truss.

Check for static determinacy:

$$2J = M + R$$

$$2 \cdot 6 = 10 + 3$$

$$12 = 13$$

FIGURE 12-41: Statically indeterminate truss.

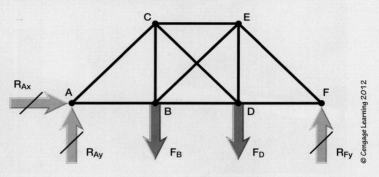

SETTING STANDARDS

The following standards have been established for the purpose of consistency throughout this chapter.

1. Measure all angles from horizontal so that we can always use $F \sin \theta$ to solve for Y-components and $F \cos \theta$ to solve for X-components.
2. When drawing free-body diagrams, assume all internal member forces to be in tension.

Examle 12-6

Draw a free-body diagram of the entire truss in Figure 12-42 for the purpose of calculating the reaction forces that occur at the simple supports. Once this is done, point free-body diagrams of each truss member joint will be drawn for the purpose of calculating the internal member forces.

FIGURE 12-42: This simple truss has one pinned support and one roller support and is subjected to one external force F_K.

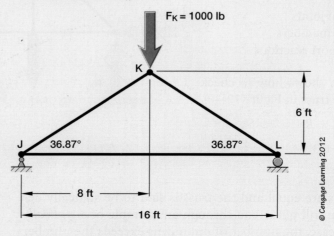

(continued)

Example 12-6 (continued)

When drawing a free-body diagram of the entire truss, treat it as a solid object. Don't worry about the internal member forces yet. In Figure 12-43, the pinned support has two reactions, and the roller support has one reaction.

FIGURE 12-43: Free-body diagram.

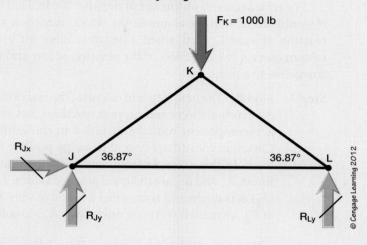

© Cengage Learning 2012

Step 1. Check for static determinacy. There are 3-joints, 3-members, and 3-support reactions in the truss.

$$2J = M + R$$
$$2 \cdot 3 = 3 + 3$$
$$6 = 6$$

Great! The truss is statically determinate, so we can move on to the next step—calculating the support reactions.

Step 2. Find R_{LY}. Choosing the right equilibrium equation can be difficult at first. We have three to choose from:

$$\Sigma M = 0 \text{ (sum of the moments)}$$
$$\Sigma F_X = 0 \text{ (sum of the forces in the X-direction)}$$
$$\Sigma F_Y = 0 \text{ (sum of the forces in the Y-direction)}$$

For consistency in this text, when solving for unknown support reactions, we will start each problem by summing the moments ($\Sigma M = 0$) about the pinned support. In this problem, the pinned support is located at joint J. Therefore, our equilibrium equation will be

$$\Sigma M_J = 0 \text{ (the sum of the moments about point J)}$$

When writing the equilibrium equation, all forces that tend to cause a rotation about point J need to go into the equation. As with the simply supported beam problems, we assume moments that occur in the clockwise direction to be negative and those that occur in the counterclockwise direction to be positive. Remember, a moment is the product of a force that acts on an object perpendicular to a moment arm distance, which originates at the fulcrum ($M = F \cdot D$). Whenever a force appears in the equation, it must be multiplied by its distance away from joint J as measured perpendicular to the direction that the force acts. The dimensions of this truss can be found in Figure 12-42.

$$\Sigma M_J = 0$$

$$-(F_K \cdot 8 \text{ ft}) + (R_{LY} \cdot 16 \text{ ft}) = 0$$

$$-(1{,}000 \text{ lb} \cdot 8 \text{ ft}) + (R_{LY} \cdot 16 \text{ ft}) = 0$$

$$R_{LY} = 500 \text{ lb}$$

We ended up with an answer of positive 500 lb. Had this number been negative, it would tell us that we assumed the wrong direction when we drew the support reaction at joint L. Also, when a negative value for a reaction force results, it is wise to correct the direction of the reaction arrow and then change the sign from a negative to a positive.

Step 3. Find R_{JY}. The next step is to calculate the reaction at joint J in the Y-direction. All external forces and support reactions that act in the Y-direction or have a Y-component must be included in this equilibrium equation. Use the Cartesian coordinate system to assign positive and negative values to each force. In this example, both reactions are pointing up. Therefore reaction forces R_{JY} and R_{LY} are assigned positive values. The external force at point K (F_K) acts down and is assigned a negative sign. By substituting in the value of F_K, we are left with one unknown (R_{JY}) to solve for.

$$\Sigma F_Y = 0$$

$$R_{JY} + (-F_K) + R_{LY} = 0$$

$$R_{JY} - 1{,}000 \text{ lb} + 500 \text{ lb} = 0$$

$$R_{JY} = 500 \text{ lb}$$

We ended up with a positive value of 500 lb for the reaction at joint J in the Y-direction. This means that we correctly assumed the direction of the reaction force.

Step 4. Find R_{JX}. Now it's time to find the last of the reaction forces at joint J in the X-direction (R_{JX}). Any external force or support reaction that acts in the X-direction or has an X-component must be included in the following equation. Fortunately, this is a very easy calculation because the only force that is applied to the truss is vertical. Because the truss is a rigid body and does not deform, the reaction at joint J in the X-direction will be zero.

$$\Sigma F_X = 0$$

$$R_{JX} = 0$$

Step 5. Calculate the internal member forces. Now that we know the values of all the support reactions, we can move to the inside of the truss to calculate the internal member forces. An internal member force is the force that a truss member is subjected to as a result of an external force(s) that is transmitted through the truss. Engineers use internal member forces to choose the proper size and material of the truss members.

Step 6. Draw a point free-body diagram of each joint. Before solving for an internal member force, we need to draw an FBD of each joint in the truss, but this time we need to show the internal member forces. You can see in Figure 12-44 that all of the arrows for internal members are pointing away from the joints, meaning those members are in tension. For consistency, we initially assume all members to be in tension. It is, of course, impossible for this to be true. However, making an assumption that all members are in tension

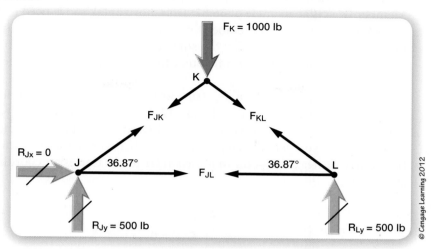

gives us a uniform way of working with force vectors through our mathematical analysis using the equilibrium equations. When we are dealing with internal member forces in this chapter, a negative answer will indicate that the member is in compression, and a positive answer means our assumption of tension was correct.

Step 7. Choose where to begin. You may recall that we assumed that internal members can only experience tension or compression. Therefore, we only have two equilibrium equations to choose from because no member can be subjected to a twisting force (or a moment). We can choose either the sum of the forces in the X-direction ($\Sigma F_X = 0$) or the sum of the forces in the Y-direction ($\Sigma F_Y = 0$).

The following table shows the number of unknowns at each joint using these two equilibrium equations. Because we can't solve for more than one unknown at a time, we can begin our calculations at either joint J or joint L by summing the forces in the Y-direction.

Joint	Equilibrium Equation	Number of Unknowns
J	$\Sigma F_X = 0$	2
J	$\Sigma F_Y = 0$	1
K	$\Sigma F_X = 0$	2
K	$\Sigma F_Y = 0$	2
L	$\Sigma F_X = 0$	2
L	$\Sigma F_Y = 0$	1

Step 8. Solve for internal member forces. For consistency in this text, we will begin our calculations from the side of the pinned support. Draw a free-body diagram of joint J, making sure to include the values of the support reactions, the angle of force F_{JK}, and labels for all of the member forces and reactions.

The equilibrium equation $\Sigma F_{JY} = 0$ means "the sum of the forces at joint J in the Y-direction equal zero." Because joint J is in equilibrium, all forces acting on it must balance. Summing the forces in the Y-direction requires that all forces or components of forces that act on joint J in the

FIGURE 12-45: *Free-body diagram of joint J.*

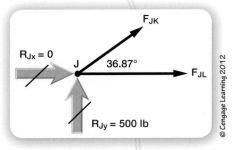

Y-direction be added together. By summing the forces in the Y-direction, we will be able to calculate the internal member force of member JK. Member JK acts at an angle of 36.87° from horizontal, so we need to find the Y-component of this member force. This can be found by multiplying the force in member JK by the sine of 36.87°. The reaction at joint J in the Y-direction (R_{JY}) is the only other force that acts in the Y-direction at joint J.

The equilibrium equation F_{JK} (sin 30°) + R_{JY} = 0 was formulated using the free-body diagram in Figure 12-45. Because force F_{JK} has a Y-component that acts upward and R_{JY} acts upward, we use positive signs for both forces. If you get stuck trying to remember whether a force should be positive or negative, you can always refer to the Cartesian coordinate system.

$$\Sigma F_{JY} = 0$$

$$R_{JY} + F_{JKY} = 0$$

$$500 \text{ lb} + F_{JK} (\sin 36.87°) = 0$$

$$F_{JK} = -833.33 \text{ lb (compression)}$$

The force in member JK (F_{JK}) was calculated to be −833.33 lb. A negative answer means that our assumption of tension in member JK was incorrect. Member JK is actually in compression. Do not switch the direction of the internal member force, simply keep the negative sign with the −833 lb to note that member JK is in compression.

We have one more unknown force to calculate at joint J: F_{JL}. Because member JL is in the X-direction, we can use the equilibrium equation $\Sigma F_{JX} = 0$ to solve for the internal member force of F_{JL}.

$$\Sigma F_{JX} = 0$$

$$R_{JX} + F_{JKX} + F_{JL} = 0$$

$$0 \text{ lb} - 833 \text{ lb} (\cos 36.87°) + F_{JL} = 0$$

$$R_{JX} = 666.67 \text{ lb}$$

FIGURE 12-46: *Free-body diagram of joint L.*

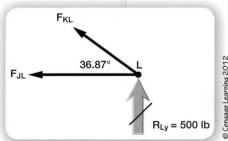

The internal member force of member JL (F_{JL}) came out to a positive 666.67 lb meaning our assumption of tension was correct (see Figure 12-46). We are now ready to move on to another joint.

Our next step in the analysis involves calculating the magnitude of the force that is carried by member KL (F_{KL}). In this case, either of joints K or L can be used to solve for F_{KL} because each joint has only one unknown to solve for. We chose to use the sum of the forces in the Y-direction at joint L ($\Sigma F_{LY} = 0$).

$$\Sigma F_{LY} = 0$$

$$R_{LY} + F_{KLY} = 0$$

$$500 \text{ lb} + F_{KL} (\sin 36.87°) = 0$$

$$F_{KL} = -833.33 \text{ lb (compression)}$$

The force in member KL (F_{KL}) was calculated to be −833.33 lb. Remember, a negative answer means that our assumption of tension in member JK was incorrect and member KL is actually in compression.

Step 9. Redraw the free-body diagram showing all forces (see Figure 12-47).

FIGURE 12-47: Finished truss problem showing all forces.

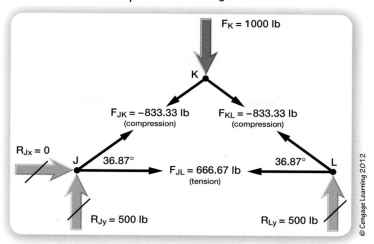

© Cengage Learning 2012

Your Turn

Draw a similar truss on a sheet of graph paper and change the applied force and the member lengths. Solve for the reaction forces and internal member forces.

Point of Interest
Nine Steps to Solve a Truss Problem

1. Check for static determinacy. There is no sense in going any further if the number of unknowns exceeds the number of equilibrium equations. A statically indeterminate truss is not unstable; it just has too many unknowns to solve for using the three equations of equilibrium.

2. Draw a free-body diagram of the entire truss. Label all joints and replace the support symbols (pinned and roller) with reaction forces.

3. Sum the moments about the pin support to determine one of the unknown reaction forces.

4. Use the X- and Y-direction equilibrium equations to determine the remaining unknown reaction forces.

5. Draw point free-body diagrams of the truss joints that show all of the internal member forces in tension. This is easily done by re-drawing the truss from step 2 in pencil. Erase a small portion from the center of each member and draw arrows pointing away from the joints to note tension.

6. Begin the analysis by selecting a joint that has only one unknown force.

7. Use the X- and Y-direction equilibrium equations ($\Sigma F_x = 0$ or $\Sigma F_y = 0$) to solve for all the uknown truss member forces.

8. Repeat the previous step for the remaining joints and solve for the remaining unknown member forces.

9. Identify all of your answers in a list or by re-drawing the free-body diagram of the entire truss and labeling every support reaction and internal member force.

Example 12-7 (see Figure 12-48)

1. Check for static determinacy

   $2J = M + R$

   $2 \cdot 4 = 5 + 3$

   $8 = 8$

2. Draw a free-body diagram of the truss showing reaction forces.

3. Find R_{DY} by summing the moments around point A.

   $\Sigma M_A = 0$

   $-(F_c \cdot 12 \text{ ft}) + (R_{DY} \cdot 17 \text{ ft}) = 0$

   $-(2{,}400 \text{ lb} \cdot 12 \text{ ft}) + (R_{DY} \cdot 17 \text{ ft}) = 0$

   $R_{DY} = 1{,}694 \text{ lb}$

4. Find R_{AY} by summing the forces in the Y-direction.

   $\Sigma F_Y = 0$

   $R_{AY} + (-F_c) + R_{DY} = 0$

   $R_{AY} + (-2{,}400 \text{ lb}) + 1{,}694 \text{ lb} = 0$

   $R_{AY} = 706 \text{ lb}$

5. Find R_{AX} by summing the forces in the X-direction.

   $\Sigma F_X = 0$

   $R_{AX} = 0$

6. Draw point free-body diagrams of the truss joints showing the internal member forces (see Figure 12-49).

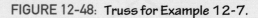

FIGURE 12-48: Truss for Example 12-7.

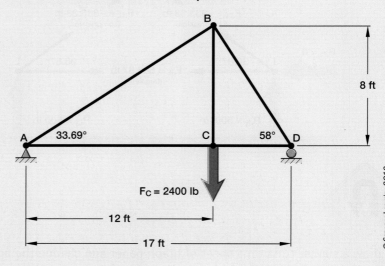

© Cengage Learning 2012

FIGURE 12-49: Free-body diagram showing internal member forces.

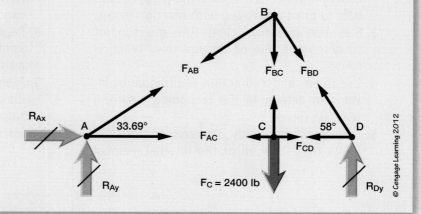

© Cengage Learning 2012

Example 12-7 (continued)

7. Solve for the internal member forces in members AB and AC from joint A.

$$\Sigma F_{AY} = 0$$

$$R_{AY} + F_{ABY} = 0$$

$$706 \text{ lb} + F_{AB} (\sin 33.69°) = 0$$

$$F_{AB} = -1,273 \text{ lb (compression)}$$

$$\Sigma F_{AX} = 0$$

$$R_{AX} + F_{AC} + F_{ABX} = 0$$

$$0 \text{ lb} + F_{AC} + (-1,273 \text{ lb} \times \cos 33.69°) = 0$$

$$F_{AC} = 1,059 \text{ lb (tension)}$$

8. Solve for the internal member forces in members BD and CD from joint D.

$$\Sigma F_{DY} = 0$$

$$R_{DY} + F_{BDY} = 0$$

$$1,694 \text{ lb} + F_{BD} (\sin 58°) = 0$$

$$F_{BD} = -1,998 \text{ lb (compression)}$$

$$\Sigma F_{DX} = 0$$

$$-F_{CD} + (-F_{BDX}) = 0$$

$$-F_{CD} + (-1,998 \text{ lb} \times \cos 58°) = 0$$

$$F_{CD} = 1,059 \text{ lb (tension)}$$

9. Solve for the internal member forces in members BC from joint C:

$$\Sigma F_{CY} = 0$$

$$F_C + F_{BC} = 0$$

$$-2,400 \text{ lb} + F_{BC} = 0$$

$$F_{BC} = 2,400 \text{ lb (tension)}$$

OFF-ROAD EXPLORATION

Download West Point Bridge Designer from http://bridgecontest.usma.edu and compete in a local West Point Bridge Design Contest.

SUMMARY

- The science of *mechanics* involves the study of forces and their effects. Two of the common subject areas within the science of mechanics are (1) *statics*, the study of objects at rest, and (2) *dynamics*, the study of objects in motion.

- A *structure* is a stable assembly of connected components that is capable of withstanding loads.

- A *truss* is a triangular arrangement of members commonly made of wood or steel that forms a rigid framework.

- Civil engineers are responsible for the design and construction of roads, bridges, buildings, airports, waterways, and water- and wastewater-treatment facilities.

- A push or pull that a structural member is subjected to from other parts of the structure is called an *internal member force*.

- *Static loads* are usually gravity-type loads. These would include the weight of the structure itself or any other load that is stationary. *Dynamic loads* are movement-based loads.

- A *scalar* quantity has magnitude only, whereas *vector* quantities have magnitude, direction, and sense.

- A *force system* is two or more forces that are collectively considered.

- Any object that does not change its shape or size when a force is applied to it is referred to as a *rigid body*.

- A *moment* is a tendency of a force to twist or rotate an object.

- A *free-body diagram*, or *FBD*, is a simplified graphic representation of a rigid body, mechanism, or particle that is isolated from its surroundings and shows all of the forces that act on the object.

- A *cantilever* is a structural member with one built-in end and one unsupported end.

BRING IT HOME

Directions: Use the diagram in Figure 12-50 to calculate the resultant force for each of the following:

1. $F_1 = 50$ N; $F_2 = 20$ N
2. $F_1 = 1,000$ lb; $F_2 = 1,000$ lb
3. $F_1 = 800$ N; $F_2 = 300$ N
4. $F_1 = 1,200$ lb; $F_2 = 1,500$ lb
5. $F_1 = 3$ tons; $F_2 = 4$ tons

Directions: Calculate the X and Y components of the force acting on point A (F_A) shown in Figure 12-51.

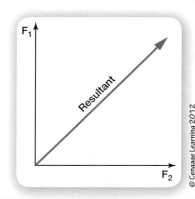

FIGURE 12-50: Use this diagram to solve problems 1–5.

© Cengage Learning 2012

BRING IT HOME

(Continued)

FIGURE 12-51: Use this diagram to solve problems 6-10.

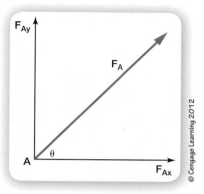

© Cengage Learning 2012

6. $F_A = 400$ lb; $\theta = 20°$
7. $F_A = 1,200$ N; $\theta = 60°$
8. $F_A = 150$ lb; $\theta = 80°$
9. $F_A = 2,800$ N; $\theta = 33°$
10. $F_A = 9,000$ lb; $\theta = 15°$

Directions: Use the analysis procedure outlined in this chapter to solve the following problems.

11. The top of a ladder rests against a wall 14 feet above the floor. If the angle formed between the ladder and the floor is 80°, calculate (a) the length of the ladder, (b) the distance from the base of the wall to the foot of the ladder, and (c) the angle between the ladder and the wall.
12. Use Figure 12-52 to (a) draw a free-body diagram of the entire beam and (b) calculate the reaction forces at supports A and B.
13. Use Figure 12-53 to (a) draw a free-body diagram of the entire beam, (b) find the X and Y components of the 1 kip force (1 kip = 1,000 lb), and (c) calculate the reaction forces at points A and B.
14. A delivery truck is positioned in the middle of a bridge. (a) Draw a free-body diagram of the entire truss. (b) Calculate the support reaction forces. (c) Calculate all of the internal member forces.

FIGURE 12-52

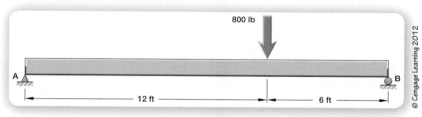

© Cengage Learning 2012

FIGURE 12-53

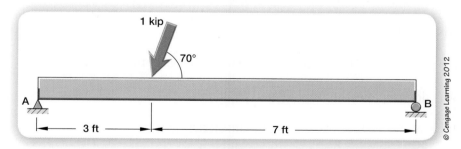

© Cengage Learning 2012

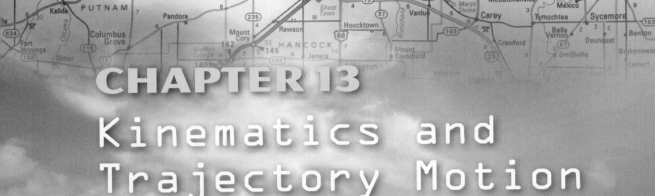

CHAPTER 13
Kinematics and Trajectory Motion

GPS DELUXE

| START LOCATION | DISTANCE | END LOCATION |

Menu

Before You Begin

Think about these questions as you study the concepts in this chapter:

1. What is mechanics?

2. What is the difference between kinetics and kinematics?

3. What are the basic kinematic terms?

4. How does a vector quantity differ from a scalar quantity?

5. Why does a projectile take a parabolic flight path?

6. In what everyday applications can kinematics be used?

I n most college-level engineering programs, aspiring engineers are required to take courses on the topic of **mechanics**, which is the study of motion, forces, and the effects of forces on objects. Mechanics is a condensation of the discoveries of classical physicists and mathematicians such as Sir Isaac Newton and Galileo. It is a topic that applies to engineering disciplines concerned with the design, function, and motion of tangible objects.

> **Mechanics:**
> the study of motion, forces, and the effects of forces on objects.

Few of us think of today's professional golfers as students of mechanics, but anyone who wants to compete on a professional tour must learn to control the flight of a golf ball with great precision and consistency (see Figure 13-1). Becoming a student of mechanics can help a golfer in this quest.

FIGURE 13-1: *Golfers can use the principles of kinematics to control the flight of the ball.*

> **Kinematics:**
> a branch of mechanics that is concerned with the study of motion without consideration of the forces that caused the motion.

FIGURE 13-2: **An angle chart showing the relationship between the club type and the club face angle.**

Standard Golf Club Loft Angle Chart	
Club	Loft Angle Range
Driver	7–13 Degrees
2 Wood	12–15 Degrees
3 Wood	12–17 Degrees
4 Wood	15–19 Degrees
5 Wood	20–23 Degrees
6 Wood	22–25 Degrees
7 Wood	25–28 Degrees
1 Iron	15–18 Degrees
2 Iron	18–20 Degrees
3 Iron	21–24 Degrees
4 Iron	25–28 Degrees
5 Iron	28–32 Degrees
6 Iron	32–36 Degrees
7 Iron	36–40 Degrees
8 Iron	40–44 Degrees
9 Iron	45–48 Degrees
Pitching Wedge	47–53 Degrees
Gap Wedge	50–54 Degrees
Sand Wedge	54–58 Degrees
Lob Wedge	58–62 Degrees

Specifically, golfers must understand a special branch of mechanics known as **kinematics**. Kinematics is the study of motion without consideration of the forces that caused the motion.

Golfers choose from a variety of club shapes, depending on the distance they want to hit the ball. They apply their knowledge of kinematics when they make that choice. Each type of club has a specific angle that produces a different flight path (Figure 13-2).

The driver is used to achieve the greatest distance. It has a steeply angled face, as little as 7 degrees off the vertical, or almost straight up and down. Conversely, the lob wedge has the flattest angle. Some wedges are as much as 63 degrees off the vertical (see Figure 13-3). The pitching wedge is used for very short shots.

In the language of kinematics, the speed of the golfer's swing provides the initial velocity of the ball. By keeping the initial velocity of the ball the same, through years of practice, a golfer can accurately predict how far the ball will go based on the club head's angle.

FIGURE 13-3: **Effect of the club angle on the distance the ball travels.**

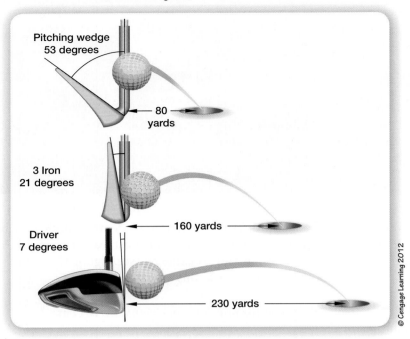

FIGURE 13-4: **The disciplines within the subject of mechanics.**

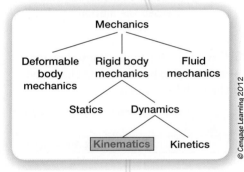

Kinematics can be applied to any sport in which something or someone is flying through the air, such as football, soccer, diving, sport shooting, freestyle motocross, water-balloon launching, and baseball. In this chapter, we will use the language of kinematics to demonstrate how objects traveling through space have different flight paths based on their initial velocities and trajectory angles. Engineers apply kinematics in the aerospace, defense, and robotic industries, just to name a few.

THE DISCIPLINES OF MECHANICS

Mechanics is a broad subject that cannot be covered in just one course. Therefore, it is typically divided into several courses that are completed during the first few years of a postsecondary engineering program.

There are three major branches of mechanics (see Figure 13-4):

1. deformable body mechanics,
2. rigid body mechanics, and
3. fluid mechanics.

In the following sections, we will focus on rigid body mechanics. The term **rigid body** describes an object that does not change in size when a force is applied to it. Much of this textbook focuses on rigid-body mechanics, which is further divided into statics (a topic you learned about in Chapter 12) and dynamics. **Dynamics** is the study of motion and the effects of forces acting on *rigid bodies* in motion. Dynamics is further divided into **kinetics**, or study of the forces that cause bodies to move, and kinematics, which is the study of motion *without* consideration of the forces that caused that motion. *Kinematics* is a Greek word meaning "to move."

BASIC KINEMATIC TERMS

Decriptive words such as fast, slow, turning, sliding, stopping, and speeding up are all used to describe the way in which an object moves. In physics, words such as

Kinetics:

a branch of dynamics that involves the study of forces that cause bodies to move.

acceleration, **speed**, **velocity**, and **distance** are used to describe motion. You need a basic understanding of these terms in order to understand kinematics.

The following terms are likely to be used in a discussion of kinematics:

▶ **Distance** is the space between two objects or points.

▶ **Speed** is the distance traveled per unit time.

▶ **Velocity** is an object's speed and direction of motion.

▶ **Acceleration** is a change in an object's speed or direction divided by the time period during which the change occured.

▶ **Gravity** is the force that pulls objects toward the center of Earth.

▶ **Gravitational acceleration** is the constant value describing the acceleration of any object falling toward Earth. This constant is 9.8 meters per second squared or 32.2 feet per second squared.

MAGNITUDE VERSUS DIRECTION

We can assign mathematical quantities to these basic kinematic terms. These mathematical quantities can then be divided into two categories: scalar and vector. A scalar quantity is one that can be described by magnitude only—for example, mass and time. As shown in Figure 13-5(a), it answers the question, "How much?" A vector quantity is one that must be described by both magnitude and direction such as velocity and force. It answers the questions "How much?" and "Which way?" (see Figure 13-5b).

FIGURE 13-5: (a) Example of a scalar quantity. This car is traveling down the road at 55 mph (shows magnitude, or speed, only). (b) Example of a vector quantity: This car is traveling west down the road at 55 mph (shows magnitude and direction).

Scalar:
a quantity that can be described by magnitude only.

Vector:
a quantity that must be described by both magnitude and direction.

(a)

© Cengage Learning 2012

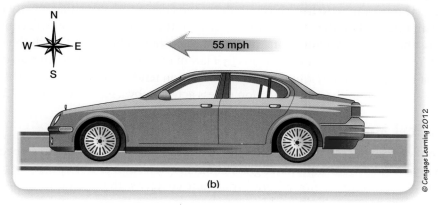

(b)

© Cengage Learning 2012

Your Turn

Identify whether the thermometer and the Navman Wind 3100 gauge in Figure 13-6 show scalar or vector quantities and explain why.

FIGURE 13-6: **(a) Thermometer showing both Celsius and Fahrenheit scales. (b) Navman Wind 3100 wind gauge.**

(a)

(b)

CHARACTERISTICS OF A PROJECTILE'S TRAJECTORY

projectile:
an object that exhibits both vertical and horizontal motion under the influence of gravity alone.

A **projectile** is an object that exhibits both vertical and horizontal motion under the influence of gravity alone. Air resistance is ignored. To analyze and predict the motion of a projectile, the vertical motion and horizontal motion must be considered separately.

Horizontal Motion

Imagine we are standing on one side of the Grand Canyon in Arizona. We are about to use a large cannon to launch you to the other side. Let's assume that gravity is absent and the cliff on the other side is lower or at the same height. You will continue across the canyon in a horizontal motion at a constant velocity and reach the other side. Your flight will observe Newton's first law of motion, which states that an object at rest tends to stay at rest and an object in motion tends to stay in motion with the same speed and in the same direction unless acted on by a force (see Figure 13-7).

Vertical Motion

If you were to step off the side of the cliff, you would free-fall straight down because of gravity. You would experience gravitational acceleration and would accelerate at a rate of 9.8 m/sec² (32 ft/sec²) until you made contact with the ground.

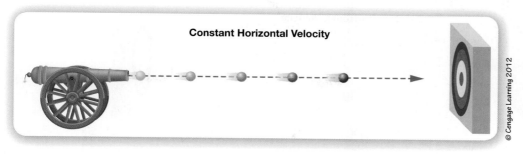

What happens if you are shot out of a cannon straight up into the air? The force of gravity would cause you to decelerate at a rate of 9.8 m/sec² until you reach a certain height at which you no longer have vertical velocity. From this point, gravity will bring you down (see Figure 13-8).

FIGURE 13-8: An illustration of the effect of gravity on vertical motion.

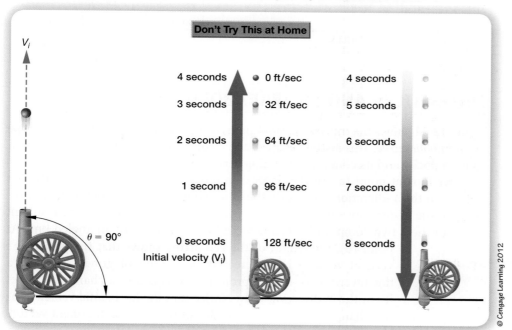

Horizontal and Vertical Motion

We have explained that gravity is a force that only affects an object's vertical motion, remembering that air resistance is also ignored. It has no effect on horizontal motion. Because there is no other force acting on the object, the horizontal velocity of a projectile remains the same throughout the time of flight. Returning to our example of a human cannon going across the Grand Canyon, the projectile (you) will have a constant horizontal velocity and a downward vertical acceleration. Therefore, you will fall to the ground. However, because you have constant horizontal velocity, your downward path will be curved (see Figure 13-9).

Figure 13-10 summarizes the forces, acceleration, and velocity experienced by a projectile in the horizontal and vertical direction.

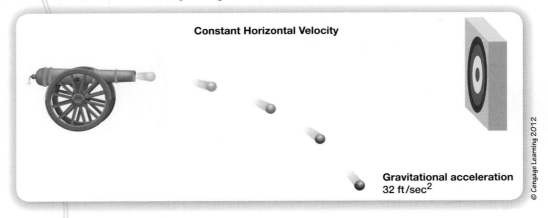

Constant Horizontal Velocity

Gravitational acceleration
32 ft/sec²

© Cengage Learning 2012

FIGURE 13-10: Horizontal and Vertical Motion

	Horizontal Motion (Ignoring Air Resistance)	Vertical Motion
Forces	None	Gravity
Acceleration	None	Gravitational acceleration downward at 32 ft/sec² or 9.8 m/s²
Velocity	Constant	Changing

PROJECTILES AND THE PARABOLIC CURVE

Figure 13-11 shows the *trajectory* (flight path) of a projectile, which takes the shape of a *parabolic curve*. This distinct shape is the result of the projectile's constant horizontal velocity and its contantly changing vertical velocity.

If we look at point (a) on the figure, the red arrow represents the projectile's velocity in the *Y*-direction (vertical). The blue arrow represents the projectile's velocity in the *X*-direction (horizontal). As you learned in Chapter 12, the resultant vector of these two component vectors would take the form of an angled line. At point (b), the length of the red arrow shows that the projectile's vertical velocity has decreased as a result of gravity exerting a downward force on the projectile. This causes a change in velocity in the *Y*-direction only. Notice that the horizontal velocity has not changed. The resultant vector of these two component vectors would also take the form of an angled line, but its angle is shallower than that of the resultant vector at point (a). This trend continues until the projectile reaches the apex of the curve. At the apex, the projectile exibits no vertical motion. Past this point, the projectile is descending. This is shown by the change in direction of the red arrows. Gravitational acceleration at 32 ft/sec² causes the vertical velocity to increase as shown by the increasing lengths of the red arrows. Notice that there is no change in the projectile's horizontal velocity or direction throughout the entire process.

FIGURE 13-11: *Parabolic path of a projectile.*

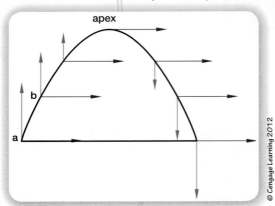

apex

b

a

© Cengage Learning 2012

ANALYZING PROJECTILE MOTION

Now that you have learned some of the characteristics of projectile motion, let's work through some projectile-motion problems. These problems will introduce several key formulas for analyzing projectile motion. Before we begin solving these problems, we will make a table like the one below to list all knowns (KNs) and unknowns (UKNs), a good technique to use when solving any mathematical problem. Finally, while working through these problems, we will assume that all projectiles will start and end on the same horizontal plane.

KNs	UKNs

Projectile-Motion Problem

In this problem, we know the launch angle and initial velocity. A projectile is fired at an angle (θ) equal to 30°. The initial velocity (V_i) of the projectile is 50 m/sec (see Figure 13-12). Solve for the following unknowns:

▶ initial horizontal velocity,

▶ initial vertical velocity,

▶ time to get to the apex of the parabolic curve,

▶ the maximum height the projectile reaches,

▶ the total time of flight, and

▶ the distance the projectile goes before it lands.

Note: The red arrow is a vector (drawn as an arrow) because it shows both magnitude and direction. On the unknown side of the table in Figure 13-13, V_{iX} means the initial velocity in the X-direction and V_{iY} means the initial velocity in the Y-direction.

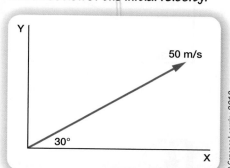

FIGURE 13-12: Showing magnitude and direction of the initial velocity.

© Cengage Learning 2012

FIGURE 13-13: Knowns and unknowns chart.

KN's	UKN's
V_i = 50 m/s	V_{iX}
Theta or Firing angle 30 deg	V_{iY}
	t_{max}
	d_Y
	Total time of flight
	d_X

SOLVING FOR THE INITIAL HORIZONTAL VELOCITY To solve for velocity in the X-direction (Figure 13-14), we use the formula derived from the Pythagorean theorem:

$$V_{iX} = V_i \cos \theta \qquad \text{(Equation 13-1)}$$

In our problem,

$$V_{iX} = (50 \text{ m/sec}) (\cos 30°)$$

$$V_{iX} = (50 \text{ m/sec}) (0.866)$$

$$V_{iX} = 43.3 \text{ m/sec}$$

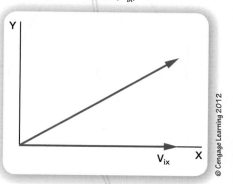

FIGURE 13-14: Solving for the horizontal velocity (V_{ix}).

© Cengage Learning 2012

A projectile is fired at an angle equal to 15°, and the initial velocity is 120 m/sec. What is the initial horizontal velocity? (Show all work.)

FIGURE 13-15: Solving for the initial vertical velocity (V_{iY}).

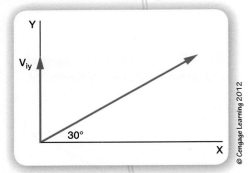

SOLVING FOR INITIAL VERTICAL VELOCITY V_{iY} To solve for velocity in the Y-direction (Figure 13-15), use Equation 13-2:

$$V_{iY} = V_i \sin \theta \qquad \text{(Equation 13-2)}$$

In our problem:

$$V_{iY} = (50 \text{ m/sec}) (\sin 30°)$$
$$V_{iY} = (50 \text{ m/sec}) (0.5)$$
$$V_{iY} = 25 \text{ m/sec}$$

Your Turn

A projectile is fired at an angle equal to 15°, and the initial velocity is 120 m/sec. Solve for initial vertical velocity. (Show all work.)

FIGURE 13-16: The amount of time that it takes a projectile to reach the apex of its trajectory is t_{max}. V_f is the velocity at the highest point, which is always equal to zero.

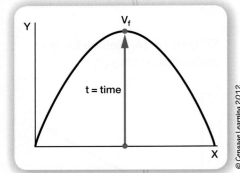

SOLVING FOR T_{max} t_{max} is the amount of time it will take for the projectile to reach its maximum height.

The vertical velocity of the projectile at the top of the curve is zero. It is denoted as V_f; $V_f = 0$ m/sec. Gravitational acceleration (g) is a vector quantity because gravity acts in a downward direction. Where an equation identifies g, its substituted value will be -9.8 m/sec² or -32 ft/sec². Where an equation identifies $-g$, its value will be 9.8 m/sec² or 32 ft/sec², as the inverse of a negative value is a positive value.

To calculate t_{max}, use Equation 13-3.

$$V_f = V_{iY} g t_{max} \qquad \text{(Equation 13-3)}$$

Isolating variable t_{max}:

$$t_{max} = \frac{(V_f - V_{iY})}{g}$$
$$= \frac{(0.0 \text{ m/sec} - 25 \text{ m/sec})}{-9.8 \text{ m/sec}^2}$$
$$= \frac{-25 \text{ m/sec}}{-9.8 \text{ m/sec}^2}$$
$$= 2.55 \text{ sec}$$

Your Turn

Using the previous example where you solved for initial vertical velocity of a projectile fired at 15°, solve for t_{max}. (Show all work.)

Solving for the Maximum Height Reached by the Projectile

Because displacement is in the vertical direction, the maximum height reached by the projectile (d_y) is solved by using Equation 13-4:

$$d_Y = V_{iY} t_{max} + \frac{1}{2} g t_{max}^{\ 2} \qquad \text{(Equation 13-4)}$$

$$= (25 \text{ m/sec}) (2.55 \text{ sec}) + (0.5) (-9.8 \text{ m/sec}^2) (2.55 \text{ sec})^2$$

$$= 63.75 \text{ m} + (-31.86 \text{ m})$$

$$= 31.89 \text{ m}$$

FIGURE 13-17: (a) Knowns and unknowns chart. (b) Showing max height, d_Y

KN's	UKN's
$V_i = 50$ m/s	d_Y = distance
t_{max} = 2.55 s	
$g = 9.8$ m/s²	
$V_{iY} = 25$ m/s	

(a)

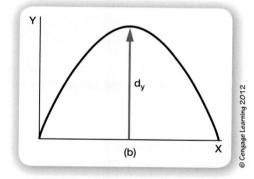

(b)

© Cengage Learning 2012

Your Turn

Using the previous example in which you solved for initial vertical velocity and t_{max} of a projectile fired at 15°, solve for maximum height. (Show all work.)

SOLVING FOR TOTAL FLIGHT TIME OF THE PROJECTILE How much time will pass before the projectile hits the ground? Because the magnitude of gravitational acceleration is constant, we can find the flight time (t) simply by doubling the time it takes the projectile to reach V_f. Previously, we calculated the time it takes the projectile to reach the apex (V_f) as 2.55 s. So, $t = 2(2.55 \text{ sec})$:

$$t = 5.1 \text{ sec (total flight time of the projectile)}$$

SOLVING FOR THE RANGE OF THE PROJECTILE How far the projectile will travel in the X-direction is called *range* (d_X) (see Figure 3-18a). We learned that the

FIGURE 13-18: (a) Calculating range, d_x; (b) knowns and unknowns table.

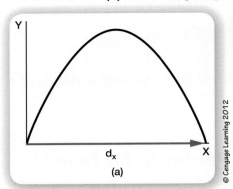

KN's	UKN's
V_{iX} = 43.3 m/sec	d_x = distance in the X direction
t = 5.1 sec	

(b)

(a)

© Cengage Learning 2012

horizontal velocity never changes and that the magnitude is the distance traveled over a period of time. Therefore, to find d_X, use Equation 13-5:

$$d_X = V_{iX}t \qquad \text{(Equation 13-5)}$$
$$= (43.3 \text{ m/sec})(5.1 \text{ sec})$$
$$= 220.83 \text{ m}$$

Your Turn

Using the previous example in which you solved for initial horizontal velocity and the time to reach the apex of the projectile's path when fired at 15°, solve for the range. (Show all work.)

We have just worked through a projectile-motion problem where we solved for six unknowns given a projectile's firing angle and initial velocity. To do this we needed to use multiple equations, in series, to solve for all the unknowns. By rearranging and combining equations, additional equations can be derived to solve for unknowns, eliminating multiple steps. Let's work through some other projectile-motion problems using different equations.

Solve for the Initial Velocity of a Projectile

A projectile is launched at an angle (θ) of 26° and travels a distance of 290 ft; gravitational acceleration equals 32 ft/sec². Calculate the initial velocity.

FIGURE 13-19: Calculating initial velocity.

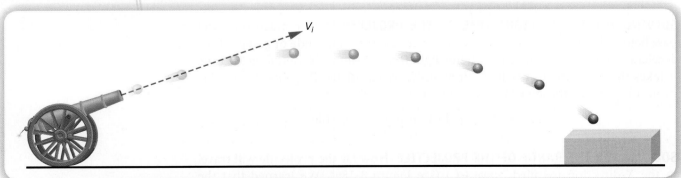

© Cengage Learning 2012

To solve for initial velocity V_i, use Equation 13-6:

$$V_i = \sqrt{\frac{-gX}{\sin 2\theta}} \qquad \text{(Equation 13-6)}$$

$$= \sqrt{\frac{32 \text{ ft/sec}^2 \times 290 \text{ ft}}{\sin(2 \times 26°)}}$$

$$= \sqrt{\frac{9{,}280 \text{ ft}^2/\text{sec}^2}{0.788}}$$

$$= 108.52 \text{ ft/sec}$$

Your Turn

A projectile is launched at an angle (θ) of 35° and travels a distance of 315 meters; gravitational acceleration equals 9.8 m/sec². Calculate the initial velocity. (Show all work.)

Solve for the Horizontal Distance of a Projectile

The initial velocity of a projectile is 500 ft/sec, and the firing angle is 15°. Calculate the horizontal distance.

FIGURE 13-20: Calculating horizontal distance.

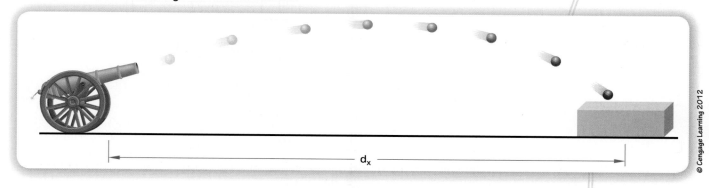

To solve for horizontal distance d_X, use Equation 13-7: Refer to figure 13-20

$$d_X = \frac{V_i^2(\sin 2\theta)}{-g} \qquad \text{(Equation 13-7)}$$

$$= \frac{(500 \text{ ft/sec})^2 \times \sin(2 \times 15°)}{32 \text{ ft/sec}^2}$$

$$= \frac{125{,}000 \text{ ft}^2/\text{sec}^2}{32 \text{ ft/sec}^2}$$

$$= 3{,}906.25 \text{ ft}$$

Solve for Firing Angle

A projectile will travel the same distance when launched at two different angles if it has the same initial velocity and the two angles deviate from 45° equally. Let's calculate a firing angle that is less than 45° and a firing angle that is more than 45° for a projectile that travels a distance of 3,906.25 feet with an initial velocity of 500 ft/sec.

To solve for an angle less than 45°, use Equation 13-8:

$$\theta_{<45°} = \frac{\sin^{-1}\left(\dfrac{-gd_X}{V_i^2}\right)}{2} \qquad \text{(Equation 13-8)}$$

$$= \frac{\sin^{-1}\left(\dfrac{32 \text{ ft/sec}^2 \times 3{,}906.25 \text{ ft}}{(500 \text{ ft/s})^2}\right)}{2}$$

$$= \frac{\sin^{-1}\left(\dfrac{125{,}000 \text{ ft}^2/\text{sec}^2}{250{,}000 \text{ ft}^2/\text{sec}^2}\right)}{2}$$

$$= \frac{\sin^{-1}0.5}{2}$$

$$= \frac{30°}{2}$$

$$= 15°$$

To solve for an angle greater than 45°, use Equation 13-9:

$$\theta_{>45°} = 90° - \left(\frac{\sin^{-1}\left(\dfrac{-gd_X}{V_i^2}\right)}{2}\right) \qquad \text{(Equation 13-9)}$$

$$= 90° - \left(\frac{\sin^{-1}\left(\dfrac{32 \text{ ft/s}^2 \times 3{,}906.25 \text{ ft}}{(500 \text{ ft/s})^2}\right)}{2}\right)$$

$$= 90° - \left(\frac{\sin^{-1}\left(\dfrac{125{,}000 \text{ ft}^2/\text{s}^2}{250{,}000 \text{ ft}^2/\text{s}^2}\right)}{2}\right)$$

$$= 90° - \left(\frac{\sin^{-1}0.5}{2}\right)$$

$$= 90° - \left(\frac{30°}{2}\right)$$

$$= 90° - 15°$$

$$= 75°$$

NOTE: In these examples, one projectile will have a lower flight path than the other.

Your Turn

Using equations introduced in this chapter, solve for t_{max} and d_Y for a projectile launched with an initial velocity of 500 ft/sec at both 15° and 75°. Sketch the flight path for both firing angles and explain the relationship of the firing angle to t_{max} and d_Y. (Show all work.)

Case Study >>>→

A Kinematic History of the Golf Ball

Golf evolved on the eastern coast of Scotland as a casual gentlemen's game. The equipment used was fashioned from what was available at the time. The clubs were made from wood, and documents dating as far back as 1550 reference the use of wooden golf balls. The evolution of the equipment used for playing has and will continue to change the game. Let's take a look at the golf ball and the changes made to it over the years.

The "featherie" was the first handcrafted golf ball in 1618 (see Figure 13-21). The ball got its name from the goose feathers that were used as its core. The feathers were soaked in water and then tightly packed into a cow- or horsehide sphere. As the wet feathers dried out, two things occurred: The leather around the feathers shrunk and tightened and the feathers inside expanded, thus making the ball hard. The ball was then painted, and the maker's mark was added. The quality varied from craftsman to craftsman; and being a handcrafted product meant the price was sometimes out of reach of the common player.

Figure 13-21: *The featherie golf ball.*

© Cengage Learning 2012

Golfers started to notice that a scuffed ball would travel further. This discovery led to the introduction of a ball made from gutta-percha, nicknamed the "guttie," in 1848. The guttie was made using the rubber sap of the gutta tree found in the tropics. The sap from the tree was heated and then easily shaped into a sphere. If the ball was damaged, it could be reheated and reshaped. The first gutties were smooth, so a scuffed featherie was still better. However, in 1880, the second-generation guttie was released. It had raised spherical bumps and was manufactured using a mold. The bumps increased the distance the ball could travel, and the mold manufacturing increased its affordability, consistency, and quality.

Coburn Haskell introduced a one-piece rubber-cored ball in 1898 (see Figure 13-22). The ball's solid rubber core was wrapped in rubber thread, which was then encased in a gutta-percha sphere. The Haskell ball was

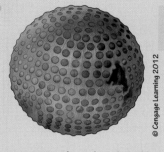

Figure 13-22: *The Haskell golf ball.*

© Cengage Learning 2012

mass produced using the W. Millison thread-winding machine. Once again, this made the golf ball more affordable to the common player.

In 1921, the ruling authorities of golf, the United States Golf Association (USGA) and the Royal and Ancient Golf Club of St. Andrews (the R and A), standardized the golf ball by imposing a weight and size constraint.

The Role of Dimples on the Golf Ball

Dimples simply make the golf ball more aerodynamic. A golf ball is subject to two types of drag. The first is drag caused by friction; this is the smaller of the two drag effects. The majority of drag comes from the ball physically separating the air as it travels; this is called "pressured drag due to separation."

The ball in Figure 13-23(a) is smooth, and the air separates at 12 and 6 o'clock. The velocity of the air behind the ball is constant; this is called *laminar flow*. By adding dimples to the ball, as shown in Figure 13-23(b), a turbulent flow is created in which the velocity of the air behind the ball varies. The dimples make the surface rough, which changes the air flow from laminar to turbulent. The turbulent flow has more energy than the laminar flow and thus stays attached longer, decreasing drag. This decreases the size of the separation behind the ball. Having a smaller separation region behind the ball means there is less pressure drag on the ball, thus allowing it to travel farther.

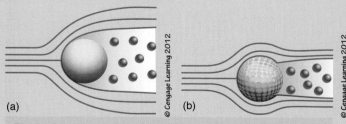

(a)　　　　　　　　　(b)

© Cengage Learning 2012

Figure 13-23: *(a) Laminar flow over a sphere and (b) turbulent flow over a sphere.*

(continued)

(continued)

The dimples on the golf ball also help control the flight of the ball. When the ball is properly hit, backspin will be created. This spin creates less pressure on the top portion of the ball, thus making it rise or fly upward. This produces two desirable effects for the golfer. First, it will allow the ball to travel lower to the ground at the beginning of the stroke, thus keeping the ball out of the wind for a longer period of time. Second, the flight path of the ball will peak out and drop more vertically, thus making the ball hold better (less bounce) (see Figure 13-24). It is possible with an iron—a greater-angled club—to produce enough backspin on the ball to stop it or even have it move back toward the golfer after it has landed.

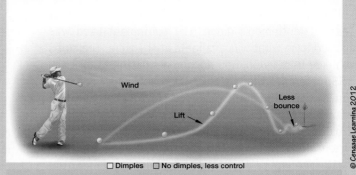

☐ Dimples   ☐ No dimples, less control

Figure 13-24: *Flight pattern with and without dimples.*

© Cengage Learning 2012

Example

Imagine that you are a professional golfer who is playing a new golf course for the first time. To promote the new course, the course owners are having a hole-in-one competition before the tournament, and you have been invited to participate. Figure 13-25 shows the competition hole. You are told the yardage of the hole and the elevation change to the pin. Your caddy suggests that you use a sand wedge, which has a loft angle of 55 degrees. Under ideal conditions, what should the golf ball's initial velocity be in order to achieve a hole-in-one?

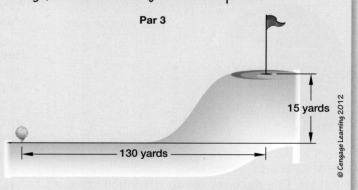

FIGURE 13-25: Par 3 hole with distance, elevation change, and initial velocity for the competition.

Par 3

15 yards

130 yards

© Cengage Learning 2012

Step 1. 1 yard is equal to 3 feet. Assuming that the tee is the origin point (0,0), identify the X/Y coordinate location of the hole along the golf ball's parabolic path in feet:

$$X = \frac{3 \text{ feet}}{1 \text{ yard}} \times 130 \text{ yards}$$

$$X = 390 \text{ feet}$$

$$Y = \frac{3 \text{ feet}}{1 \text{ yard}} \times 15 \text{ yards}$$

$$Y = 45 \text{ feet}$$

Step 2. Identify the formulas for motion in the X- and Y-axis:

$$X = (V_x)t = (Vi \cos \theta)t$$

$$Y = (V_y)t - \frac{1}{2}gt^2 = (Vi \sin \theta)t - \frac{1}{2}gt^2$$

Step 3. Algebraically manipulate the variables in the X-axis motion formula to derive a formula for time (t). Then, substitute the known values into this formula:

$$X = (V_{iX})t = (V_i \cos \theta)t \quad \rightarrow \quad t = \frac{X}{V_i \cos \theta}$$

$$t = \frac{390 \text{ ft}}{V_i \cos 55°}$$

Step 4. As time is an unknown variable, we can substitute our previously-derived formula for time into the Y-axis motion equation:

$$Y = (V_i \sin \theta)t - \frac{1}{2}gt^2$$

$$Y = V_i \sin \theta \frac{X}{V_i \cos \theta°} - \frac{1}{2}g\left(\frac{X}{V_i \cos \theta}\right)^2$$

Step 5. Substitute the known values into the new Y-axis motion equation and solve for initial velocity (V_i):

$$Y = V_i \sin \theta \frac{X}{V_i \cos \theta} - \frac{1}{2}g\left(\frac{X}{V_i \cos \theta}\right)^2$$

$$45 \text{ ft} = V_i \sin 55° \frac{390 \text{ ft}}{V_i \cos 55°} - \frac{1}{2}\left(32.2 \frac{\text{ft}}{\text{sec}^2}\right)\left|\frac{390 \text{ ft}}{V_i \cos 55°}\right|^2$$

$$45 \text{ ft} = \sin 55° \frac{390 \text{ ft}}{\cos 55°} - 16.1 \frac{\text{ft}}{\text{sec}^2}\left|\frac{390 \text{ ft}}{V_i \cos 55°}\right|^2$$

$$45 \text{ ft} = 390 \text{ ft} \frac{\sin 55°}{\cos 55° - \dfrac{16.1 \dfrac{\text{ft}}{\text{sec}^2} \times (390 \text{ ft})^2}{V_i^2 (\cos 55°)^2}}$$

$$45 \text{ ft} = 390 \text{ ft}(\tan 55°) = -\frac{16.1 \dfrac{\text{ft}}{\text{sec}^2} \times 151,100 \text{ ft}^2}{V_i^2 (\cos 55°)^2}$$

$$45 \text{ ft} - 390 \text{ ft}(\tan 55°) = -\frac{2,448,810 \dfrac{\text{ft}^2}{\text{sec}^2}}{V_i^2 (\cos 55°)^2}$$

$$512 \text{ ft} = \frac{2,448,810 \dfrac{\text{ft}^2}{\text{sec}^2}}{V_i^2 (\cos 55°)^2}$$

$$V_i^2 (\cos 55°)^2 = \frac{2,448,810 \dfrac{\text{ft}^2}{\text{sec}^2}}{512 \text{ ft}}$$

$$V_i^2 (\cos 55°)^2 = 4,783 \frac{\text{ft}^2}{\text{sec}^2}$$

$$V_i^2 = \frac{4,783 \dfrac{\text{ft}^2}{\text{sec}^2}}{(\cos 55°)^2}$$

$$V_i^2 = 14,538 \frac{\text{ft}^2}{\text{sec}^2}$$

$$V_i = \sqrt{14,538 \frac{\text{ft}^2}{\text{sec}^2}}$$

$$V_i = 121 \frac{\text{ft}}{\text{sec}}$$

SUMMARY

- Mechanics is the study of the motion of objects

- There are three major branches of mechanics: deformable body mechanics, fluid mechanics, and rigid body mechanics.

- A rigid body describes an object whose size does not change when a force is applied to it.

- Dynamics is the study of motion and the effects of forces acting on rigid bodies in motion.

- Kinematics is the study of motion without consideration of the forces that caused that motion.

- Basic kinematic terms include acceleration, velocity, and distance.

- Gravity is the force that pulls objects toward the center of Earth.

- A scalar quantity is a mathematical quantity that is limited to magnitude only.

- A vector quantity is a mathematical quantity that has both magnitude and direction.

- A projectile is an object that is moving both vertically and horizontally with gravity as the only force acting on it.

- A projectile moves at a constant horizontal velocity because there is no force acting on it.

- Gravity only acts on the vertical motion of a projectile at an acceleration of 9.8 m/s^2 or 32 ft/s^2.

- A parabolic curve is the shape that a projectile produces when the only force acting on it is gravity.

- A good technique to use when solving mathematical equations is to create a table identifying all known and unknown values.

- Some kinematic equations are based on the Pythagorean theorem.

- Several kinematic equations can be used when solving unknowns in projectile-motion problems.

1. Explain the difference between kinematics and kinetics.
2. List three vector quantities.
3. Explain why a projectile takes a parabolic path.
4. Which velocity remains constant during the flight of a projectile?
5. If an object is dropped off a cliff, how fast will it be going in 4 seconds?

6. Develop your own kinematic tale beginning with "Once upon a time, . . ." Define a setting, character, and kinematic knowns and then solve for unknowns. Show all work and explain how kinematics saved the day.
7. How do dimples improve the flight of a golf ball?

EXTRA MILE

Problem Statement: High school spirit week often culminates in a pep rally where upperclassmen engage in friendly competition through games and contests that are designed to entertain and rally the crowd's sense of school pride. The pep rally coordinator is looking for a new idea that combines engineering innovation with a traditional pep rally event.

Design Statement: Design a device that will automate the launching of water balloons during a water balloon toss event. The solution must:

▶ be easy to assemble and disassemble,

▶ utilize a mechanical form of energy transfer (air pressure and chemical combustion are strictly forbidden),
▶ launch a balloon at a uniform initial velocity at any angle,
▶ be able to adjust the water balloon's initial trajectory angle between 45 and 90 degrees.

CHAPTER 14
Introduction to Measurement, Statistics, and Quality

GPS DELUXE

Menu

| START LOCATION | DISTANCE | END LOCATION |

Before You Begin

Think about these questions as you study the concepts in this chapter:

1 Why is having a standard system of measurement important?

2 What are the two most common types of measurement systems used by engineers?

3 How are numbers expressed in scientific notation?

4 What types of linear measurement tools are most commonly used, and how does an engineer decide which measurement tool is appropriate for the task?

5 What is the difference between a data set's mean, median, mode, and range?

6 What is standard deviation, and how is it calculated?

7 What is quality, and why is it so important in the engineering world?

8 How do engineers control the quality of a product?

9 What statistical methods are used to evaluate quality?

Imagine you have purchased a flying disk that, despite your best efforts and through no fault of your own, will not fly along a straight path. On closer inspection, you notice that the disk is not perfectly round. You might assume that the product was of low quality.

The authors of *Juran's Quality Handbook*, a standard industry reference, offer two definitions for quality: (1) those product features that meet customer needs, and (2) freedom from flaws. Both of these definitions drive customer satisfaction. With the example of the flying disk, the customer is not satisfied, but is this because the product has flaws or because it was not designed with the specific features to meet customers' needs?

Another example of quality-driven customer satisfaction is a customer's selection process of a digital camera at an electronics store. Image that there are two options. The more-expensive camera has more features such as higher megapixels and optical zoom. At the other end, a less-expensive camera may be more satisfying because it meets customers' needs such as being compact and easy to use. Both cameras need to be free from flaws; otherwise, their quality will come into question. Quality in the sense of freedom of flaws is the focus of this chapter.

Quality affects the performance of objects as simple as a flying disk.

> **Quality:**
> (1) those product features that meet customer needs;
> (2) freedom from flaws.

(a) Some customers will consider the digital single-lens-reflex camera on the left to be higher quality because it has more sophisticated features. (b) Other customers will find that the compact point-and-shoot camera on the right meets their needs better. Both cameras must be free of flaws to satisfy the definition of quality for this chapter.

After World War II, the concept of a quality management system was introduced to the business world. This reinforced the idea that quality doesn't just happen, but is planned for and measured.

As part of a quality management system, data is collected during the manufacturing process and analyzed to evaluate the quality of the finished product and the methods that produced it. This analysis provides engineers with performance trends and information on whether variations in the physical dimensions of the end product are acceptable. Doing this analysis during manufacturing is valuable because it prompts changes only when needed.

In this chapter, you will learn about the standard measurement systems and precision measuring tools that are commonly used by engineers to collect data. You will also develop an understanding of quality as it relates to manufacturing and learn what engineers and manufacturers do to ensure that they are designing, producing, and selling quality products.

STANDARD SYSTEM OF MEASUREMENT

Consider how often measurement terms are used in our daily lives. We purchase a dozen bagels, give directions in miles, and complain about how much a gallon of gasoline costs. Now imagine a world in which everyone has a different understanding of these measurement terms. What impact would that have on our lives?

A lack of standardized measuring instruments would not be a big deal if all measurments were done by only one individual working independently on a job. This was often the case in ancient times. For example, the cubit once served as a unit of linear measurement. Though the method of measuring a cubit was standard, the length of a cubit was not. As shown in Figure 14-1, a cubit is the length of one man's forearm as measured from the end of the elbow to tip of the middle finger. Another historical example of a larger unit of linear measurement is the pace, which is the distance covered from the point where one foot touches the ground until that same foot touches again (see Figure 14-2).

FIGURE 14-1: A cubit is the length of one man's forearm.

Cubit

© Cengage Learning 2012

FIGURE 14-2: A pace is a unit of linear distance that is measured from the point where one foot touches the ground until that same foot touches the ground again.

Pace

© Cengage Learning 2012

The major problem with a measurement system that is based on dimensions of the human body is that no two people are exactly the same size in all respects. The need for a standardized system of measurement arose when different groups of people started trading with one another.

In the 13th century, King Edward I made one of the first steps toward a standardized measurement system when he ordered a permanent measuring stick be made of iron to serve as the standard yardstick for the kingdom. The King declared that a foot would be equal to 1/3 of the length of the yardstick, and that 1/12 of that foot would be the standard for the inch. The United States still uses these standards that were developed more than 800 years ago.

Today's Measurement Systems

In today's world, the engineer's working environment is not limited by departments, companies, or even international borders. Therefore, effective communication with others relies on knowledge of common measurement systems. The two common measurement systems are the U.S. customary system (English System) and the metric system.

The U.S. customary system of measurement, also known as the English system, is used in the United States today. The founders of the United States felt that a system of measurement was so important to the growth of their new country that they gave Congress the power to establish such standards within the U.S. Constitution.

When measuring length in the English system, there are four common units of measurement: inch, foot, yard, and mile. The main reason why the scientific community and most of the industrialized world do not use the English system is that there is no numerical pattern between these units of measurement. There are 12 inches in a foot, 3 feet in a yard, and 5,280 feet in a mile.

The metric system, also known as the International System (SI), was developed in France in the 1790s. It is a base-10 measuring system, which means that each successive unit is 10 times the size of the previous unit. For example, length in the metric system is based on the meter. A meter is divided into 10 decimeters, and each decimeter is divided into 10 centimeters. Another example is volume, which is based on the liter. A liter is divided into 10 deciliters, and each deciliter is divided into 10 centiliters.

In the metric system, different prefixes are used depending on the size of the measured value, with some prefixes serving as conventional measures (see Figure 14-3). For example, the volume of a liquid commercial product, such as a bottle of soda pop, is identified as being a specific number of milliliters.

FIGURE 14-3: Metric prefixes, decimal equivalents, and exponential equivalents.

Prefix	Decimal Equivalent	Exponential Equivalent
Pico	0.000000000001	10^{-12}
Nano	0.000000001	10^{-9}
Micro	0.000001	10^{-6}
Milli	0.001	10^{-3}
Centi	0.01	10^{-2}
Deci	0.1	10^{-1}
No prefix	1.0	10^{0}
Deka	10.0	10^{1}
Hecto	100.0	10^{2}
Kilo	1000.0	10^{3}
Mega	1,000,000.0	10^{6}
Giga	1,000,000,000.0	10^{9}

Scientific notation:

method used to express a very large or very small number as the product of a number between 1 and 10 and an appropriate power of 10 (e.g., 1.5×10^2).

Scientific Notation

You may have noticed that some of the numbers in the right-hand column of Figure 14-3 are expressed as exponents. This method is called **scientific notation**. Scientific notation is a method used to write very large or very small numbers in a more concise form. The method expresses a number as a value between 1 and 10 and an exponent, or power of 10. For example, the number 123 billion written in expanded form is 123,000,000,000. In scientific notation it would be 1.23×10^{11}. The first number, 1.23, is between 1 and 10. The second number is the base, and because we are using a base-10 number system, it must be 10. The third number is the exponent or power of 10. In our example, the exponent is 11.

Another way to think about scientific notation is that the exponent provides information on how many places and in which direction to move the decimal. A positive exponent moves the decimal to the right, whereas a negative exponent moves the decimal to the left. Zeros are used as placeholders.

An advantage of writing large or small numbers in scientific notation is that it helps to prevent errors when performing mathematical calculations. By consolidating the large number of zeros through the use of an exponent, you don't lose track of them.

Example

Problem: Express the number 0.000001 in scientific notation.

Step 1. Move the decimal place to the right to satisfy the condition of having one nonzero digit to the left of the decimal.

$$0.000001 \rightarrow 1.0$$

Step 2. Count the number of positions that the decimal place moved, noting also the direction that it moved. In this case, the decimal point moved six places to the right. Moving the decimal place to the *right* results in a *negative* exponent, whereas moving it to the *left* results in a *positive* exponent. So,

0.000001 written in scientific notation is 1×10^{-6}

Your Turn

Problem: Write the following numbers in scientific notation:

543,000
32,000,000
2,030,000
0.000065
0.104
0.00007050

MEASURING TOOLS AND TECHNIQUES

The world of engineering often requires more precise measures of distance than the inch, foot, yard, or mile. Therefore, measurement tools are needed that offer greater degrees of precision. Let's start with the basic tools that are used in both systems of measurement.

The Fractional Rule

An inch is divided into halves, quarters, eighths, sixteenths, and so on. A standard inch rule as shown in Figure 14-4 can be used to measure linear distance to an accuracy of 1/16th of an inch. This means that each inch on the rule is divided into 16 equal spaces that are separated by increment marks. If you were to start at zero and count over 4 increments, or 4/16ths of an inch, the resulting value would be simplified to 1/4 inch.

The Decimal Rule

A decimal inch rule divides the inch into 10 equal spaces that have labeled increment marks (see Figure 14-5). The distance between these increments is equal to 1/10th of an inch, or 0.1 in. Each 1/10th increment is further divided into five smaller increments. The resulting distance between each smaller increment is equal to 1/20th of an inch, or 0.02 in.

The Metric Rule

The metric rule reflects the base-10 system of measurement. The standard metric rule has a numeric label next to each centimeter increment. Each centimeter is divided into 10 equal spaces. Each of these smaller increments represents a millimeter (see Figure 14-6).

Conversion Methods and Factors

The industrial world does not yet operate on a single measurement system. For this reason, engineers must use conversion methods and factors when working with different measurement tools and communicating geometric information to manufacturers. For example, an engineer might need to convert fractional inches to decimal inches or convert values from one system to another, such as U.S. customary

FIGURE 14-4: A standard inch rule can be used to measure linear distance to an accuracy of 1/16th of an inch.

© Cengage Learning 2012

FIGURE 14-5: A decimal inch rule divides the inch into 10 equal spaces that have labeled increment marks.

© Cengage Learning 2012

FIGURE 14-6: A metric rule reflects the base-10 system of measurement and has a numeric label next to each centimeter.

© Cengage Learning 2012

decimal inches to metric millimeters. Let's look at the U.S. customary rule for inches and the metric rule together (see Figure 14-7). To convert between inches and metric millimeters, the conversion factor is 1 inch = 25.4 millimeters.

FIGURE 14-7: A comparison of the inch and metric rules.

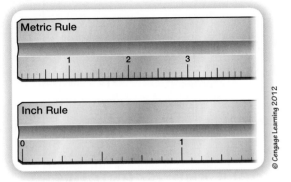

© Cengage Learning 2012

Example

Problem: Convert 4½ inches to millimeters.

Step 1. Convert any fraction to decimals:

½ = 1 ÷ 2 = 0.5 inches then add the whole number to the decimal value.

So,

4½ inches = 4.5 inches

Step 2. Multiply the number of inches by 25.4 millimeters/inch (the conversion factor):

4.5 inches × 25.4 millimeters/inch = 114.3 millimeters

Step 3. To convert back from millimeters to inches, divide 114.3 by the conversion factor:

114.3 millimeters ÷ 25.4 millimeters/inch = 4.5 inches

Many construction trades still use the fractional inch as the base unit of measurement. To convert a fractional inch value to a decimal inch value, divide the numerator (the top part of the fraction) by the denominator (the bottom part of the fraction).

$$\frac{1}{8} = 1 \div 8 = 0.125$$

Though the need to convert a decimal inch value to a fractional inch value is far less frequent, two methods are commonly employed.

METHOD 1: DIVISION Starting from the decimal point, count the decimal places. If there is one decimal place, put the number over 10 and reduce. If there are two places, put the number over 100 and reduce. If there are 3 places, put it over 1,000 and reduce, and so on.

$$0.75 = \frac{75}{100} = \frac{3}{4}$$

METHOD 2: REFERENCE CHART Instead of doing a mathematical conversion, many people choose to reference a fraction-to-decimal conversion chart like the one shown in Figure 14-8.

FIGURE 14-8: A fraction-to-decimal conversion chart.

Fraction	Decimal	Fraction	Decimal
1/64	.015625	33/64	.515625
1/32	.03125	17/32	.53125
3/64	.046875	35/64	.546875
1/16	**.0625**	**9/16**	**.5625**
5/64	.078125	37/64	.578125
3/32	.09375	19/32	.59375
7/64	.109375	39/64	.609375
1/8	**.125**	**5/8**	**.625**
9/64	.140625	41/64	.640625
5/32	.15625	21/32	.65625
11/64	.171875	43/64	.67185
3/16	**.1875**	**11/16**	**.6875**
13/64	.203125	45/64	.703125
7/32	.21875	23/32	.71875
15/64	.234375	47/64	.734375
1/4	**.25**	**3/4**	**.75**
17/64	.265625	49/64	.765625
9/32	.28125	25/32	.78125
19/64	.296875	51/64	.796875
5/16	**.3125**	**13/16**	**.8125**
21/64	.328125	53/64	.828125
11/32	.34375	27/32	.84375
23/64	.359375	55/64	.859375
3/8	**.375**	**7/8**	**.875**
25/64	.390625	57/64	.890625
13/32	.40625	29/32	.90625
27/64	.421875	59/64	.921875
7/16	**.4375**	**15/16**	**.9375**
29/64	.453125	61/64	.953125
15/32	.46875	31/32	.96875
31/64	.484375	63/64	.984375
1/2	**.50**	**1**	**1.00**

PRECISION MEASURING TOOLS

Precision is critical in engineering. It means to be accurate or exact. When an engineer is designing a project, her measurements must be precise in order to create a quality product. To ensure that they are, she must use special measuring instruments.

> **Precision:**
> to be accurate or exact.

Measuring with a Dial Caliper

Dial calipers are one of the most commonly used measuring instruments in the engineering world (see Figure 14-9). They are perhaps the most versatile and easy to read of all the precision measurement tools and are accurate to the nearest 1/1000 of an inch (0.001 in).

The dial caliper is adjusted by using your thumb to move the slider back and forth along a stationary blade until it has opened wide enough to accept the work

FIGURE 14-9: A dial caliper.

piece that you want to measure. For example, if you are measuring the length of a bolt, you would place the bolt in between the inside measurement surfaces of the caliper and slowly close the gap until the bolt is held between the two surfaces. Be careful not to apply too much pressure so that you do not damage the caliper.

As shown in Figure 14-10, there are four types of measurements that can be made with a dial caliper: step distance, hole depth, inside length and diameter, and outside length and diameter.

FIGURE 14-10: The four types of dial caliper measurements: (a) step distance, (b) hole depth, (c) inside length and diameter, and (d) outside length and diameter.

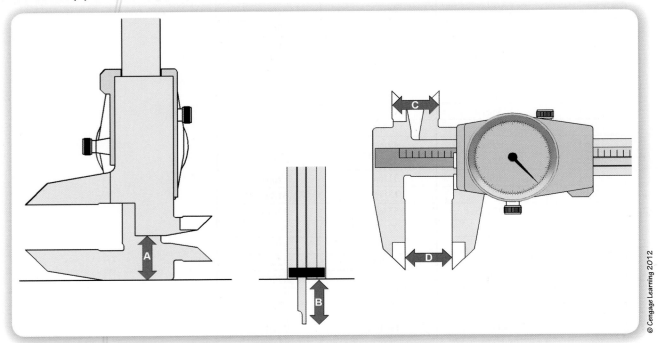

© Cengage Learning 2012

Reading a Dial Caliper

The dial caliper blade contains a blade scale that displays two different types of graduations. A standard dial caliper will measure up to 6 inches, hence the large vertical markings on the fixed blade scale will read from 0 to 6. Each inch is divided into 10 equal parts. The smaller numbers between the inch increments are labeled 1 through 9. The space between these smaller increments is equal to 1/10 of an inch, or 0.100 in. The dial face on the dial caliper is divided into 100 increments. Each increment equals 1/1000 of an inch, or 0.001 in.

The reading of the dial caliper is the same regardless of whether you are measuring an outside diameter, inside diameter, hole depth, or step measurement:

1. Identify the number of whole inch values shown on the blade scale.
2. Identify the number on 1/10th inch divisions that appear past the whole inch value.
3. Identify the number of thousandths indicated by the needle on the dial face.
4. Add the values to find the dial caliper reading.
5. Repeat the process to ensure accuracy.

Example

Problem: Find the dial caliper reading in Figure 14-11.

Step 1. Identify the number of whole inch values shown on the rack:

0.000 in

Step 2. Identify the number on 1/10 inch divisions that appear past the whole inch value:

0.500 in

Step 3. Identify the number of thousandths indicated by the needle on the dial face:

0.025 in

Step 4. Add the values to find the dial caliper reading:

0.000 in
0.500 in
+0.025 in

0.525 in

Step 5. Repeat steps 1–4 to ensure accuracy.

FIGURE 14-11: Dial Caliper.

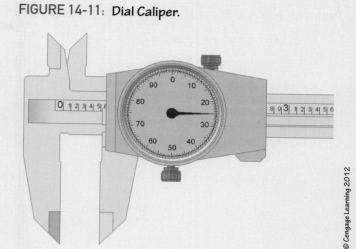

© Cengage Learning 2012

Your Turn

Problem 1: Find the dial caliper reading in Figure 14-12.

FIGURE 14-12: Dial Caliper.

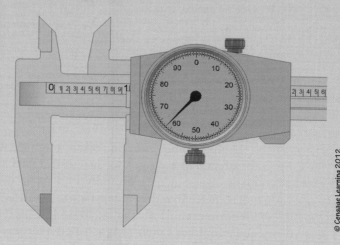

© Cengage Learning 2012

(*continued*)

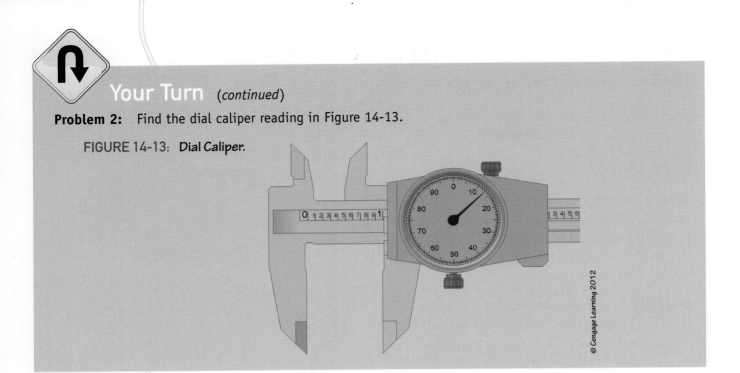

Your Turn (*continued*)

Problem 2: Find the dial caliper reading in Figure 14-13.

FIGURE 14-13: **Dial Caliper.**

© Cengage Learning 2012

Measuring with a Micrometer

A **micrometer** is used to measure the thickness of flat stock (like paper or aluminum plate) or the diameter of round objects (such as bolts and steel drill rods). The major parts of a micrometer are the frame, anvil, spindle, lock ring, sleeve (also called the barrel), and thimble (see Figure 14-14). The object being measured is placed between the anvil and the spindle, and the thimble is turned to close the gap and hold the object in place.

FIGURE 14-14: **A zero to 1-inch micrometer.**

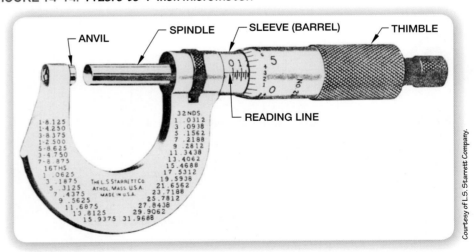

Courtesy of L.S. Starrett Company.

All micrometers can measure to an accuracy of 1/1000 of an inch. Some have the ability to measure to the nearest 1/10,000 of an inch. Micrometers come in many sizes but measure only in a range of 1 inch. For example a zero to 1-inch micrometer like the one in Figure 14-14 is a standard measuring tool commonly found in manufacturing facilities. It is used to measure objects or features that are no larger than 1 inch. A 1 to 2-inch micrometer would be used to measure an object that is between 1 and 2 inches thick.

Reading a Micrometer

The largest vertical markings on the sleeve are 0, 1, 2, 3, 4, 5, 6, 7, 8, and 9 (see Figure 14-15). These markings represent an inch that has been divided into 10 equal parts. Therefore, each division equals 1/10 of an inch (0.100 in). Located between each of these divisions are three smaller vertical markings (see Figure 14-15). These markings divide the 0.100 in into four equal parts, each having a value of 1/40 of an inch or 25/1000 (0.025 in). The user turns the thimble to either increase or decrease the gap between the spindle and the anvil. Every complete turn of the thimble equals 1/40 of an inch, which moves the spindle 0.025 in. The space between each of these marks on the beveled edge of the thimble is equal to 1/1000 of an inch or 0.001 in.

The steps for reading a micrometer are as follows:

1. Identify the largest 1/10-inch division shown on the sleeve.
2. Identify the number of quarter divisions shown past the 1/10-inch mark (each subdivision = 0.025").
3. Identify the 1/1000th increment on the thimble that is closest to the zero line on the sleeve.
4. Add the values to find the micrometer reading.
5. Repeat the process to ensure accuracy.

Example

Problem: Find the micrometer reading in Figure 14-15.

FIGURE 14-15: Zero to 1-inch micrometer.

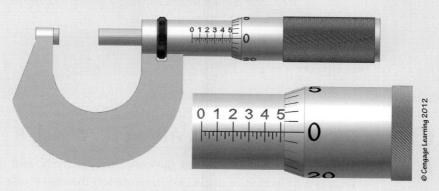

© Cengage Learning 2012

Step 1. Identify the largest 1/10 inch division shown on the sleeve:

0.500 in

Step 2. Identify the number on quarter divisions shown past the 1/10 inch mark (each subdivision = 0.025 in):

0.025 in

Step 3. Identify the 1/1000 increment on the thimble that is closest to the zero line on the sleeve:

0.000 in

(continued)

Step 4. Add the values to find micrometer reading:

$$
\begin{array}{r}
0.500 \text{ in} \\
0.025 \text{ in} \\
+0.000 \text{ in} \\
\hline
0.525 \text{ in}
\end{array}
$$

Step 5. Repeat steps 1–4 to ensure accuracy.

Your Turn

Problem: Find the micrometer reading in Figure 14-16.

FIGURE 14-16: **Zero to 1-inch micrometer.**

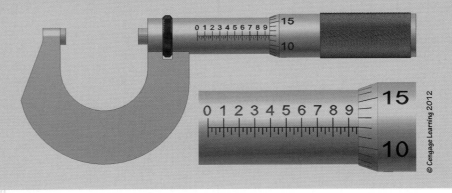

© Cengage Learning 2012

 BASIC STATISTICAL ANALYSIS OF DATA

Statistics can be described as the collection, analysis, and interpretation of data. Data is a collection of organized information that can come from observation, experience, or a set of premises. In the engineering world, this information usually is in the form of numbers obtained from measurements. A group of these numbers is called a *data set* and is often generated by quality engineers.

To generate a data set, an engineer can use either the sample method or the population method. The *sample* method involves measuring only a *percentage* of the parts manufactured, whereas the *population* method involves measuring *all* of the parts manufactured.

The majority of basic statistical analysis of data involves identifying or calculating the central tendencies of a data set. The three basic central tendencies of a data set are the mode, median, and mean (average). A manufacturing process generates a "normal" data set if the resulting data forms a bell-shaped curve when plotted on a histogram. A "normal" data set will also have the same value for its median, mean, and mode.

Mode

The **mode** is the numeric value that occurs with the greatest frequency in a data set.

Data set: 13, 7, 8, 23, 14, 13, 21, 25

Step 1. Arrange the data set values in numerical order:

7, 8, 13, 13, 14, 21, 23, 25

Step 2. Determine which number occurs most frequently:

7, 8, **13**, **13**, 14, 21, 23, 25

Mode = 13

If a data set has two or more values that occur with equal frequency and with a greater frequency than any other value, the data set is multimodal. A data set with two modes is called **bimodal**. A data set with three modes is called **trimodal**.

Example

Data set: 7, 3, 4, 34, 12, 45, 4, 34, 5, 8, 21

Step 1. Arrange the data set values in numerical order:

3, 4, 4, 5, 7, 8, 12, 21, 34, 34, 45

Step 2. Identify the values that occur with the greatest frequency:

3, **4**, **4**, 5, 7, 8, 12, 21, **34**, **34**, 45

Modes = 4, 34

In this data set, the values 4 and 34 occur with the same frequency (twice each) and with a greater frequency than any other values in the data set. Therefore, the data set is bimodal.

Median

The **median** is the numeric value that occurs in the middle of a given data set arranged in numerical order. In a data set with an *odd* number of values, there will be an equal number of values above and below the median.

Example

Data set: 12, 1, 7, 20, 2, 8, 4, 18, 6, 15, 10

Step 1. Arrange the data set values in numerical order:

1, 2, 4, 6, 7, 8, 10, 12, 15, 18, 20

(continued)

Step 2. Identify the value that occurs in the middle of the data set:

1, 2, 4, 6, 7, **8**, 10, 12, 15, 18, 20

Median = 8

If the data set has an *even* number of sample values, then the median will be half of the sum of the two middle values.

Example

Data set: 1, 6, 4, 3, 5, 2

Step 1. Arrange the data set values in numerical order:

1, 2, 3, 4, 5, 6

Step 2. Identify the two values that occur in the middle of the data set:

1, 2, **3**, **4**, 5, 6

Step 3. Add the two middle values and divide the sum in half:

3 + 4 = 7

7 ÷ 2 = 3.5

Median = 3.5

Mean:

the average of a sample of measurements.

Mean

The **mean** value of a data set is the "average." To calculate the mean, all of the numeric values are added together and then divided by the total number of values in the data set.

Example

Data set: 12, 14, 23, 7, 9, 10

Step 1. Identify the total number of values contained in the data set:

Total number of values = 6

Step 2. Add the values in the data set:

12 + 14 + 23 + 7 + 9 + 10 = 75

(continued)

Step 3. Divide the sum of the values in the data set by the total number of values:

$75 \div 6 = 12.5$

Mean = 12.5

Range

The *range* (R) of a data set is useful information, although it is not a measure of central tendency. The range is the difference between the highest and lowest values of a data set. It gives the quality engineer a quick look at how much a manufacturing process or sample varies.

Example

Data set: 12, 14, 23, 7, 9, 10

Step 1. Identify the largest and smallest values in the data set:

Largest value = 23

Smallest value = 7

Step 2. Subtract the smallest value from the largest value:

$23 - 7 = 16$

Range (R) = 16

Calculating Standard Deviation

The **standard deviation** of a data set is a value that indicates how close the values are to the mean value of that data set. A small standard deviation indicates that the data values are close to the mean, whereas a large standard deviation indicates that the data values are far from the mean.

The standard deviation is used in combination with the mean to group data values from a normal manufacturing process on a histogram. A normal manufacturing process will produce a bell-shaped curve on the histogram, wherein certain percentages of the data elements will occur within ± 1, ± 2, and ± 3 standard deviations from the mean (see Figure 14-17).

A normal distribution curve is typically drawn six standard deviations wide, with the mean (usually the target dimension value) at the center. Statisticians have found that 68% of the data will fall within ± 1 standard deviations from the mean; 95% of the data will fall within ± 2 standard deviations from the mean; and 99% of the data will fall within ± 3 standard deviations from the mean.

Standard deviation (s):
a value that identifies how close the values are to the mean value of a data set.

FIGURE 14-17: A standard deviation (or bell) curve showing that certain percentages of data elements will occur within ±1, ±2, ±3 standard deviations from the mean.

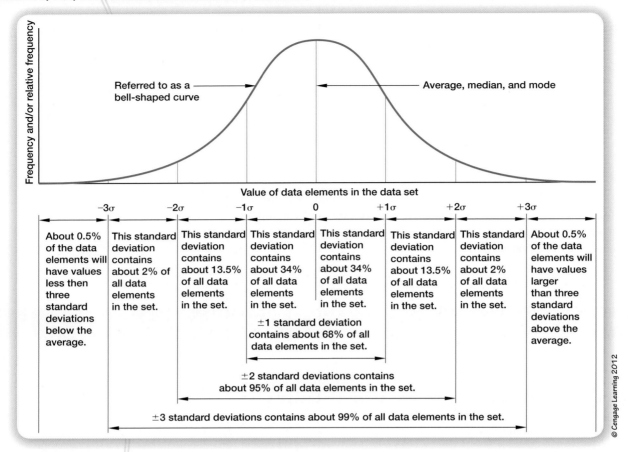

© Cengage Learning 2012

Example

Problem: Find the s standard deviation for the following data set:

$$x_1 = 5$$
$$x_2 = 6$$
$$x_3 = 8$$
$$x_4 = 9$$

Step 1. Calculate the mean ($\bar{x}$) of the data set using Equation 14-1:

$$\bar{x} = \frac{\Sigma x}{n}$$

(Equation 14-1)

Whereas:

Σx = sum of the values in the sample data set

n = number of values in the sample data set

$$\bar{x} = \frac{(x_1 + x_2 + x_3 + x_4)}{n}$$

$$\bar{x} = \frac{(5 + 6 + 8 + 9)}{4}$$

$$\bar{x} = 7$$

(continued)

Step 2. Calculate the sample standard deviation (s) of the data set using Equation 14-2:

$$s = \sqrt{\frac{\Sigma(x - \bar{x})^2}{n - 1}}$$

(Equation 14-2)

Replace variable n with the value 4:

$$s = \sqrt{\frac{\Sigma(x - \bar{x})^2}{4 - 1}}$$

Replace variable $\bar{x}$ with the value 7:

$$s = \sqrt{\frac{\Sigma(x - 7)^2}{4 - 1}}$$

Replace each x variable with its respective value from the data set:

$$s = \sqrt{\frac{(5 - 7)^2 + (6 - 7)^2 + (8 - 7)^2 + (9 - 7)^2}{4 - 1}}$$

Solve:

$$s = \sqrt{\frac{(-2)^2 + (-1)^2 + (1)^2 + (2)^2}{4 - 1}}$$

$$s = \sqrt{\frac{(-2)^2 + (-1)^2 + (1)^2 + (2)^2}{3}}$$

$$s = \sqrt{\frac{10}{3}}$$

$$s = 1.8$$

Your Turn

To illustrate the process of collecting and analyzing data, let's look at the average high temperatures in Toronto, Canada during 2007, as shown in Figure 14-18.

Problem: Identify the *mode, median, mean*, and *standard deviation* of the data set.

Month	Temperature (°F)
January	29
February	31
March	40
April	55
May	65
June	75
July	81
August	79
September	71
October	55
November	45
December	35

FIGURE 14-18: **Average high temperatures for Toronto, Canada, 2007.**

Source: www.weatherunderground.com

Career Profile

CAREERS IN ENGINEERING

Giving Health Care a Shot in the Arm

A lover of math since his childhood, Kalyan Pasupathy is always looking for patterns. And he's found some in a rather unlikely place for an engineer to set up shop: the health-care industry. Currently an assistant professor at the University of Missouri School of Medicine, Pasupathy teaches and conducts research in the school's Health Management and Informatics Department, where he specializes in operations research and systems analysis.

"It's basically a way of looking at how employees and system components are configured and how work gets done," Pasupathy says about his field, which traces its origins back to the time–motion studies that revolutionized American manufacturing at the turn of the last century. "The goal is to make the work process more efficient, more streamlined, with fewer errors."

Retail operations like WalMart and Toyota have been employing such techniques for years, Pasupathy says, but the health-care industry is lagging behind. "Traditionally, doctors, nurses and other health-care professionals are trained in their own domains in separate schools," he points out, "but they aren't necessarily trained to work with one another. We are changing that by providing interdisciplinary education to tomorrow's professionals."

On the Job

Pasupathy first entered the medical field through the American Red Cross, where he participated in a project that has saved the organization as much as $700,000 a year. "What we did was look at all the field units to see how they were operating," Pasupathy says, "and then we used that information to formulate a model that would better utilize the resources that each unit had at its disposal."

Since then, Pasupathy has applied his skill sets to a wide array of medical practices, from breast cancer screenings to the administering of drugs. "Right

Kalyan Pasupathy, Operations Research and Systems Analysis, University of Missouri School of Medicine

© Cengage Learning 2012

now, I'm working with the pharmacy department in a university hospital," he says. "We're looking at medical errors, the multiple points along the way where something can go wrong. And one of the striking things is that when you walk into a hospital pharmacy the phone is always ringing. There are all these interruptions, which can lead to errors and near-miss events. We're looking for the root causes to improve quality and safety."

Inspiration

Like many engineers, Pasupathy was a math whiz as a child, but he credits one of his high school teachers, Elizabeth Abraham, with taking the spark that was already there and setting it aflame. "She did a very good job of encouraging individual students," Pasupathy says. "She also allowed students to advance at their own individual pace, based on their skills and abilities. As an educator myself, I've used her as a role model."

Education

Pasupathy studied operations research as an undergraduate, but as a graduate student at Virginia Tech he fell under the spell of what he calls "the science of better"—that is, he studies ways to use analytical techniques to enhance an organization's performance. "The curriculum at Virginia Tech ingrained the engineering process in my brain," he says. "There's always this nagging thought, 'How do I improve this?'"

Advice for Students

If Pasupathy could pass one thing on to tomorrow's engineering students, it would be the idea that an engineer can wind up anywhere. "We're no longer just building bridges or working for manufacturing firms on the production floor," he says. "You can work in a drastically different industry or sector. Engineering and medicine are no longer divergent disciplines. There are several universities that do a lot of cool things. Pursue engineering and be ready to face surprises."

When determining standard deviation from a normal manufacturing process, a sample of the data set is typically used rather than the entire population. The sample standard deviation formula compensates for the error that is inherent in the sampling process by dividing the sum of the squared differences by a value of n-1 (See Equation 14-2). This increases the value of the standard deviation compared to a population standard deviation resulting in a wider distribution curve.

QUALITY

Manufacturers are constantly challenged to produce products faster and at the lowest manufacturing cost. At the same time, they are expected to maintain or increase product quality. The following scenarios illustrate why quality must be at the forefront of product design, process planning, and manufacturing implementation.

First, the importance of quality is sometimes dependent on the intended function. For example, producing a poorly printed newspaper may not cause a life or death situation, but producing misfiring ammunition can. During World War II, there was a problem with ammunition misfiring. To win the war and save lives, the United States needed to resolve the problem. To do that, its manufacturers examined their manufacturing processes, identified the causes of the variations that had led to the poor quality, and made changes to control and improve the manufacturing processes.

In the second scenario, think about the expression, "You get what you pay for." If two products appear to be the same but are priced differently, consumers might think that the less-expensive product is of lower quality and may choose not to buy it. This would mean fewer sales for that manufacturer that, in turn, could ultimately lead to its going out of business.

The Quality Revolution

The explosion of the quality revolution started in Japan after World War II, when manufacturers of military goods switched to producing consumer goods. At first, they produced poor-quality products. However, as time passed, they began to explore new ways of thinking about quality. They solicited the help of Dr. W. Edwards Deming and Joseph M. Juran, two American quality experts, and by the mid-1970s, Japan's reputation for producing quality automobiles and electronics began to surpass that of the United States. As a result, American manufacturers began to realize how important it was to develop processes that would monitor and control quality.

Quality-Management System

Joseph Juran developed a quality-management system known as the "Juran trilogy," which enables manufacturers to achieve the intended results:

1. quality planning,
2. quality control, and
3. quality improvement.

In other words, you need to prepare first, produce with the appropriate controls in place, and then evaluate the results to identify improvement opportunities. To illustrate the trilogy, let's apply it to baking cookies.

1. Quality planning: Decide what cookie to make and assemble the necessary ingredients, kitchenware, and appliances.
2. Quality control: While preparing the batter and baking the cookies, use the appropriate measuring cups and spoons to measure ingredients and set the oven to ensure that the correct time and temperature will be used.
3. Quality improvement: Evaluate the cookie (by a taste test) to see if it meets your expectations. On reflection, you may identify ways to improve the taste (raw materials), speed up prep and cooking time (process changes), or make the perfect cookie with the least amount of mess in the kitchen (reduce waste).

Now that we've seen a simplified version of the Juran trilogy, let's look at it in more depth.

Quality Planning

The important steps in planning for quality are:

1. Establish the project.
2. Identify the customers.
3. Discover the customers' needs.
4. Design the product.
5. Develop the manufacturing process.
6. Develop the controls and transfer them to operations.

To explore these steps further, let's imagine we are manufacturing a flying disk.

ESTABLISH THE PROJECT Establishing the project involves (1) identifying what the product will be (a flying disk) and (2) assembling a project team whose members have defined roles and responsibilities. For example, an aeronautical engineer will make sure the disk can fly, and an ergonomist will make sure it conforms to the human body and doesn't cause injury as a result of repetitive use.

IDENTIFY THE CUSTOMERS To ensure a successful product launch, the project team will need to identify their potential customers. In the case of the flying disk, there are recreational users and more serious users who play disk golf. These customers will have different expectations of what the flying disk should do and what it should look like. The project team should include market research members who can identify potential customers and determine what they are willing to pay for a product that meets their needs.

DISCOVER THE CUSTOMERS' NEEDS Once potential customers are identified, the next step is to discover their needs. The market research team members will interview customers and collect information regarding product requirements. These requirements will include what the product should look like and what customers like and dislike about the products currently available on the market. Once this information is gathered, it will need to be analyzed and prioritized before product design can begin.

DESIGN THE PRODUCT A good design will make it easier to manufacture the product and will result in increased output and decreased waste from rejected parts. This, in turn, will lower the cost of creating a quality product. Once the customers' needs are analyzed and prioritized, other issues need to be considered. To achieve a good design, a designer or engineer must have a good understanding of the process that will be used to manufacture the product. She must also take into account safety considerations, such as preventing sharp edges, as well as environmental concerns.

A product design must also adhere to regulatory agency or industry standards such as those established by the **International Organization for Standardization (ISO)**. The ISO is the world's largest publisher of standards maintained by 157 countries. This nongovernmental group serves the needs of both the public and the business sectors.

Once complete, the product design is documented in a product specification or product requirements document. It describes in detail the product's function, its appearance, and its components.

In reality, the probability is low that a product can be manufactured repeatedly with the exact same geometric dimensions. Therefore, quality planning involves identifying how much deviation is allowable in the dimensions. These deviations are referred to as *tolerances*. The range of tolerances is dependent on the product's function and the desired quality. For example, a child's toy would have a larger range of tolerances than a medical device used by a surgeon.

DEVELOP THE MANUFACTURING PROCESS The next step in quality planning is to develop the manufacturing process or series of actions or operations that will result in the production of a product. The appropriate tools, tasks, and suppliers of materials are identified. The order in which the product is manufactured must be well thought out to ensure quality.

Setting expectations and documenting these expectations through drawings or specification sheets enable appropriate materials to be purchased. Once purchased, these materials should go through a thorough inspection procedure to ensure a quality product at the end. A quality product cannot be made from inferior, substandard materials.

Manufacturing process: a series of actions or operations that result in the production of a product.

DEVELOP THE QUALITY CONTROLS AND TRANSFER THEM TO OPERATIONS Quality is no accident; it must be planned for. Designing for quality means that a manufacturing facility is capable of producing a product that meets all of the product specifications. To do this, the manufacturing areas need to follow procedures and guidelines to ensure that manufacturing processes are in control.

The last step of quality planning is to develop quality controls and transfer them to operations. There are many types of quality controls, such as the development and implementation of standard operating procedures (SOPs). These documents are like recipes, giving specific instructions on how to manufacture the product. Other types of controls include providing training to manufacturing personnel or focusing on removing the possibility of human error by using devices called *fixtures* that physically hold an object in place while it is being worked on.

Quality Control

Producing a quality product does not stop at the planning stage. Quality control involves monitoring production to identify problems that have caused quality issues and then defining and implementing an action plan to correct the problems.

To make sure a manufacturing/production process stays in control, an inspection process is typically implemented. This can take place at the end of the production process or throughout. Inspecting throughout the process is preferable and can take many forms, such as destructive testing (where the product is destroyed and analyzed), nondestructive testing (such as X-ray, ultrasound, or photographic analysis), and basic measurement. In some cases, controls can even be put in place to eliminate the need to inspect.

Quality Improvement

The last step in the Juran trilogy is quality improvement. Quality improvement includes those activities that yield a bigger change than quality control. Quality improvement involves changing the design specifications or the process to achieve a higher-quality product, increased efficiency, or lower costs.

ENGINEERING APPLICATIONS OF STATISTICS

One resolute truth exists in manufacturing: It costs just as much to make a bad part as it does to make a good part. The most important difference between a good part and a bad part is that a manufacturer cannot sell bad parts; they're waste. If company A produces too many bad parts because it cannot control its manufacturing processes, then it must compensate for its losses by increasing the per unit price of its good parts. This cost increase is transferred to the customer. If company B, a competitor of company A, can produce the same part with less waste, it can sell the part at a lower price and ultimately put company A out of business. To avoid this scenario, businesses employ statistical methods to prevent the manufacture of bad parts.

Efficiency, Accuracy, and Repeatability

In order to be competitive in both quality and price, a company needs to make sure that its production processes are efficient and cost effective. This means that little to no waste is generated with respect to time, materials, and effort. It must also produce products that are accurate, which means that they conform to predetermined standards and specifications. Lastly, a company's manufacturing processes need to be repeatable so that parts can be duplicated over and over again with little variation.

Figure 14-19 shows four targets that are often used to explain the concepts of accuracy and repeatability as they pertain to manufacturing:

▶ Target A: The holes are far away from the bulls-eye, with no noticeable grouping that would suggest repeatable performance. A manufacturing process that produced similar results would be neither accurate nor repeatable.

▶ Target B: A manufacturing process that produced similar results would show a high degree of repeatability, although the parts would not be accurate according to the blueprint target dimension.

▶ Target C: A manufacturing process that produced similar results would be accurate and repeatable.

▶ Target D: A manufacturing process that produced similar results would have a high degree of accuracy but a low degree of repeatability. If a manufacturing company produced parts in this manner, most of them would exist on the outskirts of the tolerance ranges.

FIGURE 14-19: **Four targets are frequently used to explain the concepts of accuracy and repeatability as they pertain to manufacturing processes: (a) indicates a manufacturing process that is neither accurate nor repeatable; (b) indicates a high degree of repeatability, although the parts would not be accurate, according to the blueprint target dimension; (c) indicates a process that is both accurate and repeatable; and (d) indicates a process that has a high degree of accuracy but a low degree of repeatability.**

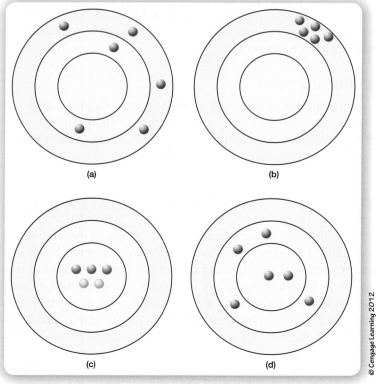

© Cengage Learning 2012

Engineers employ *statistical process control* methods to ensure that low-cost, quality products are produced efficiently and accurately using manufacturing methods that are repeatable. Therefore, a company that employs statistical methods strives for results that are analogous to the performance that is illustrated on target C.

Sources of Error in Data

Statistical information is only as valid as the values from which it is calculated. Inaccurate data is often the result of improperly calibrated measuring equipment, inappropriately selected measuring tools, or human error.

CALIBRATION One of the requirements for an ISO-recognized quality-manufacturing facility is that its measurement instruments be routinely sent out for certified third-party inspection and calibration. **Calibration** is the process of checking and adjusting a measuring device against a standard, preferably on a regular basis. For example, gauge blocks like those shown in Figure 14-20 are pieces of steel that are ground and lapped to standard lengths of extreme dimensional precision. They are often used to calibrate dial calipers and micrometers.

FIGURE 14-20: *Gauge blocks used to calibrate dial calipers and micrometers.*

Getty Images/Science and Society Picture.

Fun Fact

The surfaces of gauge blocks are so flat that when two of them are placed together, air molecules cannot occupy the space between them. The result is two metal blocks that "stick" together without glue or any other form of adhesive.

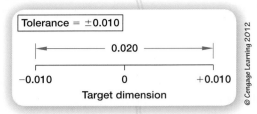

Tolerance = ±0.010

0.020

−0.010   0   +0.010

Target dimension

© Cengage Learning 2012

IMPROPER SELECTION OF MEASURING TOOL Like calibration, selecting the proper measuring tool is critical to avoid data errors. For example, if a 1/16th-inch rule were used to measure the lengths of a series of no. 2 pencils, the resulting values would be very uniform. In fact, there would probably be no variation. However, if a dial caliper were used instead of the rule, variation would be much more evident. This is because the dial caliper can accurately measure a distance that is more than 60 times smaller than the smallest increment on the standard inch rule. This illustrates that just because a measuring tool is highly accurate does not mean that it is necessarily appropriate for collecting statistical information.

A measuring tool should be accurate to within 1/10th of the *total blueprint tolerance* for the object or feature that it is measuring. A dial caliper, which can measure to an accuracy of 0.001 inch, would be an appropriate tool for measuring an object feature (such as a hole diameter) that has a tolerance of ±0.010 inch. The total blueprint tolerance for such a feature would be 0.020 inch (see Figure 14-21). 1/10th of this total blueprint tolerance is 0.002 inch. However, if the object feature had a tighter tolerance, such as ±0.002 inch, then a measuring tool that is more accurate than a dial caliper would be needed to collect statistical information.

HUMAN ERROR Using an inappropriate measuring tool or misusing a properly selected and calibrated one falls within the realm of **human error**. Human error is an incorrect act or decision by a person through either ignorance or accident. Of all the variables involved in the production of quality products, this one has the greatest potential for variation and data errors.

Human error:

an incorrect act or decision by a person through either ignorance or accident.

Variation Is Normal

Variation exists in all things, even in manufactured parts that are intended to be identical. Objects that appear to be identical in size, such as the diameters of the metal dowel pins shown in Figure 14-22, will exhibit variation when

FIGURE 14-22: Dowel pins that appear to be identical in size can exhibit variation when subject to measuring methods such as laser interferometry.

© Cengage Learning 2012

subject to extremely accurate measuring methods, such as *laser interferometry*. However, just because a machine churns out one part that is too big or too small does not mean the manufacturing process needs to be adjusted. Changes or adjustments to a process should be made only when they are based on the statistical analysis of data. For example, if the average size of a sample group (three or more parts taken in order) falls above or below a control limit (which is a calculated value, and not synonymous with the concept of a tolerance limit), then it is appropriate to make an adjustment to the machine that has produced the parts.

There are four rules that are associated with variation in manufacturing:

1. No two objects can be made to identical dimensions.
2. Variation can be measured, assuming that a well-trained person is correctly using the appropriate, well-calibrated tool.
3. Individual outcomes of a manufacturing process cannot be accurately predicted.
4. Groups of data from a normal process will form patterns that can be used to make accurate predictions regarding the behavior of that process.

> **Control limit:**
> upper and lower limits that are established through statistical calculations.

Attribute and Variable Data

There are two types of data: *attribute* and *variable*. Attribute data is yes or no, good or bad, go or no go. The degree of accuracy is not considered. For example, a material is either ferrous or it is not. One can determine this by exposing a material to a magnetic field. If an attraction exists, then the material is ferrous. If no attraction exists, then the material is not ferrous.

Variable data (or analog) falls within an acceptable range of values. This is the kind of data that is associated with dimensional measurements and tolerances as seen on engineering drawings. A tolerance is an acceptable amount of dimensional variation that will still allow an object to function correctly.

> **Tolerance:**
> is an acceptable amount of dimensional variation that will still allow an object to function correctly.

Causes of Variation in Data

Two objects that are designed to be identical will end up having some degree of variation for one of two reasons: chance causes or assignable causes. Chance causes are nonquantifiable and sometimes nonidentifiable reasons for variation in data. They are usually the result of something that has occurred at the system level. Managers and engineers are responsible for changes that address chance causes of variation.

Assignable causes are sources of quality failure that lie outside of the process and, as a result, are intermittent, unpredictable, or unstable. They can often be identified by an operator or a machine and corrected without intervention from management. For example, if an end mill breaks on a milling machine, this would be an obvious reason for why the machine is turning out parts that exhibit large amounts of variation from their intended dimensions.

Coding Data

Numbers can become cumbersome and unmanageable when they have multiple digits. Coding is a technique that is used by quality engineers to make numbers easier to work with and more manageable.

Example

Problem: A series of five measurements produced the following data set:

Actual Measured Values of Five Samples:

1.001 in

0.999 in

1.003 in

1.002 in

1.000 in

The target dimension for a series of measurements in a data set is 1.000 inch. Code this data.

Step 1. Subtract the target dimension from the actual measured value. This gives the *deviation of the sample value from the target dimension:*

    a.   1.001 in − 1.000 in = +0.001 in

    b.   0.999 in − 1.000 in = −0.001 in

    c.   1.003 in − 1.000 in = +0.003 in

    d.   1.002 in − 1.000 in = +0.002 in

    e.   1.000 in − 1.000 in = 0.000 in

Deviation of the Five Sample Values from the Target Dimension:

+0.001 in

−0.001 in

+0.003 in

+0.002 in

0.000 in

Step 2. Set the target dimension to zero and change the deviation values to whole value integers, keeping the sign:

Coded Values of the Original Five Samples Based on the Target Dimension:

+1

−1

+3

+2

0

(continued)

Simple arithmetic can be performed on coded values without the need for a calculator. For example, if you were asked to total the *actual* measured values of the five recorded samples, you would probably use a calculator to get to the final answer of 5.005 inches. However, if you were asked to total the *coded* values, you would probably be able to identify the value of 5 without the use of a calculator. Recognizing that each coded value represents 0.001 inch, and the target dimension is 1.000, it would be easy to perform the conversion without the use of a calculator and determine that the actual sum of the sample values is 5.005 inches. Coding values also makes it easier to quickly determine how many of the recorded part dimensions exist outside of a given tolerance. For example, if the tolerance for these dimensions is ±0.002 inch, then it is easy to see that four of the five sample values are within acceptable size limits.

Your Turn

Problem: A company was commissioned to produce a large number of 2" × 2" aluminum plates that have a ½" hole located at a specific distance from a standard edge (see Figure 14-23). Thirty-five samples were measured to check the accuracy of the hole locations. The values of these measurements were recorded in Figure 14-24. Code the sample part dimensions shown in Figure 14-24 using the target dimension value of 1.375.

FIGURE 14-23: An aluminum plate showing a target dimension of 1.375 inches.

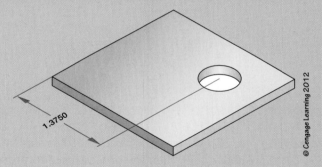

1.3750

© Cengage Learning 2012

FIGURE 14-24: A chart of sample hole center distance values.

1.378	1.376	1.377	1.376	1.376	1.378	1.377
1.378	1.372	1.376	1.375	1.379	1.377	1.380
1.376	1.374	1.374	1.375	1.375	1.372	1.379
1.375	1.375	1.370	1.377	1.371	1.374	1.374
1.375	1.373	1.373	1.373	1.374	1.375	1.375

Process Capability

How does a manufacturer know if the machines on the shop floor are capable of producing quality parts that will meet an engineer's specifications? Or how does she know if a profit could be made when taking on a production job? To answer these questions, each machine must be studied.

Each machine would be used to produce one part over and over again. A critical dimension on each part would be measured and recorded into a data set. *Process capability* would then be determined based on calculations made from the data. The term **process capability (Cp)** is a measure of the data distribution and reflects the ability of the process to produce parts within the blueprint (BP) specifications (tolerances). As was discussed earlier in the chapter, six standard deviations (three above and three below the mean) encompass 99% of the data that is collected from a process that is considered to be "normal." The formula for Cp is shown in Equation 14-3.

Process capability (Cp):
a measure of the data distribution and reflection of the ability of the process to produce parts within blueprint specifications (tolerances).

$$Cp = \frac{\text{Total BP tolerance}}{6 \text{ standard deviations}} \qquad \text{(Equation 14-3)}$$

If the tolerance for the hole center dimension shown in Figure 14-23 was ±0.010 from the target blueprint length of 1.375 in, then the range of acceptable dimension sizes would be from 1.365 in to 1.385 in. This constitutes a total blueprint tolerance of 0.020 inch. The standard deviation value for the sample measurements in Figure 14-24 is 0.0023 inch. Using this information, the process capability of the machine that drilled the holes in the metal plates can be analyzed to determine if the job is a moneymaker or money waster for the manufacturer.

Example

Problem: Find the process capability for the machine that will produce the hole center shown in Figure 14-23. Assume that the hole center dimension tolerance is ±0.010 inch, and the standard deviation for the sample data set is 0.0023 inch:

$$Cp = \frac{\text{Total BP tolerance}}{6 \text{ Standard deviations}}$$

$$Cp = \frac{0.020 \text{ inch}}{6 \times 0.0023 \text{ inch}}$$

$$Cp = \frac{0.020 \text{ inch}}{0.0138 \text{ inch}}$$

$$Cp = 1.452$$

FIGURE 14-25: How process capability (Cp) values are characterized.

Process Capability	
Above 1.33 =	Excellent
1 to 1.33 =	Not-so-Good/Not-so-Bad
Below 1 =	Poor

Cp values are characterized as shown in Figure 14-25. If a Cp value is above 1.33, then the process used to produce the parts is doing so within the blueprint tolerance. Such results are considered *excellent* because it is more than likely that the process will produce no scrap parts. In this case, any jobs that involve drilling holes in metal plates to within ±0.010 inch of a target dimension should be a moneymaker for the manufacturer.

Had the blueprint tolerance for this job been ±0.007 inch, then the Cp value would be 1.018. A process that produces a Cp value between 1 and 1.33 is characterized as *not-so-good/not-so-bad*. In other words, there is a chance that some scrap parts will be produced. A manufacturer would have to think carefully about whether or not to take on such a job.

Finally, if the blueprint tolerance for this job had been ±0.005 inch, then the Cp value would be 0.727. A Cp value below 1 is considered *poor*. This means that the machine cannot produce parts that are completely within acceptable dimensions. Scrap parts would ultimately result, which would cost the manufacturer money.

For a manufacturer to improve quality and make money, process performance needs to improve. This involves making better parts, modifying processes so that they are easier to run, generating less scrap parts, reducing dependency on mass inspection, and improving productivity. If a manufacturer discovers that the equipment is not capable of producing products that meet specifications, then the manufacturer must make a decision: modify the equipment, replace the equipment, or refuse the job and allow a competitor to take it.

CONTROL CHARTS AND STATISTICAL PROCESS CONTROL Statistical process control involves the collection of data for the purpose of process improvement, such as producing products with less variation. Less variation means less scrap, and less scrap means lower costs. *Control charts* are used in statistical process control methods to chart process performance. They can show when a process has changed or when it has to be changed. For example, a control chart will show historical data regarding a machine's performance. Over time, the machine's performance will deteriorate. By projecting the data forward, a manufacturer can predict when a machine will need to be maintenanced or replaced.

X-bar/R Chart A type of control chart used in manufacturing is the *X-bar/R chart*. This chart is used to show the average of part sample subgroups (3 to 10 samples per subgroup) and the range of those samples. A number of rules should be followed when sampling data from a machine or process to create an X-bar/R chart:

1. The process should be running optimally with no chance causes of variation.
2. An appropriate part characteristic (such as a specific part diameter) must be selected. Do not compare different geometric features.
3. The process must not be changed in any way during the study.

An X-bar/R chart indicates whether or not a process (or machine) is functioning within acceptable limits. This type of chart can only be used if the process is normal. To verify that a process is normal, coded sample data from that process must first be charted in the form of a histogram. If the data points generate a bell-shaped curve that peaks at the mean value, then the process is normal.

Once an X-bar/R chart has been generated, three circumstances may occur that would warrant adjusting the process:

1. If a plotted point falls outside of an upper or lower statistical limit, then an adjustment may be made.
2. If seven average values in a row are on one side of the process average centerline, then the process average has shifted and the process would need to be adjusted.
3. If seven average values in a row trend up or down, then an adjustment would be needed.

Equations 14-4 and 14-5 can be used to calculate the upper and lower control limits for an X-bar chart:

$$\text{UCL}_{\overline{x}} = \overline{\overline{x}} + A_2 \overline{R} \qquad \text{(Equation 14-4)}$$

$$\text{LCL}_{\overline{x}} = \overline{\overline{x}} - A_2 \overline{R} \qquad \text{(Equation 14-5)}$$

The average of all the subgroup averages is known as the *process average*. It is also referred to as X-double bar (see Equation 14-6).

$$\overline{\overline{x}} = \frac{\Sigma \overline{x}}{n} \qquad \text{(Equation 14-6)}$$

A_2 is a statistical constant that is based on the number of samples per subgroup. The value of A_2 must be looked up in a statistical chart like the one shown in Figure 14-26. The average range of all the subgroup range values is also factored into the upper and lower control limits.

FIGURE 14-26: Statistical chart showing X-bar UCL and LCL A$_2$ variable by subgroup size.

Number of Samples per Subgroup	A₂ Value
3	1.023
4	.729
5	.577
6	.483
7	.419
8	.373
9	.337
10	.308

FIGURE 14-27: A statistical chart
showing UCL$_R$ D$_4$ values based on the
number of samples per subgroup.

Number of Samples per Subgroup	D$_4$ Value
3	2.574
4	2.282
5	2.114
6	2.004
7	1.924
8	1.864
9	1.816
10	1.777

A range chart (R chart) graphically shows the amount of variation between the parts of the subgroups over time. For an R chart, only an upper control limit needs to be calculated. The formula for this limit is shown in Equation 14-7.

$$UCL_R = D_4\overline{R} \qquad \text{(Equation 14-7)}$$

D$_4$ is a statistical constant that is based on the number of samples per subgroup. The value of D$_4$ must be looked up in a statistical chart like the one in Figure 14-27. The average range of all the subgroup range values is also factored into the upper control limit. If a point falls above the upper control limit on an R chart, then there is a 99.7% chance that something has changed in the process.

An X-bar/R chart will tell the quality engineer if a process (or machine) is functioning within acceptable limits. It cannot be used to identify individual parts within the subgroups that fall outside of the blueprint tolerance and therefore must be scrapped. If a process is changed (or a machine is adjusted), then the control limits would have to be recalculated.

Putting It All Together

A quality engineer was given the task of analyzing the results of a rather old computer numerical controlled (CNC) turning machine (lathe) that had been used for mass production of cylindrical parts for many years. It had come to the plant manager's attention that the machine had been producing a large number of parts that were out of tolerance. The engineer interviewed the CNC lathe operator, who confirmed that he would routinely modify the X-axis part reference in an attempt to keep the machine from producing inaccurate part diameters. The quality engineer explained to the machine operator that variation in part sizes is to be expected and that modifying the machine parameters (though done with the best of intentions) will ultimately cause the process to exhibit greater variation. She explained how she needed the machine operator's help to analyze the CNC lathe's performance, and that the analysis would be used to determine if any adjustments to the process should be made or if the machine needed maintenance. The engineer asked the CNC lathe operator to periodically sample five cylindrical parts in order, measure their diameters (which were supposed to be made to within ±0.003 inch of a target blueprint diameter of 2.000 inches), and record the measured values in a table. The lathe operator agreed that he would not make any adjustments to the machine, even if the machine produced parts that were outside of the acceptable tolerances. An appropriate measuring tool was selected, and the CNC operator was trained how to use the tool effectively. When the CNC operator had recorded the measurements for seven sample subgroups (see Figure 14-28), he passed the information along to the engineer as requested.

Using the table of data provided by the CNC operator and the target blueprint diameter of 2.000 inches, the quality engineer coded the diameter values and

FIGURE 14-28: Subgroup sample diameter values.

Subgroup						
1	2	3	4	5	6	7
1.997	2.003	2.000	1.999	2.001	2.002	2.000
1.999	2.000	2.001	1.998	2.002	2.000	2.000
1.999	2.001	1.999	2.000	2.002	2.000	1.999
1.997	2.002	1.999	1.996	2.000	1.999	2.001
1.998	1.998	1.998	2.001	2.000	2.001	1.998

FIGURE 14-29: *Coded subgroup values.*

	Subgroup						
	1	2	3	4	5	6	7
	−3	3	0	−1	1	2	0
	−1	0	1	−2	2	0	0
	−1	1	−1	0	2	0	−1
	−3	2	−1	−4	0	−1	1
	−2	−2	0	1	0	1	−2
Total	−10	4	−1	−6	5	2	−2
Average	−2	.8	−.2	−1.2	1.4	.4	−.4
Range	2	5	2	5	2	3	3

placed them in a new table (see Figure 14-29). The totals, averages, and range values for each subgroup were then calculated and included in the new table.

Next, the quality engineer plotted the coded values on a histogram to determine if the process was generating a normal distribution curve (see Figure 14-30). If the plotted information did not reveal a normally distributed data set, then there would be no point in performing further statistical analysis on the process.

FIGURE 14-30: *A histogram of the coded data in Figure 14-29.*

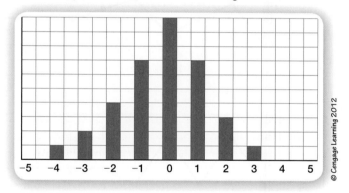

© Cengage Learning 2012

Having verified that the machining process that produced the cylindrical parts was normal, the quality engineer then calculated the process average (X-double bar) and average range (R-bar) from the subgroup averages and range values. These calculated values were then transferred to their respective charts.

$$\overline{\overline{x}} = \frac{\Sigma \overline{x}}{n} = \frac{[(-2) + (0.8) + (-.2) + (-1.2) + (1) + (0.4) + (-.4)]}{7}$$

$$= -0.23$$

$$\overline{R} = \frac{\Sigma R}{n} = \frac{(2 + 5 + 2 + 5 + 2 + 3 + 3)}{7}$$

$$= 3.14$$

The quality engineer then used the process average and range average to calculate the upper and lower control limits for the X-bar chart.

$$UCL_{\overline{x}} = \overline{\overline{x}} + A_2\overline{R}$$

$$= -0.23 + 0.577 (3.14)$$

$$= 1.6$$

$$LCL_{\bar{x}} = \bar{\bar{x}} - A_2\bar{R}$$

$$= -0.23 - 0.577\,(3.14)$$

$$= -2.0$$

The upper control limit range value was then calculated and transferred to the range chart.

$$UCL_R = D_4\bar{R}$$

$$= 2.114\,(3.14)$$

$$= 6.6$$

The quality engineer then drew horizontal lines on the charts to represent the process average, the upper and lower control limits for the chart of averages, and the upper control limit for the range chart. Finally, each coded value subgroup average and range average was plotted on its respective chart in sequence, and lines were drawn to connect the points (see Figure 14-31).

FIGURE 14-31: An X-bar/R chart based on the data in Figures 14-28 through 14-30.

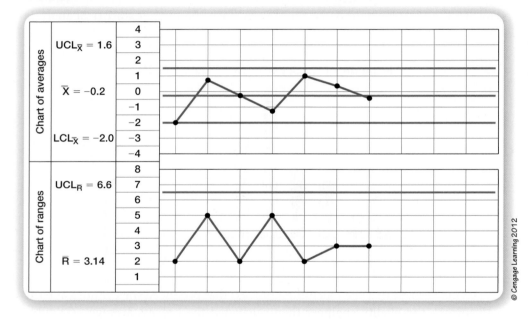

© Cengage Learning 2012

The engineer could see from the coded values that the CNC turning process had produced at least one part that was outside of the ±0.003 inch tolerance range. This meant that it was possible for other parts to be turned to a diameter that is either too small or too large and that at least one of the manufactured parts would have to be scrapped. However, the X-bar/R chart showed that the turning process was performing within acceptable limits. The subgroup averages had not formed a trend upward or downward, nor had they been concentrated to one side of the process average. Therefore, despite the fact that the process had produced at least one part that was unacceptable, there was no statistically valid reason to warrant modifying the old machine tool. The quality engineer then summarized her findings in the form of a technical report and presented the information to the company managers and the CNC operator.

The control charts in Figure 14-31 show that variation exists from one subgroup to the next. However, the amount of variation that existed in the process was contained within control limits that were calculated from statistical analysis of sample data. Also, the plotted information did not show a trend either up or down, nor did it show seven points occurring on one side of the averages centerline.

Career Profile

NUCLEAR-POWERED CAREER

William F. Bundy, Jr. is a lieutenant commander in the United States Navy and chief engineer for the USS Hampton, a nuclear submarine that's among the most advanced undersea vessels of its kind in the world. It takes a lot to keep a 360-foot-long steel tube gliding through the ocean safely and soundly, and Bundy's responsibilities include everything having to do with the ship's maintenance, repair, and operation of the Nuclear Propulsion Plant.

"Normally, we're out to sea," says Bundy about the submarine's crew, which is based in San Diego, California. "Quite often, these are local operations—two, three, or four weeks at a time. We're basically always preparing for battle, which means a lot of exercises. We've done multi-national exercises with Japan, Singapore, and Australia. And we just simulated a mock war in the Philippine Sea, taking out an aircraft carrier and destroyers. Of course, sometimes the aircraft carrier or its escorts take us out."

William F. Bundy, Jr.

© Cengage Learning 2012

On the Job

If you think Bundy spends all day twisting dials, think again. "Most of my time is spent in meetings," says the 29-year-old U.S. Naval Academy grad, who also has a Master's degree from the University of Pennsylvania's Wharton Business School. "When something on the ship breaks, there's a cross-functional team of 60 people under me that goes to work. We identify the problem, break it down on the schematics, and then get to work. There's also a significant amount of maintenance to do—thousands of items a year. And then there's just keeping it all going, the structural components and electrical systems, etc. It's a lot of work."

Inspiration

Bundy didn't have to look far for inspiration. His father, William F. Bundy, is a 30-year Navy man who now teaches at the Naval War College. But Bundy, Jr. thinks Bundy, Sr. had less of an effect than the milieu that he grew up in. "I was always hanging around sailors and officers," he says, "and they were the ones who got me seriously thinking about a Navy career."

It was the Boy Scouts' Explorer Program that got Bundy thinking about an engineering career, though. "There was this 'Get to Know Engineering' program in my community," he says, "and they introduced us to

the different engineering areas over a four-week period. They also had these really cool challenges, like 'How do you get air from here to there using only a paper clip and a rubber band?' I loved the idea that you could come up with an idea and then execute it."

Education

Nearly 10 years later, Bundy is still impressed with what he learned at the Naval Academy, where he majored in systems engineering. "It was just a phenomenal experience," he says, "and I learned a lot of valuable lessons. One of the lessons I learned, which I apply to this day, is to keep it simple. Competing with other schools, I was on a team that designed and built an unmanned, underwater vehicle. We had a gyroscope, depth sensors, and all this stuff. But the team that won, a high-school team, used a Lego Mindstorm processor. Their control system was basically a propeller and rudder."

Advice for Students

Bundy thinks everyone can benefit from studying engineering. "Whether you decide to become an engineer or not, engineering teaches you how to think through a problem," he says. "My statics course in college could have been easy; the material is pretty simple for an engineering course. But we were forced to go all the way through the engineering problem-solving process—questions, assumptions, equations, and solution. It's all about how you approach the problem. Once you understand that, you can solve any problem in the world."

SUMMARY

- Two common standard measurement systems used today are the U.S. customary system (English system) and the metric system.

- Writing very large or very small numbers in scientific notation helps to prevent errors when performing mathematical calculations.

- Although both devices measure in inches, a fractional rule uses fractions to describe values between whole numbers (e.g., 1½ inches) whereas the decimal rule uses decimals (1.5 inches).

- Conversion factors are used to convert from one measurement system to another.

- Dial calipers and micrometers are examples of precision measuring tools.

- Basic statistical analysis of data involves identifying or calculating the mode, median, and mean (average) of a data set.

- The mode is the numeric value that occurs with the greatest frequency in a data set.

- The median is the numeric value that occurs in the middle of a given data set arranged in numerical order.

- To calculate the mean (average), all of the numeric values are added together and then divided by the total number of values in the data set.

- The range is the difference between the largest and smallest values of a data set.

- The standard deviation of a data set is a value that identifies how close the values are to the mean value of that data set.

- The larger the data set that is analyzed, the more precisely the data can be interpreted.

- Businesses began to pay attention to quality after World War II when they shifted their manufacturing from military goods to consumer products.

- A quality-management system involves planning, control, and improvement activities.

- Quality control involves monitoring production to identify problems that have caused quality issues and then defining and implementing an action plan to correct the problems.

- Quality improvement involves changing the design specifications or the process to achieve a higher-quality product, increased efficiency, or lower costs.

- Engineers employ statistical process control methods to ensure that low-cost quality products are produced efficiently and accurately using manufacturing methods that are repeatable.

- Process control methods involve calibrating equipment, properly choosing measuring tools, and minimizing the probability of human error.

- Variations can be measured and are acceptable if they occur within part tolerance limits.

- Variation can result by chance or by an assignable cause.

- Coding is a technique that is used by quality engineers to quickly determine how many samples exist outside of a given tolerance.

- Process capability (Cp) is a measure of the data distribution and the ability of the process that produced the data to potentially fit inside the tolerances or blueprint specifications.

- Control charts are used in statistical process control methods to show when a process has changed or when a process needs to be changed. A typical control chart used in manufacturing is the X-bar/R chart.

1. Why is it important to have a standard for measurement?
2. Identify a historical example of an attempt to establish a standard measurement system.
3. Why do scientists use scientific notation?
4. What advantage does a dial caliper have over an inch ruler?
5. What is the of mode of a data set?
6. Give an example of a bimodal data set.
7. What is the value that identifies how close values are to the mean value of a data set?
8. What is a histogram?
9. How did Japan change its reputation for producing low-quality products?
10. How do we determine what tools to use when measuring?
11. Which form of data is "go or no go"?
12. How does coding data help eliminate errors?
13. What does the term *process capability* (Cp) refer to?
14. What is the difference between a contol limit and a tolerance?

EXTRA MILE

Problem Statement: You have been asked to gather data for company X. The company wants to know if its machining process is in control and capable of making good parts.

Activity: Using a dial caliper, measure 50 "identical" building blocks such as LEGO. Determine the Cp using Equation 14-3.

GLOSSARY

1st-class lever: a lever in which the fulcrum is positioned between the load and the effort.

2nd-class lever: a lever in which the load is positioned between the fulcrum and the effort.

3rd-class lever: a lever in which the effort is positioned between the fulcrum and the load.

A

Absolute pressure (p_{abs}): the total pressure exerted on a system, including atmospheric pressure; identified as psia.

Absolute temperature: temperature that is measured or calculated on scale (such as the Kelvin scale) that is based on a hypothetical absolute zero temperature at which matter is devoid of all thermal energy.

Absolute zero: lowest point on the kelvin scale; defined as zero (0 K).

Acceleration: a change in an object's speed or direction over a certain period of time.

Actual effort force: the amount of input force that is needed to move a load under circumstances in which a machine operates at less than 100% efficiency.

Actual mechanical advantage (AMA): the ratio between the load force and the actual effort force.

Actual stress limit: a limit beyond which the stress within a structural component will cause permanent deformation or failure; usually the yield stress for a material that must operate within the elastic region of an engineering stress–strain diagram.

Actuator: (1) a device that transfers electrical, pneumatic, or hydraulic energy into mechanical energy that moves or displaces something. (2) a device that converts fluid pressure into mechanical motion for the purpose of moving a load.

Aesthetic: relates to appearance, especially when concerning beauty.

Agriculture: the activity of producing crops or raising animals.

Air compressor: a type of mechanical device that is used to generate air pressure in a pneumatic circuit.

Allowable stress limit: an established stress limit for a structural component that is usually determined by a factor of safety; for a material that must operate within the elastic region of an engineering stress–strain diagram, it will be a fraction of the material's yield stress.

Alloy: a mixture of two or more elements, with one element being a metal.

Analog signal: a continuous electrical signal that varies in intensity over time and is capable of attaining an infinite number of values or levels within a given range.

Analog-to-digital converter (ADC): conversion of an analog signal to a digital quantity such as binary.

Anode: a negatively charged electrode.

Area (A): a measure of the two-dimensional space that is enclosed by a shape.

Area moment of inertia (I_{xx}): a mathematical property of a cross-sectional shape that indicates that shape's capacity to resist rotating in the plane of its centroidal axis; also referred to as the *second moment of an area.*

Assembly: a group of machined or handmade parts that fit together to form one unit.

Assignable causes: a source of quality failure that lies outside the process and thus is intermittent, unpredictable, or unstable.

Atmospheric pressure (p_{atm}): the pressure that is exerted by the weight of the atmosphere above the point of measurement; standard atmospheric pressure is 14.7 psia at sea level.

Atom: the smallest indivisible unit of matter and the most basic building block of an element that still retains the properties of that element.

Atomic number: the number of protons that exist within an atom's nucleus.

Attribute data: binary in nature and no analysis can be performed on it: yes or no, good or bad, go or no go.

B

BASIC Stamp: a microcontroller that runs a single board computer using the Parallax PBASIC language. The BASIC Stamp has the ability to interface with other integrated circuits and operate networks.

Battery: a device that produces electricity through a chemical reaction between two different metal electrodes separated by a chemical solution called an *electrolyte*.

Bearing: a mechanical device that is used to reduce friction where two surfaces meet and slide against each other as a result of linear or rotary motion.

Bernoulli's Principle: the velocity of a fluid increases as the pressure exerted by that fluid decreases.

Bimodal: a data set that contains two values that occur with the greatest identical frequency.

Biomass power (biopower): power that uses the energy stored within biomass to generate electricity.

Block: a pulley that contains multiple wheels that spin independently but move in unison.

Block and tackle: an arrangement of two or more pulleys that are strung together to lift or move a load.

Boring: a machining process that accurately enlarges the diameter of a hole.

Boyle's Law: the absolute pressure of a confined gas is inversely proportional to its volume, provided its temperature remains constant.

Brainstorming: any technique that is used by a design team to generate ideas.

Breakeven point: the point at which the profit from products sold equals the total cost of manufacturing the parts.

Brittleness: a mechanical property that causes a material to fracture at relatively low strain values.

BTU: British Thermal Unit; the energy required to raise the temperature of 1 pound of water 1°F.

C

Calibration: the process of checking and adjusting a measuring device against a standard.

calorie (cal): the amount of heat needed to raise the temperature of one gram of water by one degree Celsius.

Cam and follower: a mechanism that changes continuous rotary motion into intermittent linear motion.

Cantilever: a structural member that projects beyond a supporting column or wall, leaving it unsupported on one end.

Casting: a procedure for producing metal parts by filling a mold with a molten material, waiting for it to solidify, and then removing the part from the mold.

Catalyst: a material that is used to initiate a chemical reaction.

Cathode: a positively charged electrode.

Celsius scale: temperature scale that sets the boiling point of water at 100° and the freezing point of water at 0°.

Centrifugal casting: a manufacturing process that uses centrifugal force to hold the molten metal against the inside of a rotating mold.

Centroid: the center of a shape.

Centroidal axis: an imaginary line that passes through the centroid and is appropriately aligned with one of the three principle Cartesian coordinate axes (X, Y, or Z), depending on the orientation of the structural member in three-dimensional space.

Ceramics: a diverse class of materials based on nonmetallic, inorganic solid compounds that are prepared by the action of heating and subsequent cooling.

Chain: a series of links that are uniform in length and fitted together to form a continuous band.

Chain drive ratio: the ratio between the rotational speeds of the input (driver) and output (driven) sprockets in a chain drive mechanism.

Chain pitch: the distance between centers on one chain link.

Chance causes: nonquantifiable and sometimes non-identifiable reasons for variation in data.

Channel Tunnel: a tunnel under the English Channel that enables railway travel between France and England.

Charles' Law: the volume of a confined gas is proportional to its absolute temperature, provided its pressure remains constant.

Check valve: a type of one-way valve that allows fluid to flow in one direction only.

Closed circuit: provides a path for electrical flow if voltage is applied to it.

Closed system: system that allows energy but nothing else to flow through its boundary.

Closed-loop system: a control system that considers the output of a system and makes adjustments based on that output.

Closed thermodynamic system: system that does not allow mass to cross the boundary from the system to its surroundings (or vice versa), but it does allow energy to be transferred across the boundary in the form of work or heat.

Codes: standards that are enforceable by law.

Coke: a low-sulfur bituminous coal that has been baked to remove water, coal-gas, and coal-tar, leaving mostly carbon. Used in the production of steel.

Comment symbol: used when an additional explanation or comment is required. Usually connected to the symbol it is explaining by a dashed line.

Component: a single handmade or machined part that belongs to a larger whole.

Composite material: a solid that is comprised of two or more distinctly different materials that are intentionally combined in layers or mixtures to exploit and enhance certain material properties.

Compound: materials that are composed of two or more elements.

Compound gear train: a gear train that has at least one idler gear shaft that carries two gears that rotate at the same speed.

Compound machine: a combination of two or more simple machines that work together to accomplish a task.

Compression: the act of two forces pushing on an object in opposite directions, causing that object to compact.

Computer-aided design (CAD): (1) any software program that assists in the process of designing the geometry of an object. (2) a computer program that creates engineering drawings.

Computer-aided manufacturing (CAM): a manufacturing process that makes products using engineering drawings from CAD and a numerical control system.

Concurrent force system: the lines of action of all forces intersect at a common point.

Conductance (G): a measure of the ease with which electricity can pass through a material; measured in siemens.

Conductivity: a material's ability to transmit heat or electricity.

Conductor: a material through which electrons easily flow.

Coniferous tree: an evergreen tree that bears exposed seeds in the form of cones and typically has needle-shaped leaves.

Connector symbol: represents the exit to, or entry from, another part of a flowchart; usually used to break a flow line that will be continued elsewhere.

Consensus: a general agreement.

Conservation of energy: the scientific law that states that energy can be neither created nor destroyed.

Constraint: a general limit that is imposed on a design project such as a project deadline, a budget, materials, or manufacturing processes.

Contact force: any force that is transmitted from one physical object to another object through physical contact.

Control limits: (1) a detailed and predetermined set of parameters that are acceptable during the manufacturing of a product. (2) upper and lower limits calculated by three standard deviations greater than and lower than the mean.

Control system: a set of components working together to perform a given task under the direction of a processor or computer.

Convection: the transfer of heat through the movement of warmed matter through a fluid.

Coplanar concurrent force system: the lines of action of all forces act along the same plane and pass through a common point.

Coplanar force system: the lines of action of all forces within a system lie in the same plane.

Coplanar nonconcurrent force system: the lines of action of all forces act along the same plane but do not pass through a common point.

Coulomb: the basic unit of electrical charge (Q); designated by the capital letter C.

Crank: a bar that has one fixed pivot point and is allowed to rotate 360°.

Criteria: specific standards against which a design will be judged acceptable or unacceptable.

Critique: a detailed analysis and assessment of a person's work.

Cross-section: a plane made by cutting across something perpendicular to its length.

Cross-sectional area: the surface area of a cross-section.

Curing: the process by which a material transforms from a liquid or plastic state to a permanent solid state.

Current: the movement of charge carriers, such as free electrons, holes, and ions.

D

Deciduous trees: trees that lose their leaves at the end of a growing season and bear their fruit in the form of berries or nuts.

Decision matrix: a chart used by designers to quantify their opinions of two or more design ideas by assessing each idea according to a series of important considerations.

Decision symbol: indicates a junction where a decision must be made. A single entry may have any number of alternative solutions, but only one can be chosen.

Deflection: a type of deformation that occurs in beams, wherein one side of the beam is placed in compression and the other side is placed in tension.

Deformable-body mechanics: a branch of applied physics that is concerned with the deformations of solid bodies as a result of externally applied forces.

Deformation: a change in dimension(s) that is caused by the application of force, pressure, heat, or any other physical phenomenon.

Design brief: a concise information tool that summarizes the most important information about a design project, serves as an agreement between the engineer and the client, and is used as a standard for assessing a solution's validity at any point in the development process.

Design process: a systematic problem-solving method for generating and developing ideas into solutions.

Design statement: a part of a design brief that challenges the designer, describes what a design solution should do without describing how to solve the problem, and identifies the degree to which the solution must be executed.

Destructive testing: a type of material test that involves loading a test specimen to the point of structural failure while collecting data that indicates the material's mechanical properties.

Dial caliper: a precision measurement device accurate to at least 1/1000th of an inch; used to measure inside and outside dimensions and depths of materials with the output read from a dial.

Diametral pitch: a value that represents the number of gear teeth that occur in one inch of arch length along the pitch circle. This value is also used to represent the size of a gear tooth.

Die: a mold used in manufacturing processes.

Die casting: a manufacturing process similar to injection molding in which molten metal is injected directly into a metal die, rapidly cooled, and quickly ejected from the die.

Digital signal: an electrical signal that has an integral number of discrete levels or values within a given range.

Digital thermometer: a device that uses a sensor called a *thermoresistor* (or *thermistor*) to measure temperature.

Digital-to-analog converter (DAC): converts a digital signal to an analog equivalent such as a voltage.

Direct shear: the act of two offset forces acting on an object in opposite directions, trying to split the object in two along a plane that is parallel to the direction of those forces.

Directional control valve (DCV): any device that is controlled by an operator for the purpose of changing the path that a fluid takes through a circuit.

Displacement: a measure of how far an object moves as measured along a straight line from start to finish; vector quantity.

Distance: the space between two objects or points but is not confined to a straight-line path.

Document symbol: used to represent any type of hard copy input or output (e.g., in a report).

Double-acting cylinder: a common type of linear actuator that is controlled by fluid pressure in both directions (extends and retracts).

Drawing: a manufacturing process where metal is forced through or around a die by stretching, compressing, or bending.

Drilling: the simplest way to make a hole in a material.

Ductility: (1) a mechanical property that describes a material's ability to experience plastic deformation through tension without breaking. (2) a physical property that allows a material to be drawn out into a long, thin wire without breaking.

Dynamic load: objects that are in motion on or within a structure.

Dynamics: the study of objects in motion.

E

Efficiency: the ratio between a machine's actual and ideal mechanical advantage.

Effort cable: the segment of a rope or chain in a pulley system to which an effort force is applied.

Effort force: an input force that is generated by a person, motor, engine, magnetic field, spring, moving water, wind, or by any other phenomena that serves to impart energy into a machine or system.

Elastic deformation: the change in size or shape of an object under an applied load from which the object will return to its dimensions on unloading.

Elastic region: the region of an engineering stress–strain diagram that lies between the origin and the yield point, denoting the stresses and strains that are associated with a material's elastic deformation.

Elasticity: a mechanical property that allows a material to return to its original dimensions once its load has been removed.

Elastomers: natural or synthetic polymers that have the ability, at room temperature, to stretch to at least twice their length and return back to their original size after unloading.

Electrical circuit: a closed path through which electricity can flow between the terminals of a power source and through one or more electrical components.

Electrical energy: the flow of tiny charged particles called electrons, commonly referred to as electricity.

Electrically erasable programmable read only memory (EEPROM): rewritable memory chip that does not need power to keep its content.

Electricity: the transfer of energy through the flow of electrons along a conductor.

Electromagnet: a device consisting of a coil of conductive wire that is wrapped around an iron core that generates a magnetic field when current passes through the coil.

Electromagnetic induction: the generation of voltage within a conductor as a result of passing the conductor through a magnetic field.

Electromagnetic radiation: radiation that has the ability to transfer energy through empty space, unlike convection and conduction, which need matter to transfer energy.

Electron: an extremely small negatively charged subatomic particle that orbits the nucleus of an atom.

Element: a substance that is composed of only one type of atom and cannot be broken down by chemical means into a simpler form.

Energy: the ability to cause matter to move or transform.

Engineer: a person trained and skilled in the design and development of technological solutions to human problems.

Engineer's notebook: a type of journal that serves as an archival record of engineering research and as a repository for solution ideas that are generated throughout a design process.

Engineering stress–strain diagram: a graph that shows the relationship between stress (vertical axis) and strain (horizontal axis) for a specific material based on a constant cross-sectional area.

Entropy: a measurement of the amount of unusable energy in a system.

Ethanol: an alternative fuel produced from renewable and sustainable resources.

Extensometer: a mechanical gauge or electromechanical sensor that continuously measures the amount of axial deformation that occurs during a destructive test.

Extrusion: a manufacturing process used to create parts with uniform cross-sections.

F

Factor of safety (*n*): ratio between a material's actual stress limit and its allowable stress limit.

Fahrenheit scale: temperature scale that sets the boiling point of water at 212° and the freezing point of water at 32°.

Fatigue: material weakness that results from repeated loading and unloading.

Feasibility study: tells a company how practical it would be to produce a certain product.

Feedback: information returned to the input of a system in order to provide self-corrective action.

Filter: a fluid-conditioning device that is used to remove particulate matter that can damage the inner workings of moving components within a fluid power circuit.

Finite element analysis (FEA): a computer-based design analysis tool that allows the user to apply virtual forces and pressures to a 3-D CAD model in order to determine deflections, stress concentrations, and other effects.

Fixed-displacement pump: a type of hydraulic pump that generates a constant flow rate.

Flow meter: a device used to measure the flow rate of a fluid.

Flow rate (*Q*): the volume of fluid that moves past a given point in a system per unit time.

Flow velocity: the ratio between the distance that a drop of fluid travels and the amount of time that it takes to travel that distance.

Flow-control valve: a type of valve that is used to control the volume of fluid as it flows in one direction only; often used to control the speed of an actuator.

Fluid mechanics: the study of the properties of gases and liquids that are at rest or in motion.

Fluid power: the use of a confined fluid flowing under pressure to transmit power from one location to another.

Force: a push or pull that acts on an object to initiate or change its motion or to cause deformation.

Force-displacement diagram: a graph that shows the relationship between force (vertical axis) and deformation (horizontal axis) for a specific material as a result of a destructive test.

Force system: two or more forces that are collectively considered.

Forging: a form of metal forming that shapes a metal part through controlled plastic deformation.

Forming: a manufacturing process that involves permanently changing the shape of the material by applying pressure.

Fossil fuel: formed from the remains of tiny plants and animals that died millions of years ago.

Foundry: a workshop in which metal castings are produced.

Four-bar linkage: a common type of linkage system that contains three movable bars, two fixed pivot joints, and two pin joints. The fourth bar is fixed and often a part of a machine's frame.

Fracture: the act of breaking apart; total structural failure.

Free-body diagram (FBD): a simplified graphic representation of a rigid body, mechanism, or particle that is isolated from its surroundings and on which all applied forces are shown. Friction: a type of resistance force that results when two objects move against each other.

Fulcrum: the point around which a lever, wheel, or linkage pivots.

G

Gage length: the original length of a test sample as identified by two marks spaced a standard distance apart.

Gauge pressure (p_{gauge}): the pressure value that is identified by a pressure gauge that is attached to a fluid system; identified as psig.

Gay-Lussac's Law: the absolute pressure of a confined gas is proportional to its absolute temperature, provided its volume remains constant.

Gear: a toothed wheel that is used to transmit rotary motion and torque from one shaft to another.

Gear ratio: the ratio between the number of revolutions of the input (driving) gear and output (driven) gear in a gear train.

Gear train: two or more gears in mesh that are used to increase or decrease rotational speed and torque and change or maintain rotational direction between the input and output shafts of a mechanical system.

Generator: a device that converts mechanical energy into electrical energy.

Graphical flowcharting: a computer programming language using pictorial representation of a process.

Gravitational acceleration: the constant value describing the acceleration of any object falling toward Earth. This constant is 9.8 meters per second squared or 32.2 feet per second squared.

Gravity: the force that pulls objects toward the center of Earth.

Guidelines: a set of directions to follow.

H

Hardener: a catalyst that causes polymer macromolecules in a resin to cross-link to each other.

Hardness: a mechanical property that gives a material the ability to resist surface penetration, scratching, or wear.

Heat: the flow of, or potential flow of energy from, a higher-temperature substance to a lower-temperature substance.

Heat exchanger: a device that efficiently transfers heat from one substance to another.

Hooke's Law: the law of elasticity that was formulated by English scientist Robert Hooke in 1678, which states that stress and strain exhibit a linear relationship up to the proportional limit for materials in tension.

Horsepower (hp): a non-SI unit of power that is equivalent to moving a 550-lb load a distance of 1 foot in 1 second.

Human error: an incorrect act or decision by a person through either ignorance or accident.

Hydraulic amplification of force: the amplification of an input force that results when dissimilar size pistons are connected together in a closed hydraulic system; the process of generating a large output force from a small input force in a hydraulic system.

Hydraulics: the physical science and technologies associated with liquids that are at rest or flowing under pressure.

Hydrocarbon: a molecule that contains only carbon and hydrogen atoms.

Hydrodynamics: the study of fluids that are in a state of motion.

Hydrostatics: the study of the properties of fluids that are in a state of static equilibrium (at rest).

Hygrometer: an instrument used to measure the water vapor content in the air.

I

Ideal effort force: the amount of input force that is needed to move a load under ideal circumstances in which a machine operates at 100% efficiency.

Ideal mechanical advantage (IMA): the ratio between the distance across which an effort force acts and the resulting distance across which a load acts.

Idler gear: any gear that is used to bridge the gap between the input and output gears of a gear train. Idler gears are used to change or maintain the direction of rotation between the input and output gears.

Inclined plane: a stationary, flat surface that is set at an angle and used to bridge two planes that are offset by some vertical distance.

Inert: condition of having limited ability to react chemically with other elements.

Infrared thermometer: thermometer commonly used in high-temperature applications such as furnaces and kilns, where using other types of thermometers would not be possible.

Input: information fed into a data-processing system or computer.

Input/output symbol: represents data that is available for input or that results from processing (e.g., customer database records).

Intermittent: something that occurs at irregular intervals.

Internal member force: a force that is transmitted through a structural member.

International Organization for Standardization (ISO): the world's large publisher of standards.

International System of Units: abbreviated SI; the modern form of the metric system that is most widely used throughout the world in both science and commerce.

Investment casting: a manufacturing process commonly known as *lost wax casting* because the wax pattern is melted away in the process of making the mold.

Ion: a single atom or molecule that has gained or lost one or more electrons, making it positively or negatively charged.

Irrigation: the application of water to crop-producing lands through artificial means.

Isolated system: system in which no mass and energy can flow freely through its boundaries.

J

Joule: (1) SI unit for work and energy (2) the energy required to apply a force of 1 newton over a distance of 1 meter.

K

Kelvin scale: temperature scale widely used by scientists and engineers. One kelvin is equal to one Celsius degree.

Kinematics: the branch of mechanics that is concerned with the study of motion without consideration of the forces that caused the motion.

Kinetic energy: energy due to motion.

Kinetics: a branch of dynamics that involves the study of forces that cause bodies to move.

Kirchhoff's Current Law: states that the current entering a point must equal the current exiting that point.

Kirchhoff's Voltage Law: states that the sum of the individual voltage drops in a single loop and that the applied voltage must equal zero.

L

Ladder logic (or ladder diagram): a computer-programming language that resembles a ladder with two vertical lines and horizontal rungs that is often used for a Programmable Logic Controller.

Laminar flow: fluid flow that is characterized as smooth and steady.

Law of charges: opposite charges attract and like charges repel.

Lever: a rigid bar that is allowed to rotate at some angle about a pivot point called a *fulcrum*. It is used to move a load when an effort force is applied to the bar.

Linkage: an assembly of rigid mechanical components within a mechanism that are linked together for the purpose of transmitting force and controlling motion.

Load: (1) the weight of the object that needs to be moved or the *resistance* that a mechanical device must overcome to accomplish its task (2) an external force that is applied to an object; examples include tension, compression, shear, and torsion.

Long-range force: force that acts on an object without physical contact such as gravity and magnetism.

Lubricator: a device that mixes tiny drops of oil with compressed air for the purpose of lubricating the components in a pneumatic circuit.

M

Machine: any device, either fixed or moving, that helps you accomplish a task.

Macromolecule: a very large natural or synthetic molecule that is comprised of chains of smaller molecules.

Malleability: a mechanical property that allows a material to experience plastic deformation through compression without breaking.

Malleable: capable of being shaped or formed without breaking by means of hammering or some other application of pressure.

Manufacturing process: a series of actions or operations that result in the production of a product.

Market research: the study of the purchasing habits of various groups of people.

Materials: the tangible substances that make up physical objects.

Materials processing: the procedure of converting raw materials into finished products.

Matrix: a binder of continuous phase that surrounds a reinforcing material and holds it in place.

Mean: the average of a sample of measurements.

Mechanical advantage: the ratio of the output force (load) produced by a machine to the input force (effort) that is applied to the machine; also a ratio between the distance over which the effort force acts to the distance moved by the load.

Mechanical properties: attributes that all materials exhibit to some degree when subject to external forces, including ductility, elasticity, hardness, malleability, strength, and toughness.

Mechanics: the study of motion, forces, and the effects of forces on objects.

Mechanism: an assemblage of moving mechanical components that are supported by a rigid structure.

Mechatronics: the study of the combination of computer engineering, electronic engineering, and mechanical engineering.

Median: the numeric value that occurs in the middle of a given data set arranged in numerical order; one-half the values will be greater than and one-half the values will be less than the median.

Member: a component of a structure.

Mercury thermometer: a device that uses mercury (a metal that is liquid at room temperature) encapsulated in glass to measure temperature.

Metal: a solid material that is typically hard at room temperature, is shiny, and possesses good thermal and electrical conductivity.

Metric system: measuring system in which the fundamental units are the meter, liter, and kilogram.

Microcontroller: an electronic device that consists of a microprocessor, memory unit, input and output ports, a crystal oscillator for timing, ADCs, and DACs all on one circuit board.

Micrometer: a precision measuring instrument used to measure the thickness of flat stock or the diameter of round objects.

Microprocessor: a single-chip computer element that contains a control unit, central processing circuitry, and all the necessary logic functions to serve as a central processing unit (CPU) of a microcomputer or as a dedicated automatic control system.

Microscopy: the use of a microscope for investigation.

Middle Ages: the time of history between 400 to 1500 C.E.

Mode: the number that occurs most often in a data set.

Model: a detailed three-dimensional representation of a design that is used to communicate, explore, or test an idea.

Modulus of elasticity: the slope of the engineering stress–strain curve within the elastic region of the graph that occurs between the origin and the proportional limit.

Modulus of resilience (U_R): a measure of the amount of strain energy per unit volume that a material will absorb without experiencing permanent deformation. The value of the modulus of resilience is equal to the area under a material's engineering stress–strain curve between the origin and the yield point.

Modulus of toughness (U_T): the amount of work that can be done on a material per unit volume without causing fracture. The value of the modulus of toughness is equal to the area under the entire engineering stress–strain curve.

Molecule: the simplest structural unit of an element or a compound.

Moment: (M) a tendency of a force to twist or rotate an object.Motion control: an open or closed system that has moving components.

Motion energy: the kinetic energy contained in an object or a substance that is moving.

Multimeter: a measuring instrument that combines three separate meters: an ammeter to measure current, a voltmeter to measure voltage, and an ohmmeter to measure resistance.

Multiview: a type of drawing that portrays an object as a series of two or more two-dimensional views that are arranged in a specific pattern.

N

Necking: a visible decrease in cross-sectional area at the location where a material will eventually fracture.

Necking region: the region on an engineering stress–strain diagram between the ultimate stress and the point of fracture.

Neutral atom: an atom with no net electrical charge.

Neutral plane: an imaginary plane that separates the regions of tension and compression in a beam and along which no stress occurs.

Neutron: a subatomic particle with no charge; located in the nucleus of an atom.

Newton's First Law of Motion: states that a body at rest will stay at rest and a body in motion will stay in motion unless acted upon by an outside force.

Newton's Second Law of Motion: states that force is the product of mass and acceleration.

Newton's Third Law of Motion: states that every action has an equal and opposite reaction.

Nominal size: the size by which an object is identified, which may be different from its actual size.

Nonconcurrent forces: the lines of action do not share a common point.

Noncoplanar concurrent: the lines of action of all forces are not acting in the same plane but do pass through a common point.

Noncoplanar force system: the lines of action of all forces are not acting along the same plane.

Noncoplanar nonconcurrent force system: the lines of action of all forces are not acting along the same plane and do not pass through a common point.

Nonrenewable energy source: an energy source that cannot be replenished in our lifetime.

Normal force: a force such as tension or compression that acts along the central axis of a structural member.

Normally closed (NC): the contact of a relay that is closed when the coil is deenergized.

Normally open (NO): the contact of a relay that is open when the coil is deenergized.

Nuclear fission: a nuclear reaction in which a nucleus is split into smaller nuclei, resulting in the release of energy.

Nuclear fusion: a nuclear reaction in which smaller nuclei are fused together, resulting in the release of energy

Nucleus: located in the center of an atom; composed of protons and neutrons.

Numerical control: a control system that digitally communicates to machine tools, such as mills and lathes, to produce a part.

O

Offpage connector symbols: used to indicate that a flowchart continues on another page.

Offset method: the process used to identify the yield stress for a material that shows no discernible point on its engineering stress–strain curve where yielding is evident. The process involves drawing a line parallel to the linear portion of the engineering stress–strain curve, beginning at the 0.002 in/in strain value (or 0.2%) on the X-axis.

Offset yield stress: the point of intersection between an engineering stress–strain curve and a 0.2% offset line drawn parallel to the linear portion of that curve within the elastic region; also called the *proof stress*.

Ohm's law: the direct current in a conductor is directly proportional to the voltage applied across the conductor and inversely proportional to the conductor's resistance.

Open circuit: a circuit that does not provide a complete path for electrical flow.

Open system: system in which mass and energy flow freely through the boundary.

Open-loop system: a control circuit in which the system output has no effect on the control.

Open thermodynamic system: system that allows mass and energy to flow freely through the boundary.

Ore: a raw material that contains a desired metal element in a compounded state with other elements in sufficient quantities that warrant commercial extraction through mining, smelting, or other processes.

Output: the information produced by a computer.

P

Parallel circuit: provides two or more paths for electricity to flow.

Parallelogram: a four-sided figure that consists of straight lines, with the opposite sides being parallel to each other.

Pascal's Law: pressure exerted on a confined fluid is transmitted equally and perpendicular to all of the interior surfaces of the fluid's container.

Penstock: a pipe or sluice that controls the flow of water from behind a dam.

Percent elongation: a measure of the amount of stretching that occurs within the gage length of a tensile test sample.

Percent reduction in area: a measure of the amount of necking that occurs in a tensile test sample at the cross-sectional area where fracture takes place.

Perfect Gas Laws: the mathematical relationships between a gas's volume, absolute pressure, and absolute temperature within a closed system, which are defined when one of the three variables is kept constant; consisting of Charles' Law, Gay-Lussac's Law, and Boyle's Law.

Phototransistor: a digital switch that produces an output when light falls on it.

Photovoltaic (PV) cell: a semiconductor device that generates a voltage as it absorbs photo of light.

Physical property: any aspect of a material that can be measured or perceived without changing its chemical identity.

Pictorial: a type of drawing that gives the illusion of three dimensions by showing an object's height, width, and depth in a single view.

Piezoelectric effect: a phenomenon in which a material produces a small voltage when subjected to mechanical pressure.

Pilot line: a transmission line that is used to transport pressurized fluid for the purpose of controlling a valve.

Piston: a solid cylindrical component within an actuator that moves under the influence of fluid pressure; also the component within an air compressor that is used to compress and move air in a pneumatic circuit.

Pitch (p): the linear distance traveled in one revolution of a screw.

Pitch circle: an imaginary circle around which the teeth on a gear or any other uniformly toothed device are evenly spaced

Plastic deformation: deformation that occurs beyond a material's yield stress and continues to the point of fracture.

Plastics: a group of polymer materials that, while solid in the finished state, were liquid at some stage in the manufacturing process and may be formed into various shapes by the application of heat and pressure.

Pneumatics: the physical science and technologies associated with the mechanics of pressurized gases.

Polarity: the condition of being electrically positive or negative.

Polymer: a high-molecular-weight organic compound, either naturally occurring or synthetic, that is comprised of at least 100 repeating chains of simple molecules called *monomers*.

Polymerization: the process by which monomers join together to form a polymer.

Potential chemical energy: energy that is stored in the chemical bonds of molecules.

Potential energy: the amount of energy stored in a body at rest.

Potential gravitational energy: energy stored because of an object's position or place in a gravitational field.

Potential mechanical energy: energy that is stored in an object because of the application of a force (e.g., the catapult).

Potential nuclear energy: energy stored in the nucleus of an atom.

Potentiometer: a variable resistor.

Power density: power per unit volume.

Power (P): the rate at which electrical energy is transferred or transformed; measured in watts.

Precious metal: relatively scarce, highly corrosion-resistant metals that are valued for their color, luster, and malleability. Gold, silver, and platinum are examples.

Precision: to be accurate or exact.

Pressure (p): a type of load that occurs when a force is distributed perpendicular to the surface of an object.

Pressure regulator: a device that is used to manually adjust and control the pressure of the compressed air in a pneumatic system.

Pressure-relief valve: a type of safety valve that will vent fluid back to the reservoir in a hydraulic circuit to protect the pump and other components from damage resulting from excess pressure.

Prime mover: a device, such as an electric motor or an internal combustion engine, that is used to power a hydraulic pump or an air compressor.

Prismatic bar: a straight member of uniform cross-sectional area along its gage length.

Problem statement: a part of a design brief that clearly and concisely identifies a client's or target consumer's problem, need, or want.

Process: a systemic sequence of actions that is designed to control an output.

Process capability (Cp): a measure of the data distribution and reflection of the ability of the process to produce parts within blueprint specifications (tolerances).

Process control: system that will monitor an industrial process so that a consistent uniform product will be produced.

Process symbol: the most frequently used symbol in flowcharting; represents any process, function, or action.

Programmable logic controller (PLC): a specialized heavy-duty computer system used for process control in factories, chemical plants, and warehouses. Closely associated with traditional relay logic. Also called a *programmable controller* (PC).

Projectile: an object that exhibits both vertical and horizontal motion under the influence of gravity alone.

Proportional limit: the point on a stress–strain curve beyond which proportionality between stress and strain no longer exists.

Proton: a subatomic particle with a positive charge; located in the nucleus of an atom.

Prototype: a one-of-a-kind working model of a solution that is developed for testing purposes.

Pulley: a free-spinning wheel (usually grooved) around which a rope, chain, or belt is passed.

Pump: a device that is used to introduce flow or pressure into a fluid system.

Q

Quality: (1) Those product features that meet customer needs; (2) freedom from flaws.

R

Radiant energy: electromagnetic energy that travels in waves. Examples of radiant energy are light, radio waves, microwaves, and X-rays.

Rapid prototyping (RP): a collection of CAD data–driven physical model construction technologies that use additive manufacturing processes.

Reaming: manufacturing process done after a drilling operation to machine a hole to a standard diameter.

Receiver tank: a device that holds compressed air in a pneumatic system.

Recrystallization temperature: the temperature at which the strength of a particular metal decreases and is more easily formed.

Reed switch: an electromagnetically operated digital switching device.

Refractories: nonmetallic ceramic materials that are applied in situations where extreme temperatures (above 1,000°F) are encountered.

Reinforcement: the structural component of a composite material in the form of fibers, whiskers, or particles, which are encased by a matrix material that exists in a different phase.

Renewable energy source: source of energy that can be replenished in short order.

Research: the systematic study of materials and sources in order to establish facts and reach new conclusions.

Reservoir: a holding tank for nonpressurized hydraulic fluid that helps protect the fluid from outside contamination; also serves as a heat exchanger.

Resin: a gumlike solid or semisolid viscous substance that contains polymer macromolecules that crosslink when combined with a hardener.

Resistance: (1) a force that a mechanical device must overcome to accomplish its task. (2) the opposition to the flow of electricity; measured in ohms.

Rigid body: an object that does not experience a change in shape or size when a force is applied to it.

Rocker: a bar that has one fixed pivot point around which it may rotate through a fixed angle range.

Rotational speed: a measure of how fast an object rotates about an axis. Represented by the lowercase Greek letter omega (ω).

S

Scalar: a quantity that can be described by magnitude only.

Schematic: a diagram that uses symbols to represent components of a system.

Schematic symbol: a simplified graphic representation of an electrical, a mechanical, or a fluid power system component.

Scientific notation: method used to express a very large or very small number as the product of a number between 1 and 10 and an appropriate power of 10 (e.g., 1.5×10^2).

Scientist: a person who gains knowledge of the physical or material world through observations and experimentation.

Screw: a cylinder around which a helical groove or thread is wound. It is used to convert rotary motion into linear motion and is capable of generating large output forces.

Sensor: a device that responds to a physical stimulus (such as heat, light, sound, pressure, magnetism, or a particular motion) and transmits a resulting signal (as for measurement or operating a control).

Series circuit: an electrical circuit that provides only one path for electricity to flow through two or more electrical components.

Servomechanism: a feedback system that consists of a sensing element, amplifier, and a servomotor.

Shear force (*V*): the force that is transmitted through an object that experiences shear.

Shear stress (τ)): the amount of shear force per unit of shear area.

Shearing: a manufacturing process used to punch shapes out of sheet stock or plate stock.

Shutoff valve: a simple two-way valve that is used to turn all or part of a fluid power system on or off.

Shuttle valve: a type of fluid valve that is used when the control of an actuator must be shared between two independently operated directional control valves.

SI: an abbreviation that stands for "Système International d'Unités," French for International Standard of Units. The SI system is the modern form of the metric system.

Simple gear train: two or more gears in mesh that have only one gear per shaft.

Simple machine: a mechanical device that is used to change the direction or the magnitude of a single applied force. The six mechanical devices that are considered to be simple machines include the inclined plane, the wedge, the lever, the wheel and axle, the pulley, and the screw.

Simple support: pinned and roller supports.

Simply supported beam: a beam that has a pinned support and a roller support.

Single-acting cylinder: a common type of linear actuator that is controlled by fluid pressure in one direction only; a return stroke occurs automatically.

Sketch: a rough, handmade drawing that represents the main features of an object or scene.

Slag: undesirable and unwanted impurities that result from a smelting process.

Smelt: to extract a metal element from an ore by applying high, sustained heat in the presence of a fluxing material such as limestone.

Soleniod: an electromechanical device that uses the principles of electromagnetism to control the motion of an actuator.

Sound energy: energy that travels in longitudinal waves that expand and compress the substances it travels through.

Specific heat capacity (c): a measure of a substance's ability to absorb thermal energy.

Speed: the distance traveled per unit time.

Speed ratio: the ratio between the rotational speed of the input gear and output gear in a gear train.

Spreadsheet: a table consisting of rows and columns in which numbers and text are input to organize information and perform mathematical analysis.

Sprocket: a toothed wheel that is used in conjunction with a continuous chain to transfer rotational speed

and torque from an input shaft to an output shaft in a mechanical system.

Standard: a reference developed by an authority or through general consent and used as a basis for comparison and verification.

Standard deviation (σ): a value that identifies how close the values in a data set are to the mean value of that set.

Static equilibrium: a condition wherein all of the forces acting on an object are in a state of balance, causing the object to remain stationary.

Static loads: loads that include the weight of the structure itself.

Statics: the study of objects at rest.

Statistical process control: involves the collection of data for the purpose of process improvement, such as producing products with less variation.

Statistics: the collection, analysis, and interpretation of quantitative data.

Stepladder method: a group brainstorming technique in which group members are first given time to think of ideas individually; then two group members come together to share their ideas and continue to brainstorm. Group members join one by one to present their ideas and build off one another's.

Strain (*e*): how much a material deforms for every unit of original size; also called unit deformation.

Strength: a general term that refers to a material's ability to resist applied loads.

Stress: a reaction that occurs within an object that provides resistance to dimensional changes brought on by an applied load.

Structure: a stable assembly of connected components that is designed to withstand loads.

Supply and demand: the relationship between the number of products of a certain type on the market and the number of consumers ready to purchase them.

Support cable: a segment of a rope or chain in a pulley system that carries all or part of the load force.

Sustainable design: occurs when designers account for the ecological impacts of the design solution; also called *green* or *eco-design*.

Switch: a device for making, breaking, or changing the connections in an electrical circuit.

Synthetic: the quality of having been prepared or generated by human-instigated processes such as chemical reactions.

T

Tapping: a machining process used to produce internal threads.

T-connector: a fluid power component that is used to join three separate transmission lines in a pneumatic or hydraulic system.

Temperature: a measure of the average kinetic energy of the molecules of a substance.

Tension: the act of two forces pulling on an object in opposite directions, which causes the object to stretch.

Thermal conduction: occurs when heat (thermal energy) is transferred within a substance through particle-to-particle collisions.

Thermal energy: the internal energy of a body or a substance resulting from the vibration and movement of atoms and molecules.

Thermal radiation: the transfer of heat (thermal energy) from one place to another through electromagnetic waves.

Thermocouple: a device that consists of two dissimilar metals that, when heated, will generate a small voltage across the device.

Thermodynamic system: a collection of objects on which heat flow can be observed or predicted.

Thermodynamics: a branch of physics that is based on the fundamental laws of nature concerning energy and mechanical work.

Thermoplastic: plastic that can be reheated and reformed.

Thermosetting plastics: plastics that become "set" in their solid state once they are molded; they cannot be remolded.

Threading: a machining process used to produce external threads.

Tolerance: the difference between the maximum and minimum dimensions allowed within the design of a product.

Torque: the rotational equivalent of the concept of force. Also, the measure of the tendency of a force to rotate a body on which it acts around an axis. Represented by the lowercase Greek letter tau (τ).

Torque ratio: the ratio between the torque values on the input and output gears.

Torsion: a rotational effect that occurs when two forces, separated by some distance, act on an object in opposite directions, causing the object to twist along its longitudinal axis.

Toughness: a mechanical property that allows a material to absorb energy that results in permanent deformation.

Transistor: a switching device.

Transmission line: a pipe or tube that serves as a connection between two fluid power components.

Trimodal: a data set that contains three values that occur with the greatest identical frequency.

Truss: a triangular arrangement of members commonly made of wood or steel that forms a rigid framework.

Turbulent flow: fluid flow that exhibits random fluctuations in speed, direction, and pressure.

U

U.S. customary system (English system): measurement system in which the fundamental units are the inch, foot, yard, and mile.

Ultimate stress (s_u): the maximum amount of stress that a material can withstand without failing.

Uniform axial stress (s): the intensity of a normal force (F) per unit cross-sectional area (A) within a structural member.

United States customary system of units: the most commonly used system of measurement in the United States. Exceptions are the military and the fields of science and medicine, where the SI system is preferred.

Universal testing machine: a multipurpose machine that is capable of performing several different types of destructive stress tests, including tensile tests, compression tests, direct shear tests, and flexure tests.

V

Vacuum generator: a pneumatic device that incorporates a Venturi tube to generate suction by accelerating the flow of compressed air.

Valence shell: the outermost shell of an atom.

Valve: any device that is used to control the flow of fluid.

Variable data: data that falls within an acceptable range of values.

Variable-displacement pump: a type of hydraulic pump that allows the user to increase or decrease the fluid flow rate.

Vector: (1) a quantity that must be described by both magnitude and direction. (2) a representation of a quantity that can be broken into components.

Velocity: an object's speed and direction of motion.

Venturi effect: the reduction in fluid pressure that occurs as a fluid flows through the constricted section of a pipe.

Viscosity: a measure of a fluid's thickness or resistance to flow.

Voltage (V): the difference in charge between two points; measured in volts.

Volume: the amount of space occupied by a three-dimensional object or enclosed within a container.

W

Watt (W): SI unit of power.

Wedge: a rigid object with sloping sides that moves against or through a load to exert a force.

Wheel and axle: a rotating device that consists of a wheel or crank that is attached to a smaller-diameter axle.

Work: a measure of the amount of energy that is transferred when a force acts on an object and causes it to move or deform.

Work harden: to repeatedly apply pressure to a metal so that it permanently deforms and becomes stronger and harder.

Working line: a transmission line that is used to transport fluid to and from an actuator or any other device that performs work in a fluid power system.

Y

Yield stress (s_y): the point on an engineering stress–strain curve beyond which a material will begin to experience measurable permanent deformation.

Young's modulus: another term for the concept of modulus of elasticity; named in honor of English scientist Thomas Young, who was the first to identify the concept by name.

INDEX

H

Hacksaw, 383
Hammer forging, 368
Hammurabi, 5
Hardener, 309
Hardwood, 297–298
HDPE. *See* High-density polyethylene (HDPE)
Heat, 145
Heat exchanger, 163
Heat transfer, 146–147
Heating and cooling system, 275
Helical gear, 120
High-carbon steel, 285
High-density polyethylene (HDPE), 307
High-temperature geothermal energy, 162–163
Historical overview. *See* Overview and history of engineering
HMI. *See* Human-machine interface (HMI)
Hole producing, 384–387
Hooke, Robert, 11, 327
Hooke's law, 327
Horsepower (hp), 66–68, 202
Hose fittings, 222
Hot-chamber die-casting machine, 364, 365
Hot forging, 367
Hot forming metal, 366
Hot roll forging, 370
Human error, 468
Human-machine interface (HMI), 271
Hybrid plant, 163
Hydraulic actuator, 267
Hydraulic amplification of force, 238
Hydraulic engineer, 400
Hydraulic fluid reservoir, 228
Hydraulic oil, 240
Hydraulic oil filter, 223
Hydraulic system components, 226–231
Hydraulic *vs.* pneumatic systems, 218–219
Hydraulics, 215
Hydrocarbons, 167, 305
Hydrodynamics, 240–246
Hydroelectric power plant, 161
Hydropower, 160–161
Hydrostatics, 236–240
Hygrometer, 11

I

Idea generation, 37
Ideal effort force, 68
Ideal mechanical advantage (IMA)
 basic equation, 68, 69
 compound machine, 95
 defined, 68
 inclined plane, 72

 lever, 76
 pulley, 84, 86
 screw, 92
 summary (equations, listed), 98
 wedge, 73
 wheel and axle, 81
Identical resistor method, 207
Idler gear, 124
IMA. *See* Ideal mechanical advantage (IMA)
Incandescent *vs.* fluorescent bulb, 153–154
Inch (in), 65
Indirect extrusion, 371
Industrial engineering, 22
Infrared radiation, 148
Infrared thermometer, 148–149
Injection molding, 380, 381
Input, 260
Integral calculus, 12
Internal member forces, 399
Internal spur gear, 119–120
International Organization for Standardization (ISO), 464
International System (SI), 447
Interstate highway system, 14
Investment casting, 360–361, 362
Involute curve, 116
Ion, 182
Iron, 282–284
Iron meteorite, 282
Iron ore, 283
Irrigation, 6
ISO. *See* International Organization for Standardization (ISO)

J

Jaquet-Droz, Pierre, 109
Jefferson, Thomas, 448
Jet aircraft, 157
Joule, 65, 151–152, 203
Joule, James Prescott, 151, 203
Juran, Joseph M., 463
Juran trilogy, 463–465

K

Kelvin (K), 247
Kelvin, Lord, 247
Kelvin scale, 149, 247
Kerf, 382
Kiln, 298
Kilopascal (kPa), 218
Kilowatt-hours (kWh), 203
Kinematics, 427
Kinetic energy, 144–145
Kinetics, 428